Fatigue and Fracture of Metals and Alloys: Numerical and Experimental Study

Fatigue and Fracture of Metals and Alloys: Numerical and Experimental Study

Editors

Jaroslaw Galkiewicz
Lucjan Śnieżek
Sebastian Lipiec

MDPI • Basel • Beijing • Wuhan • Barcelona • Belgrade • Manchester • Tokyo • Cluj • Tianjin

Editors
Jaroslaw Galkiewicz
Kielce University of Technology
Poland

Lucjan Śnieżek
Military University and Technology
Poland

Sebastian Lipiec
Kielce University of Technology
Poland

Editorial Office
MDPI
St. Alban-Anlage 66
4052 Basel, Switzerland

This is a reprint of articles from the Special Issue published online in the open access journal *Materials* (ISSN 1996-1944) (available at: https://www.mdpi.com/journal/materials/special_issues/Fatigue_Fracture).

For citation purposes, cite each article independently as indicated on the article page online and as indicated below:

LastName, A.A.; LastName, B.B.; LastName, C.C. Article Title. *Journal Name* **Year**, *Volume Number*, Page Range.

ISBN 978-3-0365-7600-8 (Hbk)
ISBN 978-3-0365-7601-5 (PDF)

Cover image courtesy of Lucjan Sniezek

Contents

Preface to "Fatigue and Fracture of Metals and Alloys: Numerical and Experimental Study"

Dear Readers,

We aimed to present a new book that includes a set of 19 works. We have selected articles from all over the world that show modern research directions in fatigue and fracture mechanics. The papers are addressed to scientists who are involved in the assessment of fatigue strength and fracture toughness. Creating any book requires the cooperation of authors, reviewers, and editors. We would like to use this opportunity to warmly thank all involved in the editorial process for this book.

Have an excellent read.

Jaroslaw Galkiewicz, Lucjan Śnieżek, and Sebastian Lipiec
Editors

materials

MDPI

Article

Application of Operational Load Monitoring System for Fatigue Estimation of Main Landing Gear Attachment Frame of an Aircraft

Michal Dziendzikowski [1,*], Artur Kurnyta [1], Piotr Reymer [1,2], Marcin Kurdelski [1], Sylwester Klysz [1,3], Andrzej Leski [2,4] and Krzysztof Dragan [1]

1 Airworthiness Division, Air Force Institute of Technology, ul. Ks. Boleslawa 6, 01-494 Warszawa, Poland; artur.kurnyta@itwl.pl (A.K.); piotr.reymer@wat.edu.pl (P.R.); marcin.kurdelski@itwl.pl (M.K.); sylwester.klysz@itwl.pl (S.K.); krzysztof.dragan@itwl.pl (K.D.)
2 Faculty of Mechanical Engineering, Military University of Technology, ul. gen. S. Kaliskiego 2, 00-908 Warszawa, Poland; andrzej.leski@wat.edu.pl
3 Faculty of Technical Sciences, University of Warmia and Mazury in Olsztyn, ul. M. Oczapowskiego 2, 10-719 Olsztyn, Poland
4 Institute of Aviation, Lukasiewicz Research Network, al. Krakowska 110/114, 02-256 Warszawa, Poland
* Correspondence: michal.dziendzikowski@itwl.pl

Abstract: In this paper, we present an approach to fatigue estimation of a Main Landing Gear (MLG) attachment frame due to vertical landing forces based on Operational Loads Monitoring (OLM) system records. In particular, the impact of different phases of landing and on ground operations and fatigue wear of the MLG frame is analyzed. The main functionality of the developed OLM system is the individual assessment of fatigue of the main landing gear node structure for Su-22UM3K aircraft due to standard and Touch-And-Go (T&G) landings. Furthermore, the system allows for assessment of stress cumulation in the main landing gear node structure during touchdown and allows for detection of hard landings. Determination of selected stages of flight, classification of different types of load cycles of the structure recorded by strain gauge sensors during standard full stop landings and taxiing are also implemented in the developed system. Based on those capabilities, it is possible to monitor and compare equivalents of landing fatigue wear between airplanes and landing fatigue wear across all flights of a given airplane, which can be incorporated into fleet management paradigms for the purpose of optimal maintenance of aircraft. In this article, a detailed description of the system and algorithms used for landing gear node fatigue assessment is provided, and the results obtained during the 3-year period of system operation for the fleet of six aircraft are delivered and discussed.

Keywords: aircraft load monitoring; fatigue estimation; structural health monitoring; landing gear monitoring

Citation: Dziendzikowski, M.; Kurnyta, A.; Reymer, P.; Kurdelski, M.; Klysz, S.; Leski, A.; Dragan, K. Application of Operational Load Monitoring System for Fatigue Estimation of Main Landing Gear Attachment Frame of an Aircraft. *Materials* **2021**, *14*, 6564. https://doi.org/10.3390/ma14216564

Academic Editors: Jaroslaw Galkiewicz and Lucjan Śnieżek

Received: 30 July 2021
Accepted: 25 October 2021
Published: 1 November 2021

Publisher's Note: MDPI stays neutral with regard to jurisdictional claims in published maps and institutional affiliations.

1. Introduction

Developments of the industrial revolution, in particular rapid growth of rail transport, led to the discovery that failure of the material can occur under a stress level much lower than static tensile strength [1]. Limited durability of structural elements was the main driver for the pioneer experiments carried out among others by Wöhler in the field of S–N curves or Palmgren and Miner who developed the linear damage cumulation—which provided the basis for estimating safe-life of structures [2]. In aerospace, the beginnings of system solutions focused on a high level of safety date back to the early post-war years, when the safe-life design concept was introduced, in order to preserve human health and life as well as mitigating potential material losses [2,3]. Due to the advancement of aircraft structures, complexity of different types of connections and variable geometry of structural elements, the most reliable method for determining the service life is the subcomponent or full scale fatigue tests of aircraft structures [4–6]. After the F-111 aircraft crash in 1969 [7,8], a new design and maintenance concept was introduced—the damage tolerance approach [9,10]. In the new methodology, a mandatory requirement for assuring structural integrity is the introduction of Non-Destructive Testing (NDT) procedures and definition of inspection intervals of aircraft critical structural elements [11]. Non-destructive inspections were proposed in order to assure possibility of damage detection, before their development could jeopardize the safety of aircraft operation. Introduction of the damage tolerance concept resulted in significant decrease of the risk of air accidents due to fatigue damage. Despite remarkable advances in the field of fracture mechanics and the development of numerical modeling methods for complex physical processes, it is impossible to predict all factors that may increase the risk of an accident. In particular, the design load spectrum of the structure determines the intervals between subsequent NDT inspections as well as safety margins for aircraft operation. However, the way in which a particular aircraft is operated after its introduction into service does not necessarily fit to its pre-assumed profile, especially for combat aircraft, therefore the load and fatigue spectra can change during long-term operational use due to numerous factors [12]. Thus, it is necessary to introduce an Individual Aircraft Tracking (IAT) program in order to monitor actual load spectrum of every aircraft in operation [10,13,14]. A modern approach to meet this requirement is to implement Operational Loads Monitoring (OLM) systems [15] as a necessary component of aircraft avionics. The approach for this purpose has evolved from g-counters to modern on-board parametric systems incorporating several aspects, e.g., material fatigue, measurement techniques, signal analysis, modeling, etc., and utilizing a network of sensors, e.g., strain gauges or Fiber Bragg Gratings (FBG) [16–19] which are permanently mounted in the aircraft structure and measure local strain in predefined aircraft locations. Regarding the above, introduction of an Individual Aircraft Tracking (IAT) program with use of proper OLM system is essential to ensure safety and extend aircraft service life [13,14,20,21]. Development of OLM systems can also significantly increase potential use of ageing aircrafts [12], as after performing the necessary overhaul, such technology may allow their further operation beyond initially designed service life, if fatigue wear or condition of critical structures can be precisely monitored based on indications from network of integrated sensors.

In this article, an approach to fatigue estimation of the Main Landing Gear (MLG) attachment frame due to vertical landing forces based on Operational Loads Monitoring (OLM) system developed for Su-22UM3K aircraft is presented. In Section 2, the approach to OLM system design is presented and algorithms for signal analysis and fatigue assessment are defined. In Section 3 the results obtained during the 3-year period of system operation for the fleet of six aircraft are delivered and discussed, and in the last section the article is concluded.

2. Methodology for Load Analysis of Landing Gear Attachment Node

2.1. Full Scale Fatigue Test and Load Monitoring System of Su-22 Aircraft

Su-22 is a variable wing sweep angle fighter–bomber aircraft which was introduced into operation in Polish Armed Forces (PLAF) in mid-1980's. There are two versions of the aircraft: single-seated combat version M4 and a two-seated trainer version UM3K. Based on the PLAF decision, a service life extension program was launched in 2014 in order to prolong the designed operation period guaranteed by the manufacturer. As some of the aircraft were close to the original limits, the actual service life, adequate to the flight profile in Polish Air Force, had to be validated. Full-Scale Fatigue Tests (FSFTs) have been carried out in order to verify that the required service life is available with respect to adequate load spectrum [22]. Furthermore, for two-seated Su22UM3K aircraft, the development of an OLM system providing reliable and detailed data for Individual Aircraft Tracking (IAT) program was required, due to their increased usage compared to M4 version, in particular an increased number of standard and Touch-And-Go landings [14].

FSFT was designed to be a four-stage test due to the variable sweep angle of the aircraft wings [22]. This was decided in order to minimize the laborious and time consuming wing sweep changes and actuator reconfigurations. Different wing sweep angles are used for takeoff and landing (30°), subsonic flights (45°) and supersonic flights (63°). The final goal of the FSFT was determination of the total durability of the structure in order to reach 3200 flight hours and 6000 landings. Based on analysis of flight profiles of Polish Su-22, four aircraft configurations for subsequent stages were determined (Table 1). Landing loads and loads for flight with extended flaps were represented in separate stages due to different hydraulic actuators arrangement under wings. The final stage was designed to represent flights with the highest vertical overloads from Stage 2 until appearance of critical damage of the structure. In order to represent the load distribution in each stage, an array of hydraulic actuators was designed for each stage (Figure 1) taking into account load and displacement range for each considered load node as well as wing sweep angle and aircraft configuration.

Table 1. Configurations of aircraft for Full-Scale Fatigue Test.

Stage	Loads	Wing Sweep Angle
I	Landing	30°
II	Flight	45°
III	Flap	30°
IV	Flight with high overloads	45°

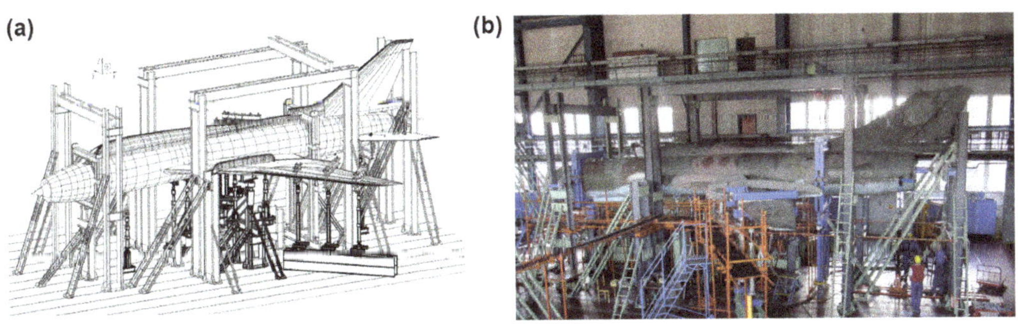

Figure 1. Full-Scale Fatigue Test rig: (**a**) schematic view and (**b**) actual view.

This article is devoted to the development of a methodology for fatigue estimation of Main Landing Gear (MLG) attachment frame due to aircraft landing; therefore, we will focus on Stage I of the test while more detailed description of the approach to FSFT of Su-22 aircraft and other test Stages can be found in [22,23]. For Stage I, the wing sweep was set to 30° and the structure was fixed with clamps on the fourth and thirty-fourth fuselage frame as well as by two rods mounted in the engine bay (highlighted in blue in Figure 2). Loads were exerted on the structure by means of 16 hydraulic actuators. Six actuators (highlighted in red in Figure 2) were used in order to represent front (1 actuator acting in the vertical direction) and main landing gear loads (2 actuator on each side representing vertical and longitudinal landing forces and 1 actuator in the middle representing transverse taxing forces), whereas 4 actuators were acting on the variable sweep wing part (2 each side) corresponding to lift and inertia forces during landing. The remaining 6 actuators were distributed along the fuselage in order to represent inertial forces. In this study, fatigue wear of the attachment frame due to it being carried by MLG is considered as one of the key factors determining remaining service life of an aircraft. Those forces were represented by actuators denoted as no. 6 in Figure 2 (one actuator on each side). Different types of full stop landings were represented during Stage I with the same landing mass of the aircraft but different vertical load levels during touchdown. Simulated vertical load sequences on the actuators in terms of the equivalent weight carried by main landing gear node is shown in Figure 3, and the number of different type of landings represented in FSFT load spectrum is provided in Table 2 below. Each full stop landing was represented by 14 load levels exerted on the frame of the aircraft in linear sequences (Figure 3).

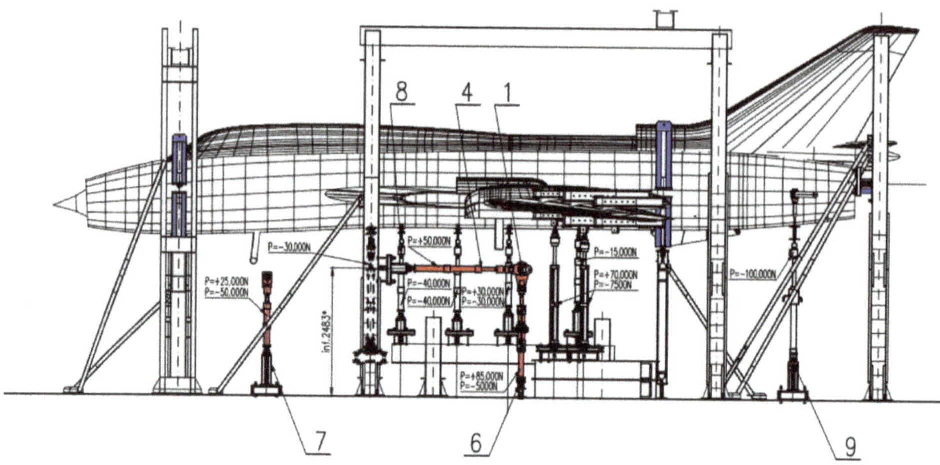

Figure 2. Configuration of actuators for Stage I of Full-Scale Fatigue Test execution.

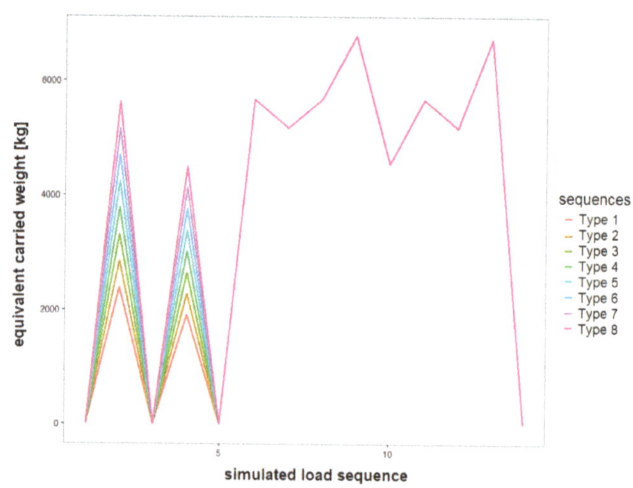

Figure 3. Simulated load sequences exerted on landing gear attachment node during the Full-Scale Fatigue Test representing different types of full stop landings.

Load envelope and landing statistics were estimated based on dedicated flight tests program and historical flight data. The information about stress level occurring in various areas of interest during flight was collected with use of a network of strain gauges installed on a test aircraft. After sensor installation, a dedicated flight test program was performed in order to acquire sufficient data about load distribution during flight for the purpose of FSFT load spectra preparation for different test stages. In total, 40 strain gauges were installed on the test aircraft selected for in flight measurements (Figure 4):

- 15 on the left wing (and symmetrically on the right wing):
  - 4 on the wing fuselage joint,
  - 2 on the main pivot joint,
  - 9 on main landing gear and in the compartment and
- 10 on the fuselage in 5 selected Sections (2 each).

  The same installation was used for test structure monitoring during FSFT.

Table 2. Number of different landings type represented during Full-Scale Fatigue Test.

Landing Type	Landing Count in Load Spectrum
Type 1	1926
Type 2	5603
Type 3	1849
Type 4	589
Type 5	157
Type 6	36
Type 7	9
Type 8	1

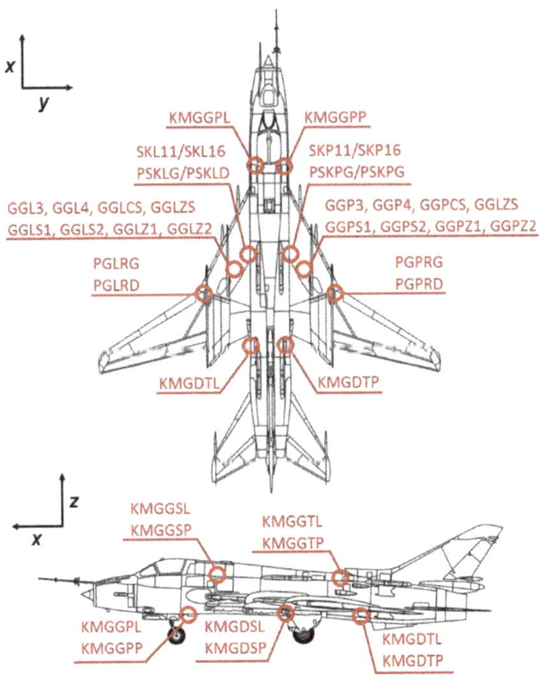

Figure 4. General view of the strain gauge network on the test aircraft (due to the symmetry of the sensor network only one side of the aircraft is shown).

The strain gauge measurement array was designed to monitor the following loads:

- bending momentum in the wing fuselage joint (one section each side);
- bending momentum in the main pivot joint (one section each side);
- bending momentum in the fuselage (5 sections);
- main landing gear loads:
  - vertical force along z axis;

- bending momentum along y axis;
- bending momentum along x axis.

All of the strain gauges were in the form of tee rosette configuration (Figure 5b), which allowed for strain measurements in the primary direction with temperature compensation due to the secondary perpendicular strain gauge. In Figure 5a, localization of strain gauges installed in the main landing gear compartment is presented.

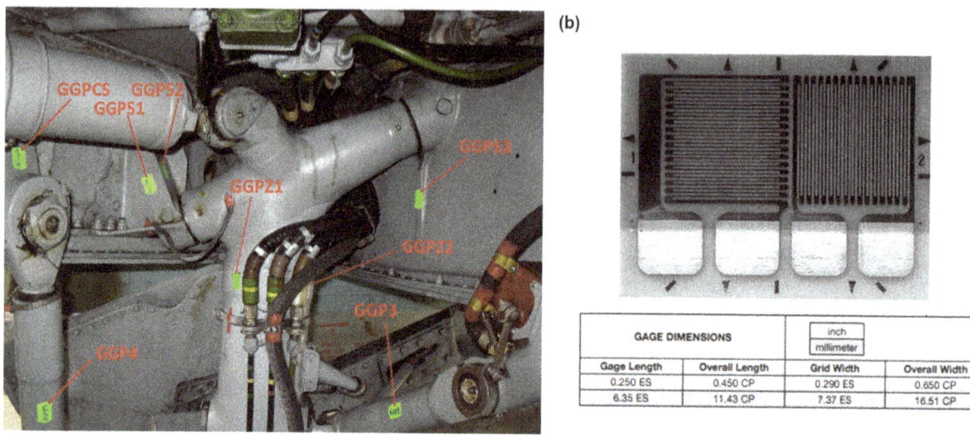

GAGE DIMENSIONS		inch / millimeter	
Gage Length	Overall Length	Grid Width	Overall Width
0.250 ES	0.450 CP	0.290 ES	0.650 CP
6.35 ES	11.43 CP	7.37 ES	16.51 CP

Figure 5. Localization and type of strain gauges installed on test aircraft: (**a**) localization of sensors in MLG compartment and (**b**) tee rosette strain gauges used (250UT manufactured by Vishay).

In addition, a load monitoring system with a reduced number of strain gauges was installed on six selected Su-22UM3K aircraft, in order to track their individual usage and remaining service life. In particular, as aircraft selected to be equipped with a Operational Load Monitoring (OLM) system are used for pilot training and certification, it was necessary to develop a methodology for main landing gear attachment node fatigue wear monitoring, as those aircraft perform more landings (especially Touch-And-Go landings) than combat one-seat aircraft. There exist many approaches to Individual Aircraft Tracking [10] depending on the type of data available for load estimation in critical locations. A through revision of different methods for fatigue assessment and comparison of different monitoring techniques is provided in [24]. One of the most common approaches is to use the vertical overload parameter recorded by Flight Data Recorder (FDR) in order to determine stress levels at critical locations. This parameter, in particular in combination with some other flight parameters, e.g., angle of attack and aircraft mass distribution, is especially efficient for fatigue assessment of wing spars or wing to fuselage attachment nodes, as vertical acceleration is the key parameter determining fatigue wear of such structural elements. Based on acceleration measurements, it is moreover possible to define many useful damage sensitive signal features for the purpose of Structural Health Monitoring [25]. However, the bandwidth and sensitivity of accelerometers used in FDR are sometimes not sufficient for proper determination of fatigue due to landings, in particular it can be hard to detect and properly assess touchdown of an aircraft based on that parameter. Furthermore, accelerometers cannot be used for determination of takeoff weight which in the case of landing gear attachment node can significantly contribute to Ground-Air-Ground cycle, therefore strain gauges were used for OLM purposes.

The sensor network of the system developed for the IAT program is reduced with respect to the system installed on the test aircraft (Figure 4). It contains eight strain gauges which are installed symmetrically in the most relevant structural elements of the aircraft on both wings and main landing gear frame. In Figure 6 general location of strain gauges of the reduced system is presented. The limitation to eight strain gauges was due to the hardware requirements of the FDR used. Four sensors are placed on the main wing spar, on lower (sensors denoted as T1 and T5) and upper (sensors denoted as T2 and T6) flanges. For main landing gear monitoring, the sensor denoted as GGPZ1 in Figure 5a was selected to measure backward bending moment of the landing gear (denoted as T3 and T7 in the reduced system), and sensor GGPS2 (Figure 5a) is used for estimation of vertical force during landing (denoted as T4 and T8 in the reduced system). Strain gauge selection was determined based on signal-to-noise ratio measured during flight tests as well as the results of sensors calibration on test specimen during FSFT. All the selected strain gauges depended linearly on the corresponding relevant forces. In particular, sensor T4 (Figure 6b) was linearly dependent on vertical force applied to main landing gear in full range of the load spectrum while it was barely sensitive to force applied in perpendicular direction. Therefore, for the purpose of fatigue wear estimation of the MLG attachment node due to vertical landing forces, the strain values ε measured by sensors T4 (left node) and T8 (right node) were used.

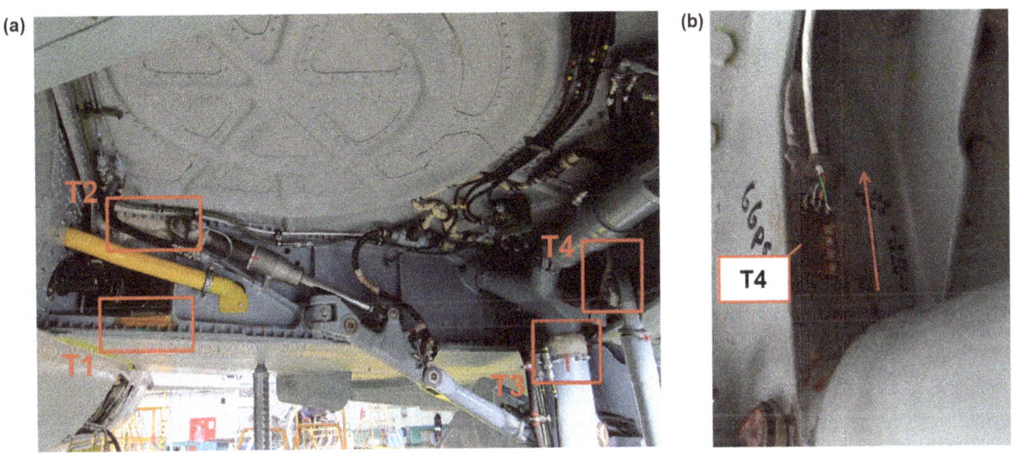

Figure 6. Reduced system for load monitoring: (**a**) general location of strain gauges of the system (left MLG bay) and (**b**) view of T4 sensor with indication of measurement direction.

2.2. Methodology of Signal Analysis for Landing Operations Extraction

In order to determine fatigue of landing gear attachment node, it is necessary to extract signals corresponding to on-ground operations from full record of a given flight. For the detection of on-ground segments recorded by the system, it was assumed that the threshold ε_T for detection of significant deformations originating from landing gear loads corresponding to landings is 20% of the base deformation, which is determined before the takeoff of the plane, using the loads at full stop. In Figure 7, an example of signal acquired during landing with multiple touchdowns preceding final touchdown and deceleration is presented. As can be seen, signal can exceed the estimated threshold during before final deceleration and load transfer to landing gear node; therefore, for a proper determination of signal corresponding to landings, an additional algorithm is required.

In the presented algorithm, the first step of on-ground operations detection and classification is to select all the signal samples $\varepsilon(t)$ below the determined threshold ε_T, i.e., satisfying the condition

$$\varepsilon(t) < \varepsilon_T, \tag{1}$$

where t is time of a given sample acquisition. Then, starting from the first signal sample satisfying this condition, consecutive disjoints sets of data are determined:

$$\tau_1 = \{\varepsilon(t_{i,1}), \ldots \varepsilon(t_{f,1})\}, \ldots, \tau_l = \{\varepsilon(t_{i,l}), \ldots \varepsilon(t_{f,l})\}, \tag{2}$$

such that the difference between the acquisition time $t_{i,k}$ of the initial signal value $\varepsilon(t_{i,k})$ from a given set τ_k and acquisition time $t_{f,k-1}$ of the final signal value $\varepsilon(t_{f,k-1})$ from the preceding set τ_{k-1} is not less than 3 s. For proper on-ground connected signal segments detection, datasets τ_j lasting no longer than 2 s, i.e., for which

$$t_{f,j} - t_{i,j} \leq 2 \tag{3}$$

are disregarded, in order to remove eventual artificial events due to natural signal disturbances, e.g., due to short power outages or drops of pressure in hydraulic blocks during flight. In the next step, for every segment τ_j, all the signal samples $\varepsilon(t)$ satisfying

$$t_{i,j} - 3 \leq t \leq t_{f,j} + 3 \tag{4}$$

are added to a given set τ_j in order to analyze full data records corresponding to subsequent on-ground operations, i.e., each part of the signal corresponding to such operation is extended by additional 3 s time offset of signal before and after the operation, in order to track all load cycles exerted on landing gear node. The first of such extended sets, τ_1, contains signal acquired during aircraft takeoff and taxiing before takeoff, and the last set τ_l corresponds to full stop landing and taxiing after landing, whereas all sets in between are classified as Touch-And-Go (T&G) landings.

In Figure 8, the outcome of the proposed algorithm for landing presented in Figure 7 is delivered. The signal corresponding to full stop landing was extracted from raw signal with use of the proposed algorithm. In particular, all the aircraft touchdowns are included in the extracted signal, and an additional 3 s time offset provides data corresponding to no load condition on landing gear node; therefore, based on such data, it is possible to capture all relevant load cycles exerted on the node due to landing.

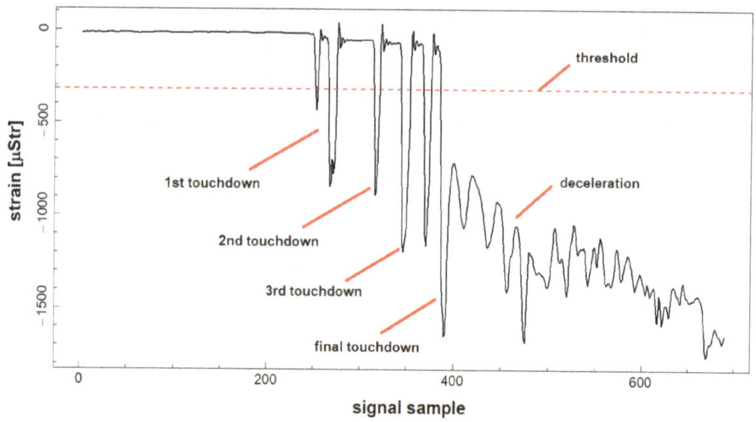

Figure 7. An example of signal acquired by strain gauge located at main gear attachment node during full stop landing with indication of signal threshold for landing detection.

All the parameters, e.g., threshold ε_T and offset levels, were decided based on algorithm results on database of reference signals, where signals corresponding to on-ground operations were manually determined, so the number and duration of landings was compared between expert and automated analysis for different adjustments of the parameters. Furthermore, the performance of the algorithm is periodically verified, i.e., number of landings is compared with maintenance data, but also validity and performance of the algorithm is evaluated by the experts based on randomly selected flights from a given period.

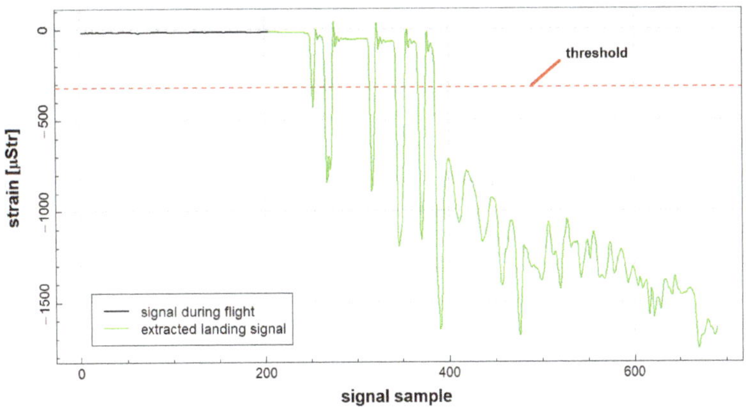

Figure 8. An example of automated extraction of signal corresponding to full stop landing.

In Figure 9a, an example of an OLM system record with indication of subsequent on-ground operations is presented. The approach is efficient in detection of on-ground operations, in particular for Touch-And-Go landings (Figure 9b). Detection of such events based on flight parameters acquired by a standard flight data recorder was very inaccurate, as many simulated landing approaches, but without touchdown, are performed during pilot training process. In Figure 10, an example of system records for a flight with three simulated attempts to landing is presented. For every such maneuver, low pass flight over runway was performed (Figure 10b) with released landing gear, which caused a slight change of strain values recorded on the node (Figure 10a); however, no actual touchdown occurred. This is correctly recognized by the proposed algorithm, as only takeoff and full stop landing were detected in that flight (Figure 10a). As both records—the altitude parameter as well as indicator of landing gear release command—were similar as for proper Touch-And-Go landings, such events were often misclassified with use of algorithms based on only flight parameters records, as sensitivity and signal sampling rate of the standard g-load sensor used on this type of aircraft are not sufficient for such events detection.

An interesting example of algorithm output is presented in Figure 11. The recorded strain signal shows a clear example of a bounced landing. In this case, after the first touchdown, the plane took off from the ground for about two seconds, and full stop landing was performed afterwards. The algorithm detected a Touch-And-Go landing event and subsequent proper landing. Such events, when detected by the system with high confidence, can provide an automated tool for human error assessment, which could be beneficial for pilot training programs.

In Figure 12, an example of a strain gauge signal recorded during full stop landing is shown, with indication of the characteristic stages of this process. One of the parameters used to assess the stress level of the main landing gear node structure during full stop and Touch-And-Go landings is the amplitude of the maximum deformation cycle recorded during the touchdown (Figure 12). A characteristic feature of the touchdown is the rapid deformation change related to the slowing down of the descent speed of the aircraft and the dissipation of energy on the elements of the landing gear node [26]. Two warning levels were defined for the strain signals characterizing hard landings, which are presented in Table 3. In addition to structural load monitoring purposes, the distribution of landings corresponding to different warning levels can also be used for the assessment of pilot training advancements. The warning levels were determined based on landing statistics determined for a certain period of time. Another approach could be based on relevant material data and depend on the stress values at the critical point for the main landing gear attachment node, yet the adopted statistical approach is better suited for pilot training purposes.

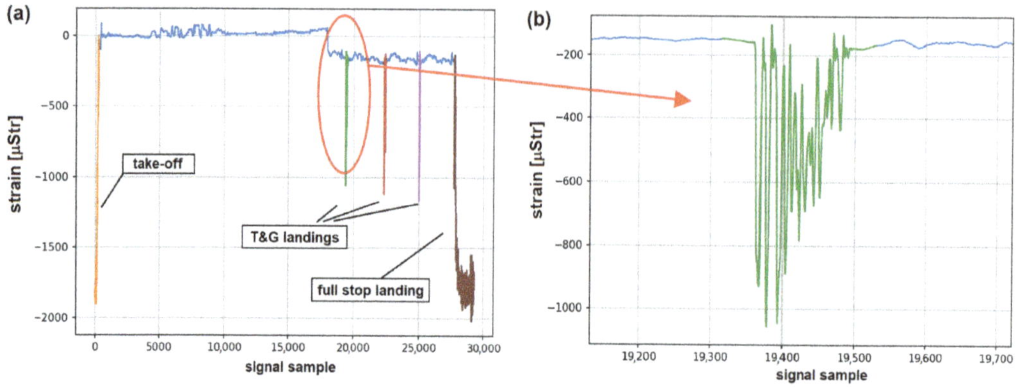

Figure 9. An example of load monitoring system records: (**a**) full record with indication of all detected on-ground operations; (**b**) signal acquired during first detected Touch-And-Go landing (marked in green).

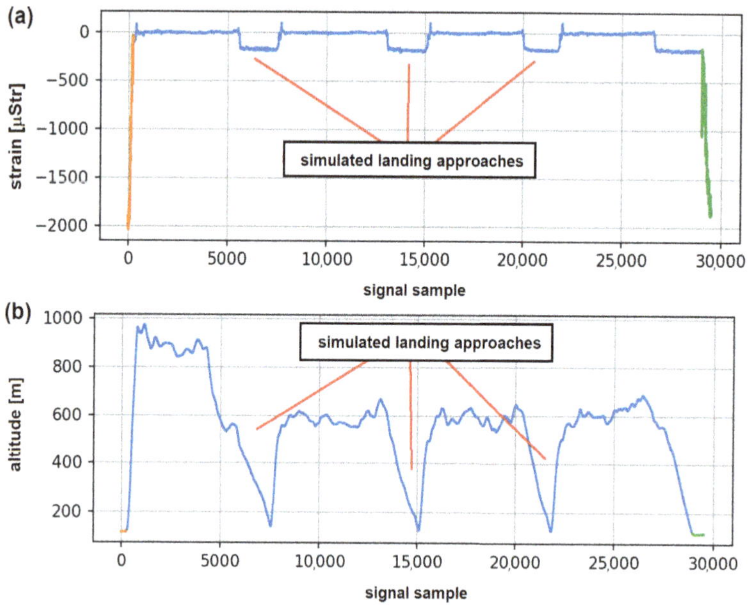

Figure 10. An example of flight records with simulated landing approaches: (**a**) strain on landing gear attachment node with indication of take-off and full stop landing; (**b**) altitude (barometric).

Table 3. Warning levels for maximum strain amplitude [μStr] during aircraft touchdown.

	Min. Touchdown Amplitude	Max. Touchdown Amplitude
standard landing	0	1100
first warning level	>1100	1500
second warning level	>1500	-

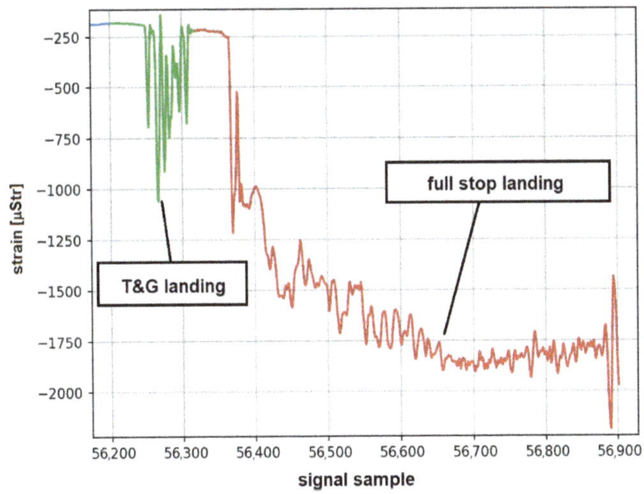

Figure 11. An example of system record with subsequent Touch-And-Go and full stop landing.

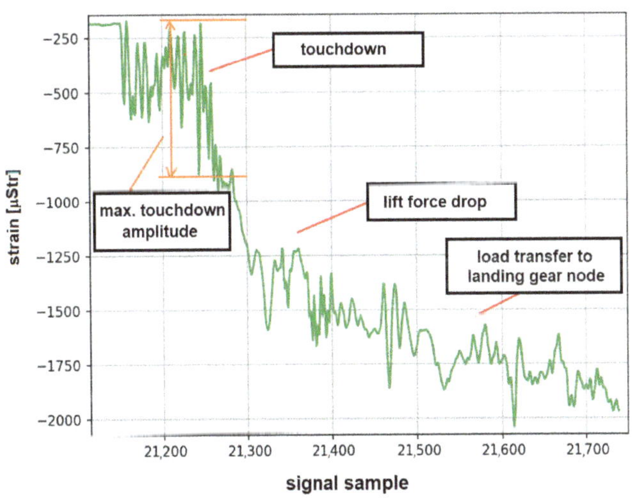

Figure 12. An example of strain record for full stop landing with indication of subsequent stages of this maneuver.

2.3. *Methodology of Landing Gear Node Fatigue Estimation*

In this paragraph, a method for fatigue estimation of Main Landing Gear (MLG) attachment frame due to vertical landing forces based on OLM system records is delivered. The classic approach to fatigue estimation is based on the S–N fatigue curve of a given material and linear cumulative Palmgren–Miner hypothesis. The S–N curve determines the relationship between the equivalent stress amplitude σ^{eq} of the zero-to-tension load cycle and the number of cycles N needed to fracture of an element, subjected to load cycles of this amplitude. The S–N curve is usually described by the relationship [27]

$$\log \sigma^{eq} = A + B \log N. \tag{5}$$

Cumulative fatigue D of a set of load cycles $\{\sigma_1^{eq}, \ldots, \sigma_M^{eq}\}$ is given by the expression

$$D = \sum_{i=1}^{M} \frac{1}{N_i} = C \sum_{i=1}^{M} \left(\sigma_i^{eq}\right)^{-\frac{1}{B}}, \tag{6}$$

where C is a constant dependent on material constants A and B.

Based on classic S–N and cumulative fatigue equations presented above, an approach to fatigue wear estimation of main landing gear attachment frame can be defined as follows. In order to determine the fatigue wear of a given structural element, a linear relation between stress level σ^c at critical point of the element and some physical parameter p can be assumed:

$$\sigma_c = \alpha p. \tag{7}$$

As the parameter p, records of strain gauge ε installed in a point of the structure where stress level is linearly proportional to σ^c can be used, or p can be a function of flight parameters determining load values of a given structural element, e.g., weight and g-load factor in the case of fuselage wing attachment. Equation (6) can be rewritten as follows:

$$D = C \sum_{i=1}^{M} \left(\sigma_{c,i}^{eq}\right)^{-\frac{1}{B}} = \tilde{C} \sum_{i=1}^{M} p_i^{-\frac{1}{B}}, \tag{8}$$

where $\tilde{C} = C\alpha^{-\frac{1}{B}}$ and p_i is value of physical parameter corresponding to equivalent zero-to-tension stress amplitude of the i-th load cycle.

The material constant C and proportionality parameter α can be omitted if reference fatigue for a given set of cycles is known, for instance, if data from Full Scale Fatigue Test are available. In that case,

$$1 = D_{FSFT} = \tilde{C} \sum_{j=1}^{M_{FSFT}} p_j^{-\frac{1}{B}}, \tag{9}$$

where summation is carried over load cycles p_j exerted on the structure during Full-Scale Fatigue Test (FSFT) and M_{FSFT} is number of load cycles during FSFT. Therefore,

$$D = \frac{D}{D_{FSFT}} = \frac{\sum_{i=1}^{M} p_i^{-\frac{1}{B}}}{\sum_{j=1}^{M_{FSFT}} p_j^{-\frac{1}{B}}}, \tag{10}$$

can be calculated if material constant B and reference data from FSFT is known.

There exist several conditions which need to be satisfied in order to apply the presented approach based on Miner's law. First, if only laboratory material data are available with no reference fatigue exerted on a real test structure, then the correspondence between stress level in the critical location of the structure and the parameter used for fatigue evaluation must be known in exact form and Miner's Equation (6) needs to be used directly instead of the Equation (10). Another requirement is a linear relation between the physical parameter used for fatigue assessment and stress level σ_c in the critical location. As mentioned in Section 2.1, strain gauge readings used in the study are linearly dependent on vertical landing force in full range of admissible loads of the aircraft, as confirmed during FSFT rig calibration. Furthermore, in the case of Su-22 aircraft, it is assumed in load spectra design that the aircraft will be operated within the linear elastic regime of the materials, as if high overloads during flight or very harsh landings occur, special procedures are introduced in order to evaluate aircraft condition (e.g., plastic deformations) and its airworthiness. In addition, no looseness in highly loaded joints is allowed, therefore linearity between vertical force and stress level in critical location is legitimate assumption in our case. Finally, fatigue limit [1] of the material needs to be considered for fatigue estimation based on Equation (10), as load cycles below the fatigue threshold in a given point can be above fatigue limit in critical location. In our case, all strain data corresponding to physical load cycles exerted on main landing gear attachment are taken into account for the purpose of fatigue assessment, therefore all the load cycles in critical location are considered as well.

For the purpose of this study, the equivalent weight w carried by landing gear node was adopted as a physical parameter needed for fatigue wear calculation in accordance with the Equation (8). Equivalent weight was assessed by strain gauge reading and results of linear physical scaling, i.e., values of strain for lifted aircraft with released landing gear were related to no load condition and values of strain readings for aircraft on-ground were related to its measured weight carried by a given landing gear node. Furthermore, data from actuators used for the Full-Scale Fatigue Test were rescaled in those units and used in the denominator of the Equation (10) for fatigue calculation. The parameter B was estimated based on laboratory fatigue tests of material used for landing gear attachment node manufacturing.

For easier interpretation of the obtained results, a notion of Landing Fatigue Equivalent (LFE) can be introduced. LFE represents the relative fatigue of landing gear node due to loads exerted during a given landing D_l with respect to mean fatigue due to simulated landings during FSFT:

$$LFE = \frac{\sum_{i=1}^{M_l} w_i^{-\frac{1}{B}}}{\frac{1}{N}\sum_{j=1}^{M_{FSFT}} w_j^{-\frac{1}{B}}}, \tag{11}$$

where:

- w_i denotes amplitude of equivalent zero-to-tension load cycles exerted on landing gear node during landing (in terms of equivalent weight carried by the node as measured by strain gauge),
- M_l denotes number of load cycles recorded during landing,
- w_j denotes amplitude of equivalent zero-to-tension load cycles exerted on landing gear node during FSFT (in terms of equivalent weight carried by the node) and
- N denotes number of simulated landings during FSFT.

In order to account properly Ground–Air–Ground cycle when calculating LFE for full stop landing, the records of strain gauge obtained for takeoff and landing are joined (Figure 13) prior to determination of load cycles with use of Range-Pair Counting algorithm [28]. Furthermore, the noise level of the recorded signal during on-ground operations was estimated, and for LFE calculation, only relevant recorded cycles, i.e., corresponding to physical load of the structure and higher than the level of noise, were considered. LFE provides a direct measure to estimate the fatigue wear of the landing gear attachment node for a given flight if:

- $LFE < 1$ then loads exerted on landing gear attachment node during a given landing was less severe than mean landing profile during FSFT; or
- $LFE > 1$ then loads exerted on landing gear attachment node during a given landing was more severe than mean landing profile during FSFT.

Cumulative LFE obtained for a given aircraft can be considered as a limiting condition for possibility of further aircraft operation instead of total estimated fatigue for landing gear attachment.

For Touch-And-Go landings, all the load cycles are due to aircraft touchdown. For full stop landing, additional information about this process can be obtained by distinction of load cycles occurring during different stages of landing (Figure 12) and calculation of the corresponding fatigue. An algorithm for the classification of structure load cycles recorded by strain gauges during the normal landings and other ground loads during take-off and full stop landing was developed. It was assumed that the cycles are classified into the following categories:

- cycles recorded during touchdown—P;
- Ground–Air–Ground cycle—GAG;
- cycles recorded during braking—W;
- cycles recorded during taxing before take-off—KSR;

- cycles recorded during taxing after landing—DKL;
- other cycles recorded during takeoff—OS;
- other cycles (not classified elsewhere)—O.

In Figure 13, the general concept of division of a single take-off and landing operation into above defined cycles is shown—the height of the shown color boxes correspond the assumed minimum and maximum values of certain types of cycles for a given flight.

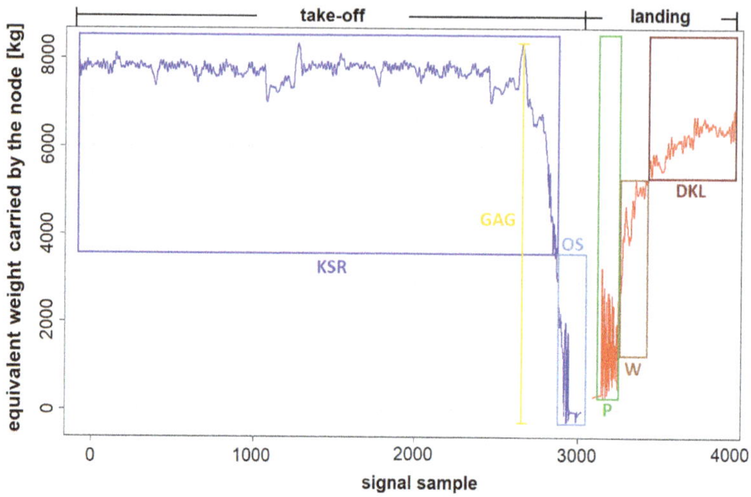

Figure 13. An example of strain record for full stop landing with indication of subsequent stages of this maneuver.

Since:

$$\sum_{i=1}^{M_l} w_i^{-\frac{1}{B}} = w_{GAG}^{-\frac{1}{B}} + \sum_{c \in P} w_c^{-\frac{1}{B}} + \sum_{c \in W} w_c^{-\frac{1}{B}} + \sum_{c \in KSR} w_c^{-\frac{1}{B}} + \sum_{c \in DKL} w_c^{-\frac{1}{B}} + \\ + \sum_{c \in OS} w_c^{-\frac{1}{B}} + \sum_{c \in O} w_c^{-\frac{1}{B}}, \tag{12}$$

where c denotes cycles corresponding to different types, also fatigue equivalent obtained for full stop landing can be factored accordingly:

$$LFE = LFE_{GAG} + LFE_P + LFE_W + LFE_{KSR} + LFE_{DKL} + LFE_{DKL} + \\ + LFE_{OS} + LFE_O. \tag{13}$$

Based on investigation of the defined contributions to fatigue equivalent, additional conclusions can be drawn, e.g., with respect to pilot training program or with respect to condition of the runway, as will be shown further in the text.

3. Results

In the period 2018–2021, a total number of 1563 full stop landings and 350 Touch-And-Go landings of six Su-22 aircraft equipped with Operational Load Monitoring system were recorded. In Figure 14, boxplots of LFE for full stop landing and Touch-And-Go landings are presented.

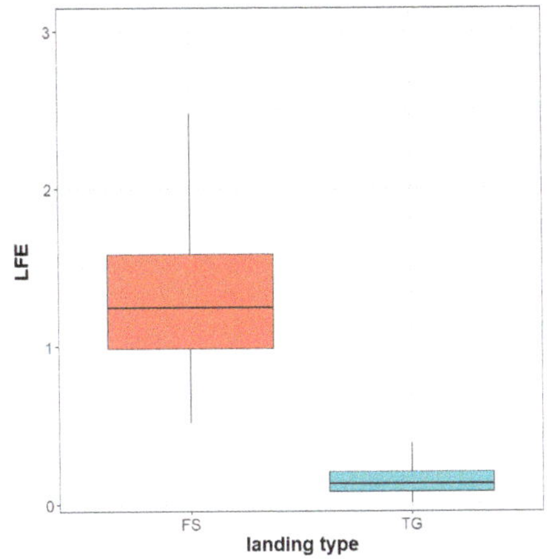

Figure 14. Boxplot of Landing Fatigue Equivalent for full stop (FS) and Touch-And-Go (T&G) landings.

It is demonstrated that the fatigue wear of the landing gear attachment node is significantly higher for full stop landings compared to Touch-And-Go landings. In the former case, the median value of *LFE* is 1.24 compared to 0.13 obtained for the latter landings type. This result is of particular importance for estimation of remaining fatigue life of landing gear attachment node for aircraft not equipped with a load monitoring system, as usually service life is evaluated with respect to number of landings without distinction between full stop and Touch-And-Go landings. Median value of fatigue corresponding to full stop landing is about 24% higher mean fatigue value of simulated landings represented during Full Scale Fatigue Test. Partially, it can be related to number of touchdown cycles represented for simulated landings. For simulated landings, only two load cycles were represented for touchdown as shown in Figure 3, in order to meet time constraint requirements for the test completion. In reality, the number of load cycles during aircraft touchdown is significantly higher as can be seen in Figure 9b or Figure 12.

In Table 4, the distribution of *LFE* to a different type of fatigue cycle is presented. Higher fatigue wear of the landing gear node obtained for full stop landing is mostly due to the Ground–Air–Ground cycle. The median value of the *LFE* for the GAG cycle is 1.04, which is approximately 84% of total median fatigue equivalent. This indicates that increased simulated take off weight of aircraft during FSFT (Figure 3) would represent better true distribution of operational take-off aircraft weight. Considerable contribution to *LFE* is also due to touchdown loads, i.e., median $LFE_P = 0.083$.

Table 4. Contribution of different types of cycles to Landing Fatigue Equivalent (*LFE*).

Cycles Type	GAG	P	D	KL	KS	Other Cycles
median *LFE*	1.040	0.083	0.004	0.030	0.003	0.016

Typically, higher *LFE* is due to increased takeoff weight or due to hard touchdown during landing (Figure 15). In Figure 15a, takeoff and full stop landing for a flight with $LFE = 3.27$ is shown. In this flight, the performed touchdown was smooth with $LFE_P = 0.25$; nevertheless, take-off weight of the aircraft was about 2000 kg higher than assumed for simulated landings during FSFT. This caused significant growth of the GAG cycle contribution to the fatigue of landing gear attachment node. The obtained LFE_{GAG} for this flight was 2.84.

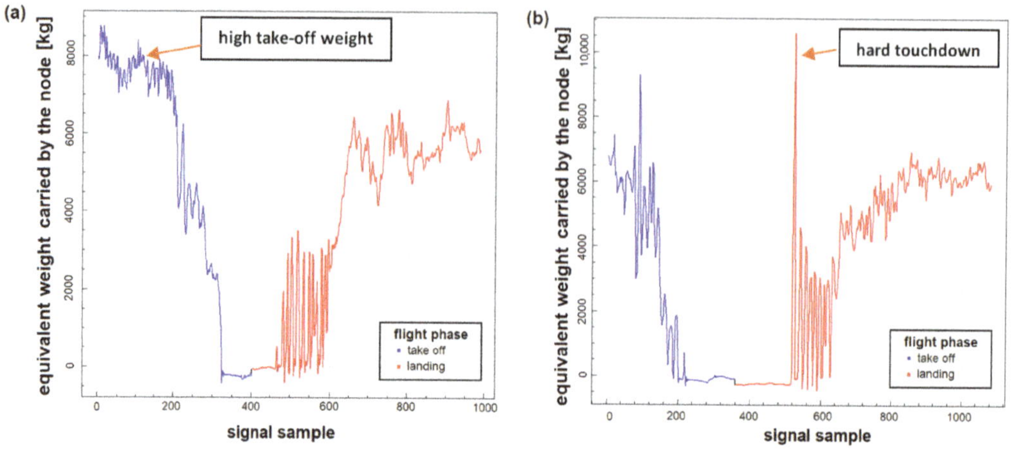

Figure 15. An example of system records for flights with high values of Landing Fatigue Equivalent obtained for full stop landing: (**a**) flight with increased take-off aircraft weight; (**b**) flight with hard touchdown.

The second example is a flight with hard touchdown during landing with very high $LFE = 8.37$. Take-off weight in this case was comparable, as assumed for FSFT, which resulted in $LFE_{GAG} = 1.30$. However, very high load cycle was exerted on the structure during touchdown with the amplitude about 10,000 kg of equivalent carried weight. Thus, LFE_P obtained for the flight was 6.95. Such events can be due to a sudden rapid gust of wind or due to the premature release of the drogue parachute (Figure 16), which is sometimes trained as simulation of landing on an airfield with a very short runway.

Furthermore, interesting cases of increased LFE were observed for aircraft operations performed from a particular air base. In Table 5, the total mean LFE with contributions of selected type of cycles is presented. Significant increase of fatigue of landing gear attachment node was obtained for aircraft operations from one of the airfields (Figure 17).

Table 5. Mean Landing Fatigue Equivalent values corresponding to different types of cycles with respect to operations performed from different air bases.

Localization	GAG	P	KL	KS	Total
air base No. 1	0.95	0.17	0.04	0.07	1.17
air base No. 2	0.99	0.20	0.03	0.06	1.24
air base No. 3	2.41	0.86	0.18	1.21	4.30

Total mean LFE obtained for landings on that airfield was 4.30 compared to fatigue equivalent 1.17 obtained for operations from the home air base. Significant increase of contributions to LFE due to different type of cycles was observed for Base 3, in particular LFE of cycles due to taxing before take-off, was more than 18 times higher than in regular conditions. In fact, abnormal variation of strain during aircraft take-off was observed as shown in Figure 18, which can be caused by particularly bad condition of runway.

Figure 16. Examples of landings with prematurely opened drogue parachute [29–31].

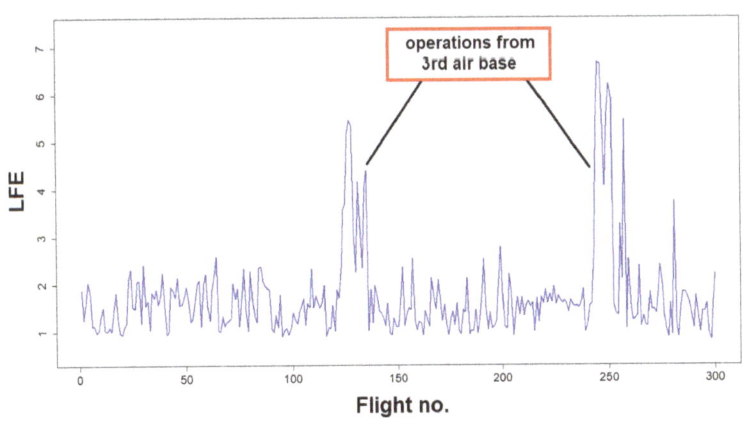

Figure 17. Landing Fatigue Equivalent for subsequent flights of an aircraft in a given period.

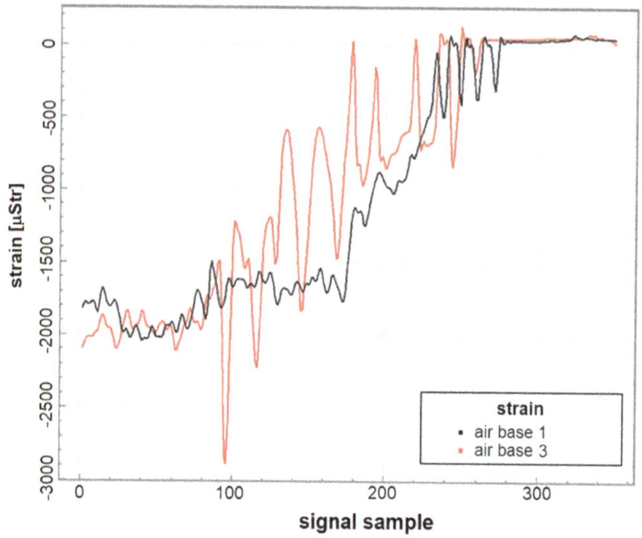

Figure 18. Records of strain during take-off for selected flights from the home base and air base no. 3.

4. Conclusions

In this paper, application of operational load monitoring system of an aircraft to fatigue equivalent estimation of landing gear attachment node was presented. In particular, system description algorithms for signal processing and fatigue estimation were provided. The notion of Landing Fatigue Equivalent (LFE) was introduced as a measure representing the relative fatigue of landing gear node due to loads exerted during a given landing with respect to mean fatigue of simulated landings during Full-Scale Fatigue Test of the structure. It was demonstrated that the LFE obtained for full stop landings is significantly higher than LFE for Touch-And-Go landings. The contribution of different types of load cycles to LFE was also investigated. Predominant contribution to LFE, i.e., over 80%, is due the Ground–Air–Ground (GAG) cycle; furthermore, loads exerted on landing gear attachment node during aircraft touchdown have considerable effect on fatigue wear of the structure. Particular examples of flights with high values of LFE were presented and discussed.

Author Contributions: Conceptualization, M.D.; methodology, M.D., M.K. and P.R.; software, M.D.; validation, M.K. and A.K.; formal analysis, M.K. and K.D.; investigation, M.D. and M.K.; resources, M.D. and A.K.; data curation, A.K. and M.D.; writing—original draft preparation, M.D. and P.R.; writing—review and editing, S.K., A.L. and K.D.; visualization, M.D. and P.R.; project administration, M.K., A.L. and K.D.; scientific consultation, S.K., A.L. and K.D.; supervision, S.K. and K.D. All authors have read and agreed to the published version of the manuscript.

Funding: This research received no external funding.

Institutional Review Board Statement: Not applicable.

Informed Consent Statement: Not applicable.

Data Availability Statement: Data used in this study are available on-demand from the corresponding author.

Acknowledgments: In the memory of Colonel Andrzej Lesniczak.

Conflicts of Interest: The authors declare no conflict of interest.

Abbreviations

The following abbreviations are used in this manuscript:

FDR	Flight Data Recorder
FSFT	Full Scale Fatigue Test
GAG	Ground–Air–Ground (cycle)
IAT	Individual Aircraft Tracking
LFE	Landing Fatigue Equivalent
MLG	Main Landing Gear
OLM	Operational Load Monitoring
T&G	Touch-And-Go (landing)

References

1. Milella, P.P. *Fatigue and Corrosion in Metals*; Springer: Milano, Italy, 2013.
2. Schütz, W. A history of fatigue. *Eng. Fract. Mech.* **1996**, *54*, 263–300. [CrossRef]
3. Riddick, H. *Safe-Life and Damage-Tolerant Design Approach for Helicopter Structures Applied Technology Laboratory*; US Army Research and Technology Laboratories (AVRADCOM): Newport News, VA, USA, 1984.
4. Schijve, J. Fatigue of aircraft materials and structures. *Int. J. Fatigue* **1994**, *16*, 21–32. [CrossRef]
5. Leski, A.; Kurdelski, M.; Reymer, P.; Dragan, K.; Sałaciński, M. Fatigue life assessment of PZL-130 Orlik structure–final analysis and results. In Proceedings of the 28th ICAF Symposium, Helsinki, Finland, 1–5 June 2015; pp. 294–303.
6. Daverschot, D.; Mattheij, P.; Renner, M.; Ardianto, Y.; De Araujo, M.; Graham, K. Full-Scale Fatigue Testing from a Structural Analysis Perspective. In *ICAF 2019—Structural Integrity in the Age of Additive Manufacturing*; Springer: Cham, Switzerland, 2019; pp. 788–800.
7. Schijve, J. Fatigue damage in aircraft structures, not wanted, but tolerated? *Int. J. Fatigue* **2009**, *31*, 998–1011. [CrossRef]
8. Wanhill, R.; Molent, L.; Barter, S. Milestone Case Histories in Aircraft Structural Integrity. In *Reference Module in Materials Science and Materials Engineering*; Elsevier Science: Amsterdam, The Netherlands, 2016.
9. Kim, Y.; Sheehy, S.; Lenhardt, D. *A Survey of Aircraft Structural-Life Management Programs in the US Navy, the Canadian Forces, and the US Air Force*; Rand Corporation: Santa Monica, CA, USA, 2006; Volume 370.
10. *MIL-STD-1530D, Department of Defense Standard Practice Aircraft Structural Integrity Program (ASIP)*; US Department of Defense: Washington, DC, USA, 2016.
11. Grandt, A.F., Jr. *Fundamentals of Structural Integrity: Damage Tolerant Design and Nondestructive Evaluation*; John Wiley & Sons: Hoboken, NJ, USA, 2003.
12. Dilger, R.; Hickethier, H.; Greenhalgh, M.D. Eurofighter a safe life aircraft in the age of damage tolerance. *Int. J. Fatigue* **2009**, *31*, 1017–1023. [CrossRef]
13. Rui, L.; Xiaofan, H.; Yuhai, L. Individual aircraft life monitoring: An engineering approach for fatigue damage evaluation. *Chin. J. Aeronaut.* **2018**, *31*, 727–739.
14. Reymer, P.; Kurdelski, M.; Leski, A.; Leśniczak, A.; Dziendzikowski, M. Introduction of an Individual Aircraft Tracking Program for the Polish Su-22. *Fatigue Aircr. Struct.* **2017**, *9*, 101–108. [CrossRef]
15. Boller, C.; Staszewski, W.J. Aircraft Structural Health and Usage Monitoring. In *Health Monitoring of Aerospace Structures*; John Wiley & Sons, Ltd.: Hoboken, NJ, USA, 2003; Chapter 2, pp. 29–73.
16. Kressel, I.; Dorfman, B.; Botsev, Y.; Handelman, A.; Balter, J.; Pillai, A.; Prasad, M.; Gupta, N.; Joseph, A.; Sundaram, R.; et al. Flight validation of an embedded structural health monitoring system for an unmanned aerial vehicle. *Smart Mater. Struct.* **2015**, *24*, 075022. [CrossRef]
17. Nicolas, M.J.; Sullivan, R.W.; Richards, W.L. Large scale applications using FBG sensors: Determination of in-flight loads and shape of a composite aircraft wing. *Aerospace* **2016**, *3*, 18. [CrossRef]
18. Iele, A.; Leone, M.; Consales, M.; Persiano, G.; Brindisi, A.; Ameduri, S.; Concilio, A.; Ciminello, M.; Apicella, A.; Bocchetto, F.; et al. Load monitoring of aircraft landing gears using fiber optic sensors. *Sens. Actuators A Phys.* **2018**, *281*, 31–41. [CrossRef]
19. Ma, Z.; Chen, X. Fiber Bragg gratings sensors for aircraft wing shape measurement: Recent applications and technical analysis. *Sensors* **2019**, *19*, 55. [CrossRef] [PubMed]
20. Molent, L. A review of a strain and flight parameter data based aircraft fatigue usage monitoring system. In Proceedings of the USAF ASIP Conference, San Antonio, TX, USA, 3–5 December 1996.
21. Molent, L.; Barter, S.; Foster, W. Verification of an individual aircraft fatigue monitoring system. *Int. J. Fatigue* **2012**, *43*, 128–133. [CrossRef]
22. Leski, A.; Reymer, P.; Kurdelski, M.; Zieliński, W.; Jankowski, K. Full scale fatigue test of the Su-22 aircraft–Assumptions, process and preliminary conclusions. *AIP Conf. Proc.* **2016**, *1780*, 020002.
23. Reymer, P.; Kurdelski, M.; Leski, A.; Jankowski, K. The Definition of the Load Spectrum for SU-22 Fighter-Bomber Full Scale Fatigue Test. *Fatigue Aircr. Struct.* **2015**, *1*, 28–33. [CrossRef]
24. Molent, L.; Aktepe, B. Review of fatigue monitoring of agile military aircraft. *Fatigue Fract. Eng. Mater. Struct.* **2000**, *23*, 767–785. [CrossRef]

25. Civera, M.; Ferraris, M.; Ceravolo, R.; Surace, C.; Betti, R. The Teager-Kaiser Energy Cepstral Coefficients as an Effective Structural Health Monitoring Tool. *Appl. Sci.* **2019**, *9*, 5064. [CrossRef]
26. Kurdelski, M.; Leski, A.; Krzysztof, D. Fatigue life analysis of main landing gear pull–rod of the fighter jet aircraft'. In Proceedings of the 28th International Congress of the Aeronautical Sciences, Brisbane, Australia, 23–28 September 2012.
27. *DOT/FAA/AR-MMPDS-01: Metallic Materials Properties Development and Standardization (MMPDS)*; Office of Aviation Research: Washington, DC, USA, 2003.
28. ASTM. *E1049–Standard Practices for Cycle Counting in Fatigue Analysis*; ASTM International: West Conshohocken, PA, USA, 2017.
29. Su-22—Hamowanie Spadochronem Przed Przyziemieniem. Available online: https://www.youtube.com/watch?v=g4jMkJiCTLE (accessed on 7 October 2021).
30. RIAT 2015 | Su-22 Arrives. Available online: https://www.youtube.com/watch?v=iyLowoSilgY (accessed on 7 October 2021).
31. Su-22 Drogue-Chute Landing Gone Wrong. Available online: https://www.youtube.com/watch?v=c3DlE_GJFMw (accessed on 7 October 2021).

materials

MDPI

Article

Experimental Study on Forged TC4 Titanium Alloy Fatigue Properties under Three-Point Bending and Life Prediction

Bohan Wang *, Li Cheng and Dongchun Li

Aeronautics Engineering College, Air Force Engineering University, Xi'an 710038, China; cheng_qiaochu@foxmail.com (L.C.); lidongdong313@163.com (D.L.)
* Correspondence: dream0729ing@foxmail.com; Tel.: +86-029-8478-7225

Abstract: Ultrasonic fatigue tests of TC4 titanium alloy equiaxed I, II and bimodal I, II obtained by different forging processes were carried out in the range from 10^5 to 10^9 cycles using 20 kHz three-point bending. The results showed that the S-N curves had different shapes, there was no traditional fatigue limit, and the bimodal I had the best comprehensive fatigue performance. The fracture morphology was analyzed by SEM, and it was found that the fatigue cracks originated from the surface or subsurface facets, showing a transgranular quasi-cleavage fracture mechanism. EDS analysis showed that the facets were formed by the cleavage of primary α grains, and the fatigue cracks originated from the primary α grain preferred textures, rather than the primary α grain clusters. From the microstructure perspective, the reasons for better equiaxed high-cycle-fatigue properties and better bimodal ultra-high-cycle-fatigue properties were analyzed. The bimodal I fatigue life prediction based on energy was also completed, and the prediction curve was basically consistent with the experimental data.

Keywords: forging; TC4 titanium alloy; three-point bending; very high cycle fatigue; microstructure; life prediction

Citation: Wang, B.; Cheng, L.; Li, D. Experimental Study on Forged TC4 Titanium Alloy Fatigue Properties under Three-Point Bending and Life Prediction. *Materials* **2021**, *14*, 5329. https://doi.org/10.3390/ma14185329

Academic Editors: Jaroslaw Galkiewicz and Lucjan Śnieżek

Received: 24 August 2021
Accepted: 13 September 2021
Published: 15 September 2021

1. Introduction

With the continuous improvement of aviation equipment reliability and life index, the ultra-high-cycle-fatigue (10^7–10^{12} cycles) problem has attracted increasingly more attention, and has become the focus of fatigue research [1–5]. Ultra-high-cycle fatigue belongs to micro-scale fatigue, which differs from traditional high-cycle fatigue (10^5~10^7 cycles) not only in the life length, but also in the more complex crack initiation and initial propagation mechanism, and also in the great difference due to different materials [6–8]. Titanium alloys have become the most widely used metal materials in the modern aviation industry because of their high specific strength, low density, excellent corrosion resistance, and good heat resistance, and the ultra-high-cycle fatigue problems are also the most prominent. Among them, TC4 titanium alloy is widely used in the manufacture of aero-engine compressor blades. The engine rotating speed is as high as tens of thousands of revolutions per minute, and it is very easy for rotating parts such as blades and discs to cause high-frequency flutter under the internal flow disturbance and rotor imbalance action, and ultra-high-cycle-fatigue fracture problems are easy to occur when accumulating for a long time [9,10]. This shows that the safety of aero-engine compressor design cannot be ensured according to the traditional 10^7 fatigue limit design theory.

Due to the observation method limitations and the theoretical analysis complexity, ultra-high-cycle-fatigue studies basically adopt phenomenological analysis methods based on experimental results, including macroscopic analysis of the S-N curve and Goodman diagram based on experimental results, and microscopic analysis of crack initiation and initial propagation based on fracture morphology. The proportion of crack initiation life is more than 95% in the high-cycle-fatigue and ultra-high-cycle-fatigue regimes [11,12]; therefore, the crack initiation mechanism and its relationship with fatigue life are very important. A large number of studies have shown that [13–15] the remarkable crack initiation zone sign for titanium alloys is the smooth facet morphology, especially for failure on the subsurface. The reasons for the facet formation involve three different mechanisms: cleavage [16,17], slip [18,19], and twins [20], which are highly controversial at present. Researchers have also made in-depth analyses based on the test results, constructed many ultra-high-cycle-fatigue fracture analysis models of titanium alloys, and predicted the life [21,22]. However, most studies have focused on the axial tension-compression loading mode, which is not consistent with the bending vibration mode of aero-engine blades, so whether these analytical models are suitable for three-point bending loading remains to be further studied.

Forging is the most widely used method for the plastic forming of titanium alloys, and it is also a key link in the blade manufacturing process. The forging performance is mainly determined by the forging process. Once a bad microstructure is formed in the forging process, the subsequent heat treatment is difficult to improve. In addition, titanium alloys are very sensitive to the forging process, in which temperature determines the solid phase transformation behavior of titanium alloys, and the deformation degree and deformation rate also affect the morphology, size, proportion, and distribution of α and β phases [23,24]. Tanaka et al. [25] studied the quantitative relationship between the Ti-17 titanium alloy fatigue properties and microstructure at different forging temperatures. It was found that the fatigue strength is closely related to the microstructure factors such as the volume fraction of the equiaxed α phase, and it is one of the crack initiation sites during low-temperature solid solution aging. The strength difference between the acicular α phase and fine $\alpha + \beta$ phase is the main reason for crack initiation after high-temperature solid solution aging. Nikitin et al. [26,27] studied the ultra-high-cycle-fatigue crack initiation mechanism of forged VT3-1 titanium alloy under tension and torsion. It was found that the traditional fatigue limit does not exist and the crack initiation position changes from surface to subsurface. When the stress ratio is $R = -1$, the subsurface cracks initiate at the strong defects, "macro region" boundaries, quasi-facets, and facets, while when the stress ratio is $R = 0.1$, the cracks mainly initiate at the "macro region" boundaries. Sinha et al. [28] studied the relationship between the fatigue life scatter and fracture mechanism of the forged Ti-6242Si alloy, and clarified the microcrack initiation and propagation process by the quantitative fatigue fracture characterization. Combined with the crystallographic characteristics analysis of the crack initiation facet and adjacent facet, it was concluded that the difference in cracking degrees along the primary α grains base plane leads to great specimen fatigue life variability.

In this paper, the forged TC4 titanium alloy three-point bending fatigue properties were studied. Equiaxed I, II and bimodal I, II were obtained by four forging processes. In addition, 20 kHz room-temperature ultrasonic fatigue tests were carried out, and the corresponding S-N curves and fatigue fractures were obtained. The crack initiation mechanism was revealed by fracture SEM + EDS analysis, and the microstructure effect on fatigue properties was analyzed by the changing trend of S-N curves. The bimodal I life prediction based on the energy method was also carried out. The work performed was helpful to determine the forging process with the best anti-fatigue ability of aero-engine blade materials.

2. Materials and Methods

2.1. Materials

The raw material was aviation-grade TC4 bar with a diameter of 28 mm, and its chemical composition is shown in Table 1. The TC4 titanium alloy phase transition temperature measured by the continuous temperature-increasing metallographic method was 998 °C. On this basis, four process combinations of two temperatures (950 °C and 985 °C) and two deformations (39.3% and 69.6%) were selected for plate die forging. The single piece blanking size was 150 mm, and the bars were heated to the specified temperature for 15~20 min and then flattened along the radial direction. After forging, the bars were annealed at 720 °C × 1 h + AC.

Table 1. Chemical composition of TC4 titanium alloy (wt%).

Al	V	Fe	O	C	H	N	Ti
6.27	4.08	0.048	0.021	0.020	0.004	0.031	Bal.

Figure 1 shows the microstructures obtained by four forging processes. It was found that the forging temperature has a great influence on the microstructure morphology, while the effect of deformation is relatively small. The equiaxed primary α phase can be clearly observed, and the remaining phases are the β transformation microstructure, including the fine acicular secondary α phase and residual β phase. No obvious defects such as crack, inclusion, segregation, and folding were found. The primary α content, grain size, and standard deviation were measured, and the results are shown in Table 2. The equiaxed I, II and bimodal I, II were obtained.

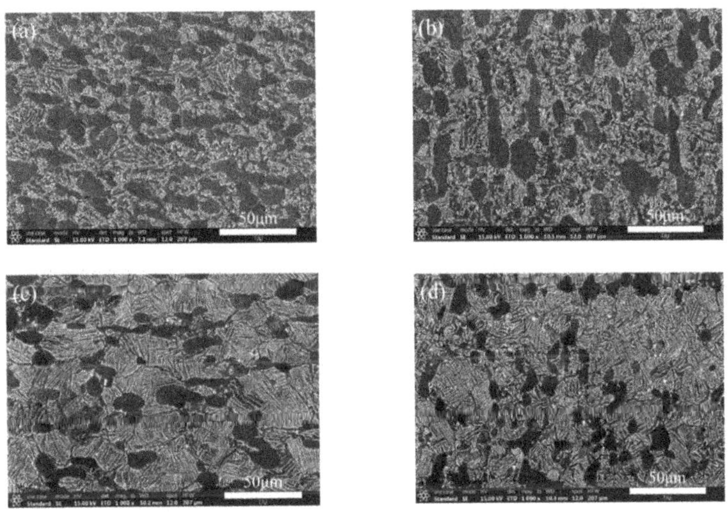

Figure 1. Microstructures of different forged TC4 titanium alloys. (**a**) Equiaxed I; (**b**) equiaxed II; (**c**) bimodal I; (**d**) bimodal II.

Table 2. Measurement results of primary α phase content, grain size, and standard deviation.

Forging Process	Forging Type	Content (%)	Grain Size (μm)	Standard Deviation (μm)	Microstructure
950 °C/39.3%	α + β	45.24	42.54	13.75	equiaxed I
950 °C/69.6%	α + β	48.74	34.68	16.80	equiaxed II
985 °C/39.3%	near β	27.54	23.62	19.92	bimodal I
985 °C/69.6%	near β	23.78	24.68	10.43	bimodal II

The room-temperature tensile properties of four kinds of forgings were obtained by a WDW-300 universal testing machine, and the test results were averaged for 3 times. The stress–strain curves were drawn according to the relationship between tension and displacement during loading, as shown in Figure 2. Finally, the tensile strength σ_R, yield strength $\sigma_{0.2}$, elongation A, section shrinkage Z, elastic modulus E, and Poisson's ratio v were obtained, as shown in Table 3.

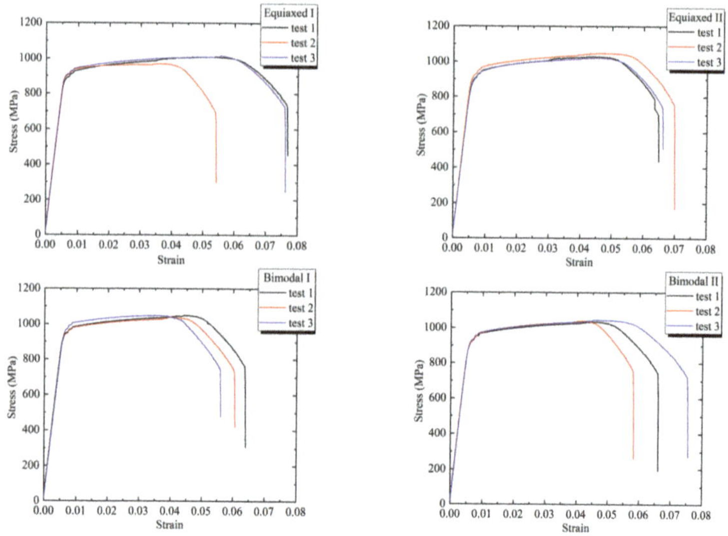

Figure 2. Stress–strain curves of different forged TC4 titanium alloys.

Table 3. Room-temperature tensile properties of different forged TC4 titanium alloys.

Microstructure	σ_R/MPa	$\sigma_{0.2}$/MPa	*A*/%	*Z*/%	*E*/GPa	*v*
equiaxed I	967	898	16.2	46.8	115.85	0.317
equiaxed II	1003	951	15.8	47.6	116.83	0.329
bimodal I	986	907	14.2	51.1	120.29	0.332
bimodal II	998	924	15.2	49.5	118.91	0.324

2.2. Methods

Finite element modal analysis was used to carry out the specimen design, the linear perturbation analysis step was selected after the three-dimensional model was introduced, the first 10 free vibration modes were extracted, the stress and displacement were set as output variables, and the structured hexahedral grid was divided. Finally, the modal analysis results of the 7th-order three-point bending vibration were obtained, as shown in Figure 3. The modal frequency was 19,954 Hz, which is close to the design frequency of 20 kHz. The middle area of the specimen bottom bore the maximum tensile stress, which is the expected fracture position, and the two displacement standing points were symmetrical with respect to the middle section, which are the fulcrum positions. By modifying the specimen length to optimize the resonant frequency, it was determined that the four forged specimen lengths *L* were 31.5, 31.6, 31.8, and 31.7 mm respectively, as shown in Figure 4.

The tests were carried out by using the HC-DF2020GD-K2 multi-function ultrasonic fatigue testing machine (HC SONIC, Hangzhou, China). The loading frequency was 18.5~20.5 kHz, the static loading range was 0~10 kN, and the force sensor control accuracy was 0.5%. The horn output amplitude was 10~150 μm, and the amplitude control precision was 1 μm. The three-point bending loading process is shown in Figure 5. The middle position of the specimen bottom bore both bending stress σ_m and vibration stress σ_a, and the stress ratio $R = (\sigma_m - \sigma_a)/(\sigma_m + \sigma_a)$. The relationship between σ_m and static pressure *P* was calibrated by a strain gauge. σ_a is proportional to the horn output amplitude A_0, and the ratio was obtained from the stress–displacement nephogram in Figure 3. The fatigue test started from the high-stress area, and the stress amplitude was gradually reduced by 15 MPa until the cycle reached 1×10^9, and each stress level was subjected to 1 to 2 samples. The output amplitudes under different ultrasonic powers were measured by an LV-S01 laser vibrometer (SOPTOP, Ningbo, China) and used to calibrate the control software. In order to restrain the temperature rise caused by high frequency and ensure the stability of the fatigue test, the specimens were cooled by water mist.

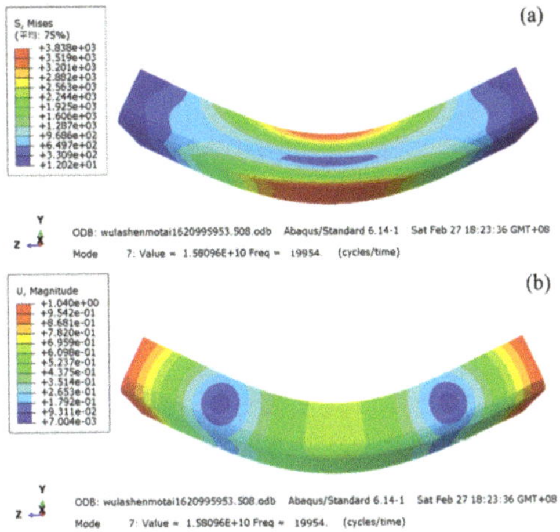

Figure 3. Modal analysis nephogram of equiaxial I. (**a**) Stress distribution; (**b**) displacement distribution.

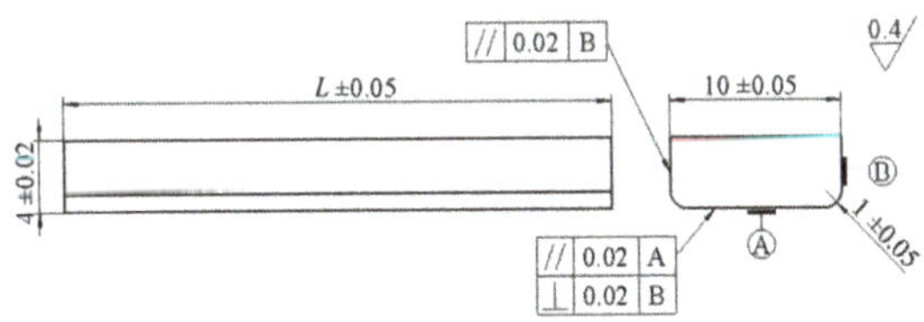

Figure 4. Design drawing of three-point bending specimen.

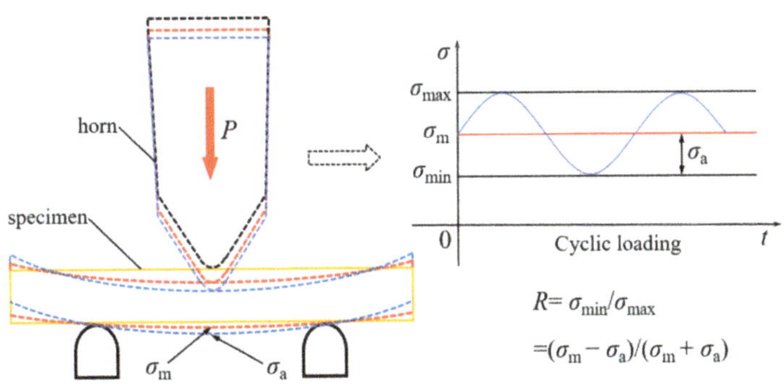

Figure 5. Schematic diagram of three-point bending loading process.

The Apreo S SEM (FEI, Hillsboro, OR, USA) was used to observe the fracture morphology, and through the energy spectrum module installed on it, EDS analysis of the ultra-high-cycle fatigue crack origin zones of the four microstructures was carried out.

3. Results

3.1. S-N Curves

As the number of loading cycles increases, the crack initiation changes from surface to subsurface, which shows that the difference between ultra-high-cycle-fatigue and high-cycle-fatigue lies not only in lower stress and longer life, but also in the fatigue failure mechanism. The S-N curves of the four microstructures include the single-step decline type, double-step decline type, and straight-step decline type, and there is no traditional fatigue limit. The equiaxed high-cycle-fatigue properties are obviously better than those of bimodal microstructures, while the equiaxed I ultra-high-cycle-fatigue property is worst. The equiaxed II ultra-high-cycle-fatigue property is similar to that of bimodal I, but the fatigue life scatter is larger, so it is considered that the bimodal ultra-high-cycle-fatigue properties are better than those of equiaxed microstructures. From the analysis in Figure 6, it seems that the high-cycle-fatigue and ultra-high-cycle-fatigue performances cannot be obtained at the same time. In the range of 10^5 to 10^9 cycles, bimodal I has better comprehensive fatigue performance, and 985 °C with a 39.3% deformation degree is determined to be the best forging process.

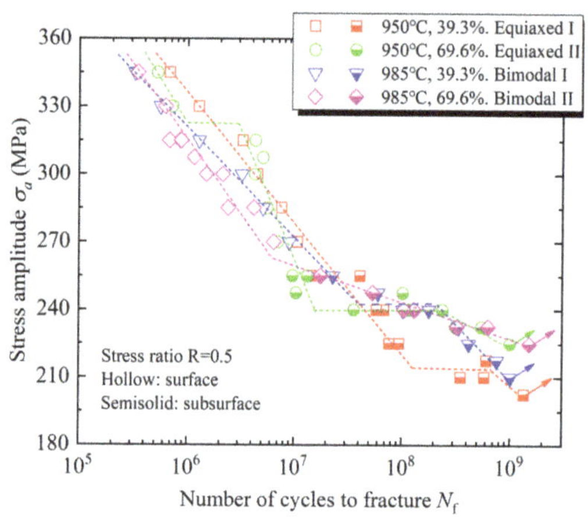

Figure 6. S-N curves of different forged TC4 titanium alloys.

3.2. Fatigue Fracture Morphology

Figure 7a–d show the high-cycle-fatigue crack origin zone SEM morphology of different forged TC4 titanium alloys at low magnification. It is observed that the river-like crack propagation patterns converge on the specimen surface, in which the bimodal I has a bright and continuous main crack, and the propagation direction is almost vertical. There are no obvious inclusion traces in the origin zones, which is significantly different from the fisheye morphology in high-strength steel. Figure 7e–h show the high-cycle-fatigue crack origin zone SEM morphology at high magnification, and many smooth facet features are observed, especially on the surface, which are considered to be the crack origins. There are a large number of cleavage steps and microplastic tear (MPT) morphologies around the facets. Overall, there is no obvious difference in the crack origin zone SEM morphology for the four kinds of microstructures, and all of them are mainly transgranular brittle fracture. Combined with the crack propagation path, it is determined that the crack first initiates from the surface facet, initially propagates with the cleavage steps and MPT characteristics, and then converges with the subsurface facet cracks to form the main crack, which finally leads to fatigue fracture, which is shown as the transgranular quasi-cleavage fracture mechanism.

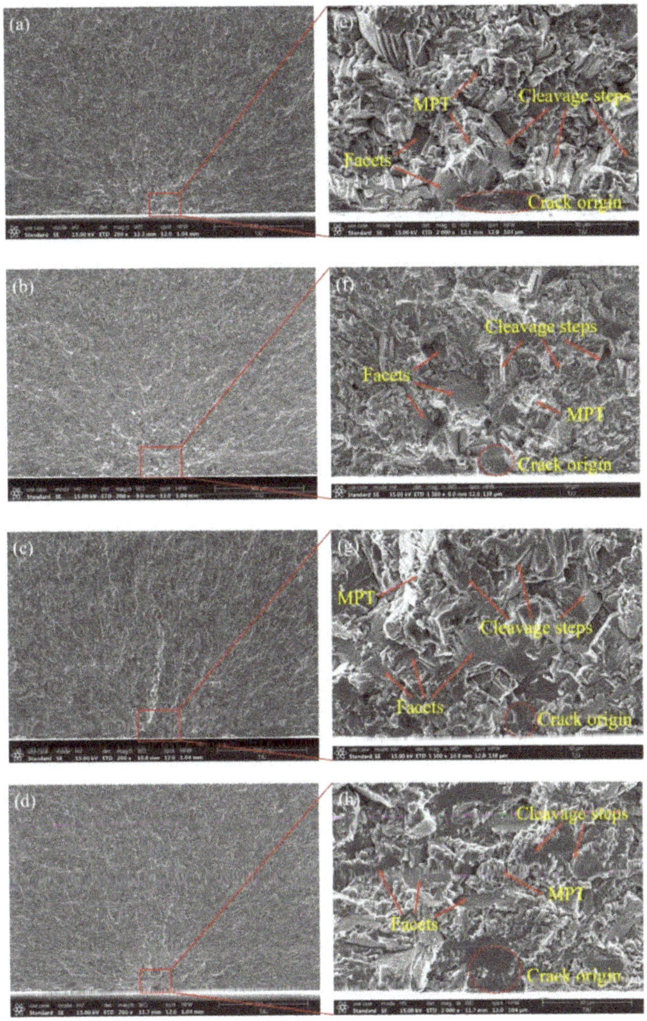

Figure 7. SEM morphology of high-cycle-fatigue crack origin zone. (**a**,**e**) Equiaxed I, σ_a = 330 MPa, N_f = 1.26 × 10^6; (**b**,**f**) equiaxed II, σ_a = 300 MPa, N_f = 4.25 × 10^6; (**c**,**g**) bimodal I, σ_a = 285 MPa, N_f = 5.08 × 10^6; (**d**,**h**) bimodal II, σ_a = 270 MPa, N_f = 6.39 × 10^6.

Figure 8a–d show the ultra-high-cycle-fatigue crack origin zone SEM morphology of different forged TC4 titanium alloys at low magnification. The crack still propagates in a river shape, but converges on the specimen subsurface, in which the equiaxed II main crack patterns are the most obvious, and the propagation direction is oblique 45°. Figure 8e–h show the ultra-high-cycle-fatigue crack origin zone SEM morphology at high magnification, and the origin zone center is the facet cluster morphology, not the inclusion characteristics. There are a large number of cleavage steps and MPT features around the facets. The analysis shows that the near small facets converge into large facets through the cleavage steps, while the distant large facets are connected by MPT characteristics, and the whole is still dominated by transgranular brittle fracture. The distance between the subsurface facet and the surface of the four kinds of microstructures is about 48, 39, 45, and 43 μm. Combined with the crack propagation path, it is concluded that the crack first originates from the subsurface facets, and then the faceted cracks with different heights and orientations connect and propagate each other through the cleavage steps and MPT features, and finally merge with the main crack to lead to fatigue fracture, which is still the transgranular quasi-cleavage fracture mechanism.

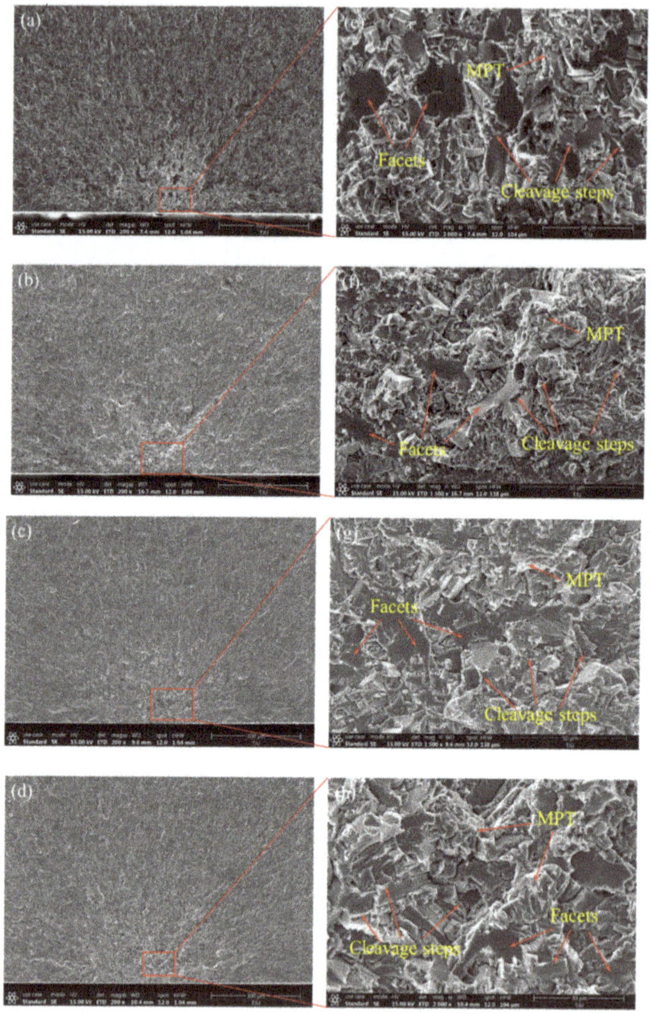

Figure 8. SEM morphology of ultra-high-cycle-fatigue crack origin zone. (**a**,**e**) Equiaxed I, σ_a = 218 MPa, N_f = 6.06 × 10^8; (**b**,**d**) equiaxed II, σ_a = 233 MPa, N_f = 3.16 × 10^8; (**c**,**f**) bimodal I, σ_a = 225 MPa, N_f = 4.15 × 10^8; (**g**,**h**) bimodal II, σ_a = 233 MPa, N_f = 3.25 × 10^8.

The fatigue crack initial propagation depth is about between 71 μm~273 μm, and then it will enter the slow and stable propagation stage. The crack propagation region is relatively flat, and the slow propagation zone shows a mixed shape of dense shear tear edges and secondary cracks. Due to the low crack propagation rate, the fatigue band is not obvious, but there are a large number of secondary cracks, indicating that the crack direction changes after it encounters the second phase or grains with different orientations in the crack propagation process. The stable propagation region is composed of many small fault blocks with different sizes and heights, on which there exist thin and short fatigue striations, as shown in Figure 9. A series of basically parallel and slightly curved fatigue striations are clearly visible. Each fatigue striation represents the crack tip position under the cycle, and the number of fatigue striations is roughly equal to the number of cycles and perpendicular to the local crack propagation direction. The high-cycle-fatigue striation spacings of the four microstructures are 0.097, 0.139, 0.219, and 0.140 μm, respectively, and the ultra-high-cycle-fatigue striation spacings are 0.095, 0.087, 0.118, and 0.086 μm, respectively. For the same microstructure, the distance between the fatigue striation observation position and the origin zone is basically the same, so the fatigue striation spacing in the local microzone basically reflects the crack propagation rate. The ultra-high-cycle-fatigue striation spacing is smaller than that of high-cycle-fatigue, so the crack propagation life is longer.

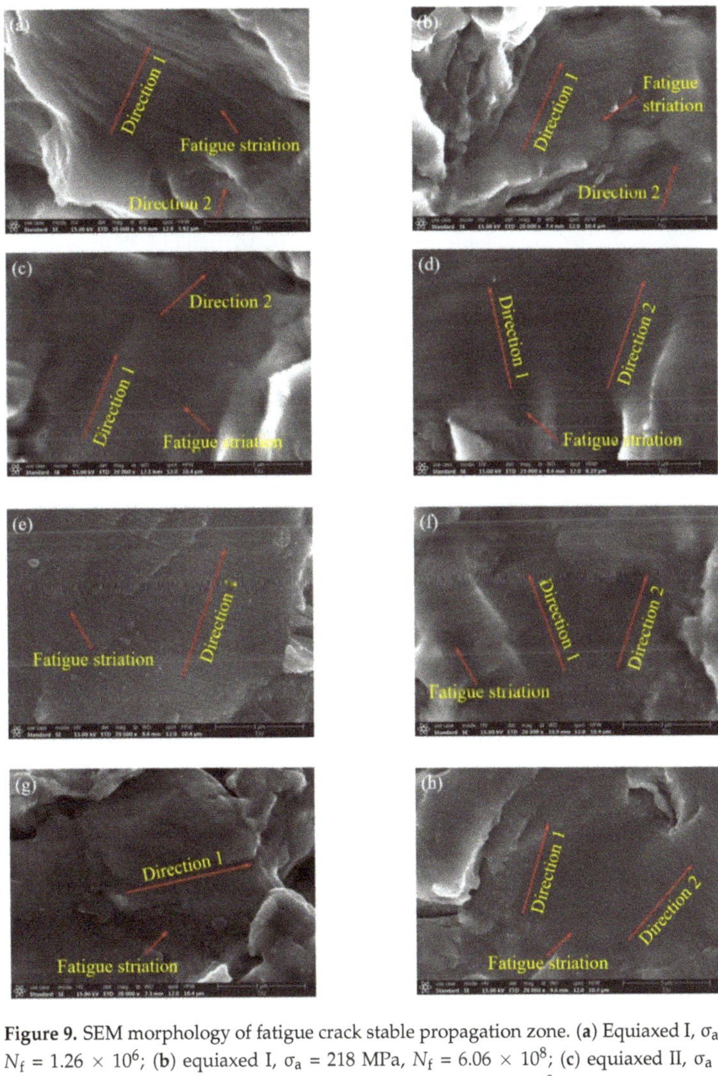

Figure 9. SEM morphology of fatigue crack stable propagation zone. (**a**) Equiaxed I, σ_a = 330 MPa, $N_f = 1.26 \times 10^6$; (**b**) equiaxed I, σ_a = 218 MPa, $N_f = 6.06 \times 10^8$; (**c**) equiaxed II, σ_a = 300 MPa, $N_f = 4.25 \times 10^6$; (**d**) equiaxed II, σ_a = 233 MPa, $N_f = 3.16 \times 10^8$; (**e**) bimodal I, σ_a = 285 MPa, $N_f = 5.08 \times 10^6$; (**f**) bimodal I, σ_a = 225 MPa, $N_f = 4.15 \times 10^8$; (**g**) bimodal II, σ_a = 270 MPa, $N_f = 6.39 \times 10^6$; (**h**) bimodal II, σ_a = 233 MPa, $N_f = 3.25 \times 10^8$.

4. Discussion

4.1. Analysis of Crack Initiation Mechanism

For TC4 titanium alloy, Al and V are the stable elements of the α and β phase, respectively, that is, the Al content in the α phase should be higher than that of the matrix, and the V content should be lower. EDS analysis was carried out on the facets in the ultra-high-cycle-fatigue crack origin zones of the four kinds of microstructures, and the results are shown in Figure 10. Among the 24 randomly selected facets, the V content is lower than the matrix level, and the Al content in most facets is higher than the matrix level, while the Al content in a few facets is lower, and the missing part is replaced by the Ti element. Considering that the facets are about tens of microns, slightly smaller than the primary α grain size, it is concluded that the facets in the crack origin zone are formed by the cleavage of primary α grains. The formation of these primary α grains is mainly attributed to the Al element-promoting effect, while a few of them hardly need the Al element assistance. Combined with SEM and EDS analysis, it is concluded that the crack initiation mechanism is the cleavage of primary α grains on the surface or subsurface, corresponding to high-cycle-fatigue and ultra-high-cycle-fatigue, respectively, and the fatigue failure mechanism is mainly transgranular quasi-cleavage brittle fracture.

The facets in the crack origin zone appear in the form of clusters, rather than the single facet. Chandran [29] believed that the spatial α grain agglomeration leads to the formation of fatigue cracks in Ti-10V-2Fe-3Al titanium alloy. In order to investigate whether the TC4 titanium alloy fatigue cracks originate from a single facet or facet clusters, four kinds of microstructure ultra-high-cycle-fatigue origin zone facet clusters were analyzed by EDS. If the fatigue cracks originate in facet clusters, the Al content should be significantly higher than the matrix level, and the V content should be lower. As shown in Figure 11, the three element contents are very close to the matrix level, so it is concluded that facet clusters are not the cause of fatigue crack formation. Kun et al. [30] obtained a similar conclusion by analyzing the Ti-8Al-1Mo-1V titanium alloy inverse pole diagram along the loading direction. It was found that the grain orientation is random on the whole, but the adjacent grain orientation is basically the same in the local microzone, indicating that the fatigue cracks originate from the preferred texture of primary α grains rather than clusters. To sum up, the fatigue crack initiation mechanism is the cleavage of primary α grains with specific spatial orientation and crystal orientation for the four kinds of microstructures.

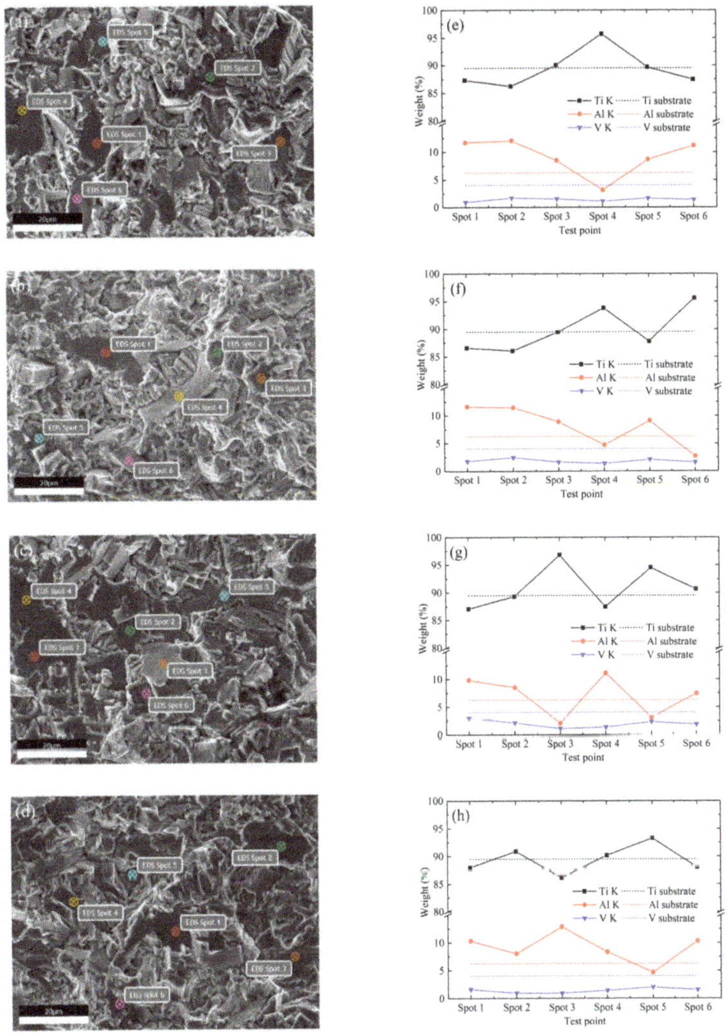

Figure 10. EDS analysis of facets in ultra-high-cycle-fatigue crack origin zone. (**a**,**e**) Equiaxed I, $\sigma_a = 218$ MPa, $N_f = 6.06 \times 10^8$; (**b**,**d**) equiaxed II, $\sigma_a = 233$ MPa, $N_f = 3.16 \times 10^8$; (**c**,**f**) bimodal I, $\sigma_a = 225$ MPa, $N_f = 4.15 \times 10^8$; (**g**,**h**) bimodal II, $\sigma_a = 233$ MPa, $N_f = 3.25 \times 10^8$.

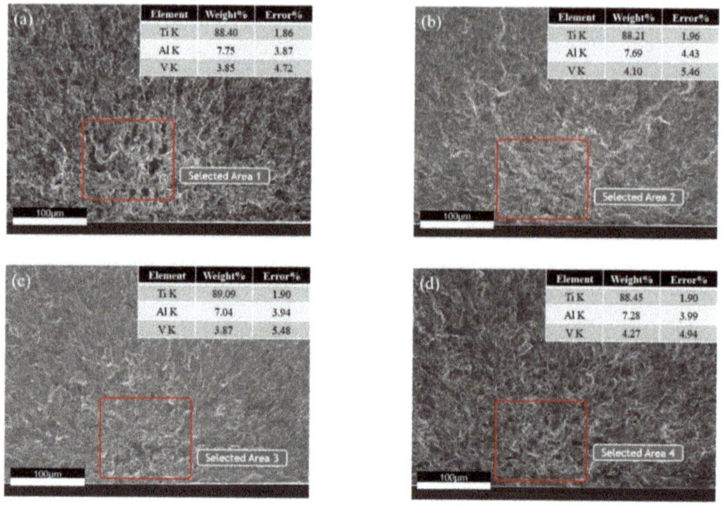

Figure 11. EDS analysis of facet clusters in ultra-high-cycle-fatigue crack origin zone. (**a**) Equiaxed I, σ_a = 218 MPa, N_f = 6.06 × 10^8; (**b**) equiaxed II, σ_a = 233 MPa, N_f = 3.16 × 10^8; (**c**) bimodal I, σ_a = 225 MPa, N_f = 4.15 × 10^8; (**d**) bimodal II, σ_a = 233 MPa, N_f = 3.25 × 10^8.

4.2. Analysis of Microstructure Influence

The titanium alloy fatigue properties are mainly affected by the microstructure such as primary α grains and are closely related to the forging process. On the one hand, the increased forging temperature will strengthen the primary α phase diffusion behavior, swallow the surrounding fine α phases, and cause the primary α grains to grow. On the other hand, the increased forging temperature will promote the α→β phase transformation behavior, resulting in the decrease in primary α grain size and content. The two mechanisms compete with each other under the deformation degree influence, resulting in changes in the primary α content and grain size, as shown in Table 2.

As the α phase is more brittle than the β phase is, it is easier to form dislocation accumulation at the α-β phase boundaries or primary α grain boundaries under fatigue loading. The larger primary α grain boundaries increase the dislocation slip length and aggravate the stress concentration at the tip of dislocation accumulation, thus promoting the primary α grain cleavage failure [31]. Stanzl et al. [32] believed that the accumulated irreversible slip in the process of low stress fatigue leads to the α grain cleavage fractures in the slip surface with a high Schmidt factor, and the subsequent process of faceted crack nucleation and propagation form a rough origin zone. Chai et al. [33] studied the subsurface defect-free crack initiation mechanism and obtained a similar conclusion: the primary α phase is the weak crack initiation part, and the longer primary α grain boundaries play the role of internal notches.

The analysis shows that the fatigue life is related to the microstructure influence on the high-cycle-fatigue and ultra-high-cycle-fatigue mechanism. The traditional high-cycle-fatigue belongs to macro-scale fatigue, the material composition can be approximately regarded as a uniform distribution, the fatigue failure is mainly controlled by surface stress, and the microstructure influence is small. At the same stress level, the primary α phase content in the equiaxed microstructure is higher than that in the bimodal microstructure, which can tolerate more slip deformation and enhance the resistance to crack initiation, so the high-cycle-fatigue performance is better. In contrast, the ultra-high-cycle-fatigue belongs to micro-scale fatigue, material composition is no longer uniformly distributed, fatigue failure is controlled by both surface stress and internal defects, and the microstructure influence is greater. The lower primary α phase content in the bimodal microstructure means fewer defects, which reduces the number of stress concentration zones caused by dislocation accumulation at the second phase interface, makes the slip deformation more uniform, and increases the crack initiation life. In addition, the β transition tissue content in the bimodal microstructure is higher and the creep resistance is greater. As shown in Figure 1, the strip-like and fine needle-like secondary α phases are arranged longitudinally and horizontally in the β matrix, isolating the primary α grains from each other, making the crack propagation path more tortuous, blocking the facet crack propagation, converging and combining into the main crack, and increasing the crack propagation life, so that the bimodal ultra-high-cycle-fatigue properties are better.

4.3. Fatigue Life Prediction Based on Energy

In the linear elastic fracture mechanics, the stress intensity factor amplitude ΔK at the crack tip is the main crack propagation control parameter, which can be calculated by the Murakami model [34], as shown in Formula (1):

$$\Delta K_{f,orFC} = n \cdot \sigma_a \left(\pi \sqrt{area_{f,orFC}}\right)^{1/2} \tag{1}$$

For surface cracks, n = 0.65; for subsurface cracks, n = 0.5; σ_a is the stress amplitude; and $area_{f,orFC}$ represents the projected area of facet or facet clusters in the principal stress direction. Based on the measurement of the facet characteristics in the bimodal I crack origin zone, it is found that there is no obvious difference in most facet sizes, with an average of about 17.5 μm, while the facet cluster size increases with the fatigue life. The stress intensity factors ΔK_f and ΔK_{FC} of facets and faceted clusters are calculated according to Formula (1). As shown in Figure 12, ΔK_f decreases with the fatigue life, which is much smaller than the stress intensity factor threshold value for macroscopic crack propagation (ΔK_{th} = 5~6 MPa·m$^{1/2}$). The ΔK_{FC} remains basically unchanged with the increased fatigue life, and the ΔK_{FC} is about 4.53 MPa·m$^{1/2}$ for surface failure and 5.49 MPa·m$^{1/2}$ for subsurface failure, which is mainly attributed to the adverse environmental effects. Surface cracks germinate and propagate in air and water fog, while subsurface cracks do so in vacuum, so the ΔK_{FC} is lower when surface failure occurs. To sum up, the amplitude of the stress intensity factor ΔK_{FC} of faceted clusters is analogized to the threshold ΔK_{th}, which controls the macroscopic stable crack propagation.

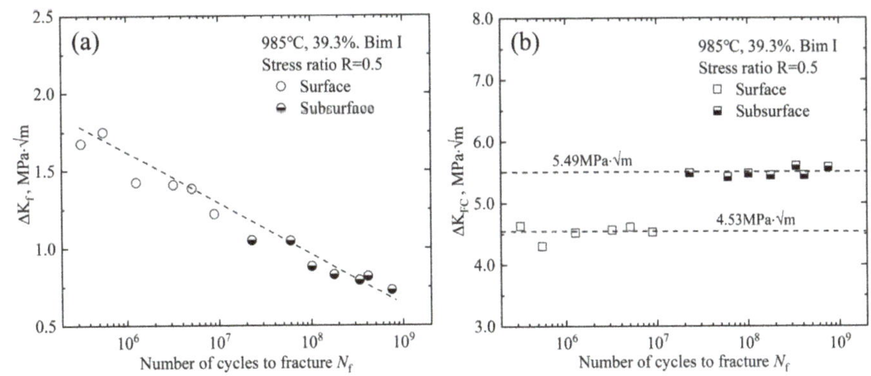

Figure 12. Calculation of stress intensity factor amplitude. (**a**) ΔK_f and N_f; (**b**) ΔK_{FC} and N_f.

For high-cycle-fatigue and ultra-high-cycle-fatigue regimes, the crack initiation and initial propagation stages occupy the vast majority, so the fatigue life can be estimated by the energy-based crack nucleation life model, and the calculation formula is [35]:

$$N_i = \frac{4\pi\mu^3 h^2 M^3 c}{0.005d^3(\Delta\sigma - 2Mk)\left[\pi^2\Delta\sigma^2(1-v)^2 + \xi M^2\mu^2\right]} \tag{2}$$

where μ is the shear homogeneous matrix modulus, h is the slip band width, d is the primary α grain size, $\Delta\sigma$ is the stress amplitude, 2 Mk is defined as the fatigue strength at 10^9 cycles, M is usually taken as 2, v is Poisson's ratio, and ξ is a numerical constant. It is known from Figure 12 that the stress intensity factor amplitude of the faceted clusters ΔK_{FC} is approximately constant, so the crack size c is represented by the facet cluster size $area_{FC}{}^{1/2}$. According to the S-N curve, the unknown parameters ξ and h are fitted nonlinearly, and the results are 1.02 and 8.25×10^{-4}, respectively. Finally, the bimodal I fatigue life prediction curve is obtained, as shown in Figure 13.

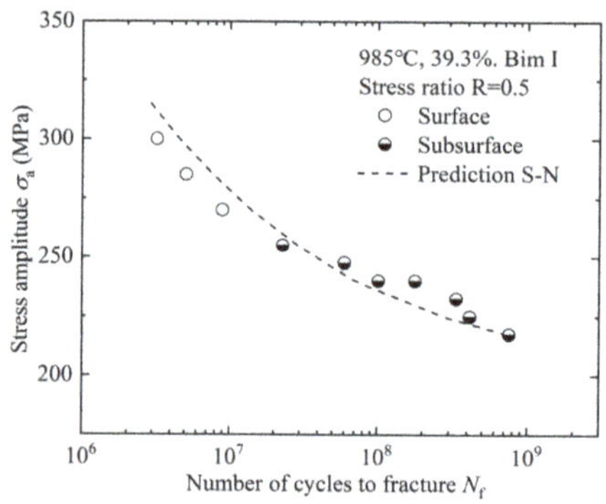

Figure 13. Results of the fatigue crack life prediction.

The life prediction curve related to the facet cluster size is basically consistent with the fatigue data, which verifies that the fatigue failure process is mainly consumed in the facet cluster formation stage, and only a small part of the propagation zone and final rupture zone. It is also shown that the energy-based life prediction model is not only suitable for axial tension–compression loading, but also for three-point bending loading. The analysis shows that the crack origin zone shows similar facet morphology under the two loading modes, and there is no essential difference in fatigue failure mechanism, so the relevant characteristic parameters in the life prediction model are universal. As far as the current research is concerned, it is reasonable to apply the axial tension–compression fatigue analysis model to three-point bending fatigue, and the author will further analyze and demonstrate this statement in the follow-up research.

5. Conclusions

In this paper, forged TC4 titanium alloy ultrasonic fatigue properties under three-point bending were studied, and the main conclusions are as follows:

1. As the number of loading cycles increases, the crack initiation changes from surface to subsurface initiation. The S-N curves have different shapes, and there is no traditional fatigue limit. The equiaxed high-cycle-fatigue performance is better, while the bimodal ultra-high-cycle-fatigue performance is better. The best forging process combination is 985 °C with a 39.3% deformation degree.
2. The crack origin zones in the four kinds of microstructures show the mixed morphology of facets, cleavage steps, and MPT characteristics. The crack originates from the surface or subsurface facet, showing a transgranular quasi-cleavage fracture

mechanism. The ultra-high-cycle-fatigue striation spacing is smaller than that of high-cycle-fatigue for the same microstructure, which reflects a higher crack propagation life.

3. The facet in the crack origin zone is formed by the cleavage of primary α grains. Most of these primary α grains are attributed to the Al element-promoting effect, while a few of them hardly need the Al element assistance. The fatigue cracks originate from the preferred texture of primary α grains, rather than the primary α grain clusters.
4. High-cycle-fatigue belongs to macro-scale fatigue, the fatigue failure is mainly controlled by surface stress, and the influence of the microstructure is small. While ultra-high-cycle-fatigue belongs to micro-scale fatigue, the fatigue failure is controlled by both surface stress and internal defects, and the influence of the microstructure is great. This makes the equiaxed and bimodal microstructure have better high-cycle-fatigue and ultra-high-cycle-fatigue performance, respectively.
5. The stress intensity factor amplitude ΔK_{FC} of faceted clusters can be analogized to the threshold ΔK_{th} to control the macroscopic stable crack propagation, and the life prediction curve related to the faceted cluster size is basically consistent with the fatigue data, which shows that it is reasonable to apply the axial tension–compression fatigue analysis model to three-point bending fatigue, which is attributed to the similar fatigue failure mechanism.

Author Contributions: B.W. and L.C. designed the experiments; B.W. and D.L. performed the experiments and funding acquisition; L.C., B.W. and D.L. analyzed the experimental results; B.W. wrote the paper. All authors have read and agreed to the published version of the manuscript.

Funding: This research was funded by the National Key Basic Research and Development Program of China (973), grant number 2015CB057400.

Conflicts of Interest: The authors declare no conflict of interest.

References

1. Zimmermann, M. Diversity of damage evolution during cyclic loading at very high numbers of cycles. *Int. Mater. Rev.* **2012**, *57*, 73–91. [CrossRef]
2. Mayer, H. Recent developments in ultrasonic fatigue. *Fatigue Fract. Eng. Mater. Struct.* **2016**, *39*, 3–29. [CrossRef]
3. Oguma, H.; Nakamura, T. Formation mechanism of specific fracture surface region in the sub-surface fracture of titanium alloy. *Adv. Mater. Res.* **2014**, *891–892*, 1436–1441. [CrossRef]
4. Günther, J.; Krewerth, D.; Lippmann, T.; Leuders, S.; Tröster, T.; Weidner, A.; Biermann, H.; Niendorf, T. Fatigue life of additively manufactured Ti-6Al-4V in the very high cycle fatigue regime. *Int. J. Fatigue* **2017**, *94*, 236–245. [CrossRef]
5. Uematsu, Y.; Kakiuchi, T.; Hattori, K. EBSD assisted fractography of subsurface fatigue crack initiation mechanism in the ultrasonic shot peened βeta-type titanium alloy. *Fatigue Fract. Eng. Mater. Struct.* **2018**, *41*, 2239–2248. [CrossRef]
6. Furuya, Y.; Nishikawa, H.; Hirukawa, H.; Nagashima, N.; Takeuchi, E. Catalogue of NIMS fatigue data sheets. *Sci. Technol. Adv. Mater.* **2019**, *20*, 1055–1072. [CrossRef] [PubMed]
7. Hong, Y.; Sun, C.; Liu, X. A review on mechanisms and models for very-high-cycle fatigue of metallic materials. *Adv. Mech.* **2018**, *48*, 1–65.
8. Sakai, T.; Nakagawa, A.; Oguma, N.; Nakamura, Y.; Ueno, A.; Kikuchi, S.; Sakaida, A. A review on fatigue fracture modes of structural metallic materials in very high cycle regime. *Int. J. Fatigue* **2016**, *93*, 339–351. [CrossRef]
9. Bao, X.; Cheng, L.; Ding, J.; Chen, X.; Lu, K.; Cui, W. The effect of microstructure and axial tension on three-point bending fatigue behavior of TC4 in high cycle and very high cycle regimes. *Materials* **2020**, *13*, 68. [CrossRef]
10. Yang, K.; Zhong, B.; Huang, Q.; He, C.; Huang, Z.Y.; Wang, Q.; Liu, Y.J. Stress Ratio and Notch Effects on the Very High Cycle Fatigue Properties of a Near-Alpha Titanium Alloy. *Materials* **2018**, *11*, 1778. [CrossRef]
11. Hong, Y.; Lei, Z.; Sun, C. Propensities of crack interior initiation and early growth for very-high-cycle fatigue of high strength steels. *Int. J. Fatigue* **2014**, *58*, 144–151. [CrossRef]
12. Pan, X.; Hong, Y. High-cycle and very-high-cycle fatigue behaviour of a titanium alloy with equiaxed microstructure under different mean stresses. *Fatigue Fract. Eng. Mater. Struct.* **2019**, *42*, 1950–1964. [CrossRef]
13. Oguma, H.; Nakamura, T. The effect of microstructure on very high cycle fatigue properties in Ti-6Al-4V. *Scr. Mater.* **2010**, *63*, 32–34. [CrossRef]
14. Jha, S.K.; Szczepanski, C.J.; Golden, P.J.; Porter, I.I.I.W.J.; John, R. Characterization of fatigue crack-initiation facets in relation to lifetime variability in Ti-6Al-4V. *Int. J. Fatigue* **2012**, *42*, 248–257. [CrossRef]

15. Liu, X.; Chen, E.; Zeng, F. Mechanisms of interior crack initiation in very-high-cycle fatigue of high-strength alloys. *Eng. Fract. Mech.* **2019**, *212*, 153–163. [CrossRef]
16. Liu, X.; Sun, C.; Hong, Y. Faceted crack initiation characteristics for high-cycle and very-high-cycle fatigue of a titanium alloy under different stress ratios. *Int. J. Fatigue* **2016**, *92*, 434–441. [CrossRef]
17. Wu, Y.; Liu, J.; Wang, H.; Guan, S.; Yang, R.; Xiang, H. Effect of stress ratio on very high cycle fatigue properties of Ti-10V-2Fe-3Al alloy with duplex microstructure. *J. Mater. Sci. Technol.* **2018**, *34*, 1189–1195. [CrossRef]
18. Huang, Z.Y.; Liu, H.Q.; Wang, H.M.; Wagner, D.; Khan, M.K.; Wang, Q.Y. Effect of stress ratio on VHCF behavior for a compressor blade titanium alloy. *Int. J. Fatigue* **2016**, *93*, 232–237. [CrossRef]
19. Yang, K.; He, C.; Huang, Q.; Huang, Z.Y.; Wang, C.; Wang, Q.; Liu, Y.J.; Zhong, B. Very high cycle fatigue behaviors of a turbine engine blade alloy at various stress ratios. *Int. J. Fatigue* **2017**, *99*, 35–43. [CrossRef]
20. Furuya, Y.; Takeuchi, E. Gigacycle fatigue properties of Ti-6Al-4V alloy under tensile mean stress. *Mater. Sci. Eng. A* **2014**, *598*, 135–140. [CrossRef]
21. Nie, B.; Zhao, Z.; Chen, D. Effect of basketweave microstructure on very high cycle fatigue behavior of TC21 titanium alloy. *Metals* **2018**, *8*, 401. [CrossRef]
22. Li, W.; Li, M.; Sun, R.; Xing, X.; Wang, P.; Sakai, T. Faceted crack induced failure behavior and micro-crack growth based strength evaluation of titanium alloys under very high cycle fatigue. *Int. J. Fatigue* **2020**, *131*, 105369. [CrossRef]
23. Deng, R.; Yang, G.; Mao, X.; Zhang, P.; Han, D. Effects of Forging Process and Following Heat Treatment on Microstructure and Mechanical Properties of TC11 Titanium Alloy. *Mater. Mech. Eng.* **2011**, *35*, 58–62.
24. Pan, Q.; Yu, X.; Qi, Y.; Huang, Q.; Pan, G. Effect of Thermal Tensile Deformation and Solution Aging Treatment on Microstructure of TC4-DT Titanium Alloy. *Mater. Mech. Eng.* **2019**, *43*, 21–26.
25. Tanaka, S.; Akahori, T.; Niinomi, M.; Nakai, M. Relationship between Microstructure and Fatigue Properties of Forged Ti-5Al-2Sn-2Zr-4Mo-4Cr for Aircraft Applications: Engineering Materials and Their Applications. *Mater. Trans.* **2020**, *61*, 2017–2024. [CrossRef]
26. Nikitin, A.; Bathias, C.; Palin-Luc, T. A new piezoelectric fatigue testing machine in pure torsion for ultrasonic gigacycle fatigue tests: Application to forged and extruded titanium alloys. *Fatigue Fract. Eng. Mater. Struct.* **2015**, *38*, 1294–1304. [CrossRef]
27. Nikitin, A.; Palin-Luc, T.; Shanyavskiy, A. Crack initiation in VHCF regime on forged titanium alloy under tensile and torsion loading modes. *Int. J. Fatigue* **2016**, *93*, 318–325. [CrossRef]
28. Sinha, V.; Pilchak, A.L.; Jha, S.K.; Porter, W.J.; John, R.; Larsen, J.M. Correlating Scatter in Fatigue Life with Fracture Mechanisms in Forged Ti-6242Si Alloy. *Metall. Mater. Trans. A* **2018**, *49*, 1061–1078. [CrossRef]
29. Chandran, K.S.R. Duality of fatigue failures of materials caused by Poisson defect statistics of competing failure modes. *Nat. Mater.* **2005**, *4*, 303–308. [CrossRef]
30. Yang, K.; Zhong, B.; Huang, Q.; He, C.; Huang, Z.Y.; Wang, Q.; Liu, Y.J. Stress ratio effect on notched fatigue behavior of a Ti-8Al-1Mo-1V alloy in the very high cycle fatigue regime. *Int. J. Fatigue* **2018**, *116*, 80–89. [CrossRef]
31. JZuo, H.; Wang, Z.G.; Han, E.H. Effect of microstructure on ultra-high cycle fatigue behavior of Ti-6Al-4V. *Mater. Sci. Eng. A* **2007**, *473*, 147–152.
32. Stanzl-Tschegg, S.; Mughrabi, H.; Schoenbauer, B. Life time and cyclic slip of copper in the VHCF regime. *Int. J. Fatigue* **2007**, *29*, 2050–2059. [CrossRef]
33. Chai, G.; Zhou, N.; Ciurea, S.; Andersson, M.; Peng, R.L. Local plasticity exhaustion in a very high cycle fatigue regime. *Scr. Mater.* **2012**, *66*, 769–772. [CrossRef]
34. Murakami, Y.; Kodama, S.; Konuma, S. Quantitative evaluation of effects of non-metallic inclusions on fatigue strength of high strength steels. I: Basic fatigue mechanism and evaluation of correlation between the fatigue fracture stress and the size and location of non-metallic inclusions. *Int. J. Fatigue* **1989**, *11*, 291–298. [CrossRef]
35. Li, W.; Xing, X.; Gao, N.; Wang, P. Subsurface crack nucleation and growth behavior and energy-based life prediction of a titanium alloy in high-cycle and very-high-cycle regime. *Eng. Fract. Mech.* **2019**, *221*, 106705. [CrossRef]

materials

MDPI

Article

Comparison of Hydrogen Embrittlement Susceptibility of Different Types of Advanced High-Strength Steels

Sangwon Cho [1,†], Geon-Il Kim [1,†], Sang-Jin Ko [1], Jin-Seok Yoo [1], Yeon-Seung Jung [2], Yun-Ha Yoo [2] and Jung-Gu Kim [1,*]

1 School of Advanced Materials Science and Engineering, Sungkyunkwan University (SKKU), Suwon 16419, Korea; jsw2811@gmail.com (S.C.); geonil1027@g.skku.edu (G.-I.K.); tkdwls121@skku.edu (S.-J.K.); wlstjr5619@skku.edu (J.-S.Y.)

2 Steel Solution Research Laboratory, POSCO Global R&D Center, Incheon 21985, Korea; ysj799@posco.com (Y.-S.J.); yunha778@posco.com (Y.-H.Y.)

* Correspondence: kimjg@skku.edu; Tel.: +82-31-290-7360

† These authors contributed equally to this study.

Abstract: This study investigated the hydrogen embrittlement (HE) characteristics of advanced high-strength steels (AHSSs). Two different types of AHSSs with a tensile strength of 1.2 GPa were investigated. Slow strain rate tests (SSRTs) were performed under various applied potentials (E_{app}) to identify the mechanism with the greatest effect on the embrittlement of the specimens. The SSRT results revealed that, as the E_{app} increased, the elongation tended to increase, even when a potential exceeding the corrosion potential was applied. Both types of AHSSs exhibited embrittled fracture behavior that was dominated by HE. The fractured SSRT specimens were subjected to a thermal desorption spectroscopy analysis, revealing that diffusible hydrogen was trapped mainly at the grain boundaries and dislocations (i.e., reversible hydrogen-trapping sites). The micro-analysis results revealed that the poor HE resistance of the specimens was attributed to the more reversible hydrogen-trapping sites.

Keywords: advanced high-strength steel; hydrogen embrittlement; hydrogen trapping; thermal desorption spectroscopy

Citation: Cho, S.; Kim, G.-I.; Ko, S.-J.; Yoo, J.-S.; Jung, Y.-S.; Yoo, Y.-H.; Kim, J.-G. Comparison of Hydrogen Embrittlement Susceptibility of Different Types of Advanced High-Strength Steels. *Materials* **2022**, *15*, 3406. https://doi.org/10.3390/ma15093406

Academic Editors: Lucjan Śnieżek and Jaroslaw Galkiewicz

Received: 15 April 2022
Accepted: 7 May 2022
Published: 9 May 2022

1. Introduction

The need to reduce environmental harm is a growing global concern. Accordingly, the automotive industry is striving to improve fuel efficiency and reduce carbon dioxide emissions to protect the environment. Globally, the industry is pushing for fuel efficiency improvements via two routes: high-efficiency engines and lightweight body designs [1–3]. A vehicle's body weight accounts for 40% of its fuel efficiency factors; therefore, reducing this weight has the greatest impact on improving fuel efficiency. Generally, a 100-kg reduction in body weight lowers carbon dioxide emissions by 7.5 to 12.5 g/km, significantly enhancing the fuel efficiency. To reduce the weight of cars, manufacturers may use nonferrous materials (e.g., resin, aluminum alloy, and magnesium alloy) [4–8]; specific methods (e.g., the miniaturization of parts); or different types of high-strength steels [9–11]. Although nonferrous materials used in automotive structures such as aluminum alloy and magnesium alloy are lighter than steel, they are also weaker, and their thicknesses must be increased to maintain body stiffness. Additionally, lightweight materials must be used in combination with other materials, such as carbon fiber-reinforced plastic, to maintain the required body stiffness. Therefore, research and development into various types of advanced high-strength steels (AHSSs) are currently underway.

Generally, the mechanical properties of steel are enhanced using methods such as solid solution, grain refinement, or precipitation; however, in the case of AHSS, phase transformation-based methods are also used. Enhancing the mechanical properties of an

AHSS increases its corrosivity and sensitivity to a delayed fracture, i.e., stress–corrosion cracking (SCC) and hydrogen embrittlement (HE), which are the main problems associated with AHSS [12].

Over the past decades, a lot of research on HE in AHSSs has been conducted: effects of the strength, microstructure, and types of defects of AHSSs. V. G. Khanzhin et al. [13,14] studied the influence of precipitate and mechanical properties on HE. According to the studies, the higher density of precipitates in a structure, the lower HE resistance, since the secondary phase particles influence both the stage of initiating hydrogen cracks and the crack growth kinetics to a critical value. Additionally, mechanical properties, their strength and toughness, affects the nucleation of hydrogen cracks, possibility of their propagation, and the kinetics of growth to a critical size.

In AHSS, the delayed fracture phenomenon is caused mainly by HE. Hydrogen inside the steel is preferentially trapped in lattice defects, such as voids, dislocations, and grain boundaries, as well as in various carbides and precipitates [15,16]. Additionally, after entrapment, hydrogen is concentrated in certain areas by stress, leading to the propagation of internal cracks and, eventually, to delayed fractures. However, the exact cause and mechanism of the delayed fracture phenomenon have not been identified to date. This is because, in addition to HE, a delayed fracture can result from the combined effects of other variables, including the stresses acting on the steel, the microstructures, mechanical properties, surface conditions, and internal cracks. Further research is required to determine the exact cause of HE. Therefore, this study uses slow strain rate tests (SSRTs), a microstructural analysis, and a thermal desorption spectroscopy (TDS) analysis to investigate the HE mechanisms of two different types of AHSSs with the tensile strength of 1.2 GPa.

2. Materials and Methods

This study used two different types of AHSSs with a tensile strength of 1.2 GPa. Table 1 provides their chemical compositions. Figure 1 presents the specimens' microstructural images, and Figure 2 shows the mechanical properties obtained by the tensile test. Steel A was comprised of fine grains with complex phases of ferrite, bainite, martensite, and a small fraction of retained austenite, with Ti and/or Nb precipitates for enhancing the tensile strength and ductility. Since this steel was cooled slowly after soaking in the austenite region, its main phases were bainite and martensite. Steel B was also a multiphase AHSS comprising ferrite, bainite, a relatively higher fraction of retained austenite, and a small portion of martensite. Under an applied stress, the phase transformation of the retained austenite increased the ductility of Steel B. As shown in Figure 2c, Steel B showed a uniform strain-hardening rate range, which is evidence of a transformation-induced plasticity effect.

Table 1. Chemical compositions of the advanced high-strength steels for use in automobiles (wt.%).

Component	C	Si	Mn	Cr	Ti	Nb	Fe
Steel A	0.11~0.18	0.4~0.7	2.2~2.7	0.0~0.1	0.01~0.02	0.01~0.02	Bal.
Steel B	0.11~0.18	1.2~1.9	2.4~2.7	0.20~0.45	0.01~0.02	-	Bal.

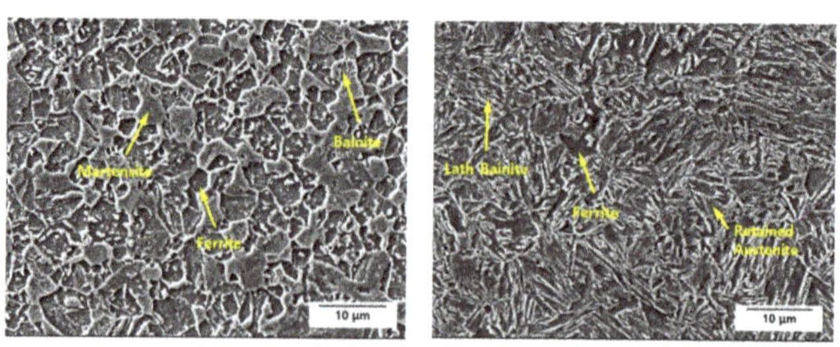

(**a**) (**b**)

Figure 1. Scanning electron microscopy images of the microstructure of (**a**) Steel A and (**b**) Steel B etched with a 2% nital solution.

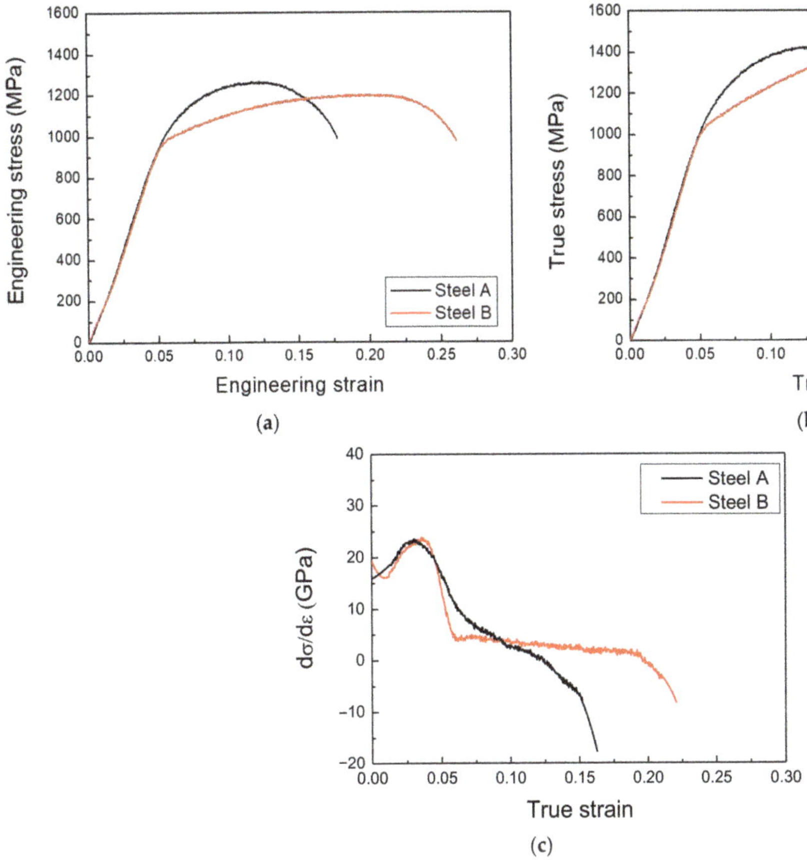

Figure 2. (**a**) Engineering stress–strain curves, (**b**) true stress–strain curves, and (**c**) strain-hardening rate vs. true strain curves for Steel A and Steel B obtained by tensile tests.

2.1. Electrochemical Tests

Potentiodynamic polarization test was performed to analyze the corrosion behavior of AHSSs and determine the applied potentials in the SSRTs. For the electrochemical tests, the specimens were cut into dimensions of 1.5 × 1.5 cm^2, abraded up to #600 with silicon–carbide paper, degreased with ethanol, and dried with N_2 gas. The electrochemical test environment used a modified Society of Automotive Engineers' (M-SAE) solution at 25 °C (room temperature) (see Table 2 for the chemical compositions). All the electrochemical experiments were performed with a triple-electrode electrochemical cell, as shown in Figure 3. The counter electrode was a graphite rod, and the reference electrode was a saturated calomel electrode (SCE). The open-circuit potential (OCP) was established over 3 h. Potentiodynamic polarization tests were conducted with a potential sweep of 0.166 mV/s in accordance with ASTM G5. After the samples were stabilized in an M-SAE solution at room temperature for 1 h, SSRTs were conducted under applied potentials (E_{app}) of −600, −750, and −1500 mV_{SCE} based on the potentiodynamic polarization test.

Table 2. Chemical composition (wt.%) of the modified Society of Automotive Engineers' solution.

NaCl	$CaCl_2$	$NaHCO_3$	$(NH_4)_2SO_4$	pH
0.5	0.1	0.075	0.35	7.3

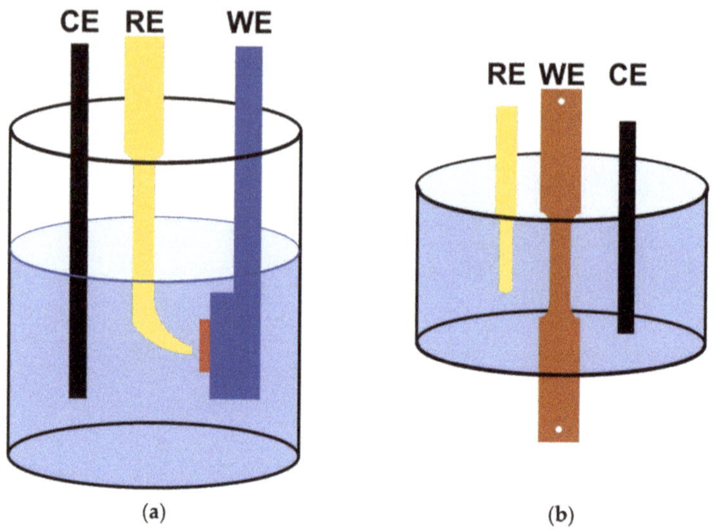

Figure 3. Schematic diagram of the three-electrode cell configuration used in (**a**) potentiodynamic polarization test and (**b**) SSRTs. CE, RE, and WE refer to counter electrode, reference electrode, and working electrode, respectively.

2.2. *Slow Strain Rate Tests*

A schematic image of the specimen for the SSRTs is presented in Figure 4. First, the critical strain rate was determined by various SSRTs in the OCP state; then, under the E_{app} values listed in Section 2.1, SSRTs were conducted at a strain rate of 10^{-5}. After the tests, each fractured specimen was cleaned with ethanol and transferred into liquid nitrogen as soon as possible. Then, the fracture surface was observed via scanning electron microscopy (SEM), and the hydrogen desorption rate was determined by a TDS analysis of the hydrogen content charged into the specimen. To enable the TDS analysis of the hydrogen content, the specimen was cut up to 10 mm from the fracture surface. To calculate the activation energy for hydrogen de-trapping, heating rates of 2 °C/min and 4 °C/min were used.

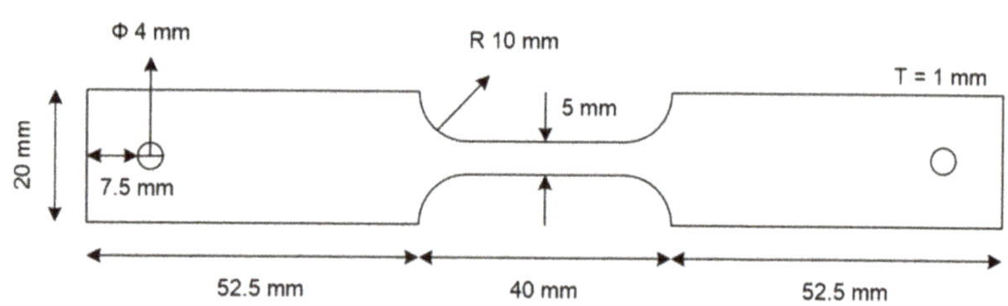

Figure 4. Dimensions of the specimen used in slow strain rate tests.

2.3. *Analyses for Hydrogen-Trapping Sites*

The grain boundary areas and austenite phase fractions of the samples were measured using electron backscattered diffraction (EBSD). X-ray diffraction (XRD) was performed at a scan rate of 1°/min, and the dislocation density was calculated using the full width at half-maximum (FWHM). Electron probe microanalysis (EPMA) and transmission electron microscopy (TEM) were used to analyze the type and characteristics of the precipitates. Before EPMA, each specimen was etched slightly with a 2% nital solution for 5 s.

3. Results and Discussion

3.1. Potentiodynamic Polarization Test

This study conducted potentiodynamic polarization tests to analyze the elec-tro-chemical properties of the specimens; the results are presented in Figure 5 and Table 3. Mild steel with a tensile strength of 270 MPa was used as a comparison material for the AHSS. The AHSSs exhibited higher corrosion rates than the mild steel. The anodic polar-ization curves of the AHSSs and the mild steel material were similar in shape; however, the ca-thodic polarization curve of the AHSSs shifted more to the right compared with the mild steel. This was because hydrogen evolution reactions are more likely in AHSS than in mild steel due to the higher levels of precipitates, carbides, and grain boundary densi-ties. Additionally, there are more phase types in AHSS compared with mild steel, result-ing in higher corrosion rates due to the large micro-galvanic effect between phases [17].

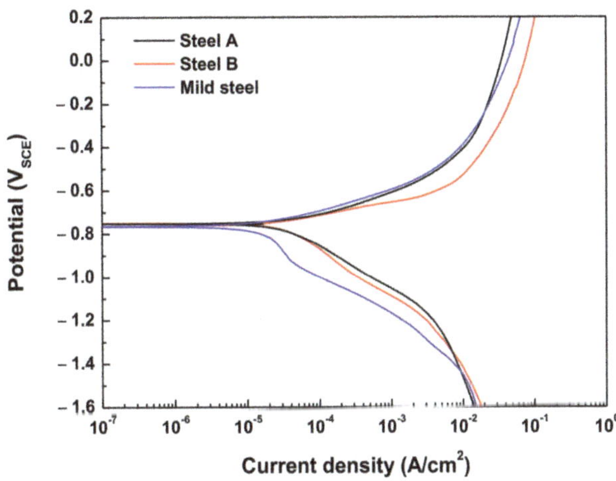

Figure 5. Potentiodynamic polarization curves in a modified Society of Automotive Engineers' solution.

Table 3. Potentiodynamic polarization test results.

Specimen	E_{corr} (V_{SCE})	I_{corr} (A/cm^2)	β_a (mV/Decade)	β_c (mV/Decade)	Corrosion Rate (mm/y)
Steel A	−0.748	3.54×10^{-5}	92	205	0.41
Steel B	−0.755	3.38×10^{-5}	61	241	0.39
Mild steel	−0.765	1.70×10^{-5}	89	419	0.20

The electrochemical properties of both AHSSs, e.g., corrosion potential and corrosion current density, were almost identical. From the polarization curve of Steel A, the redox reaction of hydrogen (Equation (1)) produced a higher hydrogen equilibrium potential ($E^0_{H_2O/H_2}$, $-0.672\ V_{SCE}$) than corrosion potential (E_{corr}, $-0.728\ V_{SCE}$). Therefore, hydrogen was also generated at the corrosion potential.

$$2H_2O + 2e^- \rightarrow H_2 + 2OH^-,\ E^0_{H_2O/H_2} = -0.672V_{SCE}. \quad (1)$$

3.2. Slow Strain Rate Tests

During an SSRT, the applied strain rate will cause differences in the occurrence of SCC and HE behaviors [18]. When the strain rate is too high, there is insufficient time for SCC to occur, resulting in only a tensile rupture. Conversely, at a low strain rate, the re-passivation of the film before the propagation of a crack by anodic dissolution means that SCC does not occur. However, HE does not require the breakdown of the passive film; instead, it is caused by hydrogen trapping inside the steel. Therefore, the lower strain rate requires more time for the hydrogen to be entrapped in the steel, making it more susceptible to HE. Accordingly, when conducting SSRTs to determine the HE characteristics, an optimal strain rate should be applied in consideration of the hydrogen trapping time. To determine the critical strain rate, the SSRTs were performed at strain rates of 2×10^{-4}/s, 10^{-4}/s, 10^{-5}/s, 10^{-6}/s, and 5×10^{-7}/s in M-SAE solution under an OCP state (see Figure 6 for the results). When the strain rate was 10^{-5}/s, the SSRT results revealed a relatively low elongation for both AHSSs. Therefore, the final SSRTs were performed under a strain rate of 10^{-5}/s.

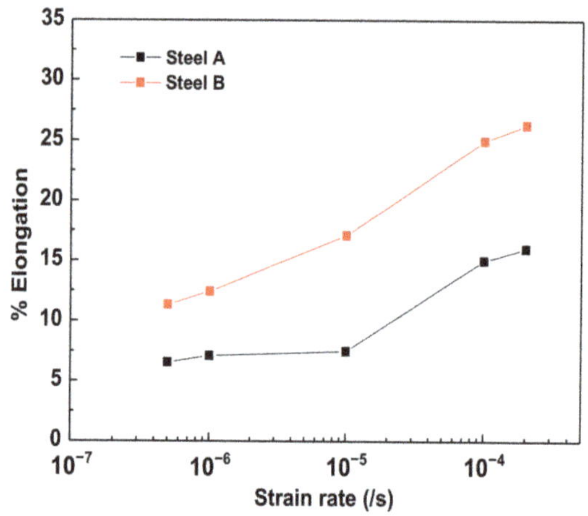

Figure 6. Elongation vs. strain rate in a modified Society of Automotive Engineers' solution under an open-circuit potential state.

Both specimens showed a decreasing elongation with a decreasing strain rate, which is representative of a typical HE elongation–strain rate curve, in which a ductility minimum is not expected [18]. A low strain rate provides sufficient time for lattice diffusion, which allows hydrogen to easily enter the trapping sites.

To minimize the hydrogen generation reactions, the applied anodic potential should be higher than the hydrogen reduction potential. Accordingly, as is shown in Figure 7 and Table 4, the SSRTs in this study were conducted using various E_{app} values. The amounts of hydrogen evolution for the E_{app} values of -1500 mV_{SCE} were calculated by integrating the base area of the current–time curves (Figure 7c,d). Since the current obtained with an E_{app} above -750 mV_{SCE} was caused by corrosion, the hydrogen evolution amounts for E_{app} values of -750 mV_{SCE} were calculated using Faraday's law as follows [18]:

$$m = \frac{I_{red,H_2O/H_2} \times t \times a}{n \times F} \quad (2)$$

where m is the reaction mass (hydrogen evolution amount, in grams), I_{red} is the current of the reduction reaction at each E_{app} (A), t is the time to fracture (s), a is the atomic weight (g/mol), n is the number of electrons (n = 2 for Equation (1)), and F is the Faraday constant (96,500 C/mol).

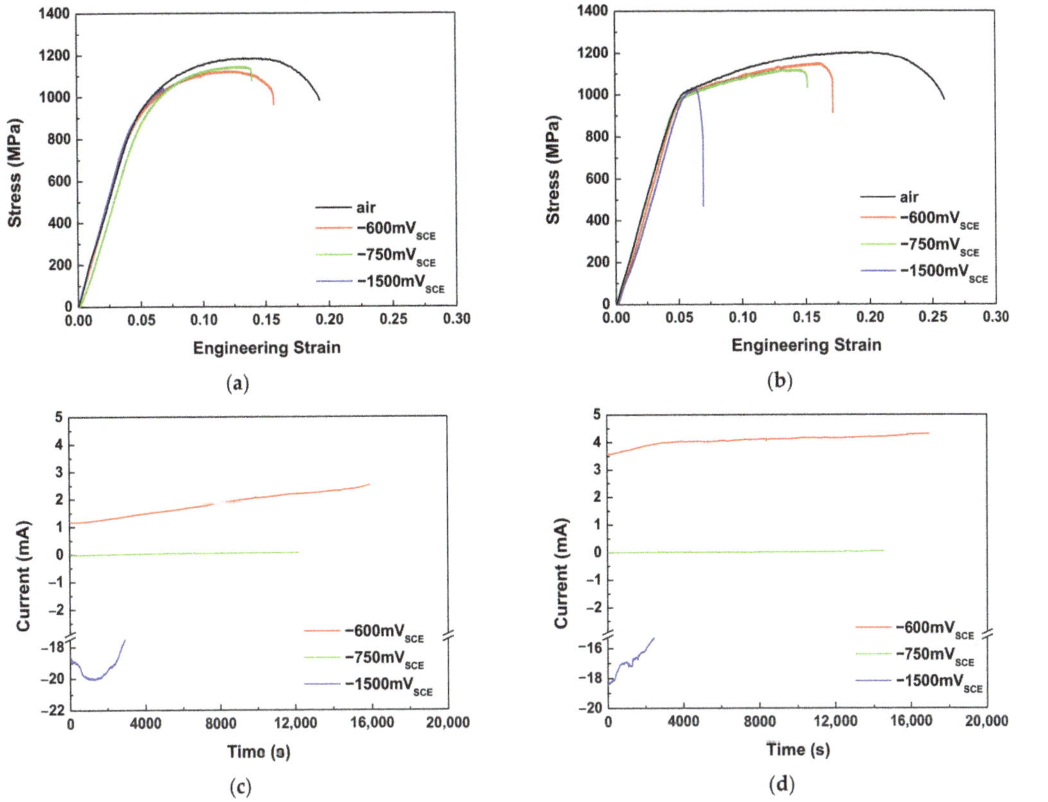

Figure 7. Stress–strain curves for (**a**) Steel A and (**b**) Steel B, and the current–time curves of (**c**) Steel A and (**d**) Steel B obtained during slow strain rate tests.

Table 4. Slow strain rate test results.

Steel	Applied Potential (mV_{SCE})	Yield Strength (MPa)	Tensile Strength (MPa)	Elongation (%)	HE Susceptibility Index, I_{HE} (%)	Hydrogen Evolution Rate (g)
Steel A	Air	781	1186	16.0	-	-
	−600	703	1124	14.1	11.9	-
	−750 (E_{corr})	699	1146	11.2	30.1	5.76×10^{-6}
	−1500	690	1041	4.6	71.0	1.23×10^{-3}
Steel B	Air	800	1202	21.6	-	-
	−600	906	1150	16.4	24.0	-
	−750 (E_{corr})	852	1119	13.3	38.3	2.62×10^{-6}
	−1500	851	1034	3.7	82.9	9.76×10^{-4}

For both AHSSs, the lower E_{app} was found to be correlated with a reduced elongation. There was an increase in the amount of corrosion with a higher anodic overvoltage, while the hydrogen evolution amount increased with the increasing cathodic overvoltage. A slightly higher amount of hydrogen was generated on Steel A compared to on Steel B.

The HE sensitivity index (I_{HE}) indicates the ductility loss of the AHSSs according to the E_{app}. Since the ductility loss of AHSSs with cathodic applied potential is related to HE, the I_{HE} was used to compare the HE resistance. The I_{HE} can be calculated using Equation (3), in which a higher I_{HE} is associated with increased HE sensitivities. The I_{HE} of

Steel B was approximately 10% higher than that of Steel A. Therefore, compared with Steel B, Steel A had a superior HE resistance.

$$I_{HE} = \frac{\varepsilon_{air} - \varepsilon_{soln.}}{\varepsilon_{air}} \times 100 \quad (3)$$

where I_{HE} is the HE sensitivity index (%), ε_{air} is the elongation tested in air, and $\varepsilon_{soln.}$ is the elongation tested under an E_{app}.

3.3. Fractography

To determine the fracture properties of the AHSS samples, after the SSRTs were conducted, the fracture surfaces and sides of the specimens were observed by SEM. The results are presented in Figures 8 and 9. Cracks were initiated and propagated from the sides in all the specimens. In Steel A, uniform pitting corrosion was observed on the sides at -600 mV$_{SCE}$, while there was no changes at -750 and -1500 mV$_{SCE}$ (Figure 8b,f,j). Dimples were observed at the crack initiation site at -600 mV$_{SCE}$, cleavage occurred at the crack initiation site at -750 mV$_{SCE}$, and transgranular fracturing was noted at -1500 mV$_{SCE}$ (Figure 8c,g,k). All the specimens exhibited dimpling at the center of their fracture surfaces (Figure 8d,h,l). Steel A only exhibited ductile fracturing at -600 mV$_{SCE}$, and the lower E_{app} values resulted in more brittle fracture behavior. Even under a potential of -1500 mV$_{SCE}$, the center of the specimen exhibited ductile fracture behavior. Therefore, hydrogen did not diffuse into the center of the specimen.

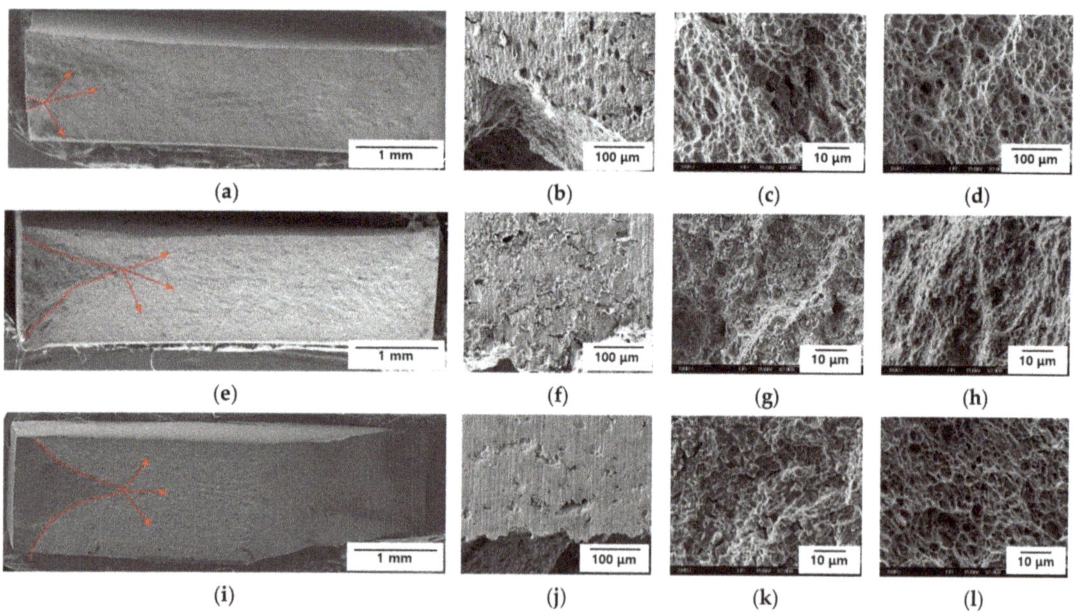

Figure 8. Fractography of Steel A at (**a–d**) -600 mV$_{SCE}$, (**e–h**) -750 mV$_{SCE}$, and (**i–l**) -1500 mV$_{SCE}$. (**a**,**e**,**i**) Entire sample, (**b**,**f**,**j**) side view, (**c**,**g**,**k**) crack initiation site, and (**d**,**h**,**l**) center. Red arrows indicate the initiation of cracks and direction of propagation.

Figure 9. Fractography of Steel B at (**a–d**) −600 mV_{SCE}, (**e–h**) −750 mV_{SCE}, and (**i–l**) −1500 mV_{SCE}. (**a**,**e**,**i**) Entire sample, (**b**,**f**,**j**) side view, (**c**,**g**,**k**) crack initiation site, and (**d**,**h**,**l**) center. Red arrows indicate the initiation of cracks and direction of propagation.

In Steel B, uniform corrosion and cracks occurred on the side of the specimen at −600 mV_{SCE}. Cracks without any corrosion were observed at −750 and −1500 mV_{SCE} (Figure 9b,f,j), and the lower E_{app} values were correlated with a higher density of cracks. Cleavage was observed at the crack initiation site at −600 mV_{SCE}, while mixed intergranular and transgranular fractures were seen at the crack initiation sites of −750 and −1500 mV_{SCE} (Figure 9c,g,k). The intergranular fracture was more obvious at −1500 mV_{SCE}, and in all the specimens, dimples occurred at the center of the fracture surfaces (Figure 9d,h,l). Steel B exhibited brittle fractures at −600 mV_{SCE}, and the lower E_{app} values were associated with more obvious brittle fracture behaviors. At −1500 mV_{SCE}, the center of Steel B demonstrated ductile fracture behavior. Thus, like Steel A, hydrogen did not diffuse into the center of the specimen. Under the same E_{app}, Steel B exhibited more brittle fracture behavior than Steel A. The fractography results confirmed that, compared with Steel A, Steel B was more susceptible to delayed fractures.

3.4. *Hydrogen Trapping and Desorption Behaviors*

To investigate the desorption behavior of diffusible hydrogen, the SSRT specimens were analyzed by TDS at the E_{app} values of −600, −750, and −1500 mV_{SCE}. The results are presented in Figure 10 and Table 5. To quantitatively analyze the desorbed hydrogen, the area below the desorption rate vs. the temperature curve was integrated [19] (see Table 5 for the results). Just 0.05 ppm of diffusible hydrogen was released in the as-received specimens. Most of the diffusible hydrogen that accumulated during the steel manufacturing process (e.g., during acid cleaning) appeared to be released during machining and storage. However, when the potential was applied, the lower E_{app} was associated with the higher hydrogen desorption rate. Under the same E_{app} values of both AHSSs, the desorbed diffusible hydrogen content of Steel B was higher than in Steel A, except for −1500 mV_{SCE}. In that case, Peak 3 of Steel A and Peak 2 of Steel B (located at approximately 220 °C) originated from the deformation field around the dislocation. In this study, as tensile deformation was considered an error, the hydrogen de-trapping from these peaks was negligible.

Theoretically, the production of hydrogen did not occur at −600 mV_{SCE}, although diffusible hydrogen was detected. It is assumed that the hydrogen was accumulated from the 1-h stabilizing process before the SSRTs were conducted.

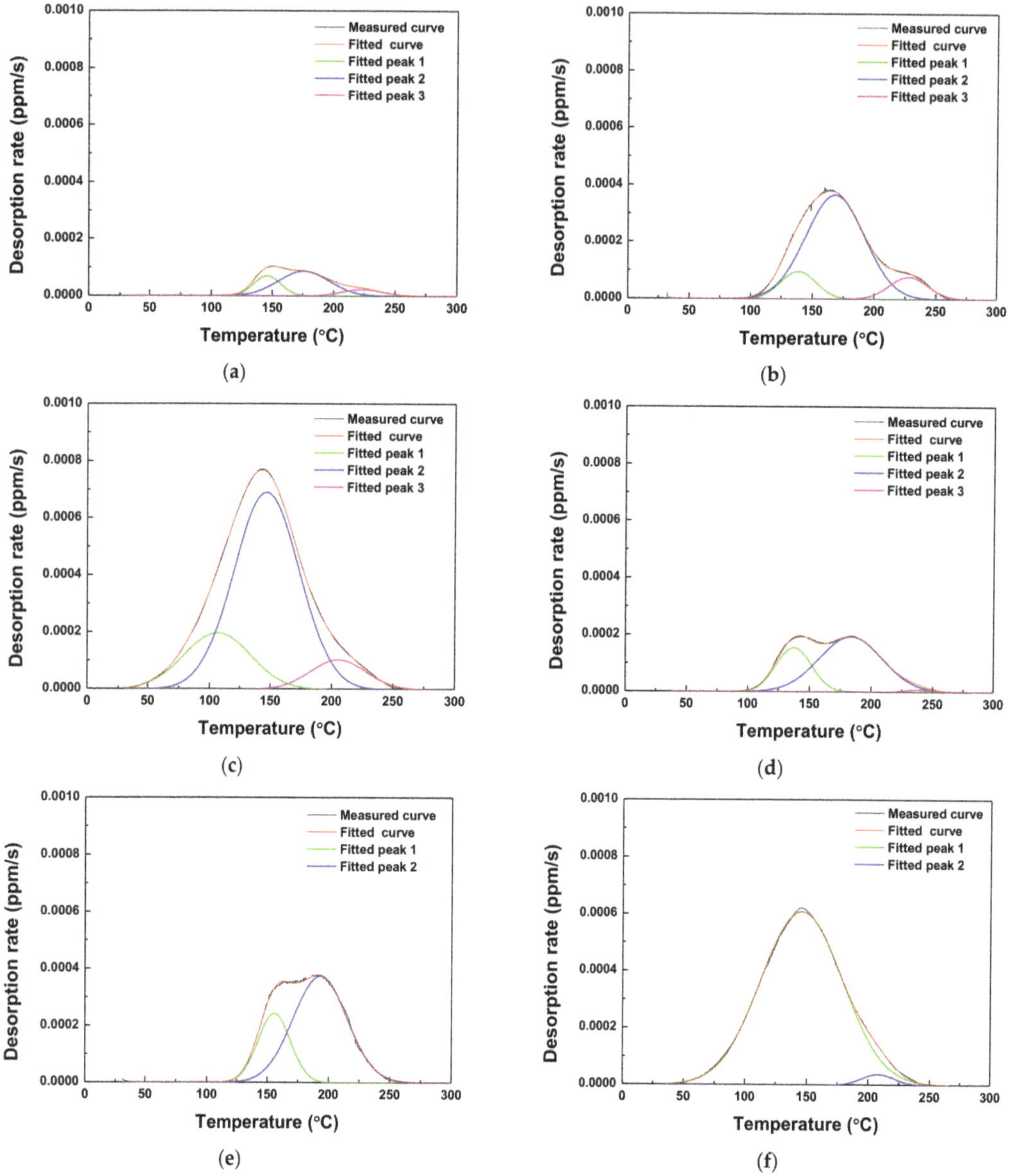

Figure 10. *Cont.*

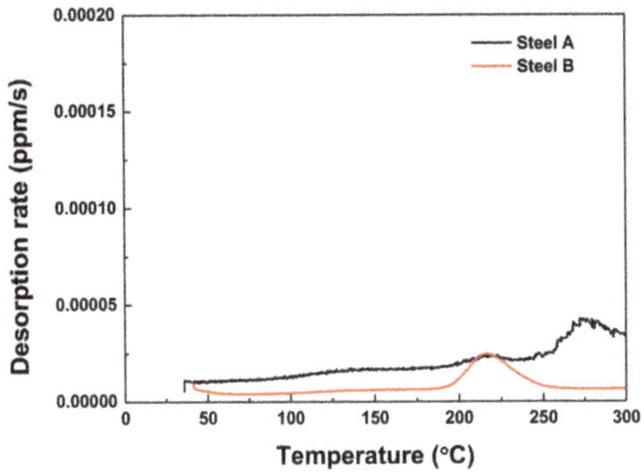

(g)

Figure 10. Hydrogen desorption rates obtained by thermal desorption spectroscopy at a heating rate of 4 °C/min in a fractured specimens of (**a–c**) Steel A and (**d–f**) Steel B at (**a**,**d**) −600, (**b**,**e**) −750, (**c**,**f**) −1500 mV_{SCE}, and (**g**) the as-received condition.

Table 5. Desorbed hydrogen contents for each peak.

Steel	Applied Potential (mV_{SCE})	Peak 1 (ppm)	Peak 2 (ppm)	Peak 3 (ppm)	Sum of Peaks (ppm)
Steel A	As-received	0.0569	-	-	0.0569 ± 0.0323
	−600	0.0622	0.1511	0.0311	0.2485 ± 0.1262
	−750	0.0579	0.3693	0.0514	0.4787 ± 0.0145
	−1500	0.2264	0.7735	0.0978	1.0977 ± 0.0968
Steel B	As-received	0.0506	-	-	0.0506 ± 0.0268
	−600	0.1388	0.3122	0.0057	0.4568 ± 0.2070
	−750	0.1748	0.4445	-	0.6193 ± 0.1280
	−1500	0.9622	0.0246	-	0.9868 ± 0.0052

To analyze the hydrogen-trapping sites in the steel specimens, the activation energy for hydrogen de-trapping was calculated using Equation (4), as proposed by Kissinger [20–22]:

$$\frac{\partial \ln(\varphi/T_c^2)}{\partial(1/T_c)} = -\frac{E_{aT}}{R} \tag{4}$$

where T_c is the temperature (K) at which the hydrogen desorption rate is maximal, φ is the heating rate (K/min), E_{aT} is the activation energy for hydrogen de-trapping (kJ/mol), and R is the ideal gas constant (8.314 J/K·mol).

As is shown in Figure 10, the desorption curves were deconvoluted into two or three peaks of Gaussian curves, indicating that diffusible hydrogen accumulated at more than two or three trapping sites. According to the Kissinger equation, the slope of $\ln(\varphi/T_c{}^2)$ vs. $1/T_c$ curve for each peak represents the activation energies (see Figure 11 for the results). The activation energies for Steels A and B corresponding to each peak are illustrated in Table 6.

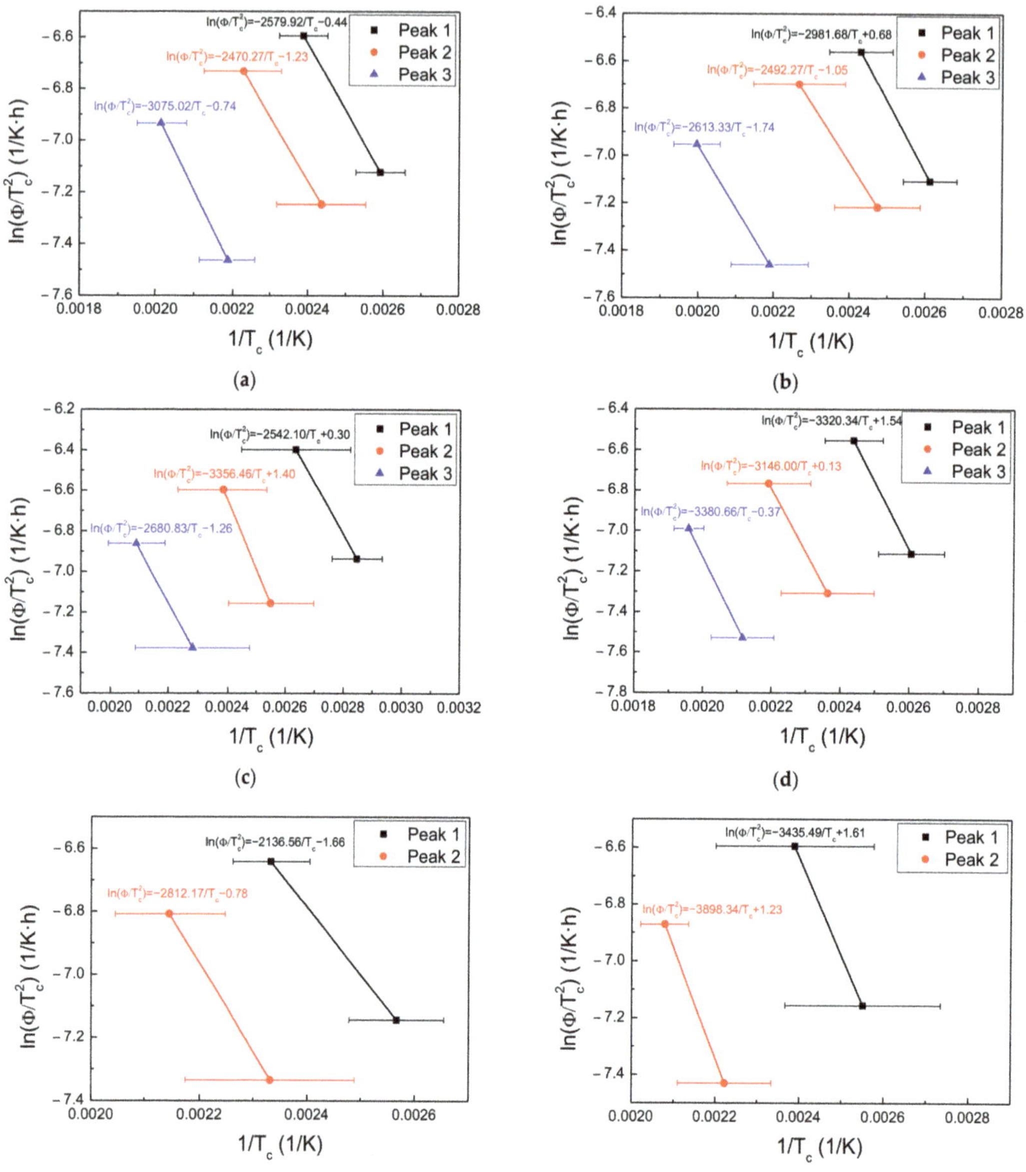

Figure 11. $\ln(\varphi/T_c{}^2)$ vs. $1/T_c$ curve for (**a–c**) Steel A and (**d–f**) Steel B at (**a**,**d**) −600, (**b**,**e**) −750, and (**c**,**f**) −1500 mV_{SCE}.

Table 6. Calculated activation energies for hydrogen de-trapping.

Steel	Applied Potential (mV_{SCE})	Peak 1 (kJ/mol)	Peak 2 (kJ/mol)	Peak 3 (kJ/mol)
Steel A	−600	21.5	20.5	25.6
	−750	24.8	20.7	21.7
	−1500	21.1	27.9	22.3
Steel B	−600	27.6	26.1	28.1
	−750	17.8	23.4	-
	−1500	28.6	32.4	-

Table 7 summarizes the activation energies for hydrogen de-trapping reported in previous related studies. Based on the published literature, the electrochemically accumulated hydrogen corresponding to Peaks 1 and 2 in Steel A was associated with the grain boundary and dislocation. Peak 3 was associated with the mechanical deformation by tensile deformation that occurred during the SSRTs [23]. For Steel B, the hydrogen corresponding to Peaks 1 and 2 at −600 and −750 mV_{SCE}, respectively, was desorbed from the grain boundary, dislocation, and ferrite–Fe_3C interface. In that specimen, the contributions from the grain boundary and dislocation were indistinguishable in Peak 1 at −1500 mV, which means that Peak 1 was the sum of the hydrogen desorbed from the grain boundary and dislocation. Peak 3 (−1500 mV_{SCE}) was associated with mechanical deformation by tensile deformation, which occurred during the SSRTs.

Table 7. Types of reversible and irreversible hydrogen-trapping sites reported in the literature.

Type of Trap	Activation Energy (kJ/mol)	References
	Reversible hydrogen-trapping sites	
Ferrite/Fe_3C	10.9	[16]
Grain boundary	17.2	[16]
Ferrite/Fe_3C interface	18.4	[16,24]
Grain boundary, Dislocation	21–29	[25–28]
Deformation field around dislocation	29 ± 5	[23]
	Irreversible hydrogen-trapping sites	
Semi-coherent TiC	49.9	[28]
High-angle grain boundary	53–59	[29]
NbC interface	63–68	[30]
Incoherent TiC	85.7, 86.9	[28]

3.5. *Analysis of Defects Acting as Hydrogen Trapping Sites*

3.5.1. Electron Backscattered Diffraction Analysis

EBSD analysis was conducted to measure the grain boundary density and fraction of retained austenite; the results are shown in Figure 12 and Table 8. Each value was measured three times to derive the mean value. The average grain sizes measured by EBSD for Steel A and Steel B were 2.79 and 4.03 μm, respectively. Mild steel has an average approximate grain size of 22 μm [31]; therefore, these values indicate that AHSSs have a smaller grain size than mild steel.

Table 8. EBSD analysis results (relative value).

Specimen	Average Grain Size	High-Angle Grain Boundary Length	Low-Angle Grain Boundary Length	Retained Austenite Fraction
Steel A	2.79 μm	15.73 mm	1.91 mm	9.5%
Steel B	4.03 μm	13.93 mm	1.99 mm	10.9%

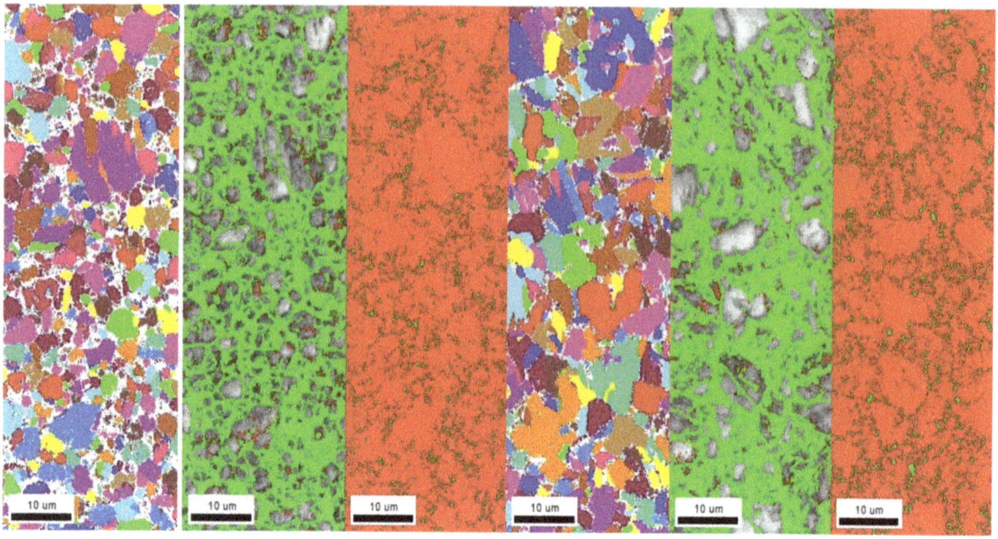

(a) (b) (c) (d) (e) (f)

Figure 12. Electron backscattered diffraction results of (**a–c**) Steel A and (**d–f**) Steel B. (**a**,**d**) Grain mapping. (**b**,**e**) Low-angle grain boundary (red) and high-angle grain boundary (green) mapping. (**c**,**f**) Face-centered cubic (red) and body-centered cubic (green) mapping.

When the misorientation of a grain boundary exceeds 15°, it is termed a high-angle grain boundary; otherwise, it is a low-angle grain boundary. In this study, Steel A had the longer high-angle grain length than Steel B, and the low-angle grain boundary lengths in both specimens were similar. The low-angle grain boundary is a reversible hydrogen-trapping site, suggesting that a longer low-angle grain boundary is more likely to induce HE [32]. Since the high-angle grain boundary is an irreversible hydrogen-trapping site, the longer high-angle grain boundary enhances the HE resistance. Therefore, in both AHSSs, the diffusible hydrogen content charged in the grain boundary is almost identical, and the non-diffusible hydrogen content charged in the grain boundary of Steel A is expected to be high.

The conducted EBSD analysis reveals that the face-centered cubic (FCC) structure reflected the retained austenite content. In Steel B, the retained austenite fraction was 10.9%, which was 1.4% higher than in Steel A. Retained austenite is an irreversible hydrogen-trapping site that enhances the HE resistance. However, in Steel B, the retained austenite fraction is not proportional to the HE resistance; this is because retained austenite with an FCC structure is transformed by tensile stress into martensite with a body-centered tetragonal (BCT) structure. Since BCT structures have a lower hydrogen solubility and faster diffusion rate than FCC structures, hydrogen accumulation via diffusion is easy in the BCT structure [33]. Thus, the hydrogen charged on the retained austenite in Steel B segregates during the tensile process and becomes susceptible to HE. Furthermore, the austenite–matrix interface is an effective diffusible hydrogen-trapping site [34]. The higher fraction of retained austenite increases the susceptibility to HE, i.e., it is expected that Steel B will be more susceptible to HE than Steel A.

3.5.2. X-ray Diffraction

The dislocation density of the samples was determined using XRD (see Figure 13 for the results). Both AHSSs were mainly comprised of α-Fe, although γ-Fe peaks were also observed. Specifically, the γ-Fe peaks were higher in intensity in Steel B compared with Steel A, which is consistent with the results of the EBSD analysis. The dislocation density is defined as the length of dislocation lines per unit volume of crystal and can be calculated using the Williamson–Smallman relationship [35], as in Equation (5) below:

$$\delta = \frac{1}{D^2} \tag{5}$$

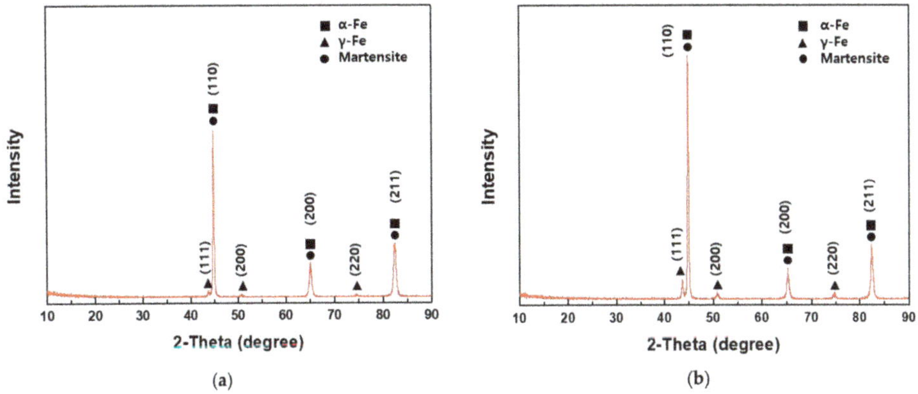

Figure 13. X-ray diffraction results for (**a**) Steel A and (**b**) Steel B.

Here, δ is the dislocation density, and D is the size of crystalline domain, which is similar to the grain size. Therefore, D can be calculated using Scherrer's equation [36], as follows:

$$D = \frac{k\lambda}{\beta \cos\theta} \tag{6}$$

in which k is the shape factor (=approx. 0.9), λ is the wavelength (Cu-K_{α} = 1.5406 Å), β is the full width at half-maximum (FWHM) value, and θ is the position of the peaks. Using the above expression, the dislocation density was calculated to be $3.488 \times 10^{14}/\text{m}^2$ and $6.263 \times 10^{14}/\text{m}^2$ for Steel A and Steel B, respectively, i.e., the dislocation density of Steel B was twice that of Steel A. Since the low-angle grain boundary areas of the two AHSSs were similar, the difference in the hydrogen content of the two types of AHSSs discharged during TDS was attributed to the difference in the dislocation density.

3.5.3. Characterization of Precipitates

To characterize the type and size of the precipitates, EPMA and TEM analyses were conducted. The results are presented in Figure 14. According to Figure 14a, the precipitate of Steel A was rich in Ti and Nb and a (Nb, Ti) precipitate surrounded the Ti-rich precipitate. However, in the precipitates of Steel B, only Ti was detected, while Nb was undetected. The precipitates of both AHSSs were approximately 1 µm in size. The TEM images and diffraction patterns for the two types of AHSSs are presented in Figure 14c,d. In Steel A, extremely fine precipitates were distributed along the grain boundary. The electron diffraction pattern and energy-dispersive X-ray spectroscopy analysis confirmed that the precipitates were amorphous Ti and Fe carbides smaller than 10 nm in size. Only the small fraction of Fe carbides was distributed randomly in the grain, and in Steel B, no TiC precipitate was observed (Figure 14d). The EPMA and TEM results revealed the presence of sub-micrometer (Nb, Ti)C and fine TiC precipitates in Steel A, although Steel B contained only a sub-micrometer TiC precipitate. The small size of the carbide produced a large effective area for hydrogen trapping [37,38]. Therefore, Steel A was able to trap considerably more hydrogen in the TiC precipitate interface compared with Steel B. Since Nb and Ti precipitates are powerful and irreversible hydrogen-trapping sites, they can positively influence HE resistance, i.e., Steel A is expected to be more resistant to HE than Steel B.

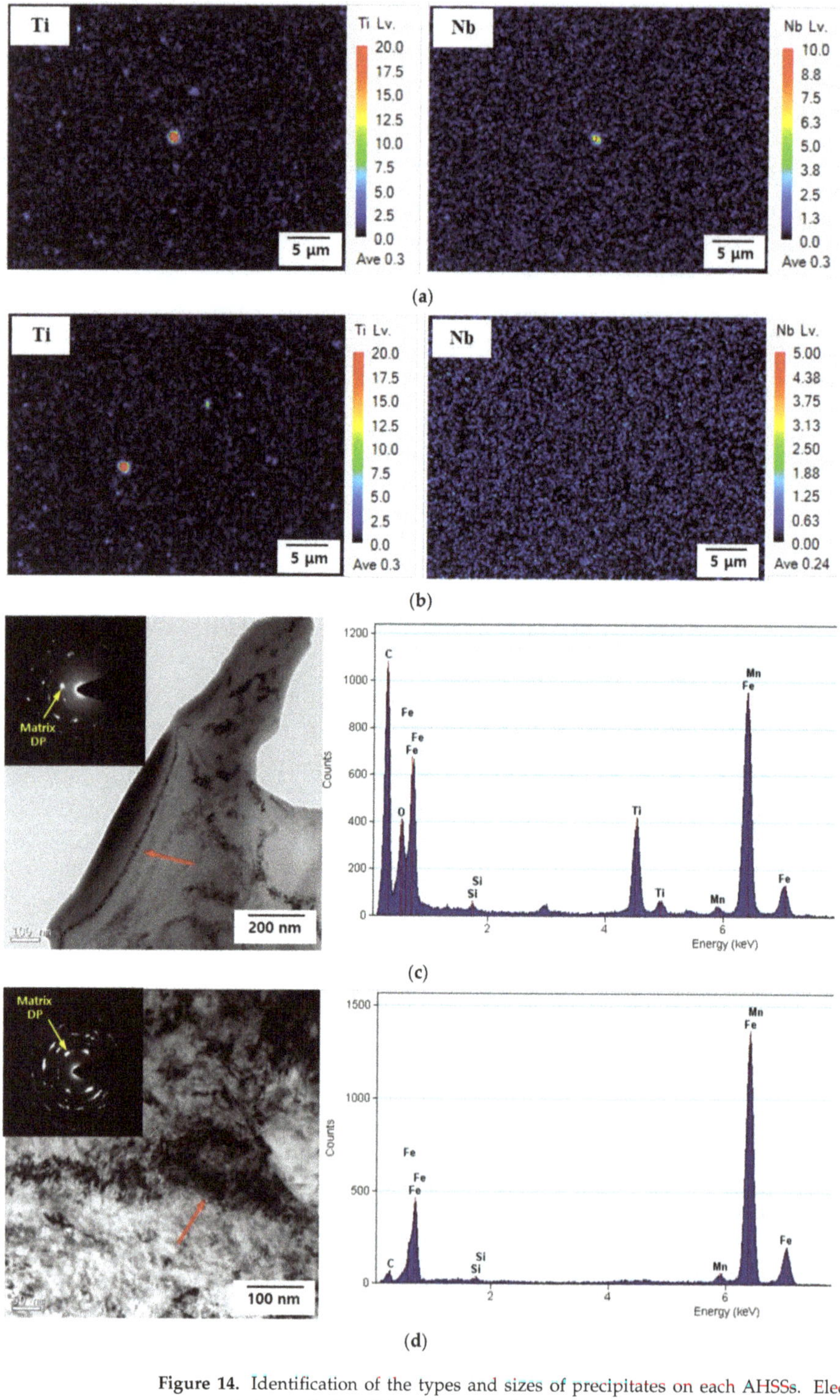

Figure 14. Identification of the types and sizes of precipitates on each AHSSs. Electron probe microanalysis results for (**a**) Steel A and (**b**) Steel B. Transmission electron microscopy results for (**c**) Steel A and (**d**) Steel B.

4. Conclusions

This study investigated the SCC and HE mechanisms of two AHSSs using SSRTs and characterized their hydrogen-trapping behaviors using TDS, EBSD, and XRD. According to the results of these investigations, the SCC and HE characteristics of the studied AHSSs can be summarized as follows:

1. For both AHSSs, elongation decreased as the cathodic overvoltage increased, i.e., both types of AHSSs were fractured by the mechanism of HE. Even when the anodic potential was applied, HE was more dominant than SCC. Although the HE sensitivity of Steel B was higher than that of Steel A, both AHSSs were more sensitive to HE than SCC.
2. In both AHSSs, the lower E_{app} was associated with a strong brittle fracture behavior. However, the center of each specimen exhibited ductile fracture behavior, because the hydrogen did not diffuse into that region. It was clear that the fracture surface of Steel B was more brittle than that of Steel A.
3. The lower E_{app} was associated with the higher rate of hydrogen desorption. In both AHSSs, diffusible hydrogen was trapped mainly at the grain boundary and dislocation.
4. The density of the irreversible hydrogen-trapping sites (high-angle grain boundaries and TiC precipitates) was higher in Steel A than in Steel B. However, the density of the reversible hydrogen-trapping sites (low-angle grain boundaries and dislocations) was lower in Steel A than in Steel B. Therefore, compared to Steel A, Steel B was more susceptible to HE.

Author Contributions: Conceptualization, S.C. and G.-I.K.; methodology, S.C. and G.-I.K.; validation, S.C., G.-I.K., S.-J.K., J.-S.Y., Y.-S.J., Y.-H.Y. and J.-G.K.; formal analysis, S.C. and G.-I.K.; investigation, S.C. and G.-I.K.; resources, S.C. and G.-I.K.; data curation, S.C., G.-I.K., S.-J.K. and J.-S.Y.; writing—original draft preparation, S.C. and G.-I.K.; writing—review and editing, S.C., G.-I.K., S.-J.K., J.-S.Y., Y.-S.J., Y.-H.Y. and J.-G.K.; visualization, S.C. and G.-I.K.; supervision, J.-G.K.; project administration, J.-G.K., and funding acquisition, J.-G.K. All authors have read and agreed to the published version of the manuscript.

Funding: This research was funded by the POSCO (Grant Number 2019Z083).

Institutional Review Board Statement: Not applicable.

Informed Consent Statement: Not applicable.

Data Availability Statement: Not applicable.

Acknowledgments: This research was supported by SungKyunKwan University and BK21 FOUR (Graduate School Innovation) funded by the Ministry of Education (MOE, Korea) and the National Research Foundation of Korea (NRF).

Conflicts of Interest: The authors declare no conflict of interest.

References

1. Miller, W.; Zhuang, L.; Bottema, J.; Wittebrood, A.J.; De Smet, P.; Haszler, A.; Vieregge, A. Recent development in aluminium alloys for the automotive industry. *Mater. Sci. Eng. A* **2000**, *280*, 37–49. [CrossRef]
2. Kulekci, M.K. Magnesium and its alloys applications in automotive industry. *Int. J. Adv. Manuf. Technol.* **2008**, *39*, 851–865. [CrossRef]
3. Cole, G.; Sherman, A. Light weight materials for automotive applications. *Mater. Charact.* **1995**, *35*, 3–9. [CrossRef]
4. Friedrich, H.; Schumann, S. Research for a "new age of magnesium" in the automotive industry. *J. Mater. Process. Technol.* **2001**, *117*, 276–281. [CrossRef]
5. Musfirah, A.; Jaharah, A. Magnesium and aluminum alloys in automotive industry. *J. Appl. Sci. Res* **2012**, *8*, 4865–4875.
6. Gándara, M.J.F. Recent growing demand for magnesium in the automotive industry. *Mater. Technol.* **2011**, *45*, 633–637.
7. Tisza, M.; Czinege, I. Comparative study of the application of steels and aluminium in lightweight production of automotive parts. *Int. J. Lightweight Mater. Manuf.* **2018**, *1*, 229–238. [CrossRef]
8. Hirsch, J. Recent development in aluminium for automotive applications. *Trans. Nonferrous Met. Soc. China* **2014**, *24*, 1995–2002. [CrossRef]

9. Baluch, N.; Udin, Z.M.; Abdullah, C.S. Advanced high strength steel in auto industry: An overview. *Eng. Technol. Appl. Sci. Res.* **2014**, *4*, 686–689. [CrossRef]
10. Galán, J.; Samek, L.; Verleysen, P.; Verbeken, K.; Houbaert, Y. Advanced high strength steels for automotive industry. *Rev. Metal.* **2012**, *48*, 118. [CrossRef]
11. Funakawa, Y.; Nagataki, Y. High Strength Steel Sheets for Weight Reduction of Automotives. *JFE Tech. Rep.* **2019**, *24*, 1–5.
12. Lovicu, G.; Bottazzi, M.; D'aiuto, F.; De Sanctis, M.; Dimatteo, A.; Santus, C.; Valentini, R. Hydrogen embrittlement of automotive advanced high-strength steels. *Metall. Mater. Trans. A* **2012**, *43*, 4075–4087. [CrossRef]
13. Khanzhin, V.; Nikulin, S.; Belov, V.; Rogachev, S.; Turilina, V.Y. Hydrogen embrittlement of steels: III. Influence of secondary-phase particles. *Russ. Metall. (Met.)* **2013**, *2013*, 790–796. [CrossRef]
14. Khanzhin, V.; Turilina, V.Y.; Rogachev, S.; Nikitin, A.; Belov, V. Effect of various factors on hydrogen embrittlement of structural steels. *Met. Sci. Heat Treat.* **2015**, *57*, 197–204. [CrossRef]
15. Hagi, H.; Hayashi, Y.; Ohtani, N. Diffusion coefficient of hydrogen in pure iron between 230 and 300 K. *Trans. Jpn. Inst. Met.* **1979**, *20*, 349–357. [CrossRef]
16. Choo, W.; Lee, J.Y. Thermal analysis of trapped hydrogen in pure iron. *Metall. Mater. Trans. A* **1982**, *13*, 135–140. [CrossRef]
17. Katiyar, P.K.; Misra, S.; Mondal, K. Comparative corrosion behavior of five microstructures (pearlite, bainite, spheroidized, martensite, and tempered martensite) made from a high carbon steel. *Metall. Mater. Trans. A* **2019**, *50*, 1489–1501. [CrossRef]
18. Jones, D.A. *Principles and Prevention of Corrosion*, 2nd ed.; Prentice Hall: Upper Saddle River, NJ, USA, 1996.
19. Yoo, J.; Xian, G.; Lee, M.; Kim, Y.; Kang, N. Hydrogen embrittlement resistance and diffusible hydrogen desorption behavior of multipass FCA weld metals. *J. Weld. Join.* **2014**, *31*, 112–118. [CrossRef]
20. Hurley, C.; Martin, F.; Marchetti, L.; Chêne, J.; Blanc, C.; Andrieu, E. Numerical modeling of thermal desorption mass spectroscopy (TDS) for the study of hydrogen diffusion and trapping interactions in metals. *Int. J. Hydrog. Energy* **2015**, *40*, 3402–3414. [CrossRef]
21. Kirchheim, R. Bulk diffusion-controlled thermal desorption spectroscopy with examples for hydrogen in iron. *Metall. Mater. Trans. A* **2016**, *47*, 672–696. [CrossRef]
22. Ryu, J.H.; Chun, Y.S.; Lee, C.S.; Bhadeshia, H.; Suh, D.W. Effect of deformation on hydrogen trapping and effusion in TRIP-assisted steel. *Acta Mater.* **2012**, *60*, 4085–4092. [CrossRef]
23. Krieger, W.; Merzlikin, S.V.; Bashir, A.; Szczepaniak, A.; Springer, H.; Rohwerder, M. Spatially resolved localization and characterization of trapped hydrogen in zero to three dimensional defects inside ferritic steel. *Acta Mater.* **2018**, *144*, 235–244. [CrossRef]
24. Hong, G.-W.; Lee, J.-Y. The interaction of hydrogen and the cementite-ferrite interface in carbon steel. *J. Mater. Sci.* **1983**, *18*, 271–277. [CrossRef]
25. Wang, M.; Akiyama, E.; Tsuzaki, K. Effect of hydrogen and stress concentration on the notch tensile strength of AISI 4135 steel. *Mater. Sci. Eng. A* **2005**, *398*, 37–46. [CrossRef]
26. Hong, G.-W.; Lee, J.-Y. The interaction of hydrogen with dislocation in iron. In *Perspectives in Hydrogen in Metals*; Elsevier: Amsterdam, The Netherlands, 1986; pp. 427–435.
27. Yamasaki, S.; Takahashi, T. Evaluation method of delayed fracture property of high strength steels. *J. Iron Steel Inst. Jpn.* **1997**, *83*, 454–459. [CrossRef]
28. Wei, F.; Hara, T.; Tsuzaki, K. Precise determination of the activation energy for desorption of hydrogen in two Ti-added steels by a single thermal-desorption spectrum. *Metall. Mater. Trans. B* **2004**, *35*, 587–597. [CrossRef]
29. Pressouyre, G. A classification of hydrogen traps in steel. *Metall. Mater. Trans. A* **1979**, *10*, 1571–1573. [CrossRef]
30. Wallaert, E.; Depover, T.; Arafin, M.; Verbeken, K. Thermal desorption spectroscopy evaluation of the hydrogen-trapping capacity of NbC and NbN precipitates. *Metall. Mater. Trans. A* **2014**, *45*, 2412–2420. [CrossRef]
31. Jangir, D.K.; Verma, A.; Sankar, K.M.; Khanna, A.; Singla, A. Influence of grain size on corrosion resistance and electrochemical behaviour of mild steel. *Int. J. Res. Appl. Sci. Eng. Technol.* **2018**, *6*, 2875–2881. [CrossRef]
32. Oger, L.; Malard, B.; Odemer, G.; Peguet, L.; Blanc, C. Influence of dislocations on hydrogen diffusion and trapping in an Al-Zn-Mg aluminium alloy. *Mater. Des.* **2019**, *180*, 107901. [CrossRef]
33. Kang, H.-J.; Kang, N.-H.; Park, S.-J.; Chang, W.-S. Evaluation of hydrogen delayed fracture for high strength steels and weldments. *J. Weld. Join.* **2011**, *29*, 33–39. [CrossRef]
34. Szost, B.A.; Vegter, R.H.; Rivera-Díaz-del-Castillo, P.E. Hydrogen-trapping mechanisms in nanostructured steels. *Metall. Mater. Trans. A* **2013**, *44*, 4542–4550. [CrossRef]
35. Shahmoradi, Y.; Souri, D.; Khorshidi, M. Glass-ceramic nanoparticles in the Ag2O–TeO2–V2O5 system: Antibacterial and bactericidal potential, their structural and extended XRD analysis by using Williamson–Smallman approach. *Ceram. Int.* **2019**, *45*, 6459–6466. [CrossRef]
36. Bykkam, S.; Ahmadipour, M.; Narisngam, S.; Kalagadda, V.R.; Chidurala, S.C. Extensive studies on X-ray diffraction of green synthesized silver nanoparticles. *Adv. Nanopart* **2015**, *4*, 1–10. [CrossRef]
37. Yamasaki, S.; Manabe, T.; Hirakami, D. *Analysis of Hydrogen State in Steel and Trapping Using Thermal Desorption Method*. Nippon Steel & Sumitomo Metal Technical Report. 2017, pp. 38–43. Available online: https://www.nipponsteel.com/en/tech/report/nssmc/pdf/116-08.pdf (accessed on 18 January 2022).
38. Turk, A.; San Martín, D.; Rivera-Díaz-del-Castillo, P.E.; Galindo-Nava, E.I. Correlation between vanadium carbide size and hydrogen trapping in ferritic steel. *Scr. Mater.* **2018**, *152*, 112–116. [CrossRef]

materials

MDPI

Article

Effect of Strain Rate and Temperature on Tensile and Fracture Performance of AA2050-T84 Alloy

Nagaraj Ekabote [1], Krishnaraja G. Kodancha [1,*], T. M. Yunus Khan [2,3] and Irfan Anjum Badruddin [3]

1 School of Mechanical Engineering, KLE Technological University, Hubballi 580031, India; ekabotenagaraj@gmail.com
2 Research Center for Advanced Materials Science (RCAMS), King Khalid University, 9004, Abha 61413, Saudi Arabia; yunus.tatagar@gmail.com
3 Department of Mechanical Engineering, College of Engineering, King Khalid University, 394, Abha 61421, Saudi Arabia; magami.irfan@gmail.com
* Correspondence: krishnaraja@kletech.ac.in; Tel.: +91-98-8659-6953

Abstract: AA2050-T84 alloy is widely used in primary structures of modern transport aircraft. AA2050-T84 is established as a low-density aluminum alloy with improved Young's modulus, less anisotropy, and temperature-dependent mechanical properties. During flights, loading rate and temperature variation in aircraft engine subsequent parts are commonly observed. The present work focuses on the effect of loading rate and temperature on tensile and fracture properties of the 50 mm thick (2-inch) AA2050-T84 alloy plate. Quasi-static strain rates of 0.01, 0.1, and 1 s^{-1} at −20 °C, 24 °C and 200 °C are considered. Tensile test results revealed the sensitivity of mechanical properties towards strain rate variations for considered temperatures. The key tensile properties, yield, and ultimate tensile stresses were positive strain rate dependent. However, Young's modulus and elongation showed negative strain rate dependency. Experimental fracture toughness tests exhibited the lower Plane Strain Fracture Toughness (K_{IC}) at −20 °C compared to 24 °C. Elastic numerical fracture analysis revealed that the crack driving and constraint parameters are positive strain rate dependent and maximum at −20 °C, if plotted and analyzed over the stress ratio. The current results concerning strain rates and temperatures will help in understanding the performance-related issues of AA2050-T84 alloy reported in aircraft applications.

Keywords: AA2050-T84 alloy; strain rate effect; plane strain fracture toughness; temperature effect; strain rate effect; constraint effect

Citation: Ekabote, N., Kodancha, K.G.; Khan, T.M.Y.; Badruddin, I.A. Effect of Strain Rate and Temperature on Tensile and Fracture Performance of AA2050-T84 Alloy. *Materials* **2022**, 15, 1590. https://doi.org/10.3390/ma15041590

Academic Editors: Jaroslaw Galkiewicz and Lucjan Śnieżek

Received: 26 January 2022
Accepted: 18 February 2022
Published: 20 February 2022

1. Introduction

Modern aircraft predominantly use lightweight structures to improve the performance-to-weight ratio. The low-density Aluminum alloy is popular among aircraft structures owing to its durable mechanical properties and ease of manufacturability [1,2]. Composites pose tough competition to Aluminum alloys due to their tailor-made properties suited for specific applications. The unpredictable behavior of composites for change in temperature and time, restricted its usage to secondary and tertiary aircraft structures [3]. Currently, the modern transport aircraft primary structures are built by Al-Li alloys. The Lithium addition to aluminum with improved manufacturing methods resulted in the enhancement of specific strength and stiffness of the alloy [4]. However, the higher cost of Al-Li alloy restricted its usage to only aerospace industries.

The apprehensions related to diversifying mechanical properties of 1st and 2nd generation Al-Li alloys directed the complete withdrawal from aerospace applications [2]. Some noteworthy limitations were anisotropic behavior, cracking during manufacturing, and thermal instability-driven lower fracture toughness [2]. The skillful and sophisticated fabrication methods steered the evolution of 3rd generation Al-Li alloys. The spars and ribs of modern transport aircraft are fabricated by a 3rd generation Al-Li alloy, AA2050-T84 [1].

AA2050-T84 alloy exhibits exceptional tensile, fatigue, and fracture toughness behavior suited for damage tolerance property requirements of wing structures [1]. Notably, the anisotropic behavior and temperature-dependent property variations of AA2050-T84 alloy were also reported [5,6]. During flights, the fuel pressure at various altitudes of an aircraft, wing lift, and drag loads at different operating conditions may result in load rate and temperature variations on spars and ribs. The sensitivity of AA2050-T84 alloy to these load and temperature variations are essential to claim its suitability to modern transport aircraft wing parts. The following paragraphs discuss the strain rate and temperature effect on various ductile and brittle material properties reported in the literature.

Mirza et al. [7] have conducted tensile tests on mild steel and aluminum at various quasi-static (lower strain rates up to 1 s^{-1}) strain rates. The results have shown negligible dependency of ductility on strain rate variations. Clausen et al. [8] have reported the negative strain rate sensitivity of AA5083–H116 through tensile tests for quasi-static strain rate variations. The strain rate dependency was related to dynamic strain aging at lower strain rates and temperatures, resulting in serrated stress-strain curves. Singh et al. [9] have reported increased flow stress with the rise in test temperature and strain rates on titanium alloys. The observed trend was attributed to the dynamic strain aging of the alloy. Through experiments and numerical analysis, Khan et al. [10] have investigated the influence of strain rate and temperature on Al2024-T351. The results inferred a strong temperature dependency and negligible strain rate effect on fracture strength of the Al2024-T351 alloy. Anderson et al. [11] have witnessed the sensitivity of DP 780 steel towards the strain rate variations. The tensile stress-strain response was steady and almost negligible concerning quasi-static strain rates. The alterations in failure surface morphology were noticed with changes in strain rates.

The experimental tensile results on DP590 and TRIP 780 steel by Roth et al. [12] have shown that ductility increases with loading speed. Rincon et al. [13] have studied the influence of temperature (between −90 °C to 270 °C) on tensile behavior of an as-cast A319 alloy and noticed the silicon dominant brittle fracturing regardless of temperature variation. Natesan et al. [14] have reported the variation in strain rate effect at different temperatures on the deformation behavior of A356-T7 cast aluminum alloys. The yield stress and strain hardening of Aluminum alloy 7075-W exhibited the positive load rate effect and negative temperature effect through plasticity experiments [15]. Hafley et al. [5] and Chemin et al. [6] have reported AA2050-T84 alloy tensile and fracture properties sensitivity to temperature variations. In summary, the material properties of various alloys of steel and aluminum generally exhibit reliance on strain rate and temperature and are noteworthy.

The dependency of fracture behavior on strain rate and temperature mainly alters the state of stress near the crack front. The state of stress variation at the crack front due to specimen type, geometry, and load type was defined by a term constraint. The constraint level at the crack component/structure drives the selection of standard test specimens for fracture toughness tests [16,17]. In Linear Elastic Fracture Mechanics (LEFM), the variation of the state of stress near the crack was measured by popular constraint parameters [18–26]. T_{11} and T_{33} are used to measure the in-plane and out-of-plane constraints in LEFM. The variations of T_{11} and T_{33} concerning specimen type, geometry, and load were well documented [16,18–26]. However, the strain rate effect at different temperatures on tensile and fracture behavior of the AA2050-T84 alloy is essential to claim its suitability to primary structures of the aircraft wing. Fracture toughness standard methods recommend the single value of fracture toughness for quasi-static load variations [27]. Notably, the above literature study shows the strong dependency of material properties on strain rate and temperature variations. Moreover, critical cracks were observed in AA2050-T84 made spars and ribs of Airbus-380 aircraft after a few flights [28]. The present literature findings demand a fracture study based on constraints near the crack of AA2050-T84 alloy at different strain rates and temperatures.

In the present work, AA2050-T84 alloy tensile behavior at different strain rates and temperatures are experimentally studied. Experimental fracture toughness tests are conducted using Compact Tension (C(T)) specimens at various temperatures. Furthermore, the effect of strain rate on fracture characterizing parameters in LEFM such as Stress Intensity Factor (K_I), T_{11}, and T_{33} are analyzed numerically for different temperatures. Finally, the AA2050-T84 alloy tensile and fracture behavior dependency on strain rates and temperatures are compared and evaluated for compatibility for aircraft wing structures.

2. Material and Test Details

2.1. AA2050-T84 Alloy

This study uses a 50 mm thick (2-inch) AA2050-T84 alloy plate to extract the test specimens. The chemical composition of AA2050-T84 alloy in wt% as obtained from the supplier is shown in Table 1. Copper is used in AA2050-T84 to provide high strength, suited for aircraft applications [29]. The Lithium addition is (<1%) restricted to balance between density reduction and increase in Young's modulus of the alloy [3,29].

Table 1. Chemical composition of 50 mm thick (2-inch) AA2050-T84 alloy plate (wt%).

Cu	Mg	Mn	Zn	Fe	Ti	Si	Li	Zr	Ag	Al
3.743	0.369	0.372	0.025	0.045	0.040	0.039	0.798	0.087	0.398	Base

2.2. Tensile and Fracture Toughness Test

ASTM E8/E8M-21, the standard test method for tension testing of metallic materials [30], was used for the tensile specimen preparation and testing of AA2050-T84 alloy. Round specimens were extracted in the rolling (along the length of the plate) direction of the AA2050-T84 plate. Figure 1 shows the tensile test specimen dimensions (in mm) used in this study. The main dimensions of the specimen are, gauge diameter (D_0 = 6 mm), gauge length (L_0 = 30 mm), and overall specimen length (L = 65 mm). The tensile specimens were designed, keeping the L_0/D_0 ratio to 5.

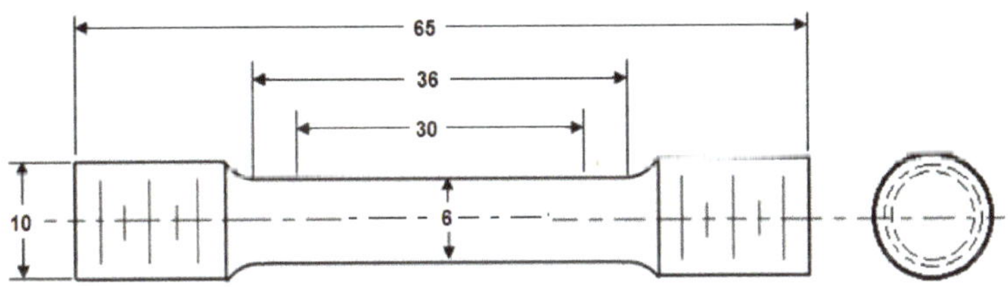

Figure 1. Tensile test specimen. All dimensions are in mm.

ASTM E399-20a, the standard test method for linear-elastic plane strain fracture toughness of metallic materials [27], was used. The commonly used fracture test specimen for primary aircraft structures is the compact tension (C(T)) specimen shown in Figure 2. The standard dimensions (in mm) are specimen width (W = 25.4 mm), specimen height (H = 2W), and specimen thickness (B) = crack length (a) = 0.5W. The C(T) specimen is extracted, ensuring the crack length in the rolling direction and load application in the transverse directions of the plate.

Flight durations and operations cause the temperature variations of the wing parts. Furthermore, the effect of these variations depends on alloy type and its ductile to brittle transition temperature [31]. However, temperature variations will be high near the aircraft engine (wing components), and the experienced load rates are dynamic. In the present study, the quasi-static strain rates considered in the tensile tests were 0.01, 0.1, and 1 s^{-1}. The temperatures considered were −20 °C (Sub-zero temperature), 24 °C (Room temperature), and 200 °C (High temperature) [32]. In the tensile and fracture toughness tests, the low-temperature chamber with liquid nitrogen and a high-temperature furnace with forced convection heating was used to maintain the sub-zero and high temperatures.

The Servo Electric Universal Testing Machine (UTM) (BISS, Bangalore, India) with 50 kN capacity was used for tensile and fracture toughness tests. In tensile testing, the applied load and deformations were recorded continuously through the load cell and extensometer, respectively. However, along with these, Crack Opening Displacement (COD) gauge (BISS, Bangalore, India) was used to record the relative displacement of two knife edges of the C(T) specimen in the fracture toughness test. These data were further processed to extract the tensile properties and fracture toughness of the AA2050-T84 alloy as per standards [27,30].

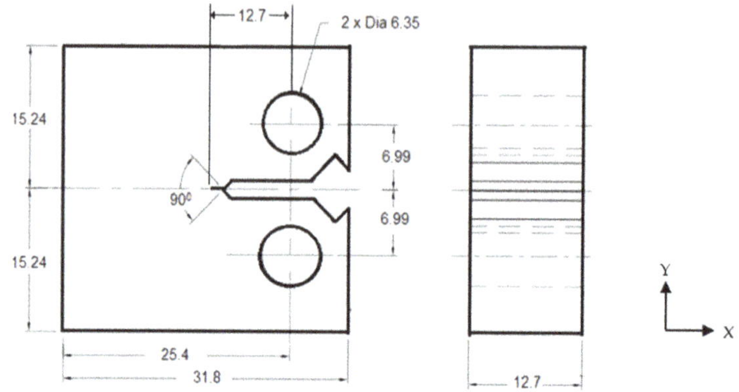

Figure 2. Compact Tension (C(T)) Specimen. All dimensions are in mm.

2.3. Finite Element Analysis

The crack driving forces and constraints of the C(T) specimen were investigated at different load rates and temperatures using 3D linear elastic finite element analysis (FEA). Half-symmetry is modeled and analyzed using Abaqus (6.14, 2014, Dassault Systemes Simulia Corp., Providence, RI, USA). The Poisson's ratio (v) and Young's modulus (E) obtained from experimentally conducted tensile results at different strain rates and temperatures are used for linear elastic fracture analysis. The output parameters viz. stress intensity factor (K_I) and constraint parameter (T_{11}) were extracted using the counter-integral method mentioned in Abaqus post-processor [33]. T_{33} is calculated by using Equation (1). In Equation (1), ε_{33} is a strain in the z-direction (thickness direction) extracted along the crack front. The material property input and the K_I extraction details were adopted as similar to the work of [22,33].

$$T_{33} = E\varepsilon_{33} + vT_{11} \tag{1}$$

Half symmetry C(T) meshed model with supports and loading is shown in Figure 3. 20-noded hexahedral elements with reduced integration were used for the meshing. A fine mesh near the crack front was used to encapsulate the crack characteristics effectively. Singularity at the crack front was emulated by shifting the mid-side nodes of crack surrounding elements towards the crack front. The crack edge (crack front) surrounded by these nodes is defined as contour integral. The output parameters are calculated along the user-defined contour integrals (in the present analysis, it is 10 contours). The detailed procedure to define the crack front and contour integrals to obtain crack driving parameters is available in the Abaqus manual [33]. Y-symmetry was imposed along the ligament (the uncracked portion in the crack plane), and the tensile load was applied through the hole to simulate Mode-I.

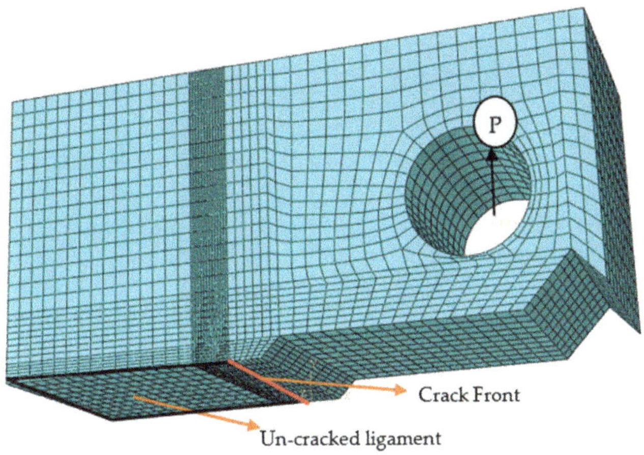

Figure 3. C(T) Meshed Model with Boundary conditions.

3. Results and Discussions

The following sections discuss the experimental and numerical analysis of the AA2050-T84 alloy.

3.1. Experimental Analysis

3.1.1. Tensile Test Analysis

The experimental tensile tests were conducted at varying temperatures and strain rates. A total of 27 tensile tests were conducted, comprising 3 experiments for each strain rate per temperature. The stress-strain curves for different strain rates and temperatures are shown in Figure 4a. The corresponding variation of average tensile properties viz. yield stress (σ_{ys}), ultimate tensile stress (σ_{ut}), Young's modulus (E), and % elongation (% et) extracted from the stress-strain curves along with error bars are presented in Figure 4b–e. The positive strain rate dependency was observed for tensile yield stress for the temperatures considered in the study, as shown in Figure 4b. The highest yield stress variation of 2% between 0.01 s^{-1} to 0.1 s^{-1} and 0.1 s^{-1} to 1 s^{-1} was observed at −20 °C. However, the lowest yield stress variation between successive strain rates, around 0.5%, was noticed at 24 °C. It was observed that the yield stress decreased from −20 °C to room temperature and further increased slightly at 200 °C, indicating the V-shaped behavior for temperature variation.

Similarly, ultimate tensile stress exhibited the positive strain rate dependency at various temperatures, as shown in Figure 4c. However, strain rate has minimal effect on ultimate tensile stress as the difference observed between successive strain rates for all temperatures is less than 1%. The ultimate tensile stress is inversely proportional to the temperature for all the strain rates and is in line with the observations of Hafley et al. [5] and Chemin et al. [6]

Figure 4d, shows the variation of Young's modulus at various strain rates and temperatures. Young's modulus showed negative strain rate sensitivity at room and higher temperatures. However, the strain rate effect on Young's modulus was negligible (around 1%) at −20 °C. The maximum Young's modulus difference of about 10% was observed at 200 °C between strain rates 0.01 and 1 s^{-1}. Furthermore, Young's modulus difference was around 5% at room temperature for successive strain rate variations. This reveals that the strain rate sensitivity towards Young's modulus was in the decreasing order of temperatures 200 °C:24 °C:−20 °C. At 200 °C. Young's modulus values were minimal and almost similar in values at −20 °C.

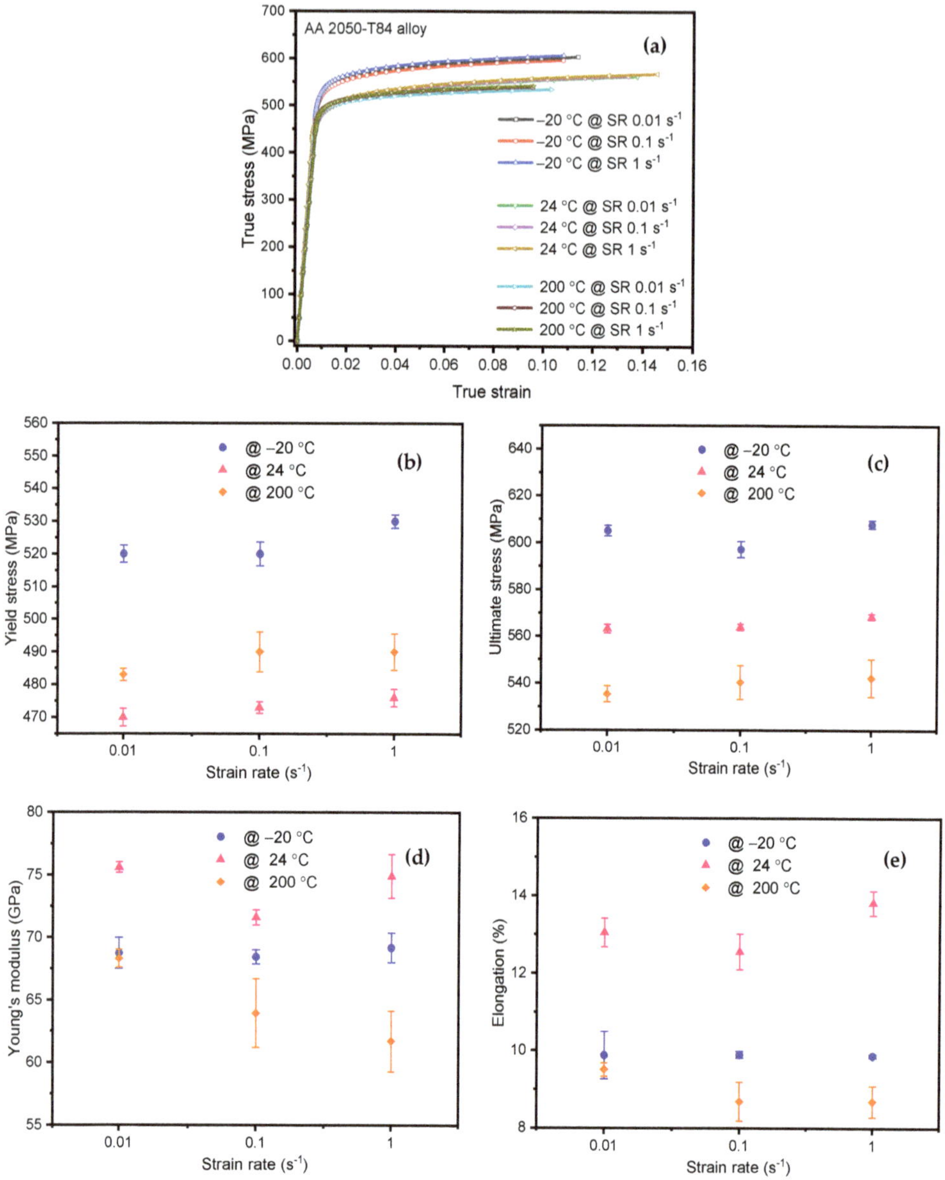

Figure 4. Strain rate and Temperature effect on: (**a**) True stress-strain curves (**b**) Yield stress; (**c**) Ultimate stress; (**d**) Young's modulus; (**e**) elongation.

Overall, a reduction between 6% and 10% is noticed in Young's modulus for the temperatures studied. The exact thickness of the plate with room temperature and −54 °C has been studied by Chemin et al. [6], revealing the same trend with a 2.5% reduction in Young's modulus. With a 100 mm (4-inch) plate, Hafley et al. [5] noticed a 9–11% reduction in Young's modulus when studied at different locations for room and −196 °C temperature.

Strong interatomic bonding between the atoms at room temperature may be the probable reason for the highest value of Young's modulus. Farraro and McLellan [34] have reported that the lower values of Young's modulus at elevated temperature indicate weakened interatomic bonding between the atoms.

Figure 4e shows the % elongation of AA2050-T84 alloy at different strain rates and temperatures. Negative strain rate dependency on % elongation as similar to Young's modulus was observed for various temperatures. The maximum % elongation was noticed at room temperature, indicating higher ductility than other temperatures.

3.1.2. Fracture Toughness Test Analysis

The fracture toughness tests were conducted as per ASTM E399-20a at different temperatures. ASTM E399-20a essentially elucidates the procedure of obtaining the single value, Plane Strain Fracture Toughness (K_{IC}), for metallic materials under quasi-static strain rates. The C(T) specimen was fatigue pre-cracked to emulate the natural crack characteristics. The effect of loading and material properties (mainly yield stress) strongly influences fatigue crack growth [35]. The pre-cracking load details are shown in Table 2. The load ratio ($\sigma_{min}/\sigma_{max}$) was maintained at 0.1 to attain the *a/W* range in between 0.45 and 0.55.

Table 2. Pre-cracking details of fracture toughness test.

Crack Length *a* (mm)	Crack Length/Width *a/W*	Maximum Stress σ_{max} (MPa)	Minimum Stress σ_{min} (MPa)	Mean Stress σ_{mean} (MPa)	Alternating Stress $\sigma_{average}$ (MPa)
12.94	0.51	250	25	137.5	112.5

The pre-cracked C(T) specimen was tested under Mode-I (opening mode) loading through tensile load application at the holes. A minimum of 3 successful fracture toughness tests was conducted at each temperature at the strain rate 0.01 s^{-1}. The K_{IC} (average of 3 test samples) obtained from experiments for −20 °C and 24 °C are 904.28 and 1059.36 MPa mm$^{1/2}$, respectively. The error bar for the K_{IC} is depicted in Figure 5.

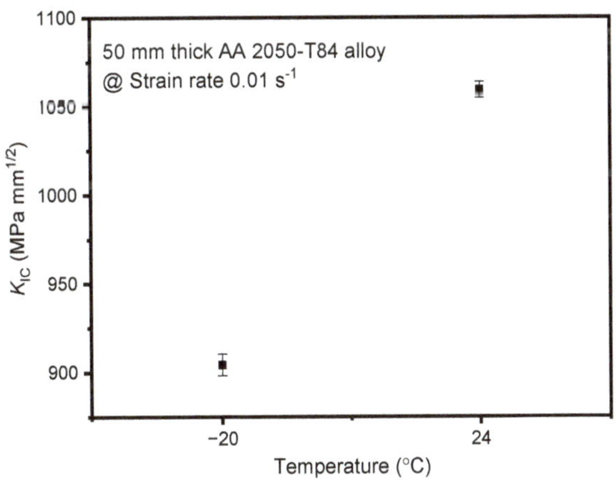

Figure 5. Plane Strain Fracture Toughness (K_{IC}) at various temperatures.

At 200 °C, all the 3 test results were invalid as the ample crack deviation was observed from the crack plane. However, the K_{IC} has reduced by about 15% from room to sub-zero temperature. Similarly, in the work of Chemin et al. [6], a reduction of the order of magnitude 16% in K_{IC} was noticed in the rolling direction of the AA2050-T84 alloy plate from room to cryogenic (−56 °C) temperature. The decrease of K_{IC} at sub-zero temperatures of AA2050-T84 alloy can be attributed to surface hardening inside the grain and validated by the grain microstructure of the alloy [6,36]. In summary, the positive temperature dependency was witnessed for K_{IC} of AA2050-T84 alloy at strain rate 0.01 s^{-1}. The economic limitations on conducting further fracture toughness tests at other strain rates impelled us to adopt the numerical analysis.

3.2. Linear Elastic Fracture Analysis

ASTM E399-20a, to predict the K_{IC} of metallic materials, recommended the single toughness value for lower strain rate variations. However, the variation of tensile properties of AA2050-T84 alloy at different lower strain rates and temperatures was substantial. The elastic fracture analysis was carried out at varied strain rates and temperatures using the Abaqus software. The current numerical procedure was adopted from Kudari et al. [22] and Kavale et al. [37] The K_I values extracted through-thickness direction of the crack at 24 °C conditions, and experimental K_{IC} are shown in Figure 6. The experimental K_{IC} value was emulated through numerical fracture analysis with less than 1% error, as observed in Figure 6.

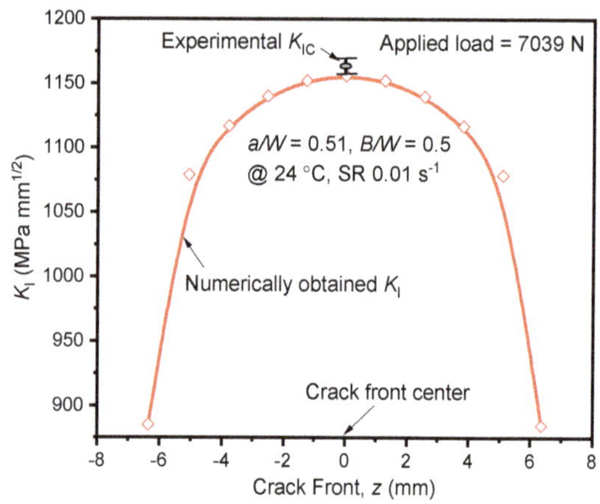

Figure 6. Stress intensity factor (K_I) along the crack front at 24 °C.

Similarly, at −20 °C, the validation of the numerical procedure was executed with less than 1% error. For all the numerical analysis, the center of the crack front was associated with the largest value of the crack characterizing parameter. Thus, the K_I values at the center of the specimen are used for analysis in further discussions.

3.2.1. Effect of Strain Rate

In linear elastic fracture analysis, the experimental load associated with K_{IC} of the alloy is the applied load at respective temperatures. However, the K_{IC} at 200 °C was unavailable, and hence for the numerical analysis, the assumed load applied up to K_I = 1200 MPa mm$^{1/2}$. The applied stress ($\sigma_{applied}$) was determined using the relationship mentioned in Equation (2) [27]. The extracted values of K_I at the crack front center for various strain rates and temperatures are plotted against the stress ratio ($\sigma_{applied}/\sigma_{ys}$) as shown in Figure 7. Positive strain rate dependency of the K_I was observed for all temperatures considered in this study. At room temperature, a steady increase of 0.6% in K_I was witnessed with the rise in strain rate. However, for strain rate 1 s^{-1}, the maximum of 1.88% increase in K_I was noticed at −20 °C. The strain rate sensitivity on K_I was found to be maximum at −20 °C and minimum at 24 °C, as the same trend was noticed for tensile yield stress values. The results of the K_I are in line with the yield stress variations of the alloy for all strain rates at different temperatures. The nominal variations of K_I (within the stress ratio of K_{IC}) indicate the negligible dependency of fracture characterizing parameters on strain rates for AA2050-T84 alloy. Moreover, ASTM

E399-20a recommendation to use the single value of K_{IC} for quasi-static strain rates seems to be justifying as the difference in numerical K_I values was minimal.

$$K_I = \frac{P_{applied}}{B\sqrt{W}} f\left(\frac{a}{W}\right) \tag{2}$$

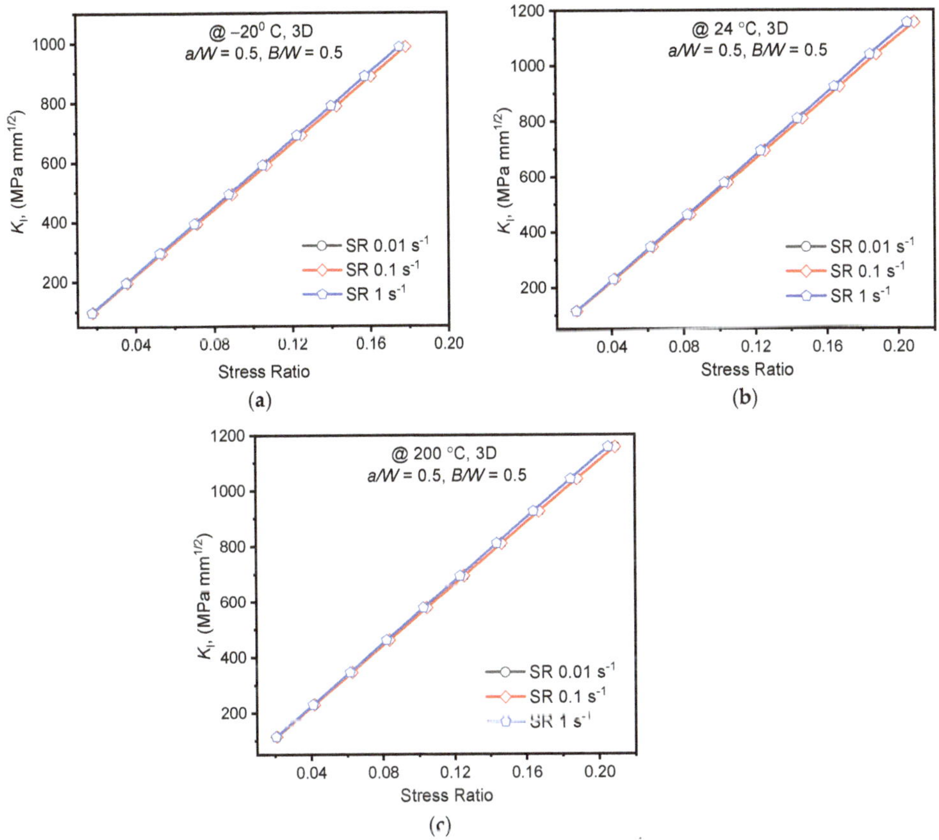

Figure 7. K_I at specimen thickness center obtained using FEA for different strain rates vs. Stress ratio (**a**) @ −20 °C; (**b**) 24 °C; (**c**) 200 °C.

Further, the effect of strain rate on crack tip/front constraints has been evaluated through T_{11} and T_{33}. The values of T_{11} and T_{33} are found to vary along the thickness similar to K_I variation and maximum being at the center of the specimen. One can infer that the crack-front constraint is high at the center than at the surface; the material may fail at the center than on surface or shows instability at the center of the specimen. As the constraint parameters do not have a unique value for the specimen thickness, maximum values at the center of the specimen are considered for further analysis. T_{11} and T_{33} variations for different strain rates within the purview of K_{IC} (or stress ratio) of the alloy at respective temperatures are plotted in Figure 8. The nature of variation was identical at all strain rates for both constraint parameters. However, the increase in stress ratio resulted in positive T_{11} and negative T_{33} values at all strain rates and temperatures. It is clear from Figure 8, that the applied stress was directly proportional to T_{11} and inversely proportional to T_{33}. The negative strain along the thickness resulted in the negative T_{33} [20]. This is in close agreement with the findings of Kudari et al. [22] for IF steel C(T) specimen.

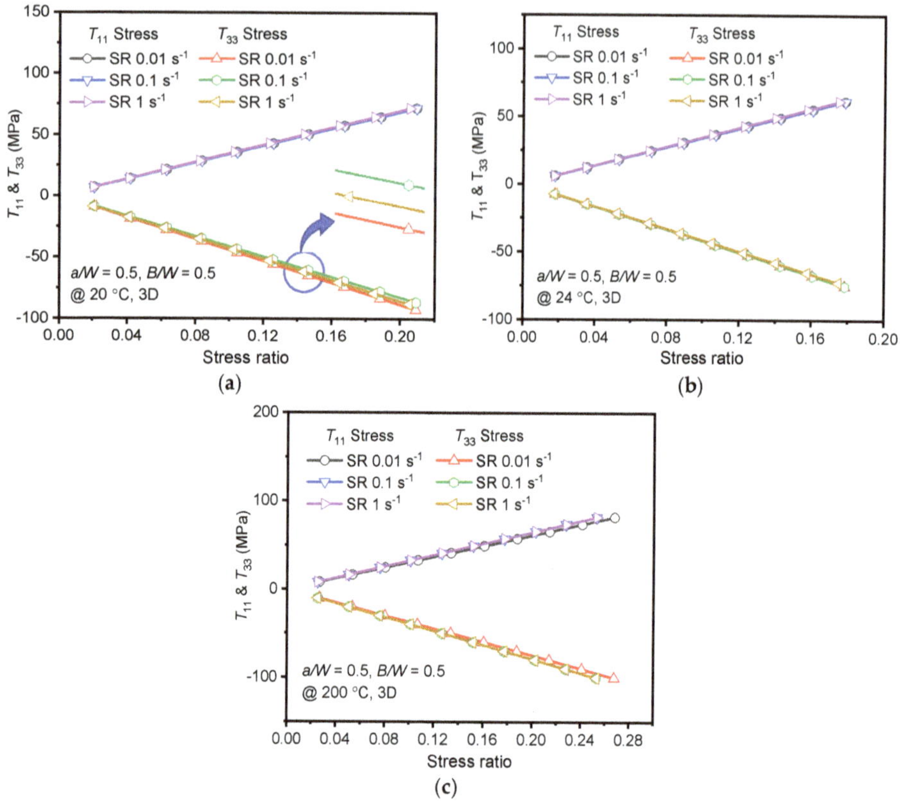

Figure 8. T_{11} and T_{33} at specimen thickness center obtained using FEA for different strain rates vs. Stress ratio (**a**) @ −20 °C; (**b**) 24 °C; (**c**) 200 °C.

At 24 °C and 200 °C, both constraint parameters were unaffected (very marginal difference) by the strain rate variation, as observed in Figure 8b,c. However, T_{33} variations between the strain rates at −20 °C were relatively substantial. At −20 °C, between the strain rates, a difference of 2.4% was found for T_{33}. T_{33} variation depended on Poisson's ratio and Young's modulus (material property) of the alloy at different strain rates. Eventually, the crack driving and constraint parameters were less sensitive to strain rate variations.

Strain rate effect on T_{11} variation is negligible as in-plane constraint depends on specimen type, geometry, and loading type only. Furthermore, the variation of hydrostatic stress along the uncracked ligament is studied at different strain rates. It is observed that no variations are found at different strain rates. The negligible variation of T_{11} can also be attributed to the uniform state of stress at the crack front and minimal variation of yield stress (or stress ratio as depicted in the graph) between the strain rates. However, the T_{33} variation is quite measurable at different strain rates for −20 °C, as shown in Figure 8a, owing to the variations in material property (both Young's modulus and yield stress). Positive T_{11} results in a lower plastic zone at the crack tip and influences the specimen's unstable crack growth. Similarly, negative T_{33} results in loss of constraint at the crack tip.

Since the crack front plasticity is restricted in LEFM regime, the state of stress may be unaltered due to strain rate variations at identical temperatures. The current serrated stress-strain curves may affect the plasticity ahead of the crack front and can be accounted in Elastic-Plastic Fracture Mechanics (EPFM) regime.

3.2.2. Effect of Temperature

The variation of K_I at different temperatures for quasi-static strain rates is plotted in Figure 9. The K_I variation is linearly increased with an increase in stress ratio as expected, and the nature of variation was the same for all temperatures considered. At the peak stress ratio, the difference between K_I at 200 °C and −20 °C is 5.1%, 5.16%, and 7.31% at strain rates 0.01, 0.1, and 1 s^{-1}, respectively. However, the variation of K_I was minimum (around 4%) for the temperatures 24 °C and 200 °C at all strain rates. Notably, the temperature effect was highest at strain rate 1 s^{-1} and in the decreasing order of 1: 0.1: 0.01 s^{-1}. The substantial variation of K_I at sub-zero temperatures indicates that for identical load conditions, the AA2050-T84 alloy is more prone to fracture failure than the other two temperatures as it possesses lower $K_{IC.}$

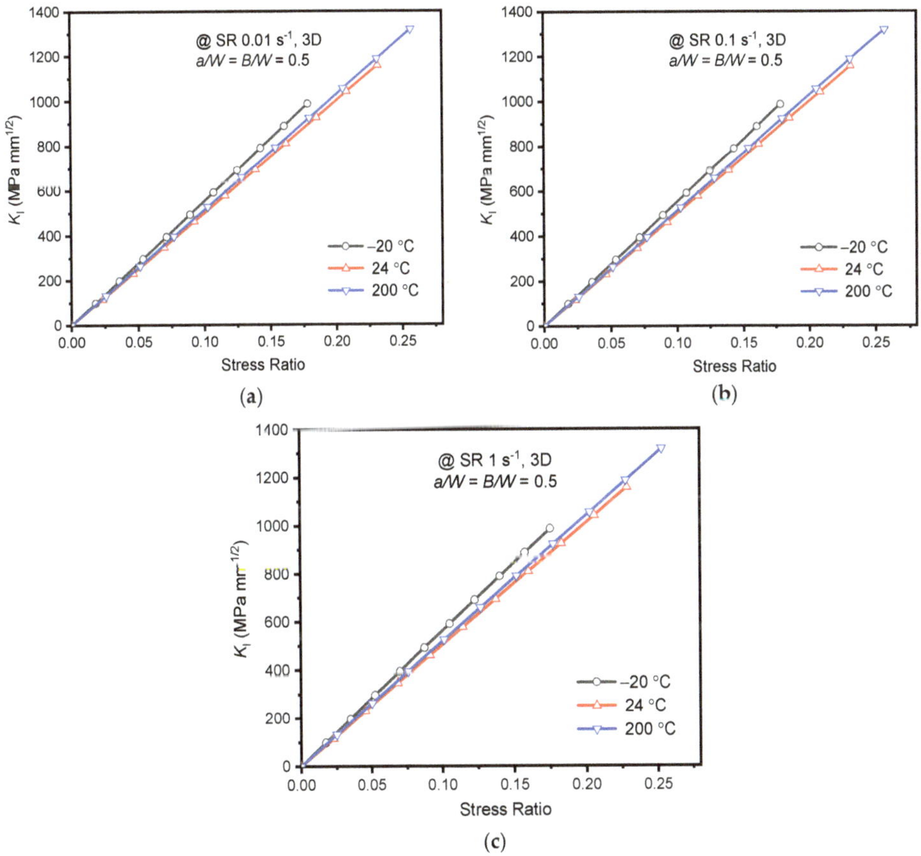

Figure 9. K_I at specimen thickness center obtained using FEA for different temperature vs. Stress ratio (**a**) SR 0.01 s^{-1}; (**b**) SR 0.1 s^{-1}; (**c**) SR 1 s^{-1}.

Similarly, Figure 10 shows the variation of T_{11} and T_{33} at different temperatures for quasi-static strain rates. The variation of T_{11} and T_{33} were almost identical at 24 °C and 200 °C at quasi-static strain rates. At 200 °C, the T_{11} was less subtle and owed lower in-plane constraints than the −20 °C, and 24 °C. Chemin et al. [6] have related dislocations gathered along the grain boundaries, led to stress concentrations under loading and promoted the lower fracture toughness of the AA2050-T84 alloy at sub-zero (−56 °C) temperature. Similarly, in the current analysis, the sensitivity of fracture toughness and in-plane constraint against the stress ratio is highest at −20 °C. The sensitivity may be associated to crack front stress concentrations at grain boundaries at −20 °C and can be accounted through constraint parameters as shown in the current numerical analysis. However, the T_{11} and T_{33} variations seem identical for strain rates 0.01 and 1 s^{-1} through Figure 10a,c at −20 °C. Eventually, T_{11} and T_{33} in combination with K_I is maximum at 1 s^{-1} compared to 0.01 s^{-1}. Thus, the highest constraint associated with numerically obtained K_I of the AA2050-T84 alloy at −20 °C is 1 s^{-1}. This behavior also resulted in the lower K_{IC} at −20 °C than 24 °C.

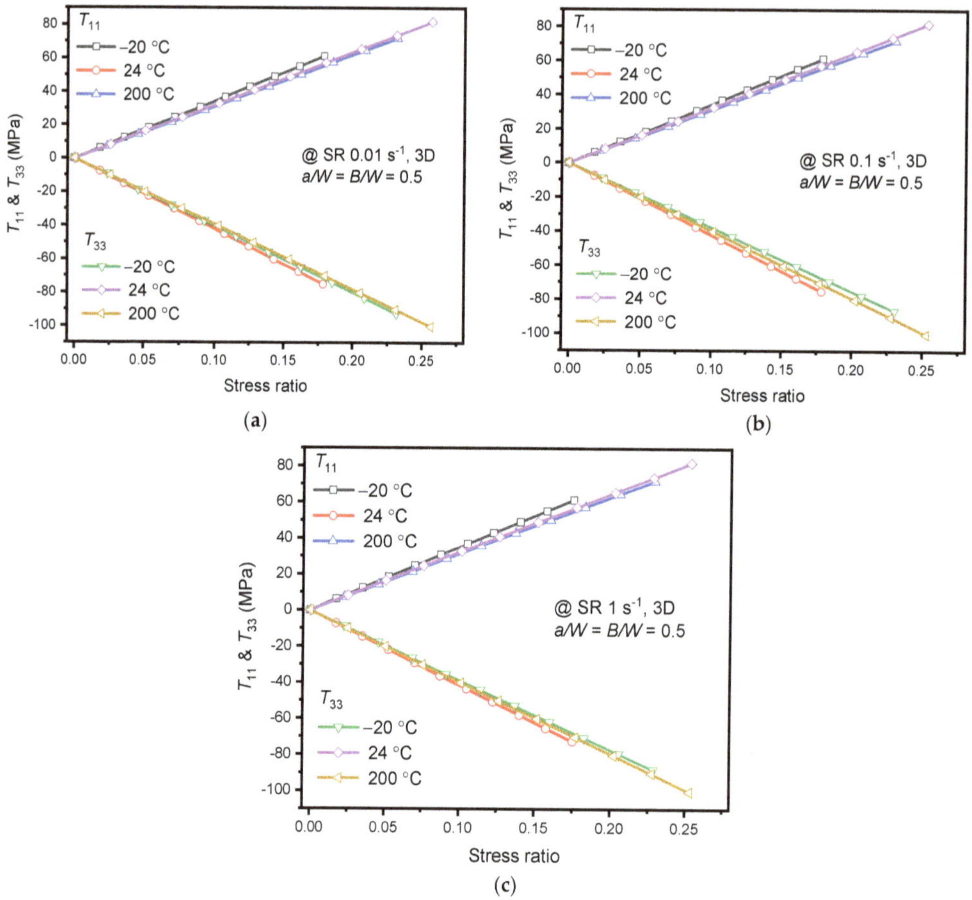

Figure 10. T_{11} and T_{33} at specimen thickness center obtained using FEA for different temperature vs. Stress ratio (**a**) SR 0.01 s^{-1}; (**b**) SR 0.1 s^{-1}; (**c**) SR 1 s^{-1}.

In summary, the major constraint loss was observed for temperature variation compared to quasi-static strain rate variations. Moreover, at −20 °C, AA2050-T84 alloy possesses lower K_{IC} with the highest in-plane crack tip/front constraint compared to the other two temperatures. This behavior of the alloy makes it vulnerable to fracture failure in cryogenic (sub-zero) applications at a strain rate 1 s^{-1}.

4. Conclusions

In the present study, the tensile and fracture behavior of the 50 mm thick (2-inch) AA2050-T84 plate was considered at various temperatures for quasi-static strain rates. Tensile tests revealed the sensitivity of mechanical properties towards the strain rates and temperatures. Positive strain rate dependency was observed for temperatures considered on yield stress and ultimate tensile stress of the alloy. A maximum of 2% increase in yield stress was noticed between strain rates at −20 °C. The lowest strain rate sensitivity of around 0.5% was witnessed at room temperature. Notably, the ultimate stress variation between the strain rates for temperatures was less than 1%. However, Young's modulus and % elongation were negative strain rate dependent. The maximum decrease of Young's modulus up to 10% was noticed at 200 °C. The minimum Young's modulus variation was witnessed at −20 °C between the strain rates.

Temperature sensitivity towards tensile behavior of the AA2050-T84 alloy was noticed. The maximum yield stress variation of 10–11% was witnessed between room temperature and −20 °C. Notably, the yield stress increase was only up to 3% between room temperature and 200 °C. Similarly, for ultimate stress, the variation was up to 7.5% between room temperature and −20 °C. However, the reduction of Young's modulus up to 18% was noticed between room temperature and 200 °C. This implies that yield and ultimate stress are quite substantial at −20 °C compared to other temperatures, making the AA2050-T84 alloy vulnerable at sub-zero temperatures. Moreover, an increase in strain rate prompts the decrease in % elongation, implying the brittle behavior of the alloy at higher strain rates.

The crack driving and constraint parameters are less sensitive to strain rate variations. However, at −20 °C, crack characterizing and constraint parameters to strain rate variations were moderately considerable. The temperature effect is highest at strain rate 1 s^{-1} and in the decreasing order of 1:0.1:0.01 s^{-1}.

Overall, the AA2050-T84 alloy tensile and fracture performance obtained through experimental and numerical analyses exhibited the dependency on strain rates and temperatures. Hence, these mechanical properties of the alloy strongly influence the damage tolerance design of spars and wings of the aircraft. The authors believe that the current results concerning strain rates and temperatures will help in understanding the performance-related issues of AA2050-T84 alloy reported in aircraft applications.

Author Contributions: Conceptualization, N.E. and K.G.K.; methodology, N.E. and K.G.K.; software, N.E.; validation, N.E. and K.G.K.; formal analysis, N.E.; investigation, N.E.; resources, N.E. and K.G.K.; data curation, K.G.K.; writing—original draft preparation, N.E.; writing—review and editing, N.E. and K.G.K.; visualization, K.G.K., T.M.Y.K. and I.A.B.; supervision, K.G.K., T.M.Y.K. and I.A.B.; project administration, K.G.K., T.M.Y.K. and I.A.B.; funding acquisition, T.M.Y.K. and I.A.B. All authors have read and agreed to the published version of the manuscript.

Funding: This research was funded by King Khalid University through grant number (R.G.P. 2/107/41).

Institutional Review Board Statement: Not applicable.

Informed Consent Statement: Not applicable.

Data Availability Statement: This study didn't report any data.

Acknowledgments: The authors extend their appreciation to the Deanship of Scientific Research at King Khalid University for funding this work through research groups program under grant number (R.G.P. 2/107/41). The authors also thank the financial help provided by KLE Technological university through the capacity buildings grants.

Conflicts of Interest: The authors declare no conflict of interest.

References

1. Wanhill, R.J.; Bray, G.H. Aerostructural design and its application to aluminum–lithium alloys. In *Aluminum-Lithium Alloys*, 1st ed.; Prasad, N.E., Gokhale, A., Wanhill, R.J.H., Eds.; Butterworth-Heinemann: Oxford, UK, 2014; pp. 27–58.
2. Lynch, S.P.; Wanhill, R.J.; Byrnes, R.T.; Bray, G.H. Fracture toughness and fracture modes of aerospace aluminum–lithium alloys. In *Aluminum-Lithium Alloys*, 1st ed.; Prasad, N.E., Gokhale, A., Wanhill, R.J.H., Eds.; Butterworth-Heinemann: Oxford, UK, 2014; pp. 415–455.
3. Edgar, A.; Starke, J.R. Historical Development and Present Status of Aluminum Lithium Alloys. In *Aluminum-Lithium Alloys*, 1st ed.; Prasad, N.E., Gokhale, A., Wanhill, R.J.H., Eds.; Butterworth-Heinemann: Oxford, UK, 2014; pp. 3–26.

4. Dorin, T.; Vahid, A.; Lamb, J. Aluminium lithium alloys. In *Fundamentals of Aluminium Metallurgy*, 1st ed.; Lumley, R.M., Ed.; Woodhead Publishing: Duxford, UK, 2018; pp. 387–438.
5. Hafley, R.A.; Domack, M.S.; Hales, S.J.; Shenoy, R.N. *Evaluation of Aluminum Alloy 2050-T84 Microstructure and Mechanical Properties at Ambient and Cryogenic Temperatures*; Technical Report for NASA; Langley Research Center: Hampton, VA, USA, August 2011.
6. Chemin, A.E.; Afonso, C.M.; Pascoal, F.A.; Maciel, C.D.; Ruchert, C.O.; Bose Filho, W.W. Characterization of phases, tensile properties, and fracture toughness in aircraft-grade aluminum alloys. *Mater. Des. Process. Commun.* **2019**, *1*, 1–3. [CrossRef]
7. Mirza, M.S.; Barton, D.C.; Church, P. The effect of stress triaxiality and strain-rate on the fracture characteristics of ductile metals. *J. Mater. Sci.* **1996**, *31*, 453–461. [CrossRef]
8. Clausen, A.H.; Børvik, T.; Hopperstad, O.S.; Benallal, A. Flow and fracture characteristics of aluminium alloy AA5083–H116 as function of strain rate, temperature and triaxiality. *Mater. Sci. Eng. A* **2004**, *364*, 260–272. [CrossRef]
9. Singh, N.; Singh, V. Effect of temperature on tensile properties of near-α alloy Timetal 834. *Mater. Sci. Eng. A* **2008**, *485*, 130–139. [CrossRef]
10. Khan, A.S.; Liu, H. Strain rate and temperature dependent fracture criteria for isotropic and anisotropic metals. *Int. J. Plast.* **2012**, *37*, 1–5. [CrossRef]
11. Anderson, D.; Winkler, S.; Bardelcik, A.; Worswick, M.J. Influence of stress triaxiality and strain rate on the failure behavior of a dual-phase DP780 steel. *Mater. Des.* **2014**, *60*, 198–207. [CrossRef]
12. Roth, C.C.; Mohr, D. Effect of strain rate on ductile fracture initiation in advanced high strength steel sheets: Experiments and modeling. *Int. J. Plast.* **2014**, *56*, 19–44. [CrossRef]
13. Rincon, E.; Lopez, H.F.; Cisneros, M.M.; Mancha, H.; Cisneros, M.A. Effect of temperature on the tensile properties of an as-cast aluminum alloy A319. *Mater. Sci. Eng. A* **2007**, *452*, 682–687. [CrossRef]
14. Natesan, E.; Ahlström, J.; Manchili, S.K.; Eriksson, S.; Persson, C. Effect of Strain Rate on the Deformation Behaviour of A356-T7 Cast Aluminium Alloys at Elevated Temperatures. *Metals* **2020**, *10*, 1239. [CrossRef]
15. Pandya, K.S.; Roth, C.C.; Mohr, D. Strain rate and temperature dependent fracture of aluminum alloy 7075: Experiments and neural network modeling. *Int. J. Plast.* **2020**, *135*, 102788. [CrossRef]
16. Ekabote, N.; Kodancha, K.G.; Kudari, S.K. Suitability of Standard Fracture Test Specimens for Low Constraint Conditions. *Conf. Ser. Mater. Sci. Eng.* **2021**, *1123*, 012033. [CrossRef]
17. Ekabote, N.; Kodancha, K.G.; Revankar, P.P. Elastic-plastic fracture analysis of anisotropy effect on AA2050-T84 alloy at different temperatures: A numerical study. *Frat. Integrita Strutt.* **2022**, *16*, 78–88.
18. Gupta, M.; Alderliesten, R.C.; Benedictus, R. A review of T-stress and its effects in fracture mechanics. *Eng. Fract. Mech.* **2015**, *134*, 218–241. [CrossRef]
19. O'dowd, N.P.; Shih, C.F. Family of crack-tip fields characterized by a triaxiality parameter—I. Structure of fields. *J. Mech. Phys. Solids.* **1991**, *39*, 989–1015. [CrossRef]
20. Hutař, P.; Seitl, S.; García, T.E.; Fernández-Canteli, A. Experimental and numerical analysis of in-and out-of plane constraint effects on fracture parameters: Aluminium alloy 2024. *Appl. Comput. Mech.* **2013**, *7*, 53–64.
21. Fernández-Canteli, A.; Giner, E.; Fernández-Zúniga, D.; Fernández-Sáez, J. A unified analysis of the in-plane and out-of-plane constraints in 3-D linear elastic fracture mechanics. In Proceedings of the 19th European Conference on Fracture, Kazan, Russia, 26–31 August 2012.
22. Kudari, S.K.; Kodancha, K.G. 3D Stress intensity factor and T-stresses (T11 and T33) formulations for a Compact Tension specimen. *Frat. Integrita Strutt.* **2017**, *11*, 216–225. [CrossRef]
23. Wang, X. Elastic T-stress for cracks in test specimens subjected to non-uniform stress distributions. *Eng. Fract. Mech.* **2002**, *69*, 1339–1352. [CrossRef]
24. Neimitz, A.; Galkiewicz, J. Fracture toughness of structural components: Influence of constraint. *Int. J. Press. Vessel. Pip.* **2006**, *83*, 42–54. [CrossRef]
25. Galkiewicz, J.; Janus-Galkiewicz, U. The Numerical Analysis of the In-Plane Constraint Influence on the Behavior of the Crack Subjected to Cyclic Loading. *Materials* **2021**, *14*, 1764. [CrossRef] [PubMed]
26. Kim, Y.K.; Oh, B.T.; Kim, J.H. Effects of Crack Tip Constraint on the Fracture Toughness Assessment of 9% Ni Steel for Cryogenic Application in Liquefied Natural Gas Storage Tanks. *Materials* **2020**, *13*, 5250. [CrossRef]
27. *ASTM E399-20*; Standard Test Method for Linear-Elastic Plane-Strain Fracture Toughness of Metallic Materials. ASTM International: West Conshohocken, PA, USA, 2020.
28. Brian Falzon. The Airbus A380 Wing Cracks: An Engineer's Perspective. Available online: https://theconversation.com/the-airbus-a380-wing-cracks-an-engineers-perspective-5318 (accessed on 22 January 2022).
29. Rioja, R.J.; Liu, J. The evolution of Al-Li base products for aerospace and space applications. *Metall. Mater. Trans. A* **2012**, *43*, 3325–3337. [CrossRef]
30. *ASTM E8/E8M-21*; Standard Test Methods for Tension Testing of Metallic Materials. ASTM International: West Conshohocken, PA, USA, 2021.
31. Mouritz, A.P. *Introduction to Aerospace Materials*; Woodhead Publishing Limited: Cambridge, UK, 2012; pp. 454–468.
32. Ekabote, N.; Kodancha, K.G. Temperature and test specimen thickness (TST) effect on tensile and fracture behavior of AA2050-T84 alloy. *Mater. Today Proc.* 2021, *in press*. [CrossRef]

33. Dassault Systemes. Abaqus Analysis User Guide. Available online: http://wufengyun.com/v6.14/books/usb/default.htm?startat=pt05ch22s02abm03.html (accessed on 22 January 2022).
34. Farraro, R.; McLellan, R.B. Temperature dependence of the Young's modulus and shear modulus of pure nickel, platinum, and molybdenum. *Metall. Trans. A* **1977**, *8*, 1563–1565. [CrossRef]
35. Borges, M.F.; Antunes, F.V.; Prates, P.A.; Branco, R.; Vojtek, T. Effect of Young's modulus on fatigue crack growth. *Int. J. Fatigue* **2020**, *132*, 105375. [CrossRef]
36. Abd El-Aty, A.; Xu, Y.; Guo, X.; Zhang, S.H.; Ma, Y.; Chen, D. Strengthening mechanisms, deformation behavior, and anisotropic mechanical properties of Al-Li alloys: A review. *J. Adv. Res.* **2018**, *10*, 49–67. [CrossRef] [PubMed]
37. Kavale, S.M.; Kodancha, K.G.; Ekabote, N. Effect of Poisson's ratio on KI, T11 and T33 for SENB and CT specimen–A FE study. *Procedia Struct. Integr.* **2019**, *14*, 584–596. [CrossRef]

materials

MDPI

Review

Void-Induced Ductile Fracture of Metals: Experimental Observations

Wiktor Wciślik [1,*] and Sebastian Lipiec [2]

[1] Faculty of Civil Engineering and Architecture, Kielce University of Technology, 25-314 Kielce, Poland
[2] Faculty of Mechatronics and Mechanical Engineering, Kielce University of Technology, 25-314 Kielce, Poland
* Correspondence: wwcislik@tu.kielce.pl

Abstract: The paper presents a literature review on the development of microvoids in metals, leading to ductile fracture associated with plastic deformation, without taking into account the cleavage mechanism. Particular emphasis was placed on the results of observations and experimental studies of the characteristics of the phenomenon itself, without in-depth analysis in the field of widely used FEM modelling. The mechanism of void development as a fracture mechanism is presented. Observations of the nucleation of voids in metals from the turn of the 1950s and 1960s to the present day were described. The nucleation mechanisms related to the defects of the crystal lattice as well as those resulting from the presence of second-phase particles were characterised. Observations of the growth and coalescence of voids were presented, along with the basic models of both phenomena. The modern research methods used to analyse changes in the microstructure of the material during plastic deformation are discussed. In summary, it was indicated that understanding the microstructural phenomena occurring in deformed material enables the engineering of the modelling of plastic fracture in metals.

Keywords: ductile fracture; material testing; void nucleation; growth; coalescence; microstructure; material characterisation

Citation: Wciślik, W.; Lipiec, S. Void-Induced Ductile Fracture of Metals: Experimental Observations. *Materials* **2022**, *15*, 6473. https://doi.org/10.3390/ma15186473

Academic Editor: Javad Mola

Received: 30 July 2022
Accepted: 14 September 2022
Published: 18 September 2022

1. Introduction

Ductile fracture criteria can be classified based on the physical sense of the quantities defining them. If the criterion includes values derived by macroscopic analysis based on mechanics of deformable bodies, without precisely defined areas of fracture initiation, a global criterion is formulated. The most known and commonly used global ductile fracture criteria (also known as phenomenological) were determined on the basis of such quantities as J integral and crack tip opening displacement δT.

Another approach involves analysis of the stress and strain (or other parameter) fields in the most stressed areas of the material (process zone). The stress and strain state in the process zone determines the strength of the whole element. Such an approach and arising criteria are local. The criterion values of local models depend only on the material properties.

An undoubted advantage of using a local approach to the analysis of the fracture is the independence of the obtained results on the geometry of the samples used. The local approach is based on actual physical phenomena (degradation of the microstructure) that occur in the material subjected to load. In general, understanding and describing these phenomena allow for their inclusion in the cracking description of any element type, and thus the development of a universal, comprehensive method of predicting the durability of structural elements and assessing the safety of their work.

Conducting an analysis according to a local approach requires the use of hybrid methods: obtaining data from experimental research (with accompanying in situ analyses, e.g., video extensometer, recording of acoustic emission signals, using computed tomography techniques) and supplementing them with numerical, metallographic and fractographic analyses.

Among the physical phenomena taken into account in the local approach to fracture, the nucleation and growth of microvoids in the structure of the material subjected to plastic deformation is of particular practical importance. Observation of this phenomenon is the basis for formulating theoretical models of nucleation, growth and coalescence of voids, which in turn enables the establishment of models of plastic materials with microdamages. These issues are therefore crucial for assessing the safety of structural elements. This article is a literature review on the microstructure observations and void growth in deformed metals. Particular emphasis was placed on ductile fracture associated with substantial plastic deformation, without taking into account cleavage cracking. The results of pioneering works from the second half of the twentieth century, as well as the latest results obtained with advanced research tools (such as microtomography), are discussed. The structure of the article is divided into sections according to the different mechanisms in ductile fracture: Section 2 provides general information on void-induced failure; further, the nucleation of voids (Section 3) and subsequent stages of their growth and coalescence (Section 4) are described.

2. Cracking of Metals by the Development of Voids

Void initiators are primarily defects in the crystalline structure (point, linear and planar defects) and second-phase particles, which can be introduced into the material microstructure intentionally (e.g., metal matrix composites) or constitute an undesirable contamination of the material. All these discontinuities, being local stress concentrators, under plastic deformation create voids with sizes in the order of tenths of a micrometre. As the plastic strain increases, the voids created in this way increase their size many times and then coalesce, forming a larger defect and leading to failure. An example of a sample structure in which a fracture developed as a result of the growth and coalescence of voids is presented in Figure 1a.

Figure 1b presents a microscopic photograph of the fracture surface of the S355 steel sample damaged due to the development of voids. The characteristic dimple microstructure is observed, in which the voids of a few μm in size are surrounded by plastically deformed ligaments (light bands in Figure 1b) [2].

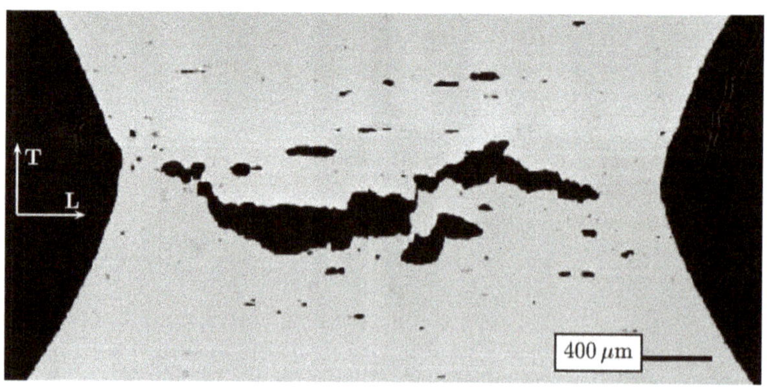

(**a**)

Figure 1. *Cont.*

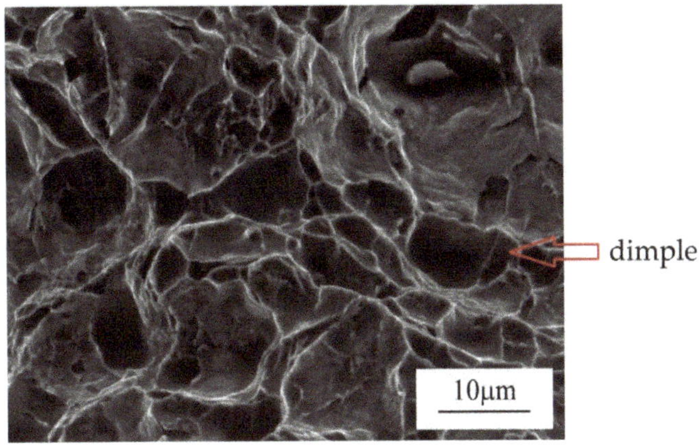

(b)

Figure 1. Void-induced failure of metals: (a) Formation of a macroscopic crack due to void growth and coalescence in X52 steel [1]; (b) Fracture surface of a tensile S355 steel specimen and dimple microstructure resulting from void development (authors' own study).

A separate issue is the macroscopic observation of fracture surfaces at void-induced failure. Ductile fracture of a round tensile bar has been relatively well-documented in the literature [3]. Initially, voids develop mainly in the centre of the sample, whereby the crack propagates approximately perpendicular to the sample axis, in the plane of the minimum cross section (in the neck area). As the crack approaches the sample surface, the influence of constraints decreases, so the crack deviates from its initial plane. As a result, the fracture of the sample takes the form of a cup–cone fracture (Figure 2), in which the central part is flat (cup) and an inclined area (cone) is formed around it.

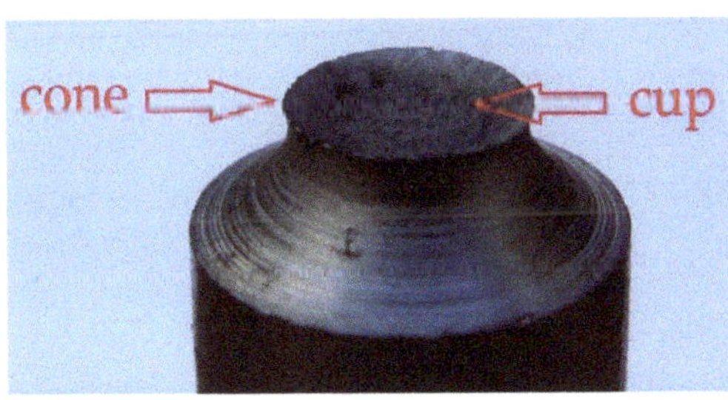

Figure 2. An example of a S355 steel specimen failure due to void development. Characteristic cup–cone shape is clearly visible [authors' own study].

An in-depth analysis of microstructures with the use of different methods allowed for the development of the systematics of the mechanisms of the formation of voids. This applies to both the nucleation of the voids and their subsequent development.

The initial experimental observations of the development of voids in plastically deformed metals took place at the turn of the 1950s and 1960s [4]. Since that time, a number of research methods have been developed to observe the expansion of voids in metals. The most common studies include the use of optical microscopy and scanning electron microscopy (SEM). The first attempts of 3D characterisation of voids involved the use of indirect methods, such as change in elastic stiffness [5,6] and change in resistivity and density of the sample [7,8]. A common disadvantage of indirect methods is the ability to measure only the volume fraction of voids, without taking into account their distribution, shape, size and other geometric features [9]. The use of X-ray tomography offers much more possibilities in this respect.

Section 3 describes the results obtained with the above-mentioned methods, taking into account different nucleation mechanisms and different types of void initiators. Section 4 presents the observation of the growth and coalescence of voids, as well as characterises the basic analytical models of these phenomena.

3. Void Nucleation in Metals

3.1. Overview

In general, voids are nucleated in the vicinity of stress concentrators on a microscopic scale [10]. In the literature, the most frequently mentioned mechanisms include the formation of voids at the intersection of the slip bands [11], at the grain boundaries [12], at twin boundaries and at vacancy clusters [13], but most importantly around inclusions and precipitates [14–16].

3.2. Nucleation with No Particles Involved

In the absence of material discontinuities, the formation of voids is observed as planar slip band decohesion or grain boundary decohesion (homogeneous nucleation). One of the first widely known studies in which the participation of dislocations in the formation of voids was experimentally confirmed was developed by Gardner et al. [12], who noticed that the dislocation structures in iron and beryllium crystals evolved into cells at high strains. The boundaries between individual cells were characterised by sufficient surface energy to create a void, even in the absence of other internal stress concentrators.

The observations of the development of voids in the Nb-Cr-Ti alloy [11] showed that the voids formed in this way are characterised by a flattened shape.

The concept of void nucleation as a result of vacancy condensation and the presence of dislocation boundaries has been known for a long time [17,18], but only modern research techniques allowed for a better understanding and documentation of this phenomenon [19]. For example, in [20], by investigating the fracture mechanism of copper containing copper oxide particles, it was found that the voids are first nucleated at the nanoscale, most often without any relation to the presence of the second-phase particles. An example of a photograph of nanovoids, made with the use of the high-angle annular dark-field scanning transmission electron microscopy (HAADF-STEM), is shown in Figure 3.

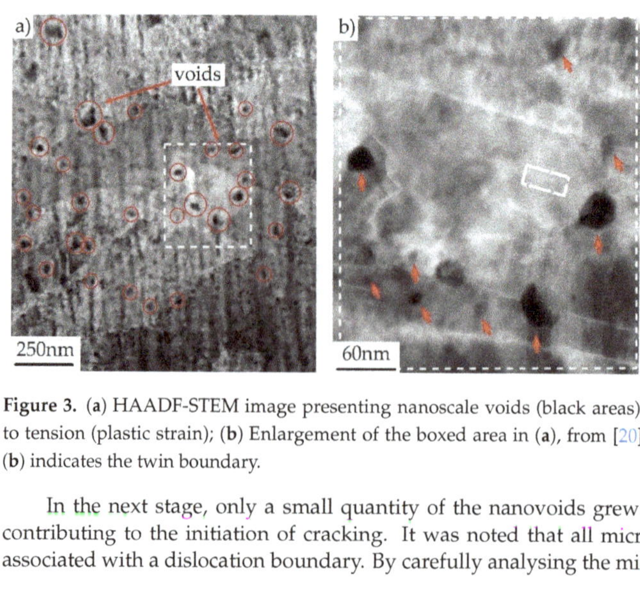

Figure 3. (**a**) HAADF-STEM image presenting nanoscale voids (black areas) in copper subjected to tension (plastic strain); (**b**) Enlargement of the boxed area in (**a**), from [20]. White rectangle in (**b**) indicates the twin boundary.

In the next stage, only a small quantity of the nanovoids grew to the microscale, contributing to the initiation of cracking. It was noted that all microscale voids were associated with a dislocation boundary. By carefully analysing the microstructure of the material in the neck area, the authors distinguished three basic groups of microvoids, depending on their location and the mechanism of formation, namely: (i) voids related to intragranular, inclusion-free dislocation boundaries; (ii) voids associated with the inclusion-free intersection between one or more dislocation boundaries and one or more grain boundaries; (iii) voids associated with an inclusion intersected by a dislocation boundary.

As mentioned before, voids can be initiated by grain separation. This mechanism is favoured by the following conditions: high share of the hydrostatic stress; small value of the void spacing to void diameter ratio; and high value of the precipitate-free zone (PFZ) thickness to void spacing ratio [21].

3.3. The Role of Second-Phase Particles

In technical alloys, voids are often initiated by the failure of second-phase particles, randomly distributed over the material matrix (heterogeneous nucleation). The presence of a hard particle locally limits elastic and plastic deformation of the matrix, which in turn causes local stress concentration. With the increase in plastic strain, the stress value also increases, which ultimately leads to the particle cracking or its separation from the matrix [22]. As the result, a void is formed (Figure 4).

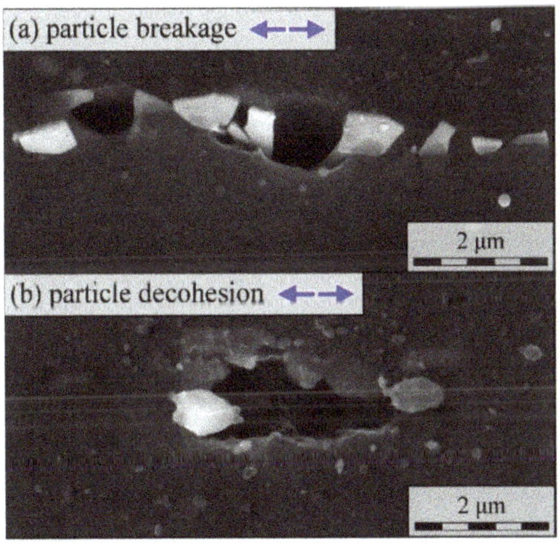

Figure 4. Two mechanisms of void nucleation in AZ31 magnesium alloy: (**a**) Particle fragmentation/fracture; (**b**) Matrix–particle separation [23].

The occurrence of one of the two mentioned mechanisms depends largely on the mechanical properties of the matrix and the particle (strength, ductility) and the strength of the matrix–particle interface. In general, the phenomenon of particle separation is primarily observed in relatively soft, ductile matrices. However, high yield stress and hardening exponent of the matrix as well as the high particle stiffness promote particle cracking.

Moreover, the nucleation mechanism depends on the particle geometry (size, shape) and its orientation with respect to the direction of the principle tensile stresses. Larger particles usually break, as do elongated particles parallel to the principal stress direction. In practice, engineering materials have different void populations; therefore, both mechanisms occur simultaneously. The local stress state is also of great importance, namely the predominance of normal stress over shear stress, which most often forces the particle fracture [24,25].

Due to the heterogeneous stress state changing with time, the nucleation of voids is continuous across the entire range of plastic strain. In other words, the development of microdamage includes the simultaneous growth and coalescence of existing voids as well as the nucleation of new ones [26].

Among the experimental studies, the process of nucleation of voids and microcrack formation by fracture of silicon particles in Al-Si alloys is relatively well-documented. The most important results described in the literature are summarised in [27]. As shown in [28,29], cracking of silicon particles and their separation from the matrix were observed already at strain values of 1–2%, while the quantity of damaged particles increased linearly with the increasing strain [30]. According to the results of observations described in [31], a maximum of 10% of the particles were damaged in the samples subjected to tension or compression. In general, larger particles crack first [32], which is often explained by a greater probability of internal defects in this type of particles. Smaller-sized particles tend to detach from the matrix, initiating voids.

As noted in [30], the phenomenon of fracture of silicon particles in the Al-7Si-0.4Mg alloy subjected to tension and bending occurs mainly in the case of elongated particles. The authors also analysed the structure of the alloy deformed to failure. In the areas with a homogeneous deformation (a few mm from the fracture surface) between 3 and 10% of the Si particles cracked. However, in the immediate vicinity of the fracture surface, the proportion of fractured particles locally increased to about 15–20%. In the coarser structures, the fracture of the particles was sudden and occurred at low strains. The development of microdamage in the finer structures was gradual.

3.4. Estimating Particle Strength

To understand the phenomenon of void nucleation by fracture of the second-phase particles, it is important to estimate their mechanical properties, especially strength. The first and best-known attempts were described in [33,34]. In [34], the microscopic maps of the locations of cracked and separated MnS inclusions were compared with the numerically determined stress and strain distributions, which gave rise to assessment of the critical fracture stress of MnS particles of about 1120 MPa. Additionally, in a similar manner, the critical stress of separation of the MnS particle and the matrix was determined, where the stress value was about 810 MPa.

A similar approach was used in [35], but this time the X-ray technique was used to assess the particle deformation. The tests were carried out on a sample made of an aluminium alloy, subjected to pure bending. Using the X-ray method, the strain values of silicon particles were determined, and then the stresses in the particles were determined as a function of the measured strains.

The typical research methods used so far did not take into account the internal defects of the particles; therefore, the obtained results are not precise.

The significant progress made in recent years in the field of methods of microstructural materials testing allowed for a more in-depth analysis of this issue. An interesting attempt to determine the strength of silicon particles in the A356 aluminium alloy is described in [36]. In order to expose the silicon particles, the ground surface of the sample was subjected to deep etching with a mixture of phosphoric, acetic and nitric acids. As a result, the particles to be tested were exposed to a height of several dozen micrometres. In the next step, using the focused ion beam (FIB) method, the geometric notch of the selected particles was cut. The prepared particles were subjected to an eccentric compression test in which the load was transferred to the particle by a tungsten needle. The test scheme is shown in Figure 5. The entire course of the process, up to the particle fracture, was recorded using a scanning electron microscope (SEM) with the video recording. In the next stage, based on the values of the measured force, the evaluation of the stress values in the particle was performed, using the simplified analytical method and with the use of the finite element method (FEM). For the defect-free particles, the particle strength was determined to be around 16 GPa. Detailed visual inspection of particles using SEM, combined with FEM modelling, made it possible to evaluate the effect of different types of particle defects on their strength. It was found that the presence of defects can reduce the strength of the silicon particles to 2–3 GPa.

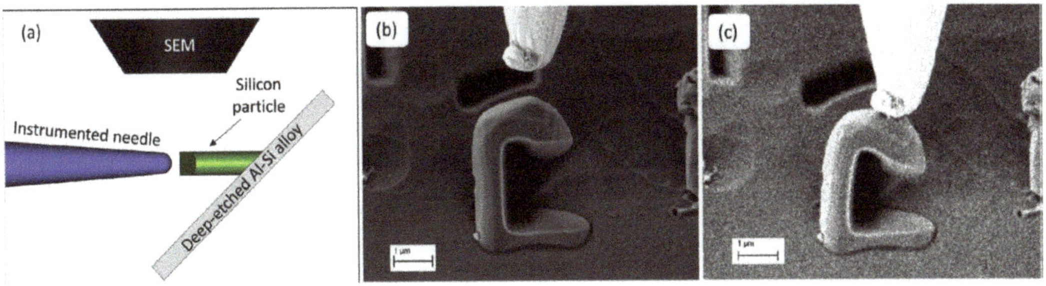

Figure 5. Strength test of silicon microparticles in A356 aluminium alloy: (**a**) Scheme of the micromechanical test; (**b**) C-shaped particle and the tungsten tip before test; (**c**) Last frame before particle fracture, from [36].

The article [37] also provides an example of particle strength assessment by means of a microscopic three-point bending test. First, the silicon particles were extracted from the Al-Si alloy by dissolving the aluminium matrix. Then, after cleaning and selecting the particles, microscopic beams were cut from them using the above-mentioned FIB technique. The beam prepared in this way was placed on a steel base with a cut-out hole and subjected to load. Figure 6 illustrates the tested sample. The results obtained during the experiment were compared with the results of analytical and numerical calculations, which resulted in determining the strength of silicon at the level of about 9 GPa, assuming no particle defects.

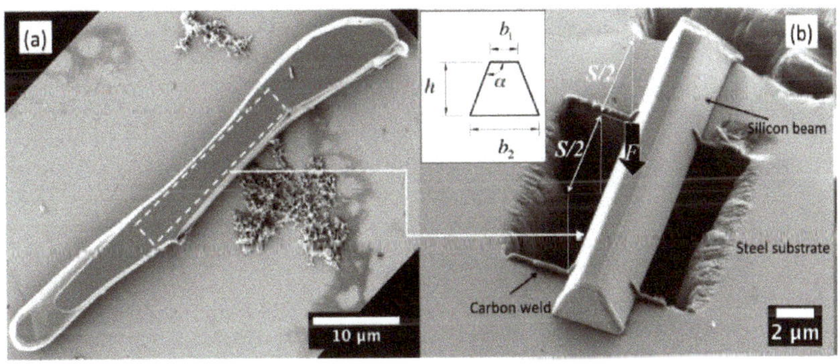

Figure 6. (**a**) Eutectic silicon particle extracted from the Al-Si alloy; (**b**) Microscopic three-point bending specimen prepared from the particle in (**a**), from [37].

A significant technical problem during the microscale strength tests is the measurement of deformations and stresses. Recently, the Raman spectroscopy technique offers a wide range of analysis of stress values on a microscopic scale, and thus also the evaluation of particle strength. The Raman effect is related to inelastic light scattering. After filtering, the light scattered on the sample goes to the spectrometer, which records its spectrum. The spectrum is presented as a function of the Raman shift, defined as the difference between the frequency of the scattered light and the input light [38]. In materials science, spectrum analysis enables the recording of the stress values, especially in the case of uniaxial stress states. However, the authors of [39] present the methodology of plane stress analysis in Si wafers on a microscopic scale.

A more complex analysis is discussed in [40], where the stress state in eutectic silicon particles in the Al-Si alloy was analysed. The alloy was tested in as-received condition under uniaxial loading. Cracking of silicon particles was observed already at the stress value of 600 MPa. Importantly, the use of Raman spectroscopy enables the assessment of the effect of particle size and its neighbourhood on the strength.

As already mentioned above, the second mechanism of void formation involves the decohesion of the matrix and the second-phase particles. Due to technical difficulties in a detailed, experimental study of this phenomenon, there are relatively few works of this type. Thus, numerical analyses are of particular importance. The solutions described in the literature are most often theoretical in nature. For example, in [41–43] a hypothetical aluminium matrix–silicon particle interface strength was determined in the range of about 4–7 GPa in tension, while the shear strength was only about 300 MPa. As this paper deals primarily with experimental research, these issues will not be described in detail here. More information on the simulation of this phenomenon, and the use of cohesive models, can be found in [44–47].

3.5. *Effect of Martensite Cracking in DP Steels*

As it has been shown in many studies, for example [48], the mechanism of void formation in dual-phase (DP) steels is only slightly based on the fracture and separation of the second-phase particles, because the fracture of martensite plays the most important role. The development of microdamage of DP1000 steel subjected to tension was analysed in detail in [48]. Strength tests were carried out inside the scanning electron microscope (SEM) chamber and stopped at regular intervals, each time photographing the microstructure of the material in the selected area. Cracking of the particles was observed at overall small strain, of the order of 2%. With the increase in plastic deformation, the voids created in this way grew, but no crack development in its vicinity was observed. As the strain value was increased further, the martensite phase cracked. Due to increasing deformation, the crack turned into a void, which, being a stress concentrator, became the cause of crack propagation in the ferrite. The high intensity of this phenomenon accounted for its dominant role in the failure of DP1000 steel.

Using advanced research methods, the authors of [48] attempted to estimate the local values of strains and stresses accompanying martensite cracking and void initiation. The obtained photographs of the microstructure were processed using the digital image correlation (DIC) technique. The photograph of the undeformed structure was divided into subset windows, then an algorithm was applied to track these areas in the photographs of the deformed structure. The results obtained in this way made it possible to determine the vectors of displacements and local deformations. Further, the experimentally determined values of displacements were used as boundary conditions in the finite element method (FEM) model of the tested sample, which allowed for the estimation of martensite cracking stress at the level of about 1700 MPa.

The dominant role of martensite cracking in the formation of voids in DP steels, mainly of the coarse structure, was also emphasised in [49]. This is the leading mechanism at low strains. At a later stage, voids were formed mainly by the decohesion of ferrite and martensite. The latter mechanism dominated in steels with the finer structure in the entire range of deformation. The occurrence of the decohesion is mainly attributed to the lower deformability of martensite.

Observations of martensite cracking in DP600 steel under uniaxial tension at low strain values were also described in [50], although as the authors point out, the importance of this mechanism in the entire process of void nucleation is not great. The process of fracture and separation of second-phase particles led to the formation of a few voids, which, however, were characterised by large dimensions, and therefore their area fraction was significant.

The dominant mechanism for the formation of voids was, as in [51], decohesion at the interface between ferrite and martensite, observed in the entire range of deformation of the tensile sample. The voids were initiated mainly at the interfaces perpendicular to tensile stresses and enlarged along the ferrite grains. The authors of [50] also noticed that with the increase in strain, the mean size of the voids decreased, which indicates a high intensity of nucleation of new voids also immediately before failure.

The authors of [52] drew similar conclusions. While examining the development of microdamage in DP600 and DP800 steels under uniaxial tension, it was noticed that at low strain, the void initiators were the globular aluminium oxide inclusions, but with higher deformations the voids were formed near to the ferrite–martensite interfaces as well as in the ferrite matrix and close to martensite islands.

3.6. Quantitative Description of Void Nucleation

Regardless of the single void nucleation analysis, it is important to evaluate the void nucleation globally, determining the number of nucleated voids as a function of remote strain and the location of the analysed area. In this case, it is particularly important to continuously track changes in the microstructure of the material throughout the deformation range up to the failure.

In recent years, the widespread use of the X-ray microtomography method gave great opportunities in this regard, and has contributed to a much better understanding of the phenomenon of failure and void development [9,53].

For example, in [54], the tomography method was used to assess changes in the microstructure of JIS SUM24L free-cutting steel under uniaxial tension. The tests were carried out on tensile specimens subjected to uniaxial stress state. The tests were interrupted at various stages, each time taking tomographic photographs of the microstructure in one selected area. In order to precisely determine the value of strain, especially after necking, changes in the width of the specimen were recorded using tomographic images. As part of the microtomographic research, in the first step, the region of interest (ROI) was distinguished, along with the voids present in the unstrained material. The photographs were then binarised, indicating base material and voids. In the next stage, a 3D labelling algorithm was applied to the binarised images, and then the volume and position of each of the detected voids was determined. In this way, over four thousand voids and the second-phase particles were marked in the ROI, which allowed for their tracking in the entire range of given deformation.

A separate issue was the development of an algorithm that allows for tracking each of the voids in subsequent stages, with increasing values of strain. The adopted procedure included the determination of a transformation matrix that was calculated by minimising the sum of distance difference between corresponding pairs of objects detected in subsequent stages of loading. Due to the heterogeneity of deformation and the different quality of individual photos, the obtained results were not fully accurate, hence the matching probability parameter M_p was introduced into the analysis. The algorithm developed in this way took into account the translation and rotation of the voids related to the occurrence of plastic deformation. A detailed description of the voids tracking algorithm is presented in [55,56].

The results of the observation of microstructure changes indicate that the nucleation of voids was continuous throughout the analysed range of deformations, up to failure. This phenomenon is well-illustrated by the graph in Figure 7, where the number of voids in the unit volume [mm^3] of ROI was determined as a function of true strain. From the very beginning, a constant increase in the number of voids is visible. Just before failure, one can see a flattening of the curve, which seemingly means a reduction in the intensity of void nucleation at this stage. In fact, as noted by the authors of [54], nucleation of voids is still present; however, at this stage, the ductile fracture process is controlled primarily by the void coalescence phenomenon, which explains the flattening of the curve in Figure 7.

Figure 8 presents tomographic cross sections of ROI, made at different strains. The vertical axis in the individual figures is equated with the direction of loading. Dark areas represent voids, while lighter areas represent the matrix. In Figure 8b (with the strain 0.23), the neck is clearly visible.

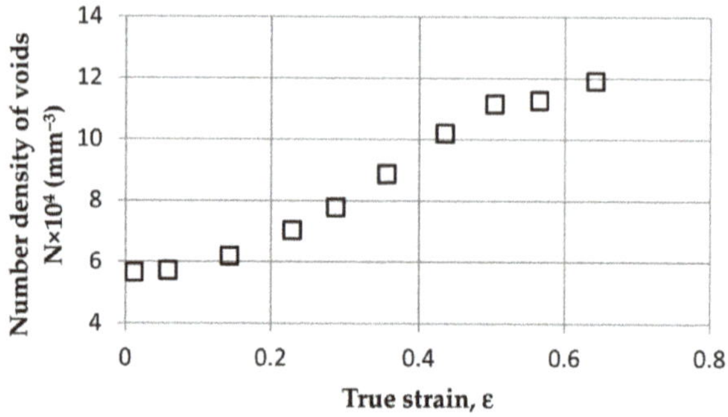

Figure 7. Experimentally determined relation between density of voids and plastic strain in JIS SUM24L free-cutting steel, from [54].

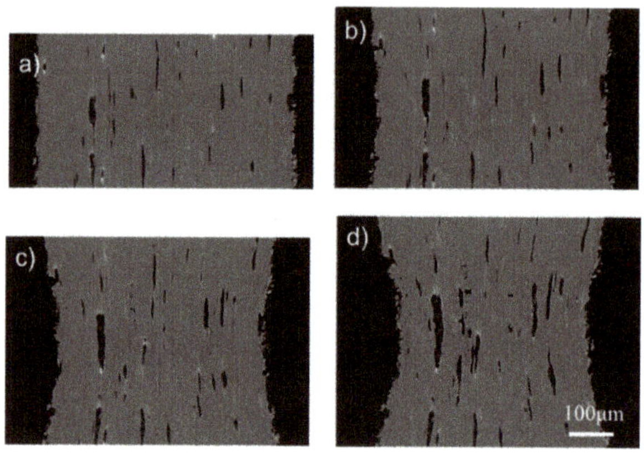

Figure 8. Free-cutting steel (JIS SUM24L grade) microstructure at different strains: (**a**) 0; (**b**) 0.23; (**c**) 0.50; (**d**) 0.64, from [54].

Detailed analysis of the individual pictures shows that the nucleation of voids occurs primarily at the interface between the matrix and the particles. The voids are also initiated at the places where the particles of the second phase break. As can be seen from the comparison of the subsequent photographs, some voids seem to disappear as the strain increases. This is due to their rotation and displacement, because the strains are not uniform throughout the tested sample. The void-tracking algorithm described above allowed for the inclusion of this phenomenon in the void development analysis.

Additionally, on the basis of the obtained results of tomographic examinations, changes in the volume and diameter of voids as a function of plastic strain were determined. While, as predicted, the total volume of the voids in the sample increased with increasing strain, the mean diameter of the voids was almost constant, regardless of the strain level. The authors of [54] indicate the nucleation of new small voids at higher strain levels as a possible reason.

The authors of [57] also emphasised the key role of small voids in the initiation of fracture. Using the 4D X-ray microtomography technique (3D + time), changes in the microstructure of SA508 steel were analysed in the entire range of tensile strains up to failure. It was observed that large Al_2O_3 and MnS_2 particles (with sizes ranging from several to tens of micrometres) cracked or separated from the matrix at zero values of plastic strain (elastic range). Considering the fact that such particles were scattered and spaced far apart, the voids they initiated did not coalesce, and therefore their contribution to the fracture initiation was insignificant. On the other hand, the elongation of voids along the tensile axis was observed, but the their contraction in the perpendicular direction was not significant. Moreover, the rotation of the elongated voids took place as a result of the increase in the value of shear stresses after the formation of the neck.

On the contrary, small particles of cementite (with a size of the order of 100–500 nm) detached from the matrix when the remote strain of the order of two was achieved. The initiation of small voids was, however, sudden. A large accumulation of small voids favoured their coalescence, which led to formation of a crack. The microtomographic image of the microstructure of the sample before failure is shown in Figure 9. In the central part of the region of interest, a cluster of small voids initiating a crack is clearly visible.

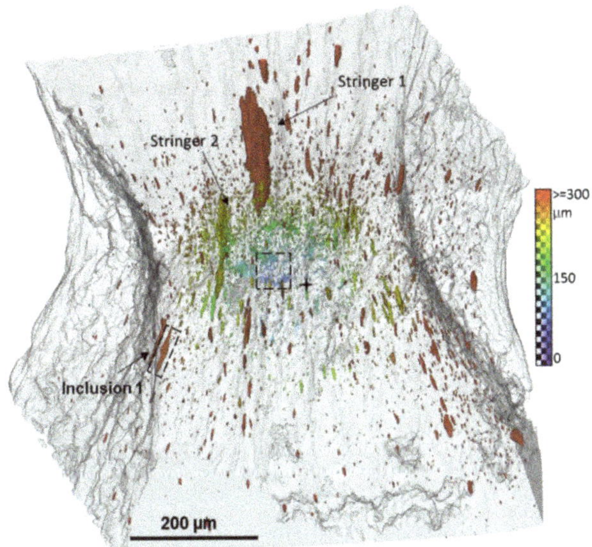

Figure 9. Microtomographic 3D image of the distribution of voids in SA508 steel, at the onset of failure, in the central part a cluster of small voids initiating a macroscopic defect is present, from [57]. The colour scale indicates the distance between void and sample centre.

Larger elongated voids predominate at a greater distance from the centre of the sample. The two largest voids (marked in the figure as stringer 1 and 2) were formed from an agglomerate of particles. Additionally, the authors of the paper carefully analysed the void nucleation around the inclusion marked as "inclusion 1" in Figure 9, indicating the mechanism of matrix–particle decohesion.

3.7. *Effect of Stress State on Void Nucleation Intensity*

In the previously mentioned work [10], the influence of the type of load on the intensity of void nucleation was determined. Figure 10 illustrates an exemplary dependence of the number of nucleated voids per unit volume of the cast Al-Si-Mg sample as a function of strain for various loading conditions. The results were obtained using the proprietary analytical void nucleation model. The lowest intensity of void nucleation was obtained for compression and torsion. As expected, among the simple load cases, the voids in the sample subjected to tension showed the highest nucleation intensity. The simultaneous action of tension and torsion resulted in the most intense nucleation of the voids.

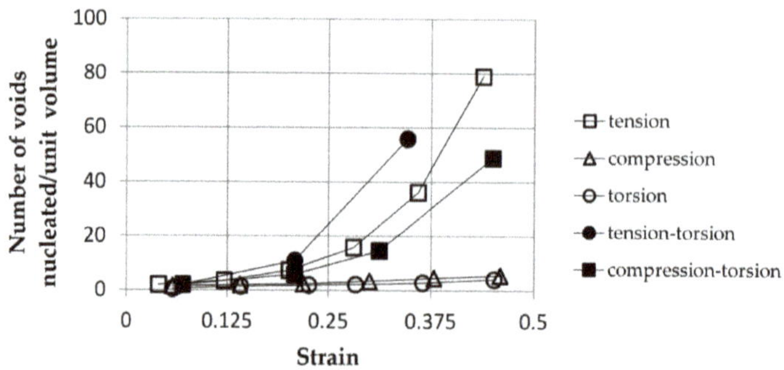

Figure 10. Effect of the loading type on the intensity of void nucleation in cast Al-Si-Mg alloy as a function of strain, based on [10].

Regardless of the above, the authors of [10] defined the material constants of void nucleation for simple loading conditions (tension/compression, torsion).

Recently, the authors of [58] conducted a thorough experimental analysis of the impact of the stress state on void nucleation in DP780 and CP800 steels. In order to obtain different components of the stress state, various strength tests were carried out: simple shear, hole tension, v-bending and biaxial tension. In each case, the tests were stopped, recording the material microstructure in the region of interest, using microtomography. Nucleation intensity was measured as the average number of nucleated voids in 1 mm^3 of material in the process zone. At failure, the highest number of voids (about 30,000/mm^3) was observed in the biaxial tension specimens. In the case of hole tension, this value was much lower and ranged from about 9500 to 18,000, depending on the material tested. The specimens subjected to shearing were characterised by the lowest nucleation intensity, i.e., at failure, values between 3000 and 4000 voids in 1 mm^3 of material were recorded.

One of the most frequently analysed issues related to the dependence of void nucleation on stress state is the influence of stress triaxiality T on the value of the strain needed to initiate the void. Stress triaxiality T describes the effect of the spherical component of the stress tensor (hydrostatic tension or compression) and is defined as the quotient of the mean stress (arithmetic mean of the principal stresses) and the Huber von Mises stress. The works published so far, for example, [59,60], unanimously indicate that the increase in triaxiality (increase in the hydrostatic pressure share) is accompanied by an exponential decrease in the value of nucleation strain.

The author of [61] drew similar conclusions, at the same time indicating the large influence of the Lode parameter on nucleation and the growth of voids. The Lode parameter takes into account the influence of the third stress tensor invariant. According to Yu [61], the value of the Lode parameter does not significantly affect the value of the void nucleation strain; however, it plays an important role, as interfacial cracks nucleate from different positions for different Lode parameters and propagate in different patterns. This is due to the fact that the Lode parameter changes the principal stress distribution, even at constant triaxiality.

Han et al. [62], studying the development of voids in QP980 steel under shear load, noticed that a large number of small voids (less than 5 μm in size) was formed at phase interfaces. In turn, a few microvoids generated from inclusions had more than 5 μm.

The phenomenon of the development of voids under shear was also analysed in detail by the authors of [63], also indicating the low intensity of nucleation in these conditions. The combination of microtomographic tests with FEM simulation allowed for the determination of the mechanism of ductile fracture of FB600 steel, initiated by separation of the matrix from CaO particles. The voids created in this way grew towards the largest local deformations, forming microcrack-like defects. As noted, a shear-band type of failure was formed on the microscopic scale even with a small volume fraction of voids, of the order of 0.015%. The void volume fraction measured before failure did not exceed 0.1%.

In recent decades, computer simulations have made a huge contribution to understanding the phenomena of void development [64]. As this article focuses primarily on experimental observations, the review of FEM results will not be discussed in detail here. However, it is worth paying attention to molecular dynamics simulation [65,66], which offers new possibilities compared to traditional continuum solutions, as it enables material modelling at the atomic level. For example, in [67], the mechanism of decohesion of the $AlCu_2$ particle and the aluminium matrix was analysed. In the first stage, the breaking of the bonds between single-inclusion and matrix atoms was observed, which initiated the particle separation. In the next stage, the crack grew steadily, with no dislocation involved. The fracture development in this case was driven by the lattice trapping phenomenon. After the fracture reached a critical size, nucleation of Shockley partial dislocations at the crack tip was observed. Then, the dislocations moved from the particle towards the matrix, whereby the rate of crack propagation increased suddenly, leading to the complete separation of the particle and the matrix. The authors called this stage of separation *dislocation-mediated delamination*.

4. Void Growth and Coalescence in Metals

4.1. Mechanisms of Growth and Coalescence of Voids

In the course of the realisation of ductile failure, after nucleation (characterised in the previous section of the article) due to plastic strain and hydrostatic stresses, the voids in the material increase [68,69]. With the action of strain, the voids grow, change their shape and move, changing their position. With stable void growth, plastic strain forms relatively uniformly in the material. However, from a certain point, the strain localises between adjacent voids. Outside the plane of strain localisation, the material undergoes elastic unloading. The occurrence of local strain localisation limits the ductility of materials. Two mechanisms for the realisation of strain localisation have been identified. The first involves the softening of strain through factors such as microstructural changes, thermal interactions and damage evolution. There is a local degradation of the load-carrying capacity of the material, resulting in strain localisation in a thin band [70–73]. The initiation of strain localisation depends on a number of factors, including stress state, material properties and material porosity [74–76].

The second mechanism of strain localisation is associated with the phenomenon of void coalescence. There is a local instability, conditioned by the interaction between adjacent voids in the material. The moment of localisation of plastic strain is identified with the beginning of the process of void coalescence. Once the coalescence process has started, the kinematics of void enlargement differ significantly from the kinematics of void growth prior to this mode of instability. For void coalescence induced by macroscopic strain localisation, the width of the localisation band is narrower. This is due to the restriction of deformation to areas (ligaments) between adjacent voids [76]. The process of void coalescence becomes the direct cause of the initiation and growth of a ductile crack in the material. There are three ways to realise void coalescence [77]:

(1) Internal necking (Figure 11a,d);
(2) Shear coalescence (Figure 11b,e);
(3) Necklace coalescence (Figure 11c).

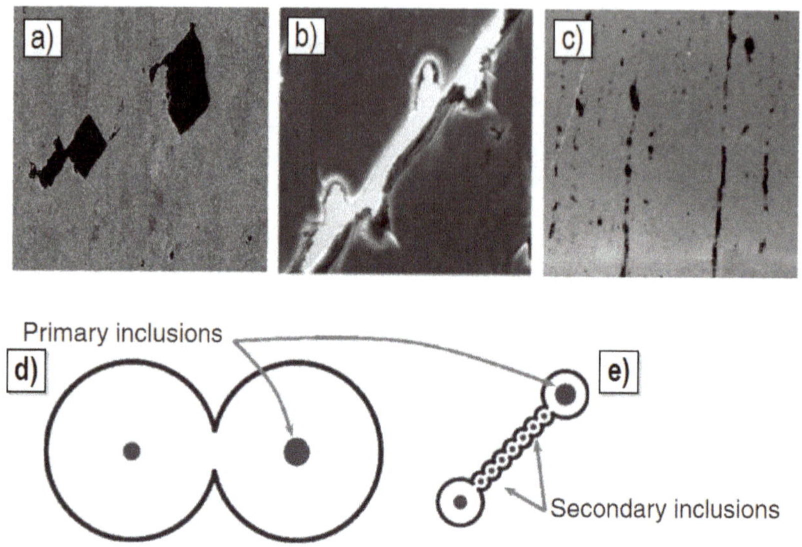

Figure 11. Methods of realising void coalescence in materials (example in steel): (**a**,**d**) Internal necking; (**b**,**e**) Shear coalescence; (**c**) Necklace coalescence [24,78,79].

The internal necking mechanism is the most commonly observed way of the material void coalescence process, initially observed by Argon [33]. With the internal necking process, the plane of localisation is almost perpendicular to the main direction of the applied load. During the process of coalescence of voids, a reduction in the area of the intervacancy ligament can be observed, which is similar to the necking phenomenon during the tensile test of the specimen under uniaxial loading [77,80–83]. A large contribution to understanding the process of coalescence of voids according to the internal necking mechanism was made by micromechanical analyses involving numerical calculations using the finite element method [71,84–87].

The second distinguished method of coalescence is the mechanism involving localised shear occurring between initial voids (of large size), when the voids are distributed along a line inclined at 45° to the main direction of loading (Figure 11b). The stress state accompanying the development of the failure process according to the void mechanism is the subject of many scientific studies. The stress state is most often defined by the stress triaxiality factor. However, the value cannot unambiguously describe the effect of stress triaxiality on void growth and coalescence. For a given value of the triaxiality factor, more than one stress state exists. The necessary information is very often extended by the magnitude of the Lode parameter. The development of the stress state in ductile fracture was the subject of the work. These quantities were determined on the basis of experimental results and numerical calculations. Selected results will be cited later in this work (Section 4.3). Coalescence in this case can occur according to the so-called "void sheeting" mechanism [88]. Plasticity is localised in shear bands containing small secondary voids; large primary voids are connected through coalescence of these secondary voids. This coalescence mechanism is often observed in high-strength materials with low-to-medium strain-hardening capacity [77,89].

A third possible mechanism for the coalescence of voids is called necklace coalescence (Figure 11c). It is the least commonly occurring in materials. The mechanism involves localisation in a direction parallel to the action of the applied load. This mechanism was observed in areas of voids, which are distributed in elongated concentrations. The mechanism of necklace coalescence is considered to be of major importance in the process of occurrence of ductile delamination cracking [77,90,91].

Many elements influence the development of the growth process and the coalescence of voids in ductile failure. These may include the state of stress in the material (which can be defined by the stress triaxiality factor, the Lode parameter), the contribution of shear stress and the level of plastic strain, taking into account the level of critical strain [92]. On the basis of the above quantities and a wide range of experimental and fractographic studies, numerical calculations, various failure models have been proposed, a selection of which will be discussed later in this paper.

4.2. Classical Models for the Growth and Coalescence of Voids

Of the classic local models describing the process of void growth, mention should be made of the solution proposed by McClintock [93]. In his work, he analysed the growth of a cylindrical void in a rigid plastic material under axisymmetric loading, assuming a plane strain condition. The model assumes that the relative volume of voids reaching a critical value will result in crack initiation. As a continuation of McClintock's study, an approximate solution for spherical void growth was proposed by Rice and Tracey [94]. The void growth model developed by Rice and Tracey allowed for the formulation of a fracture criterion, specifying that a crack will be initiated if the normalised void radius reaches a critical size. Similar results were obtained in [34,95–97]. In the cited papers, it was shown that the level of plastic strain and stress triaxiality have a significant impact on the realisation of the void growth process. The cited works formed the basis for further research and development of further local ductile fracture models.

A well-known model describing the phenomenon of void growth in ductile fracture is that by Gurson [98]. The author, with a similar methodology to the work of Rice and Tracey [94], developed an approach to analyse plastic flow in porous material assuming material continuity. Gurson included in the model the interaction between voids and the effect of void growth on material softening. A modification of Gurson's model was proposed by Tvergaard and Needleman [3]. The authors made a change in the definition of the relative volume of voids in the material and added an acceleration factor to account for the phenomenon of void coalescence. This resulted in a revised description of the plastic flow of the material in the initial stage of ductile fracture. Expanding on the ideas presented in the Gurson and GNT models, there have been many studies on the analysis of the growth and coalescence of voids during ductile fracture [71,99–106].

With analytical models for describing the process of void coalescence, it is worth mentioning Brown–Embury [107] and Thomason [71,106,108] as base models. These are models developed at the micromechanical scale. The Brown–Embury model refers to a perfectly ductile material and assumes the presence of shear bands at 45° between voids. The model relates the possibility of void coalescence depending on the diameter of the voids R and the distances between their centres (X). The criterion assumes that for a given void form factor, there is a minimum relative distance between voids, below which coalescence cannot be initiated, regardless of the stress state. Thomason's void coalescence model was developed for elastic–perfectly plastic materials, using solutions for slip lines. For the axisymmetric problem, Thomason's model assumed that the average normal stresses affect the void if the stresses reach a certain specified value.

The aforementioned classical models relating to the growth process and the coalescence of voids assume a number of simplifications. They do not address all aspects of strain and ductile failure of real materials. Thus, there is a major role for developing in situ studies of the development of voids in materials, with a particular focus on metals.

4.3. Experimental Verification of the Growth and Coalescence Process of Voids

The much more complex actual mechanism of ductile failure was highlighted in [109], taking into account the development of critical strain levels and stress triaxiality. In addition to the processes of void initiation, growth and coalescence (for tension dominated loading) and the occurrence of shear and void coalescence (for shear dominated loading), the authors point to seven types of micromechanisms that occur sequentially or are complementary in the failure process. An example of this complex type of ductile failure can be observed in the occurrence of a cup-and-cone failure on a tensile specimen, particularly for structural steels with medium-strength characteristics and high levels of ductility (Figure 12).

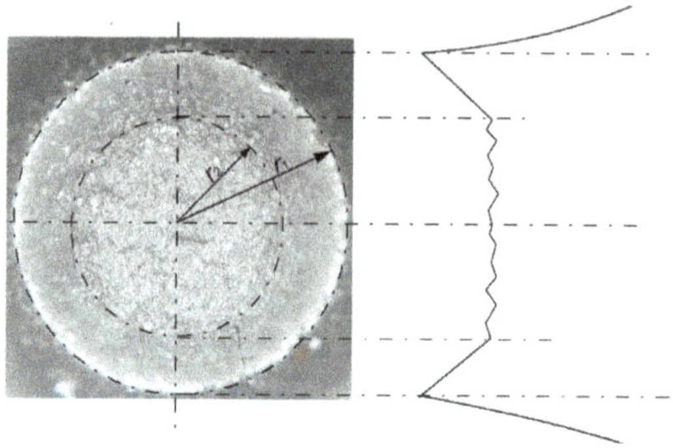

Figure 12. A cup-and-cone failure on a cylindrical, tensile specimen: view of break surface with scheme of the break plane profile (for S355 steel) [110].

Based on an analysis of the results of experimental uniaxial tensile tests on materials such as aluminium, nickel and copper, seven different types of ductile failure mechanisms were demonstrated (Figure 13) [109]:

(1) Intervoid necking (occurring with initiation, growth and coalescence of voids): for triaxiality stress factor T $\geq$ 0.33.
(2) Intervoid shearing, for which the initiation and subsequent elongation of voids along shear bands is characteristic. The consequence is the coalescence of voids and the formation of macrocracks in the planes of the shear bands. The mechanism occurs for a triaxiality stress factor T less than 0.33, located in a single plane.
(3) Void sheeting, when shear develops between existing voids in the material and the simultaneous process of nucleation and coalescence of new voids. The final ductile fracture that develops connects the existing voids in the material. The mechanism occurs on multiple planes, for stress triaxiality less than 0.33.
(4) The Orowan alternating slip (*OAS*) mechanism, which assumes the occurrence of void nucleation at the intersection of slip bands and the consequent growth of prismatic voids in the results of alternating slip along shear bands. *OAS* is characterised by its occurrence on multiple planes, for a triaxiality factor T less than 0.33.
(5) Destruction by specimen necking (T $\geq$ 0.33).
(6) Shear leading to the destruction of a specimen, occurring in a single plane. It is realised by the sliding of the material along a single slip band. Consequently, it will lead to a loss of cohesion and failure of the specimen (triaxiality factor T $\leq$ 0.33).
(7) Multiplanar shear occurring along multiple shear bands; also referred to as the slipping-off mechanism (T $\leq$ 0.33).

The nature of the interaction in the failure process between these seven mechanisms depends on, among other things: the local state of stress, the type of microstructure of the material under consideration (microstructural changes that occur during deformation can eliminate or create void nucleation sites in the material) and the strain-hardening capacity of the material (a low strain-hardening capacity favours the interlaminar shear mechanism, while a high one predisposes to the occurrence of void sheeting). For engineering materials in the occurrence of ductile failure, the first three mechanisms identified by Noell et al. [109] will be most relevant.

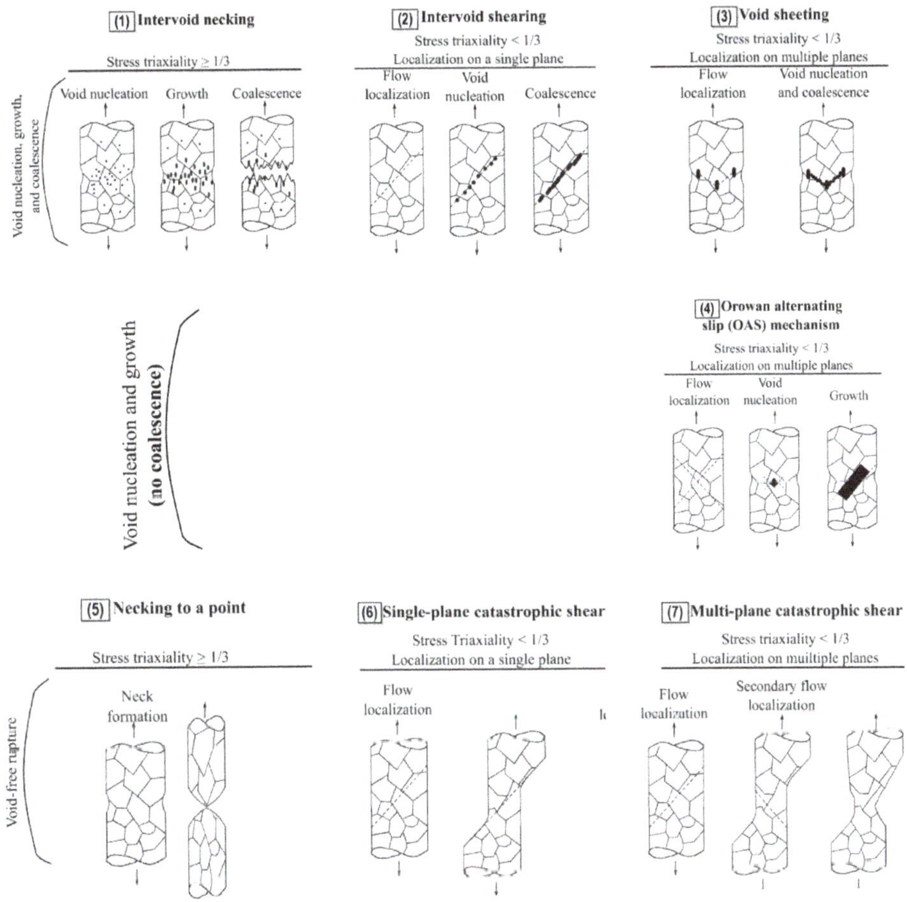

Figure 13. Types of ductile failure mechanisms: (**1**) Intervoid necking, (**2**) Intervoid shearing, (**3**) Void sheeting, (**4**) The Orowan alternating slip, (**5**) Necking to a point, (**6**) Single-plane catastrophic shear, (**7**) Multiplane catastrophic shear [109].

There are many analytical models in the literature describing ductile failure according to the void mechanism. An important aspect is the possibility of their experimental validation. Particularly difficult to validate experimentally is the phenomenon of void coalescence [111–114]. The use of X-ray computed tomography is helpful during in situ testing. In order to experimentally verify the growth process and the coalescence of voids, X-ray tomography was used for the uniaxial tensile testing of copper and Glidcop alloy in [80]. For this purpose, specimens were specially prepared for testing by laser-drilling holes in the material. Figure 14 shows the resulting images of growth and coalescence of the modelled voids. In the case of copper, the process of void coalescence occurred for true strain levels of more than 100%, while for the Glidcop alloy the strain was in the range of 50%. No secondary nucleation of voids was observed for copper. The occurrence of ripple marks on the surface of the voids was identified with the realisation of slip in the specimen. With the Glidcop alloy, coalescence was realised by nucleation of secondary voids on alumina particles between the modelled holes. Based on the results obtained, an attempt was made to verify the selected models of growth and coalescence of voids. Those models that provide quantitative verification of the results were considered. Good agreement was shown for the void growth model proposed by Rice and Tracey when taking into account the change in stress triaxiality in the specimen with progressive strain. When verifying the void coalescence process, greater discrepancies from experimental results were shown for the Brown and Embury model (differences of about 50%) than for the Thomason model (differences of 2–40%). With discrepancies in the Brown and Embury model, constraints that cause delayed void coalescence may have an impact. In the case of Glidcop alloy specimens, Thomason's model overestimates the level of critical strain at failure due to the failure to account for the secondary void nucleation occurring in the material.

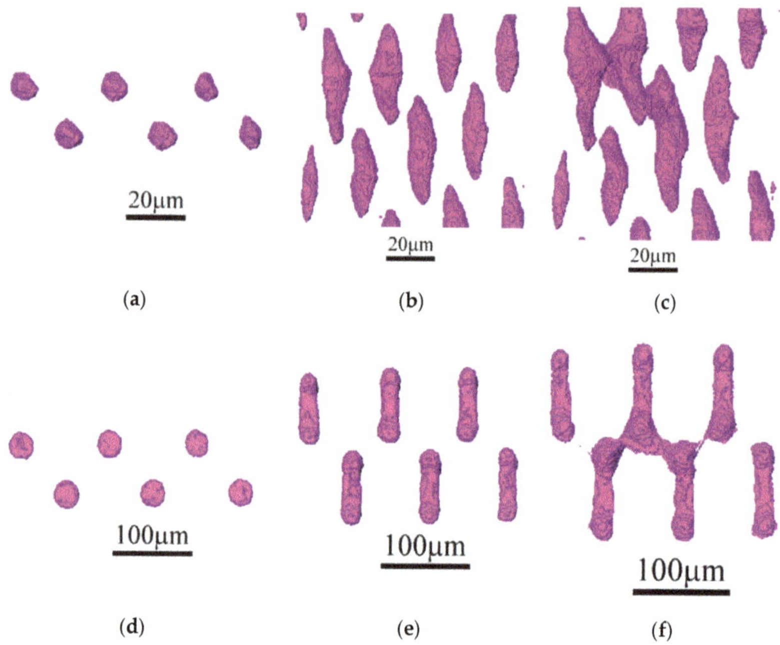

Figure 14. Growth and coalescence of modelled holes in copper specimens at true strain levels: (**a**) 0.00; (**b**) 0.77; (**c**) 1.01; in Glidcop alloy specimens at true strain levels: (**d**) 0.00; (**e**) 0.45; (**f**) 0.50 [80].

The authors of [77] proposed a criterion for the occurrence of void coalescence in material. It is based on a modification of the assumptions of the classical Thomason criterion (and its extension by Benzerga [81]). The criteria mentioned concern the analysis of the action of normal stresses, excluding the contribution of shear components. However, shear stresses can play an important role in the process of void coalescence and subsequent crack initiation in a material. An RVE model was used, which (unlike previous models) is affected by shear components. A positive calibration of the proposed criterion for void coalescence with the results of numerical calculations for the three-dimensional model was carried out. The calibration considered the values of void form factor and void distribution. The classical Thomason criterion (with a modification by Benzerga) was generalised to any loading condition. Modifications of Thomason's criterion for the void coalescence process, based on—among other things—the results of numerical calculations, are included in [82].

When analysing the growth process and the coalescence of voids in the material, an important aspect is to determine the volume contribution of the voids and their initial shape [115–117]. This allowed, among other things, for the subsequent modelling of the correct shape in the numerical model. In a number of papers, authors based on the Rice–Tracey model and the assumption of a spherically shaped void obtained oversimplified results. With the development of computed tomography methods, especially high-resolution μXCT, it has been possible to determine the true shapes of voids present in the material at the microscale level and to include them in the model describing the growth and coalescence of voids [118–123]. An example of an analysis image using the μXCT technique with the determined shape of the voids and their representation in the numerical calculation programme is shown in Figure 15 [124]. In this paper, on the basis of μXCT observations and numerical calculations, an attempt was made to determine the influence of the initial void shape (spherical, cylindrical, elliptical) on the macroscopic description of the void growth process in the material as an important stage of material destruction according to the ductile mechanism [125]. The effects of stress triaxiality, shape factor and the orientation and initial volume fraction of voids were taken into account. The influence of the initial void shape on the growth process is important at low levels of stress triaxiality. An increase in stress triaxiality reduces the difference in volume growth of voids with different shapes. For stress triaxiality above 2, almost the same increase in voids of different shapes was recorded. At low triaxiality (0.33), the volume contribution of the spherical void increases the most. The initial orientation of the voids has a strong influence on the growth process. If there is an orientation of the voids with an area located normal to the direction of stretching, greater void growth was observed. The initial volume proportion of voids showed less influence on the nature of their growth compared to the other parameters analysed [124].

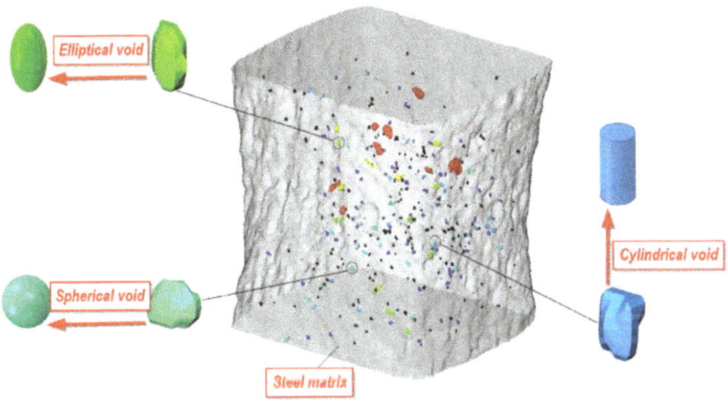

Figure 15. Shapes of voids in the material determined using computed tomography techniques (for structural steel) [124].

The influence of the shape of the void on the nature of its growth can be determined using other advanced research methods. In [126], a research methodology was used to complement the discrete dislocation plasticity (DDP) method with calculations using XFEM [127,128]. Higher stress levels, strain hardening and void growth rates occurred under biaxial loading (compared to uniaxial loading). With a constant initial proportion of void volume, it was observed that elliptical-shaped voids showed larger surface areas relative to cylindrical voids. The voids with a larger surface area in relation to volume showed a tendency to grow faster, but with a lower proportion of strain hardening.

The Rice–Tracey void growth model with Huang [97] corrections was verified on the basis of CT (compact tension) analysis. Specimens from three steels (single-phase ferritic steel, two-phase steel and steel with martensitic microstructure) subjected to tension were analysed. The analyses showed an increase in voids from a few μm to 30 μm. Attention was drawn to the change in the initially spherical shapes of the voids and to the need to account for the effect of the change in stress triaxiality in the Rice–Tracey model, depending on the microstructure and characteristics of the material under study [129]. Relationships of the initial shape of voids and the distance between voids to the level of fracture toughness were developed [130], which are shown in Figure 16.

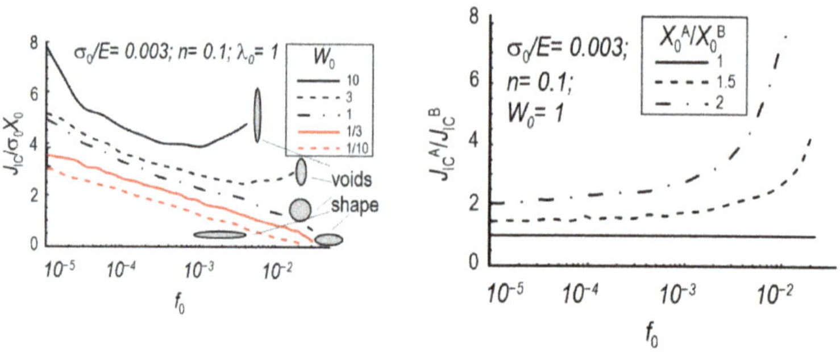

(a) (b)

Figure 16. Dependence of fracture toughness level on initial porosity: (**a**) Depending on the initial shape of the void; (**b**) Depending on the anisotropic void spacing [130].

Researchers are also using advanced computed tomography techniques to verify growth models and in situ void fusion: SRCT and SRCL [131]. The idea of measurements according to these techniques is explained in [132,133]. The use of the SRCT technique carries limitations, primarily with regard to the specimens used. Specimens must have a cross-section size of approximately 1 mm for a resolution of the order of micrometres. This has consequences in terms of the use of nonstandard specimens and the influence of specimen geometry on the results obtained (influence of the presence of a plane stress region and the development of a plastic zone). In such cases, it becomes helpful to use the SRCL technique for in situ studies. This allows for a high-quality 3D image to be obtained from specimens with larger dimensions than SRCT and the inclusion of areas on the sides of the specimens. Using the SRCL technique combined with numerical calculations of the CT (compact tension) specimen [131], parameters were determined in the GNT model relating to the realisation of failure by void mechanism in the AA6061 aluminium alloy. Accurate observation of the nucleation, growth and fusion of voids is possible through the use of TEM (transmission electronic microscope) observations in experimental studies. Compared to the SEM, the TEM technique allows for the dynamic growth of voids in the material to be captured and for the void bonding process in particular to be observed at the submicron scale level. In [134], the TEM technique was used to analyse the growth and bonding of voids in different types of materials: copper and aluminium-copper alloy.

4.4. Consideration of the Stress State in the Realisation of the Ductile Failure Process

Analysis using numerical calculations (e.g., finite element method) helps understand the growth process and the coalescence of voids in detail. Very often, an elementary cell model (representative RVE volume model) is used, which in its structure contains a void or particle with specific material characteristics. Classical studies consider the analysis of cylindrical voids, spherical voids and spherical particles. Parameters that can be taken into account in the RVE loading process are the stress triaxiality parameter (T), the Lode parameter (L) and the shear coefficient (S). There has been much work involving studies of the effect of stress triaxiality on the development on ductile failure, including a description of the void growth and coalescence process. However, it has become an important task to determine the simultaneous influence of the three aforementioned parameters of stress triaxiality, Lode parameter and shear rate when analysed using RVE. A solution to the task posed was proposed in [77]. The RVE model was used in [135] to extend Gurson's proposal to high porosity materials. The variables in the model were the volume proportion of voids and the value of the stress triaxiality factor. It was shown that the distribution of plastic deformation depends on the volume fraction of voids; a concentration of deformation occurs for a fraction with a large volume of voids. A small distance between individual voids leads to a faster coalescence process and fracture.

The contribution of the critical level of plastic strain, the stress triaxiality factor and the Lode parameter to the realisation of the ductile fracture process has been highlighted in a number of works by Wierzbicki and coworkers [136–138]. On the basis of numerous experimental studies and numerical calculations, the dependence of the critical strain at ductile fracture on the stress triaxiality was determined (Figure 17). Compression, tension and shear specimens of aluminium alloy 2024-T351 characterised by various levels of stress triaxiality factor from −0.33 to 1 were analysed. Based on the analyses, the authors concluded that shear failure occurs for compression specimens with negative values of stress triaxiality factor. For notched tensile specimens, the observed failure character was dependent on the level of stress triaxiality. For high triaxiality there was a failure mechanism involving initiation, growth and coalescence of voids. For low levels of stress triaxiality, failure was a combination between two mechanisms: void and shear. Work by Wierzbicki and colleagues and other researchers has highlighted the need to calibrate the material relationship used in the numerical calculation programme, particularly when analysing high-plasticity materials.

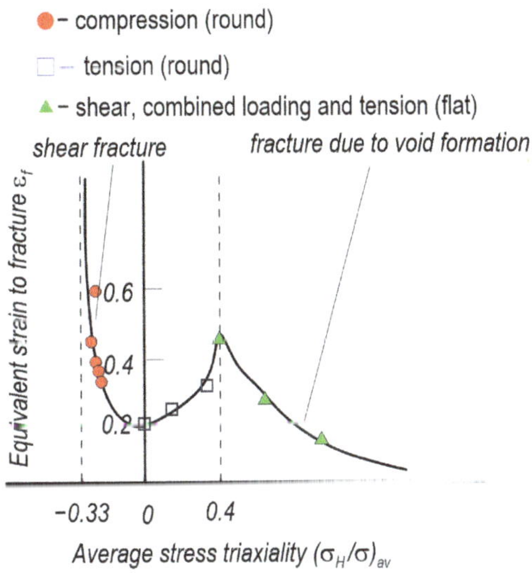

Figure 17. Numerically and experimentally determined dependence between critical plastic strain and the level of stress triaxiality factor (for 2024-T351 aluminium alloy) [139].

The influence of the Lode parameter value on the void growth and coalescence process has formed the basis of a number of papers [92,137,140–144]. Barsoum and Faleskog proposed a micromechanical model based on experimental studies and numerical calculations of a three-dimensional elementary cell containing a single spherical void [145]. The macroscopic stress state was defined by two variable quantities: the stress triaxiality factor and the Lode parameter. On the basis of the results obtained, it was determined that the effect of the Lode parameter on the change in the shape of voids and the rate of their growth increases as the level of the stress triaxiality factor decreases. For a stress triaxiality factor of T = 1, there is an increase and a change in the shape of the void due to shear strain. The void undergoes a change in shape from spherical to ellipsoidal, up to the onset of plastic localisation, so as to further obtain a 'penny' shape. For triaxiality stress level 2, there is an almost spherical increase in voids. The influence of the triaxiality parameter and Lode on the formation of plastic localisation was also determined. High values of the T parameter were accompanied by low levels of critical strain. The minimum value of critical strain occurred at a triaxiality factor of 0, while the maximum value occurred at T = 1. This marks the shear state as the most critical state from the point of view of destruction, where material failure will occur at an angle close to 45° to the plane of occurrence of the lowest principal stress [145].

5. Summary and Conclusions

Due to its great practical importance, the phenomenon of ductile failure associated with the development of voids in metals has been the subject of interest of researchers since the turn of the 1950s and 1960s. Over about 6 decades, many research techniques have been developed that have contributed to a better understanding of the nature of this phenomenon.

It was indicated that void formation in metals is induced by material discontinuities on the nano- and microscopic scale. The void initiators mainly include (i) intersection of the slip bands, (ii) grain boundaries, (iii) twin boundaries, (iv) vacancy clusters and (v) second-phase particles (inclusions and precipitates).

The void formation mechanism depends primarily on the material purity. In the absence of second-phase particles, voids are formed mainly in the vicinity of defects of the crystal structure (point, linear and planar defects). Materials that are commonly used in engineering (mainly steels and aluminium alloys) most often show the presence of inclusions, which, as stress concentrators, are void initiators. Depending on the geometrical and strength parameters of particles and matrix, voids are formed by particles' separation or fracture.

As has been demonstrated, the local values of the accompanying stresses, measured on a microscopic scale, are much higher than those measured macroscopically, and range from several hundred to several thousand MPa.

After the nucleation process, voids grow with increasing strain and change their shape and position in the material. The moment of localisation of strain in the immediate vicinity of voids is identified with the initiation of the coalescence process. Three ways of realising the coalescence of voids in the material are characterised: internal necking, shear coalescence and necklace coalescence. Selected analytical models relating to the growth process and the coalescence of voids under ductile failure mechanism are cited in this paper. An important element is the possibility of experimental verification of the classical models. This verification is possible by using advanced measurement methods (computed tomography, microstructure analysis, etc.) for in situ testing.

Despite enormous progress in this subject, there are still many unresolved issues. The course and intensity of the void development are significantly influenced by the load conditions, most importantly the stress state. However, the literature rarely mentions other factors that undoubtedly determine the development of voids, e.g., dynamic phenomena, temperature, etc.

It has not been unequivocally determined whether the stress or strain criteria must be satisfied for the void nucleation. Moreover, the literature lacks unambiguous sets of parameters describing the development of voids in materials with practical engineering application. Little is known yet about the development of voids under low-triaxiality conditions.

It should be strongly emphasised here that the observations of the development of voids are not only of cognitive importance, but above all they become the basis for the formulation of theoretical models, which further allows for the description of defective materials. It is an issue of great practical importance, thanks to which it becomes possible to perform an engineering safety assessment of structural elements containing defects, operating in pre-failure conditions. In the future, an even more accurate understanding of the ductile failure process according to the void formation mechanism will be possible with the development of advanced testing and measurement methods. These can be used for in situ research. An important aspect of supplementing the information obtained about the void mechanism will be the inclusion of the results of advanced numerical calculations.

Author Contributions: Conceptualisation, W.W. and S.L.; investigation, W.W. and S.L.; resources, W.W. and S.L.; data curation, W.W. and S.L.; writing—original draft preparation, W.W. and S.L.; writing—review and editing, W.W. and S.L.; visualisation, W.W. and S.L.; supervision, W.W. and S.L.; project administration, W.W. and S.L.; funding acquisition, W.W. and S.L. All authors have read and agreed to the published version of the manuscript.

Funding: The APC was funded by Ministry of Science and Higher Education, grant number 01.0.08.00/1.02.001SUBB.MPKM.22.001.

Institutional Review Board Statement: Not applicable.

Informed Consent Statement: Not applicable.

Data Availability Statement: No new data were created or analysed in this study.

Conflicts of Interest: The authors declare no conflict of interest.

References

1. Benzerga, A.A.; Besson, J.; Pineau, A. Anisotropic Ductile Fracture: Part I: Experiments. *Acta Mater.* **2004**, *52*, 4623–4638. [CrossRef]
2. Wcislik, W. Experimental Determination of Critical Void Volume Fraction FF for the Gurson Tvergaard Needleman (GTN) Model. *Proc. Struct. Integr.* **2016**, *2*, 1676–1683. [CrossRef]
3. Tvergaard, V.; Needleman, A. Analysis of Cup-Cone Fracture in a Round Tensile Bar. *Acta Metall.* **1984**, *32*, 157–169. [CrossRef]
4. Plateau, J.; Henry, G.; Crussard, C. Quelques nouvelles applications de la microfractographie. *Rev. Met. Paris* **1957**, *54*, 200–216. [CrossRef]
5. Schmitt, J.H.; Jalinier, J.M. Damage in Sheet Metal Forming—I: Physical Behavior. *Acta Metall.* **1982**, *30*, 1789–1798. [CrossRef]
6. Corbin, S.F.; Wilkinson, D.S. Influence of Matrix Strength and Damage Accumulation on the Mechanical Response of a Particulate Metal Matrix Composite. *Acta Metall. Mater.* **1994**, *42*, 1329–1335. [CrossRef]
7. Pardoen, T.; Delannay, F. Assessment of Void Growth Models from Porosity Measurements in Cold-Drawn Copper Bars. *Metall. Mater. Trans. A* **1998**, *29*, 1895–1909. [CrossRef]
8. El Guerjouma, R.; Baboux, J.-C.; Ducret, D.; Godin, N.; Guy, P.; Huguet, S.; Jayet, Y.; Monnier, T. Non-Destructive Evaluation of Damage and Failure of Fibre Reinforced Polymer Composites Using Ultrasonic Waves and Acoustic Emission. *Adv. Eng. Mater.* **2001**, *3*, 601–608. [CrossRef]
9. Landron, C.; Maire, E.; Adrien, J.; Bouaziz, O. Damage characterization in dual-phase steels using x-ray tomography. In *The Optical Measurements, Modeling, and Metrology*; Proulx, T., Ed.; Springer: New York, NY, USA, 2011; Volume 5, pp. 11–18.
10. Horstemeyer, M.F.; Gokhale, A.M. A Void–Crack Nucleation Model for Ductile Metals. *Int. J. Solids Struct.* **1999**, *36*, 5029–5055. [CrossRef]
11. Chan, K.S.; Davidson, D.L. Evidence of Void Nucleation and Growth on Planar Slip Bands in a Nb-Cr-Ti Alloy. *Metall. Mater. Trans. A* **1999**, *30*, 579–585. [CrossRef]
12. Gardner, R.N.; Pollock, T.C.; Wilsdorf, H.G.F. Crack Initiation at Dislocation Cell Boundaries in the Ductile Fracture of Metals. *Mater. Sci. Eng.* **1977**, *29*, 169–174. [CrossRef]
13. Nowak, Z. *The Method of Identification in Mechanics of Ductile Materials with Defects*; IFTR Reports; Institute of Fundamental Technological Research Polish Academy of Sciences: Warsaw, Poland, 2006.
14. Crussard, C.; Plateau, J.; Tamhankar, R.; Henry, G.; Lajeunesse, D. A comparison of ductile and fatigue fractures. In Proceedings of the Fracture, Swampscott, MA, USA, 12–16 April 1959; pp. 524–561.
15. Puttick, K.E. Ductile Fracture in Metals. *Philos. Mag. A J. Theor. Exp. Appl. Phys.* **1959**, *4*, 964–969. [CrossRef]
16. Rogers, H.C. The Tensile Fracture of Ductile Metals. *Trans. Metall. Soc. AIME* **1960**, *218*, 498–506.
17. Orowan, E. Fracture and Strength of Solids. *Rep. Prog. Phys.* **1948**, *XII*, 185–232. [CrossRef]
18. Koppenaal, T.J. Porosity in Plastically Deformed Cu-10 at.% Al Single Crystals. *Acta Metall.* **1961**, *9*, 1078–1079. [CrossRef]
19. Noell, P.; Carroll, J.; Hattar, K.; Clark, B.; Boyce, B. Do Voids Nucleate at Grain Boundaries during Ductile Rupture? *Acta Mater.* **2017**, *137*, 103–114. [CrossRef]
20. Noell, P.J.; Sabisch, J.E.C.; Medlin, D.L.; Boyce, B.L. Nanoscale Conditions for Ductile Void Nucleation in Copper: Vacancy Condensation and the Growth-Limited Microstructural State. *Acta Mater.* **2020**, *184*, 211–224. [CrossRef]
21. Pardoen, T.; Dumont, D.; Deschamps, A.; Brechet, Y. Grain Boundary versus Transgranular Ductile Failure. *J. Mech. Phys. Solids* **2003**, *51*, 637–665. [CrossRef]
22. Ashby, M.F.; Gandhi, C.; Taplin, D.M.R. Overview No. 3 Fracture-Mechanism Maps and Their Construction for f.c.c. Metals and Alloys. *Acta Metall.* **1979**, *27*, 699–729. [CrossRef]
23. Henseler, T.; Osovski, S.; Ullmann, M.; Kawalla, R.; Prahl, U. GTN Model-Based Material Parameters of AZ31 Magnesium Sheet at Various Temperatures by Means of SEM In-Situ Testing. *Crystals* **2020**, *10*, 856. [CrossRef]
24. Pineau, A.; Benzerga, A.A.; Pardoen, T. Failure of Metals I: Brittle and Ductile Fracture. *Acta Mater.* **2016**, *107*, 424–483. [CrossRef]
25. Wciślik, W.; Pała, R. Some Microstructural Aspects of Ductile Fracture of Metals. *Materials* **2021**, *14*, 4321. [CrossRef]

26. Cox, T.B.; Low, J.R. An Investigation of the Plastic Fracture of AISI 4340 and 18 Nickel-200 Grade Maraging Steels. *Metall. Mater. Trans. B* **1974**, *5*, 1457–1470. [CrossRef]
27. Tiryakioğlu, M. Intrinsic and Extrinsic Effects of Microstructure on Properties in Cast Al Alloys. *Materials* **2020**, *13*, 2019. [CrossRef] [PubMed]
28. Gangulee, A.; Gurland, J. On the Fracture of Silicon Particles in Aluminum-Silicon Alloys. *Trans. Metall. Soc. AIME* **1967**, *239*, 269–272.
29. Finlayson, T.; Griffths, J.; Viano, D.; Fitzpatrick, M.; Oliver, E.; Wang, Q. In stresses in the eutectic silicon particles of strontium-modified A356 castings loaded in tension. In *Shape Casting: 2nd International Symposium*; The Minerals, Metals & Materials Society: Warrendale, PA, USA, 2007; pp. 127–134.
30. Caceres, C.H.; Griffiths, J.R. Damage by the Cracking of Silicon Particles in an Al-7Si-0.4 Mg Casting Alloy. *Acta Mater.* **1996**, *44*, 25–33. [CrossRef]
31. Yeh, J.-W.; Liu, W.-P. The Cracking Mechanism of Silicon Particles in an A357 Aluminum Alloy. *Metall. Mater. Trans. A* **1996**, *27*, 3558–3568. [CrossRef]
32. Poole, W.J.; Charras, N. An Experimental Study on the Effect of Damage on the Stress–Strain Behaviour for Al–Si Model Composites. *Mater. Sci. Eng. A* **2005**, *406*, 300–308. [CrossRef]
33. Argon, A.S.; Im, J.; Safoglu, R. Cavity Formation from Inclusions in Ductile Fracture. *Metall. Mater. Trans. A* **1975**, *6*, 825. [CrossRef]
34. Beremin, F.M. Cavity Formation from Inclusions in Ductile Fracture of A508 Steel. *Metall. Mater. Trans. A* **1981**, *12a*, 723–731. [CrossRef]
35. Coade, R.W.; Griffiths, J.R.; Parker, B.A.; Stevens, P.J. Inclusion Stresses in a Two-Phase Alloy Deformed to a Plastic Strain of 1%. *Philos. Mag. A* **1981**, *44*, 357–372. [CrossRef]
36. Mueller, M.G.; Žagar, G.; Mortensen, A. In-Situ Strength of Individual Silicon Particles within an Aluminium Casting Alloy. *Acta Mater.* **2018**, *143*, 67–76. [CrossRef]
37. Mueller, M.; Fornabaio, M.; Žagar, G.; Mortensen, A. Microscopic Strength of Silicon Particles in an Aluminium–Silicon Alloy. *Acta Mater.* **2016**, *105*, 165–175. [CrossRef]
38. Grodecki, K.; Bozek, R.; Strupinski, W.; Wysmolek, A.; Stepniewski, R. Micro-Raman Spectroscopy of Graphene Grown on Stepped 4H-SiC (0001) Surface. *Appl. Phys. Lett.* **2009**, *100*, 261604. [CrossRef]
39. Harris, S.J.; O'Neill, A.E.; Yang, W.; Gustafson, P.; Boileau, J.; Weber, W.H.; Majumdar, B.; Ghosh, S. Measurement of the State of Stress in Silicon with Micro-Raman Spectroscopy. *J. Appl. Phys.* **2004**, *96*, 7195–7201. [CrossRef]
40. Harris, S.; O'Neill, A.; Boileau, J.; Donlon, W.; Su, X.; Majumdar, B. Application of the Raman Technique to Measure Stress States in Individual Si Particles in a Cast Al–Si Alloy. *Acta Mater.* **2007**, *55*, 1681–1693. [CrossRef]
41. Ward, D.K.; Curtin, W.A.; Qi, Y. Aluminum-Silicon Interfaces and Nanocomposites: A Molecular Dynamics Study. *Compos. Sci. Technol.* **2006**, *66*, 1151–1161. [CrossRef]
42. Ward, D.K.; Curtin, W.A.; Qi, Y. Mechanical Behavior of Aluminum-Silicon Nanocomposites: A Molecular Dynamics Study. *Acta Mater.* **2006**, *54*, 4441–4451. [CrossRef]
43. Noreyan, A.; Qi, Y.; Stoilov, V. Critical Shear Stresses at Aluminum–Silicon Interfaces. *Acta Mater.* **2008**, *56*, 3461–3469. [CrossRef]
44. Needleman, A. Some Issues in Cohesive Surface Modeling. *Procedia IUTAM* **2014**, *10*, 221–246. [CrossRef]
45. Galkiewcz, J. Cohesive Model Application to Micro-Crack Nucleation and Growth. *Procedia Struct. Integr.* **2016**, *2*, 1619–1626. [CrossRef]
46. Gałkiewicz, J. Microscopically Based Calibration of the Cohesive Model. *J. Theor. Appl. Mech.* **2015**, *53*, 477–485. [CrossRef]
47. Kossakowski, P.; Wciślik, W. The effect of stress state triaxiality on the value of microvoid nucleation strain in S235JR steel. *Przegląd Mech.* **2013**, *3*, 15–21.
48. Alharbi, K.; Ghadbeigi, H.; Efthymiadis, P.; Zanganeh, M.; Celotto, S.; Dashwood, R.; Pinna, C. Damage in Dual Phase Steel DP1000 Investigated Using Digital Image Correlation and Microstructure Simulation. *Model. Simul. Mater. Sci. Eng.* **2015**, *23*, 085005. [CrossRef]
49. He, X.J.; Terao, N.; Berghezan, A. Influence of Martensite Morphology and Its Dispersion on Mechanical Properties and Fracture Mechanisms of Fe-Mn-C Dual Phase Steels. *Met. Sci.* **1984**, *18*, 367–373. [CrossRef]
50. Avramovic-Cingara, G.; Saleh, C.A.R.; Jain, M.K.; Wilkinson, D. Void Nucleation and Growth in Dual-Phase Steel 600 during Uniaxial Tensile Testing. *Metall. Mater. Trans. A* **2009**, *40*, 3117–3127. [CrossRef]
51. Szewczyk, A.F.; Gurland, J. A Study of the Deformation and Fracture of a Dual-Phase Steel. *Metall. Mater. Trans. A* **1982**, *13*, 1821–1826. [CrossRef]
52. Santos, R.O.; da Silveira, L.B.; Moreira, L.P.; Cardoso, M.C.; da Silva, F.R.F.; dos Santos Paula, A.; Albertacci, D.A. Damage Identification Parameters of Dual-Phase 600–800 Steels Based on Experimental Void Analysis and Finite Element Simulations. *J. Mater. Res. Technol.* **2019**, *8*, 644–659. [CrossRef]
53. Wu, S.C.; Xiao, T.Q.; Withers, P.J. The Imaging of Failure in Structural Materials by Synchrotron Radiation X-Ray Microtomography. *Eng. Fract. Mech.* **2017**, *182*, 127–156. [CrossRef]
54. Seo, D.; Toda, H.; Kobayashi, M.; Uesugi, K.; Takeuchi, A.; Suzuki, Y. In Situ Observation of Void Nucleation and Growth in a Steel Using X-Ray Tomography. *ISIJ Int.* **2015**, *55*, 1474–1482. [CrossRef]

55. Kobayashi, M.; Toda, H.; Kawai, Y.; Ohgaki, T.; Uesugi, K.; Wilkinson, D.; Kobayashi, T.; Aoki, Y.; Nakazawa, M. High-Density Three-Dimensional Mapping of Internal Strain by Tracking Microstructural Features. *Acta Mater.* **2008**, *56*, 2167–2181. [CrossRef]
56. Toda, H.; Maire, E.; Aoki, Y.; Kobayashi, M. Three-Dimensional Strain Mapping Using in Situ X-Ray Synchrotron Microtomography. *J. Strain Anal. Eng. Des.* **2011**, *46*, 549–561. [CrossRef]
57. Guo, Y.; Burnett, T.L.; McDonald, S.A.; Daly, M.; Sherry, A.H.; Withers, P.J. 4D Imaging of Void Nucleation, Growth, and Coalescence from Large and Small Inclusions in Steel under Tensile Deformation. *J. Mater. Sci. Technol.* **2022**, *123*, 168–176. [CrossRef]
58. Pathak, N.; Adrien, J.; Butcher, C.; Maire, E.; Worswick, M. Experimental Stress State-Dependent Void Nucleation Behavior for Advanced High Strength Steels. *Int. J. Mech. Sci.* **2020**, *179*, 105661. [CrossRef]
59. Wcislik, W. Experimental and Numerical Determination and Analysis of Selected Parameters of the Gurson-Tvergaard-Needleman Model for S355 Steel and Complex Stress States. Ph.D. Thesis, Kielce University of Technology, Kielce, Poland, 2014.
60. Testa, G.; Bonora, N.; Ruggiero, A.; Iannitti, G.; Gentile, D. Stress Triaxiality Effect on Void Nucleation in Ductile Metals. *Fatigue Fract. Eng. Mater.Struct.* **2020**, *43*, 1473–1486. [CrossRef]
61. Yu, Q. Influence of the Stress State on Void Nucleation and Subsequent Growth around Inclusion in Ductile Material. *Int. J. Fract.* **2015**, *193*, 43–57. [CrossRef]
62. Han, S.; Chang, Y.; Wang, C.; Han, Y.; Dong, H. Experimental and Numerical Investigations on the Damage Induced in the Shearing Process for QP980 Steel. *Materials* **2022**, *15*, 3254. [CrossRef]
63. Tancogne-Dejean, T.; Roth, C.; Morgeneyer, T.; Helfen, L.; Mohr, D. Ductile Damage of AA2024-T3 under Shear Loading: Mechanism Analysis Through In-Situ Laminography. *Acta Mater.* **2020**, *205*, 116556. [CrossRef]
64. Shakoor, M.; Bernacki, M.; Bouchard, P.-O. Ductile Fracture of a Metal Matrix Composite Studied Using 3D Numerical Modeling of Void Nucleation and Coalescence. *Eng. Fract. Mech.* **2017**, *189*, 110–132. [CrossRef]
65. Nguyen, T.D.; Plimpton, S.J. Aspherical Particle Models for Molecular Dynamics Simulation. *Comput. Phys. Commun.* **2019**, *243*, 12–24. [CrossRef]
66. Lucchetta, A.; Brach, S.; Kondo, D. Effects of Particles Size on the Overall Strength of Nanocomposites: Molecular Dynamics Simulations and Theoretical Modeling. *Mech. Res. Commun.* **2021**, *114*, 103669. [CrossRef]
67. Zhao, Q.Q.; Boyce, B.L.; Sills, R.B. Micromechanics of Void Nucleation and Early Growth at Incoherent Precipitates: Lattice-Trapped and Dislocation-Mediated Delamination Modes. *Crystals* **2021**, *11*, 45. [CrossRef]
68. Anderson, T.L. Fracture mechanics. In *Fundamentals and Applications*, 3rd ed.; Taylor and Francis Group: London, UK, 2005.
69. Dzioba, I. *Modelowanie i Analiza Procesu Pękania w Stalach Ferrytycznych*; Politechnika Świętokrzyska: Kielce, Poland, 2012.
70. Fressengeas, C.; Molinari, A. Inertia and Thermal Effects on the Localization of Plastic Flow. *Acta Metall.* **1985**, *33*, 387–396. [CrossRef]
71. Pardoen, T.; Hutchinson, J.W. An Extended Model for Void Growth and Coalescence. *J. Mech. Phys. Solids* **2000**, *48*, 2467–2512. [CrossRef]
72. Mercier, S.; Molinari, A. Predictions of Bifurcation and Instabilities during Dynamic Extension. *Int. J. Solids Struct.* **2003**, *40*, 1995–2016. [CrossRef]
73. Aretz, H. Numerical Analysis of Diffuse and Localized Necking in Orthotropic Sheet Metals. *Int. J. Plast.* **2007**, *23*, 798–840. [CrossRef]
74. Barsoum, I. The Effect of Stress State in Ductile Failure. Ph.D. Thesis, Royal Institute of Technology, Stockholm, Sweden, 2008.
75. Needleman, A.; Rice, J.R. Limits to ductility set by plastic flow localization. In *Mechanics of Sheet Metal Forming: Material Behavior and Deformation Analysis*; Koistinen, D.P., Wang, N.-M., Eds.; Springer US: Boston, MA, USA, 1978; pp. 237–267, ISBN 978-1-4613-2880-3.
76. Reboul, J.; Srivastava, A.; Osovski, S.; Vadillo, G. Influence of Strain Rate Sensitivity on Localization and Void Coalescence. *Int. J. Plast.* **2020**, *125*, 265–279. [CrossRef]
77. Tekoglu, C.; Leblond, J.-B.; Pardoen, T. A Criterion for the Onset of Void Coalescence under Combined Tension and Shear. *J. Mech. Phys. Solids* **2012**, *60*, 1363–1381. [CrossRef]
78. Wang, J.; Liang, J.; Wen, Z.; Yue, Z. Void Configuration-Induced Change in Microstructure and Deformation Mechanisms of Nano-Porous Materials. *J. Appl. Phys.* **2019**, *126*, 085106. [CrossRef]
79. Besson, J. Continuum Models of Ductile Fracture: A Review. *Int. J. Damage Mech.* **2010**, *19*, 3–52. [CrossRef]
80. Weck, A.; Wilkinson, D.S.; Maire, E.; Toda, H. Visualization by X-Ray Tomography of Void Growth and Coalescence Leading to Fracture in Model Materials. *Acta Mater.* **2008**, *56*, 2919–2928. [CrossRef]
81. Benzerga, A.A. Micromechanics of Coalescence in Ductile Fracture. *J. Mech. Phys. Solids* **2002**, *50*, 1331–1362. [CrossRef]
82. Scheyvaerts, F.; Pardoen, T.; Onck, P.R. A New Model for Void Coalescence by Internal Necking. *Int. J. Damage Mech.* **2010**, *19*, 95–126. [CrossRef]
83. Gologanu, M.; Leblond, J.-B.; Perrin, G.; Devaux, J. Theoretical Models for Void Coalescence in Porous Ductile Solids: I—Coalescence "in Layers". *Int. J. Solids Struct.* **2001**, *38*, 5581–5594. [CrossRef]
84. Morin, L.; Leblond, J.-B.; Benzerga, A.A. Coalescence of Voids by Internal Necking: Theoretical Estimates and Numerical Results. *J. Mech. Phys. Solids* **2015**, *75*, 140–158. [CrossRef]
85. Koplik, J.; Needleman, A. Void Growth and Coalescence in Porous Plastic Solids. *Int. J. Solids Struct.* **1988**, *24*, 835–853. [CrossRef]

86. Richelsen, A.B.; Tvergaard, V. Dilatant Plasticity or Upper Bound Estimates for Porous Ductile Solids. *Acta Metall. Mater.* **1994**, *42*, 2561–2577. [CrossRef]
87. Srivastava, A.; Needleman, A. Void Growth versus Void Collapse in a Creeping Single Crystal. *J. Mech. Phys. Solids* **2013**, *61*, 1169–1184. [CrossRef]
88. Bandstra, J.; Koss, D.; Geltmacher, A.; Matic, P.; Everett, R. Modeling Void Coalescence during Ductile Fracture of a Steel. *Mater. Sci. Eng. A* **2004**, *366*, 269–281. [CrossRef]
89. Chang, Z.; Chen, J. A New Void Coalescence Mechanism during Incremental Sheet Forming: Ductile Fracture Modeling and Experimental Validation. *J. Mater. Process. Technol.* **2021**, *298*, 117319. [CrossRef]
90. Gologanu, M.; Leblond, J.-B.; Devaux, J. Theoretical Models for Void Coalescence in Porous Ductile Solids: II—Coalescence "in Columns". *Int. J. Solids Struct.* **2001**, *38*, 5595–5604. [CrossRef]
91. Pala, R.; Dzioba, I. Influence of Delamination on the Parameters of Triaxial State of Stress before the Front of the Main Crack. *AIP Conf. Proc.* **2018**, *2029*, 020052. [CrossRef]
92. Ganjiani, M.; Homayounfard, M. Development of a Ductile Failure Model Sensitive to Stress Triaxiality and Lode Angle. *Int. J. Solids Struct.* **2021**, *225*, 111066. [CrossRef]
93. McClintock, F.A. A Criterion for Ductile Fracture by Growth of Hole. *J. Appl. Mech.* **1968**, *35*, 353–371. [CrossRef]
94. Rice, J.R.; Tracey, D.M. On the Ductile Enlargement of Voids in Triaxial Stress Fields. *J. Mech. Phys. Solids* **1968**, *17*, 201–217. [CrossRef]
95. Budianski, B.; Hutchinson, J.W.; Slutski, S. *Void Growth and Collapse in Viscous Solids*; Mechanics of Solids; Pergamon Press: Oxford, UK, 1982.
96. Becker, R.; Smelser, R.E.; Richmond, O. The Effect of Void Shape on the Development of Damage and Fracture in Plane Strain Tension. *J. Mech. Phys. Solids* **1989**, *37*, 111–129. [CrossRef]
97. Huang, Y. Accurate Dilatation Rates for Spherical Voids in Triaxial Stress Fields. *J. Appl. Mech.* **1991**, *58*, 1084–1086. [CrossRef]
98. Gurson, A.L. Continuum Theory of Ductile Rupture by Void Nucleation and Growth: Part 1—Yield Criteria and Flow Rules for Porous Ductile Media. *J. Eng. Mater. Technol.* **1977**, *99*, 2–15. [CrossRef]
99. Berdin, C.; Hausild, P. Damage mechanisms and local approach to fracture part I: Ductile fracture. In *Transferability of Fracture Mechanical Characteristics*; NATO Science Series: Brno, Czech Republic, 2001.
100. Gologanu, M.; Leblond, J.B.; Devaux, J. Approximate Models for Ductile Metals Containing Nonspherical Voids—Case of Axisymmetric Prolate Ellipsoidal Cavities. *J. Mech. Phys. Solids* **1993**, *41*, 1723–1754. [CrossRef]
101. Gologanu, M.; Leblond, J.B.; Devaux, J. Approximate Models for Ductile Metals Containing Nonspherical Voids—Case of Axisymmetric Oblate Ellipsoidal Cavities. *J. Eng. Mater. Technol.* **1994**, *116*, 290–297. [CrossRef]
102. Gologanu, M.; Leblond, J.B.; Perrin, G.; Devaux, J. *Recent Extensions of Gursons Model for Porous Ductile Metals*; Continuum Micromechanics; Springer: Berlin/Heidelberg, Germany, 1995.
103. Gao, X.; Wang, T.; Kim, J. On Ductile Fracture Initiation Toughness: Effects of Void Volume Fraction, Void Shape and Void Distribution. *Int. J. Solids Struct.* **2005**, *42*, 5097–5117. [CrossRef]
104. Gao, X.; Kim, J. Modeling of Ductile Fracture: Significance of Void Coalescence. *Int. J. Solids Struct.* **2006**, *43*, 6277–6293. [CrossRef]
105. Zhang, Z.; Skallerud, B. Void Coalescence with and without Prestrain History. *Int. J. Damage Mech.* **2010**, *19*, 153–174. [CrossRef]
106. Besson, J. *Local Approach to Fracture*; Ecole des Mines de Paris: Paris, France, 2004.
107. Brown, L.M.; Embury, J.D. The initiation and growth of voids at second phase particles. In Proceedings of the Third International Conference on Strength of Metals and Alloys, ICSMA 3, Cambridge, UK, 20–25 August 1973; Institute of Metals: Cambridge, UK, 1973; pp. 164–169.
108. Thomason, P.F. A Three-Dimensional Model for Ductile Fracture by the Growth and Coalescence of Microvoids. *Acta Metall.* **1985**, *33*, 1087–1095. [CrossRef]
109. Noell, P.J.; Carroll, J.D.; Boyce, B.L. The Mechanisms of Ductile Rupture. *Acta Mater.* **2018**, *161*, 83–98. [CrossRef]
110. Dzioba, I.; Lipiec, S.; Pała, R.; Furmańczyk, P. On Characteristics of Ferritic Steel Determined during the Uniaxial Tensile Test. *Materials* **2021**, *14*, 3117. [CrossRef]
111. Babout, L.; Maire, E.; Buffière, J.Y.; Fougères, R. Characterization by X-Ray Computed Tomography of Decohesion, Porosity Growth and Coalescence in Model Metal Matrix Composites. *Acta Mater.* **2001**, *49*, 2055–2063. [CrossRef]
112. Gammage, J.; Wilkinson, D.; Brechet, Y.; Embury, D. A Model for Damage Coalescence in Heterogeneous Multi-Phase Materials. *Acta Mater.* **2004**, *52*, 5255–5263. [CrossRef]
113. Jia, S.; Povirk, G.L. Modeling the Effects of Hole Distribution in Perforated Aluminum Sheets II: Minimum Strength Failure Paths. *Int. J. Solids Struct.* **2002**, *39*, 2533–2545. [CrossRef]
114. Nagaki, S.; Nakayama, Y.; Abe, T. Relation between Damage Due to Circular Holes and Local Deformation of Perforated Sheets. *Int. J. Mech. Sci.* **1998**, *40*, 215–226. [CrossRef]
115. Cao, T.-S.; Mazière, M.; Danas, K.; Besson, J. A Model for Ductile Damage Prediction at Low Stress Triaxialities Incorporating Void Shape Change and Void Rotation. *Int. J. Solids Struct.* **2015**, *63*, 240–263. [CrossRef]
116. Komori, K. Improvement and Validation of the Ellipsoidal Void Model for Predicting Ductile Fracture. *Proc. Eng.* **2017**, *207*, 2036–2041. [CrossRef]
117. Shen, Y.; Ma, T.; Li, J. A Damage Evolution Model Considering Void Shape Effect and New Damage Level Definition. *Eng. Fract. Mech.* **2022**, *267*, 108461. [CrossRef]

118. Achouri, M.; Germain, G.; Dal Santo, P.; Saidane, D. Experimental Characterization and Numerical Modeling of Micromechanical Damage under Different Stress States. *Mater. Des.* **2013**, *50*, 207–222. [CrossRef]
119. Azghandi, S.H.M.; Weiss, M.; Arhatari, B.D.; Adrien, J.; Maire, E.; Barnett, M.R. A Rationale for the Influence of Grain Size on Failure of Magnesium Alloy AZ31: An in Situ X-Ray Microtomography Study. *Acta Mater.* **2020**, *200*, 619–631. [CrossRef]
120. Croom, B.P.; Jin, H.; Noell, P.J.; Boyce, B.L.; Li, X. Collaborative Ductile Rupture Mechanisms of High-Purity Copper Identified by in Situ X-Ray Computed Tomography. *Acta Mater.* **2019**, *181*, 377–384. [CrossRef]
121. Fabrègue, D.; Landron, C.; Bouaziz, O.; Maire, E. Damage Evolution in TWIP and Standard Austenitic Steel by Means of 3D X Ray Tomography. *Mater. Sci. Eng. A* **2013**, *579*, 92–98. [CrossRef]
122. Nemcko, M.J.; Wilkinson, D.S. On the Damage and Fracture of Commercially Pure Magnesium Using X-Ray Microtomography. *Mater. Sci. Eng. A* **2016**, *676*, 146–155. [CrossRef]
123. Song, Q.-Y.; Heidarpour, A.; Zhao, X.-L.; Han, L.-H. Experimental and Numerical Investigation of Ductile Fracture of Carbon Steel Structural Components. *J. Constr. Steel Res.* **2018**, *145*, 425–437. [CrossRef]
124. Xie, J.; Zhang, R.; Liu, T.; Zhou, C.; Jia, L.J. Effect of Initial Void Shape on Void Growth of Structural Steels Based on Micromechanical RVE Models. *J. Mater. Civil Eng.* **2022**, *34*, 04022010. [CrossRef]
125. Benzerga, A.; Leblond, J.-B. Ductile Fracture by Void Growth to Coalescence. *Adv. Appl. Mech.* **2010**, *44*, 169–305. [CrossRef]
126. Usman, M.; Waheed, S.; Mubashar, A. Effect of Shape on Void Growth: A Coupled Extended Finite Element Method (XFEM) and Discrete Dislocation Plasticity (DDP) Study. *Eur. J. Mech.-A Solids* **2022**, *92*, 104471. [CrossRef]
127. Romero, I.; Segurado, J.; LLorca, J. Dislocation Dynamics in Non-Convex Domains Using Finite Elements with Embedded Discontinuities. *Model. Simul. Mater. Sci. Eng.* **2008**, *16*, 035008. [CrossRef]
128. Liang, S.; Zhu, Y.; Huang, M.; Li, Z. Simulation on Crack Propagation vs. Crack-Tip Dislocation Emission by XFEM-Based DDD Scheme. *Int. J. Plast.* **2019**, *114*, 87–105. [CrossRef]
129. Landron, C.; Maire, E.; Bouaziz, O.; Adrien, J.; Lecarme, L.; Bareggi, A. Validation of Void Growth Models Using X-Ray Microtomography Characterization of Damage in Dual Phase Steels. *Acta Mater.* **2011**, *59*, 7564–7573. [CrossRef]
130. Pardoen, T.; Hutchinson, J.W. Micromechanics-Based Model for Trends in Toughness of Ductile Metals. *Acta Mater.* **2003**, *51*, 133–148. [CrossRef]
131. Shen, Y.; Morgeneyer, T.F.; Garnier, J.; Allais, L.; Helfen, L.; Crépin, J. Three-Dimensional Quantitative in Situ Study of Crack Initiation and Propagation in AA6061 Aluminum Alloy Sheets via Synchrotron Laminography and Finite-Element Simulations. *Acta Mater.* **2013**, *61*, 2571–2582. [CrossRef]
132. Helfen, L.; Myagotin, A.; Rack, A.; Pernot, P.; Mikulík, P.; Di Michiel, M.; Baumbach, T. Synchrotron-radiation Computed Laminography for High-resolution Three-dimensional Imaging of Flat Devices. *Phys. Status Solidi A* **2007**, *204*, 2760–2765. [CrossRef]
133. Bull, D.J.; Helfen, L.; Sinclair, I.; Spearing, S.M.; Baumbach, T. A Comparison of Multi-Scale 3D X-Ray Tomographic Inspection Techniques for Assessing Carbon Fibre Composite Impact Damage. *Compos. Sci. Technol.* **2013**, *75*, 55–61. [CrossRef]
134. Gao, B.; Xiang, Q.; Guo, T.; Guo, X.; Tang, S.; Huang, X.X. In Situ TEM Investigation on Void Coalescence in Metallic Materials. *Mater. Sci. Eng. A* **2018**, *734*, 260–268. [CrossRef]
135. Bensaada, R.; Kanit, T.; Imad, A.; Almansba, M.; Saouab, A. Void-Growth Computational Analysis in Elastic-Plastic Porous Materials. *Int. J. Mech. Sci.* **2022**, *217*, 107021. [CrossRef]
136. Wierzbicki, T.; Bao, Y.; Lee, Y.-W.; Bai, Y. Calibration of Seven Fracture Models. *Int. J. Mech.Sci.* **2005**, *47*, 719–743. [CrossRef]
137. Bai, Y.; Wierzbicki, T. A New Model of Metal Plasticity and Fracture with Pressure and Lode Dependence. *Int. J. Plast.* **2008**, *24*, 1071–1096. [CrossRef]
138. Bai, Y.; Wierzbicki, T. Application of Extended Mohr–Coulomb Criterion to Ductile Fracture. *Int. J. Fract.* **2010**, *161*, 1. [CrossRef]
139. Bao, Y.; Wierzbicki, T. On Fracture Locus in the Equivalent Strain and Stress Triaxiality Space. *Int. J. Mech. Sci.* **2004**, *46*, 81–98. [CrossRef]
140. Zhang, K.S.; Bai, J.B.; François, D. Numerical Analysis of the Influence of the Lode Parameter on Void Growth. *Int. J. Solids Struct.* **2001**, *38*, 5847–5856. [CrossRef]
141. Liu, X.; Yan, S.; Rasmussen, K.J.R.; Deierlein, G.G. Experimental Investigation of the Effect of Lode Angle on Fracture Initiation of Steels. *Eng. Fract. Mech.* **2022**, *271*, 108637. [CrossRef]
142. Danas, K.; Ponte Castañeda, P. Influence of the Lode Parameter and the Stress Triaxiality on the Failure of Elasto-Plastic Porous Materials. *Int. J. Solids Struct.* **2012**, *49*, 1325–1342. [CrossRef]
143. Dunand, M.; Mohr, D. Effect of Lode Parameter on Plastic Flow Localization after Proportional Loading at Low Stress Triaxialities. *J. Mech. Phys. Solids* **2014**, *66*, 133–153. [CrossRef]
144. Bonora, N.; Testa, G. Plasticity Damage Self-Consistent Model Incorporating Stress Triaxiality and Shear Controlled Fracture Mechanisms—Model Formulation. *Eng. Fract. Mech.* **2022**, *271*, 108634. [CrossRef]
145. Barsoum, I.; Faleskog, J. Micromechanical Analysis on the Influence of the Lode Parameter on Void Growth and Coalescence. *Int. J. Solids Struct.* **2011**, *48*, 925–938. [CrossRef]

materials

MDPI

Article

Mechanical Properties Analysis of Explosive Welded Sheet of AA2519-Ti6Al4V with Interlayer of AA1050 Subjected to Heat-Treatment

Ireneusz Szachogłuchowicz *, Lucjan Śnieżek and Tomasz Ślęzak

Faculty of Mechanical Engineering, Institute of Robots & Machine Design, Military University of Technology, ul. gen. S. Kaliskiego 2, 00-908 Warsaw, Poland; lucjan.sniezek@wat.edu.pl (L.Ś.); tomasz.slezak@wat.edu.pl (T.Ś.)
* Correspondence: ireneusz.szachogluchowicz@wat.edu.pl

Abstract: The paper presents results of investigations of welding sheets of AA2519-Ti6Al4V, a difficult-to-joint components materials, produced by explosive welding with a thin technological interlayer of AA1050. The joining process leads to the formation of intermetalics in the vicinity of joint and generates significant residual stresses. In the next step the laminate was subjected to a heat treatment process in order to improve the mechanical properties by precipitation hardening. This treatment should not be carried out before welding because of negative influence on a ductility of the aluminum alloy. Material in this state was subjected to the tests of chemical composition, microstructure, and microhardness. A tensile test was carried out with accompanying strain analysis by the digital image correlation (DIC) method. Moreover, the residual stresses were determined which were measured by using two methods, the X-ray diffraction and the hole drilling. This approach made it possible to measure the residual stresses both in the plane parallel to the surface and in the cross section of the laminate.

Keywords: explosive welding; laminate; AA2519; Ti6Al4V; heat treatment; residual stress measurement

Citation: Szachogłuchowicz, I.; Śnieżek, L.; Ślęzak, T. Mechanical Properties Analysis of Explosive Welded Sheet of AA2519-Ti6Al4V with Interlayer of AA1050 Subjected to Heat-Treatment. *Materials* **2022**, *15*, 4023. https://doi.org/10.3390/ma15114023

Academic Editor: Michele Bacciocchi

Received: 12 May 2022
Accepted: 2 June 2022
Published: 6 June 2022

1. Introduction

Explosive welding technology enable to join the materials difficult-to-join by conventional methods and it is possible to be realized on large plates also. This process is realized by explosively generated pressure wave that imparts very high velocities to the solids (welded materials). During collision the pressure reaches up to 2×10^4 MPa. Such level of pressure makes it possible to obtain physical states unattainable under conventional types of loads [1,2]. As a result, significant residual stresses appear, deviating from the primary distribution in the combined materials [3,4].

AA2519 alloys andTi6Al4V alloys are characterized by increased ballistic resistance. These alloys are used in aerospace industry. The combination of these materials using an explosive welding method should increase ballistic resistance by changing the density centers of the material and the appearance of intermetallics in the joint layer. In the welding zone of Al-Ti explosively produced laminates, the intermetallic phases Al_3Ti and Ti_3Al were revealed [5–7] which may promote delamination of the produced material and the decrease of strength properties. Additionally, significant residual stresses are present. Therefore, proper heat treatment should be carried out to decrease the negative influence of pointed factors, especially that, the heat treatment is recommended to reduce the adverse effects of the residual stresses formed in the material as a result of the explosive welding process [8]. The most important in such a case is the selection of appropriate processing parameters [9]. In considered laminate, among all components the AA2519 alloy is characterized by the lowest melting temperature, which amounts 821 K. Carefully selected heat treatment parameters, due to the addition of Cu, should provide the effect of precipitation hardening. Additionally, the alloy addition of Zr increases the resistance to secondary recrystallization and increases the tensile strength. It was revealed [10] that the observed intermediate layer is formed by the deposition of aluminum on a titanium substrate due to the interaction of high temperature and pressure. The thickness of this layer depends primarily on the reaction time of the materials at temperature above 873 K. The observed interlayer is formed by mutual diffusion until the formation of Al_3Ti intermetallic phases. In case of realization of the fusion at the temperature reaching 1073 K, initially Ti_3Al and TiAl phases are formed. Further maintaining the temperature at 1073 K results in TiAl phases gradually transforming into Ti_3Al. These transformations lead to the formation of continuous homogeneous Ti_3Al layer on the titanium surface. In case of Ti_3Al intermetallic phases, the microhardness can increase up to two times. While the research on microstructure and phase transitions are well described, this study has been undertaken on the influence of heat treatment on the global state of the material.

The purpose of this research is to investigate the influence of hardening heat treatment on the mechanical properties and state residual stresses of explosive welded sheet of AA2519-Ti6Al4V with interlayer of AA1050.

2. Materials and Equipment

The subject of the study was an explosive-welded sheet of AA2519-Ti6Al4V with interlayer of AA1050. The AA2519 aluminum alloy is characterized by good fracture toughness because of the copper content of 5.3–6.4%. An important advantage is its increased ballistic resistance and the appropriate weldability. This alloy is strengthened by precipitations through the process of supersaturation and aging. Investigated variety of aluminum alloy was produced by adding the alloying elements Zr and Sc in the amount of 0.1–0.25% for both. The addition of Sc has the effect of fragmenting the structure of the AA2519 alloy, which improves the strength properties of the material. Precipitation strengthening was obtained and the resistance to secondary recrystallization of microstructure increased. The primary effect is a greater stability of mechanical properties at elevated temperature as a result of homogeneous dispersion of Al_3Zr phase. After casting, the alloy was subjected to hot rolling and then annealing at a temperature of 693 K in air condition for 2 h. A fine-grained structure was obtained, which has a positive effect on the increase of plasticity and fatigue strength. The strength properties and chemical composition of the modified AA2519 alloy are given in Table 1.

Table 1. Strength properties and chemical composition of the AA2519 alloy.

Strength Properties			Chemical Composition [wt%]									
$R_{0.2}$ [MPa]	R_m [MPa]	A [%]	Si	Fe	Cu	Mg	Zn	Ti	V	Zr	Sc	Al
312	335	6.5	0.06	0.08	5.77	0.18	0.01	0.04	0.12	0.2	0.36	rest

The second basic material of the laminate, Ti6Al4V alloy, is used in special constructions which is caused by its high mechanical properties, corrosion resistance, and a melting point reaching 1955 K. For this reason, it is used in the production of aircraft components, including: the jet engine rotor blades or the wing caissons. The strength properties and chemical composition of the Ti6Al4V alloy are presented in Table 2.

Table 2. Strength properties and chemical composition of the Ti6Al4V alloy.

Strength Properties			Chemical Composition [wt%]							
$R_{0.2}$ [MPa]	R_m [MPa]	A [%]	O	V	Al	Fe	H	C	N	Ti
950	1020	14	<0.20	3.5	5.5	<0.30	<0.0015	<0.08	<0.05	rest

Explosive joining of materials with different mechanical properties requires the usage of an interlayer. The AA1050 alloy was used, because it has a very good adhesive property. The strength properties and chemical composition are shown in Table 3.

Table 3. Strength properties and chemical composition of the AA1050 alloy.

Scheme			Chemical Composition [wt%]							
$R_{0.2}$ [MPa]	R_m [MPa]	A [%]	Fe	Si	Zn	Mg	Ti	Mn	Cu	Al
78	168	2.9	0.4%	<0.25	<0.07	0.18	<0.05	<0.05	<0.05	rest

The layered material subjected to testing was manufactured in the Department of High Energy Technologies EXPLOMET Galka Szulc general partnership. The functional properties of explosives, including the detonation velocity, are derived from their chemical composition, the physical form of their individual components, e.g., fragmentation, the degree of pore surface development present in the structure of crystals of the ammonium nitrate used, and the quality of their mixing. The assumed plan of plating tests included the execution of joints in the AA2519-Ti6Al4V double-layer system with a separate spacer made of AA1050 aluminum alloy (Figure 1) [11].

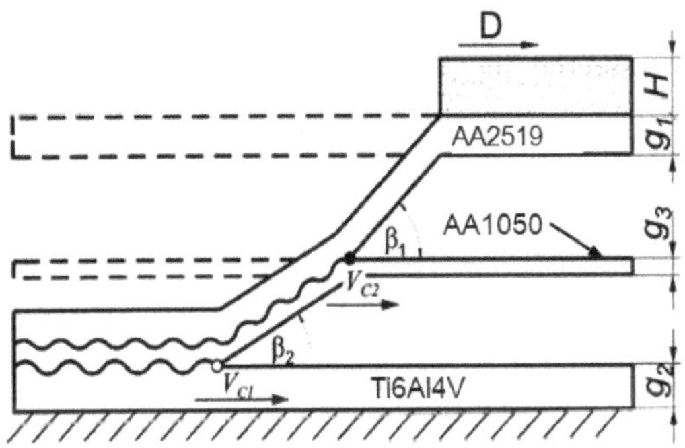

Figure 1. Schematic diagram of the collision course of explosion welded three-layer: β—angle of collision of combined materials, β_1—angle of collision of base material with interlayer material, β_2—angle of collision of overlay material with interlayer material V_c—velocity of collision point with respect to contact point of welded plates, V_{c1}—velocity of collision point of contact between base material and interlayer material, V_{c2}—velocity of the collision point of the contact point of the base material and the interlayer, D—velocity of detonation of the explosive, H—height of the explosive layer, g_1—thickness of the plate to be shot, g_2-thickness of the base plate, g_3—thickness of the interlayer.

The final products were in the form of plates with dimensions of 200 × 150 mm and thickness of 6.7 mm, approximately (Figure 2). The thickness can slightly vary because of the dynamic nature of the joining process. The plates were rolled in order to give them adequate flatness. Both described technological processes significantly deformed the material and introduced elastic-plastic residual strains. In the next step the plates were annealed at the temperature of 803–823 K for 2 h, cooled down in water at room temperature and aged at 438 K for 10 h. The main aim of this treatment was to obtain an effect of precipitation hardening. Besides the changes of microstructure and mechanical properties this treatment has seriously influenced the residual stresses state.

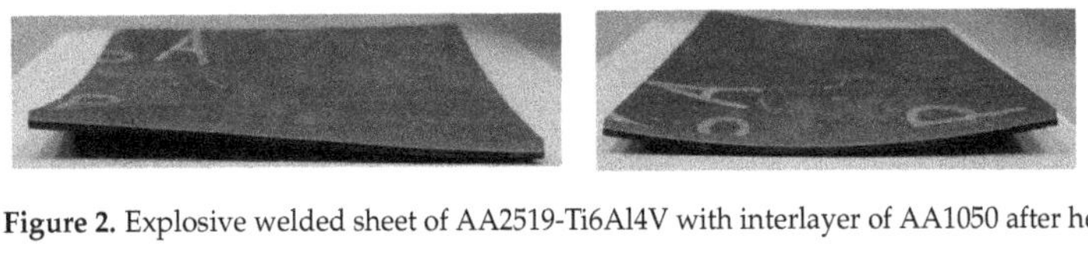

Figure 2. Explosive welded sheet of AA2519-Ti6Al4V with interlayer of AA1050 after heat treatment.

Monotonic tensile tests were carried out on specimens made of manufactured laminate after heat treatment. All the specimens subjected to axial tension had the same geometry. Five specimens were made of each material (Figure 3). Tensile testing of AA2519-Ti6Al4V laminate with AA1050 interlayer in axial tension conditions was performed according to PN-EN ISO 6892-1:2010 on an Instron 8802 monotonic and fatigue hydraulic pulser.

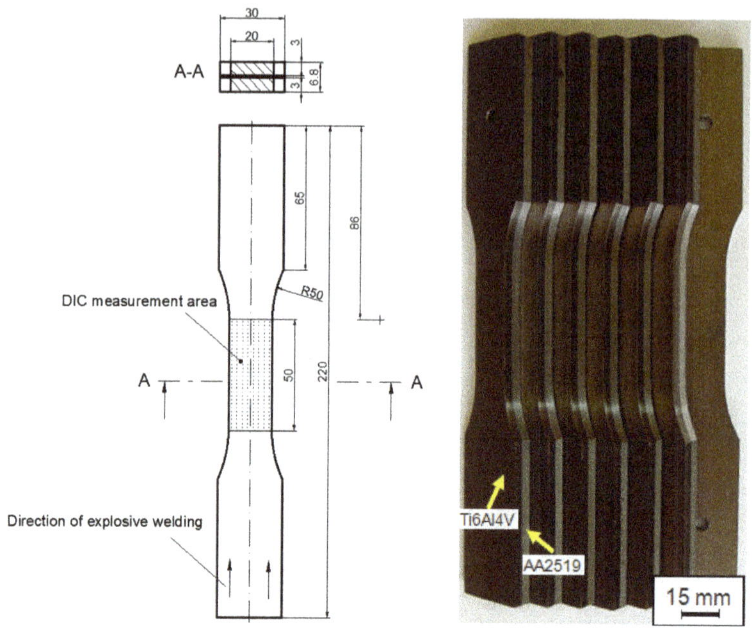

(**a**) (**b**)

Figure 3. Dimensions of samples for tensile testing of AA2519-Ti6Al4V laminate with inter-layer of AA1050 (**a**) and prepared test specimens (**b**).

Microhardness tests were carried out using a fully automatic universal laboratory hardness tester DURA SCAN 70. The microhardness test was conducted in accordance with EN ISO 6507-1. The indenter load was F = 0.9807 N. The process time to load the indenter to the nominal load was 3 s. It took 10 s to load the sample to 0.98 N.

Digital image correlation (DIC) allows the measurement of deformations and deformation of the surface of the tested object (Figure 4). The digital image correlation method is used in optical, non-contact, and three-dimensional systems for real-time measurement of displacement and deformation. The DIC system allows speckle images to be captured using CCD cameras, while further data analysis and processing are carried out using Istra 4D software.

The procedure for measuring residual stresses using the hole trepanation method is standardized by ASTM Standard Test Method E 837 [12], and the technical procedure of implementation is described in Tech Note TN-503 [13]. Micro-Measurement RS-200 milling guide was used to conduct the hole drilling process, necessary in order to measure the residual stresses. This equipment enables precise drilling of holes in the center point of strain gauge rosette which is the most important condition for making correct measurements. Figure 5 shows the distribution of gauge rosettes.

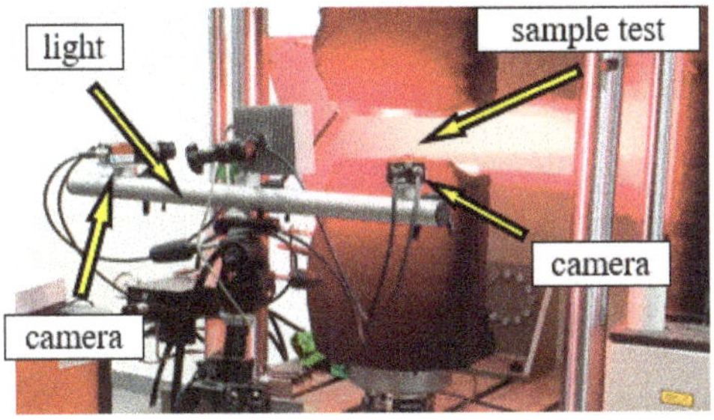

Figure 4. Digital image correlation measurement system from Dantec.

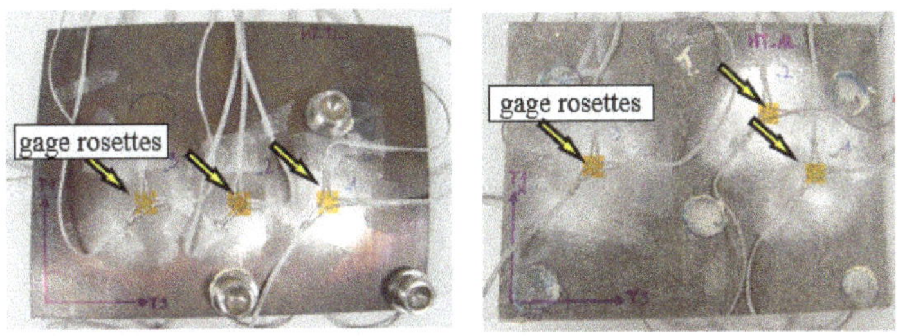

Figure 5. Arrangement of gauge rosettes on TiAl4V side 9 (**a**) AA2519 (**b**).

Furthermore, it enables to control the depth of hole and to measure its diameter. The carbide milling cutters with a diameter of 1.6 mm were used during the tests and these tools were driven by high-speed rotation air turbine supplied by compressor. Testing stand is presented in Figure 6.

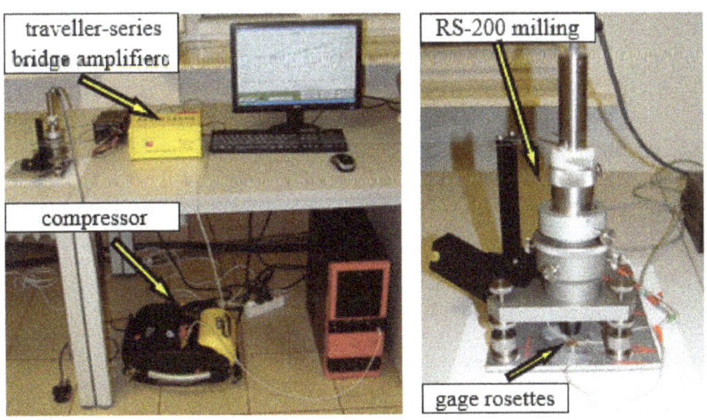

Figure 6. The overall view of testing stand (**a**) and the milling guide of RS-200 (**b**).

Surface residual stresses in the two main, perpendicularly oriented directions ("σ1, σ2") based on sin2ψ diffractometric measurements were obtained using an X-ray diffractometer -Bruker D8 Discover (Bruker Corporation, 40 Manning Road, Billerica, Massachusetts, United States of America) with a Euler wheel and a sample positioning system along the three axes. Test samples for the research were prepared using electrical discharge machining. Radiation and beam optics were characterized by CoKα filtration. Phase analysis was performed in the Crystal Impact Match software with an ICDD PDF 4+ 2019 crystal-lographic database. TARSIuS (Texture-Aided Residual Stress Investigation System) pack-age is a software developed by the Institute of Metallurgy and Materials Sciences of the Polish Academy of Sciences in Krakow and it is used to visualize the residual stresses in materials. Because this method is quite new, we decided to implement it in a short explanation of its basics with examples. The areas were near to the connection of the plates and the detailed locations are shown in the Figure 7.

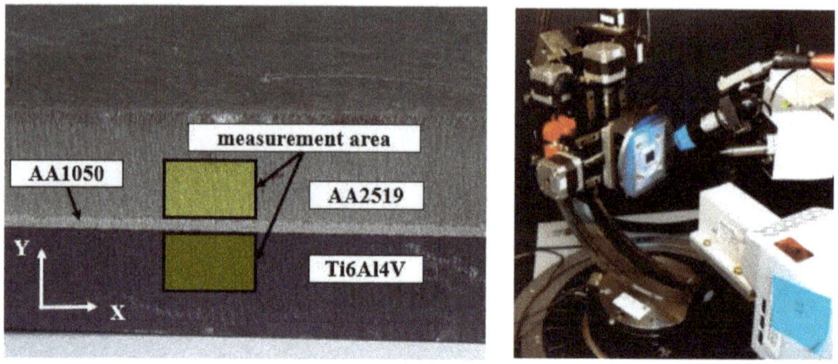

(a) (b)

Figure 7. The surface area of the sample (**a**) and Bruker D8 Discover diffractometer with Euler wheel (**b**).

3. Results

In the Figure 8 the view of plates after heat treatment and the metallographic cross-section of the composite are shown.

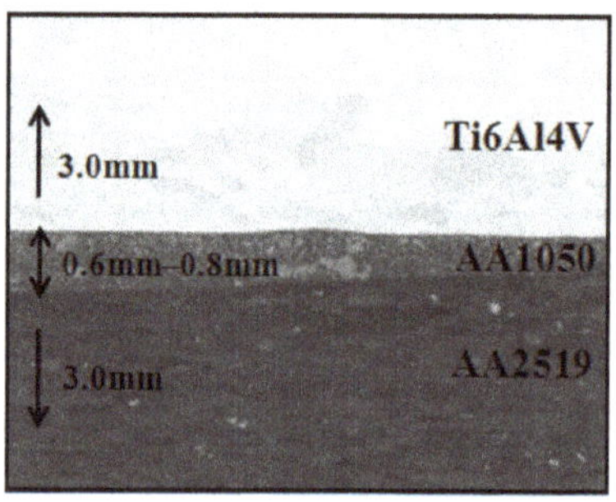

Figure 8. Explosive welded sheet of AA2519–Ti6Al4V with interlayer of AA1050) a metallographic cross-section of the composite.

The picture in the Figure 1 has shown that the plates after heat treatment are significantly distorted. This deformation is caused by the difference of the value of the coefficient of thermal expansion and the characteristic temperatures of phase transitions.

The microstructure of the heat treated of explosive welded sheet of AA2519-Ti6Al4V with interlayer of AA1050 in the vicinity of joint is shown in the photographs (Figure 9).

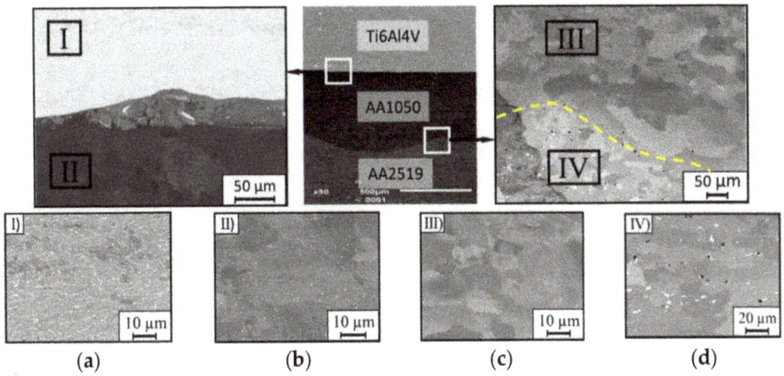

Figure 9. Metallographic cross-section of the produced explosive welded sheet of AA2519-Ti6Al4V with interlayer of AA1050 after heat treatment: (**a**) sector I–Ti6Al4V alloy; (**b**) sector II–AA1050; (**c**) sector III–AA1050; (**d**) sector IV–AA2519 alloy.

Heat treatment has not affected the structure of titanium alloy (Figure 9a). The AA1050 interlayer by the boundary with titanium is characterized by the grain growth with uniform distribution (Figure 9b). Near the AA2519 alloy, the grains of the AA1050 alloy structure have an equiaxial character (Figure 9c). For AA2519 alloy, the heat treatment process improved the solubility of copper-rich precipitates in the aluminum matrix. The effect of this process is the fine-grained structure and steady distribution of precipitates in the matrix. At the interface between AA1050 aluminum and AA2519 alloy, a sublayer of about 4–6 μm width was observed, which was formed as a result of heat treatment. It is characterized by a fragmented structure and numerous voids with diameters of about 0.3–0.5 μm (Figure 9d). The slight voids were probably produced accidentally during the supersaturation and annealing process.

3.1. Strength Properties

Tensile tests of AA2519-Ti6Al4V laminate with interlayer of AA1050 were conducted according to [11]. The tests were carried out using five samples and obtained results were repeatable. Samples were cut along the direction of explosive welding. The results obtained during the tests are presented in the form of graphs in Figure 10.

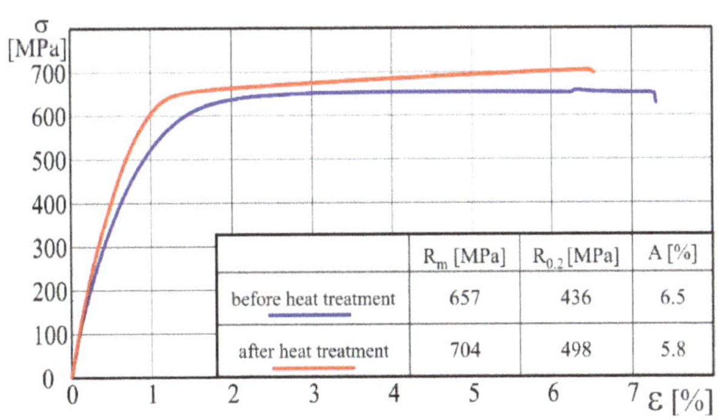

Figure 10. Stress–strain curves of AA2519-Ti6Al4V laminate with interlayer of AA1050 in production state and after heat treatment.

The heat treatment has resulted in an increase of the ultimate strength Rm from 657 MPa up to 704 MPa, which is about 7%, whereas the yield strength $R_{0.2}$ has increased from 436 MPa to 498 MPa and the increase is about 14%. On the other hand, the heat treatment has negative influence on the elongation, A, which has decreased from 6.5% to 5.8%. Based on these results, it can be stated that the ductile of heat-treated laminate decreased.

During tensile tests, an analysis of strain distribution using was applied using digital image correlation (DIC). In case of the laminate subjected to additional heat treatment, the strain distribution for the AA2519 alloy, up to the moment of a sample rupture, is characterized by high uniformity (Figure 11).

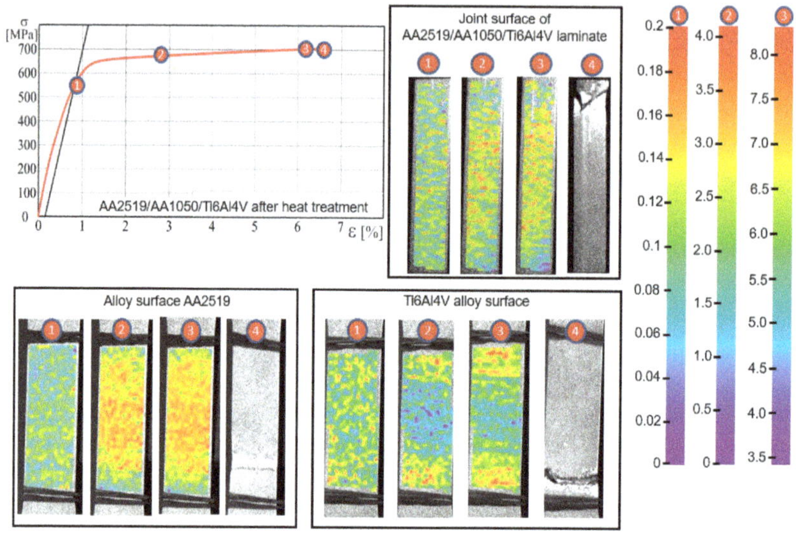

Figure 11. Strain distribution using DIC method for selected points of the stress-strain curve obtained for AA2519-Ti6Al4V laminate with interlayer of AA1050 after heat treatment: 1—The yield strength R0.2; 2—the plastic strain of 2%; 3—the ultimate strength Rm; 4—the fracture of the specimen.

In case of the Ti6Al4V alloy, after exceeding the yield point, the strain maps illustrate the occurrence of horizontal bands. DIC analysis made on the lateral surface has not revealed local or banded strain inhomogeneity, which makes it impossible to predict the location of cracking even under high loadings.

3.2. *Microhardness*

The measurement of microhardness HV0.1 was conducted according to the standard [15]. The study was performed in the immediate vicinity of the center layer. Obtained results are presented in the Figure 12, where the AA1050 intermediate layer is in grey.

The microhardness of the AA2519 alloy after heat-treatment was below 100 HV. In case of the interlayer, which is composed of the AA1050 alloy, there is noticeable a strengthening effect near the boundary with the AA2519 due to the formation of intermetallic precipitates. In case of Ti6Al4V alloy, the microhardness after heat treatment was about 350 HV and has not changed compared to the state before heat treatment [16].

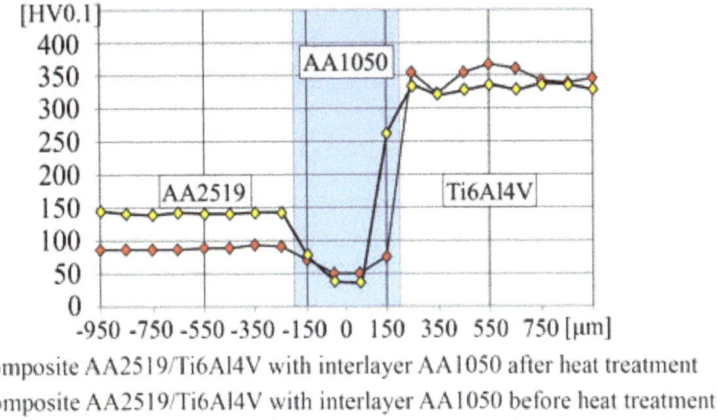

Figure 12. The results of microhardness measured in AA2519–Ti6Al4V laminate with interlayer of AA1050.

3.3. Measurements of Residual Stresses

The applied processes of explosive joining and following rolling significantly influenced the stress state of the material. It is important to define this state, therefore the residual stresses were measured in two perpendicular planes, namely in the plane parallel to the surface and in the cross section. The first measurements were conducted using hole-drilling method and the second by X-ray diffraction.

3.3.1. Hole-Drilling Method

This method was used to determine the state of surface residual stresses and their change with the thickness. The measurements were made at three randomly placed points on both sides of the plate, from the aluminum and titanium. In the presence of residual stresses, even a small hole into the material causes stress relaxation because each normal stress component perpendicular to the free surface (e.g., the hole surface) is zero. Moreover, this relaxation is related to the change in strain field in the direct vicinity of the hole according to Hook's law. This method is often referred to as a "semi-destructive" technique because in many cases a small hole does not significantly damage the structural integrity of the test object (the dimensions of the hole are usually 0.8–4.8 mm-diameter and depth). After measuring large objects, it is possible to remove the hole by careful welding and smoothing the surface with a grinder.

The obtained output voltage was transmitted from the rosette to the channels of the ESAM Traveler Plus strain gauge bridge, where it was amplified. An exemplary graph presenting the changes of voltage recorded during the test is presented in Figure 13a.

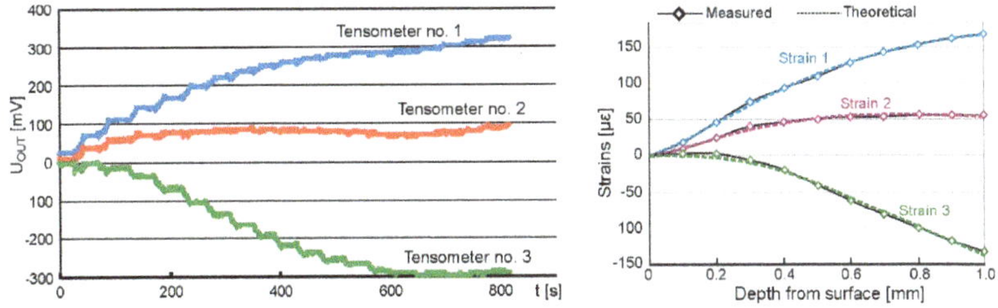

(a) (b)

Figure 13. The overall view of testing stand (**a**) and the milling guide of RS–200 (**b**).

The calculation of the strain values was made up to the depth of 1.0 mm employing Equation (1).

$$E = 4 \times U_{OUT} / (U_0 \times N \times K \times A) \times 10^6 \quad [\mu\varepsilon = 10^{-6}] \tag{1}$$

where:

U_{OUT}—output voltage [V];

U_0—supply voltage [V];

N—coefficient depending on the bridge type—for quarter-bridge N = 1;

K—strain gauge constant;

A—coefficient of output signal amplification.

As a result of the measurements and calculations carried out, the characteristics of changes in relative strain values as a function of the depth of the hole drilled were obtained. To determine the values of principal stresses and their direction, the following relations were applied.

$$\sigma_{max} = \frac{\varepsilon_1 + \varepsilon_2}{4 \cdot A} - \frac{1}{4 \cdot B}\sqrt{(\varepsilon_3 - \varepsilon_1)^2 + (\varepsilon_3 + \varepsilon_1 - 2\varepsilon_2)^2}\ [\text{MPa}] \tag{2}$$

$$\sigma_{max} = \frac{\varepsilon_1 + \varepsilon_2}{4 \cdot A} + \frac{1}{4 \cdot B}\sqrt{(\varepsilon_3 - \varepsilon_1)^2 + (\varepsilon_3 + \varepsilon_1 - 2\varepsilon_2)^2}\ [\text{MPa}] \tag{3}$$

$$\alpha = \frac{1}{2} arctg \frac{\varepsilon_1 - 2\varepsilon_2 + \varepsilon_2}{\varepsilon_2 - \varepsilon_1}\ [\text{MPa}] \tag{4}$$

where:

σ_{max}, σ_{min}—principal stresses;

ε_1, ε_2, ε_3—strains measured on strain gauges number 1, 2, and 3;

A, *B*—coefficients depending on material properties and geometry of rosette and hole;

α—angle between strain gauge no. 1 and the direction of the nearest principal stress.

The result of performed analysis is presented in Figure 13b. Calculated changes of strains (dashed lines) correspond to the measured data (diamonds). This accordance was gained under the assumption that the residual stresses vary linearly with the depth. Determination of the stress value and its character was accomplished using software H-Drill. Obtained results of measured principal residual stresses and the angle from the principal axis to defined direction, are presented in Table 4. The minus value of the angle indicates that the angular deviation is counterclockwise.

The character of stresses was analyzed up to the depth of 1.0 mm and in all cases a linear character of the stress changes was revealed. This assumption was justified by the values of statistical parameter RMS misfit which vary from 1.9 me to 2.8 me for AA2519 and from 4.6 me to 6.0 me for Ti6Al4V. If the uniform state of stresses was assumed, the RMS misfit would have vary from 6.8 me to 11.6 me and from 9.1 me to 16.2 me, respectively.

Determined values of principal stresses in aluminum alloy AA2519 vary with the depth within the ranges: σ_{min} from −102 MPa up to +90 MPa and σ_{max} from −27 MPa up to +146 MPa. On the surface, the values of σ_{min} and σ_{max} are −96.0 ± 5.9 MPa and −20.7 ± 7.6 MPa, respectively. The stresses are compressive and at different depths change to tensile stresses. Comparing the principal directions of residual stresses with the geometry of investigated plate it can be stated that σ_{min} is oriented perpendicular to the distortion, namely parallelly to the maximum gradient, and σ_{max} opposite. The values of principle stresses in titanium layer change with the ranges: σ_{min} from +8 MPa up to +376 MPa and σ_{max} from +132 MPa up to +509 MPa and they were determined at different depths. The residual stresses are tensile and increase linearly for the surface values of σ_{min} and σ_{max} equal to +36.0 ± 21.7 MPa and +141.3 ± 11.8 MPa, respectively. It must be pointed that the maximum values determined at greater depths can be subjected to high uncertainty. When drilling a hole in the titanium alloy, high cutting resistance was observed which may influence the closest vicinity of the hole, thus the near-surface values of residual stresses are most reliable. The directions of principal stresses are oriented similarly to previously described but σ_{min} and σ_{max} are swapped because on this side of the laminate the surface is convex (on the side of aluminum is concave) which changes the compressive stresses to tensile.

Table 4. The results of residual stresses measurements of the heat treated laminate AA2519-Ti6Al4V laminate with interlayer of AA1050.

	AA2519								
	Number of Measuring Point								
	# 1			**# 2**			**# 3**		
depth [mm]	0.0	0.5	1.0	0.0	0.5	1.0	0.0	0.5	1.0
	residual stresses s [MPa] and angular orientation a [°]								
s_{max}	−10	49	108	−27	59	146	−25	56	136
s_{min}	−102	−66	−31	−98	−26	45	−88	1	90
a	−81	−77	−74	−88	−82	−77	−86	−86	−87
	Ti6Al4V								
	Number of measuring point								
	# 1			**# 2**			**# 3**		
depth [mm]	0.0	0.5	1.0	0.0	0.5	1.0	0.0	0.5	1.0
	residual stresses s [MPa] and angular orientation a [°]								
s_{max}	134	285	436	132	166	201	158	329	509
s_{min}	8	120	232	39	85	128	61	223	376
a	−32	−29	−28	−19	−25	−31	−6	7	17

3.3.2. X-ray Diffraction Method

Based on the gained data, the two-dimensional maps of residual stresses were developed whose visualization enable easy interpretation of the stress state (Figure 14).

Determined values of residual stresses measured in the vicinity of the joint show that the directions of principal stresses are oriented approximately parallelly/perpendicularly to this joint. It is indicated by the orientation of the angular marks in the circles placed at the measuring points (Figure 14). Residual stresses are both compressive and tensile in AA2519, but in Ti6Al4V are only compressive. The measurement results presented in form of maps are the most reliable for the titanium alloy layer. The maximum residual is in its central part, where they reach the value of −1250 MPa. The lower values of compressive stresses in the horizontal direction may be related to the change of shape of the plate after heat treatment, which is slightly distorted.

The values of stresses in the layer of AA2519 alloy should be treated as speculative, because the values significantly exceeded the yield point of this material. This is most likely due to the formation of intermetallic phases in the aluminum and grain growth.

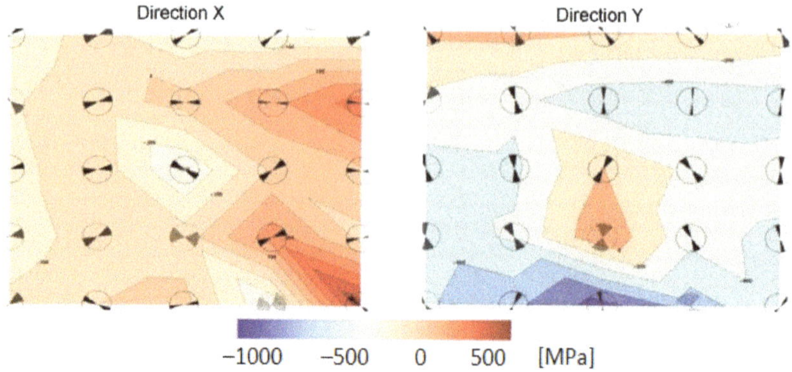

(a)

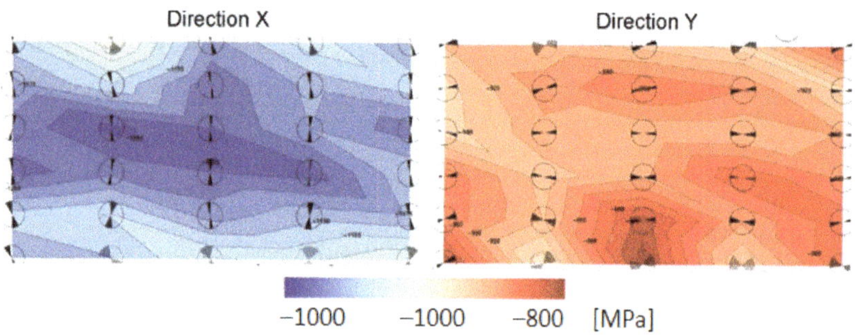

(b)

Figure 14. The maps of residual stresses measured in two perpendicular directions X and Y in the (**a**) AA2519 alloy and (**b**) Ti6Al4V; description in text.

4. Summary and Conclusions

The results of measurements the residual stresses obtained by two different methods seem to be inconsistent with each other thus this issue needs wider explanation in order to clarify doubts. First difference relates to the orientation of the measuring plane—the hole-drilling method (HDM) was applied in plane parallel to the surface contrary to X-ray method (XRM), which was applied in perpendicular plane. Second difference is related with the place of measurements and state of the material. HDM has been adopted to determine the stresses on the surface and at shallow depth of the material in the form of full-scale plate after heat treatment without any additional mechanical processing. Whereas XRM allowed to measure the stress state near the interlayer, but the small sample must be cut from the plate for proper preparation. This process can influence the global stress state. The last difference relates to the nature of the measurement. HDM allows to determine only the value of the first-order residual stresses because the resultant strains are measured in some distance from the measurement point. XRM is based on the measurement of the distance between the crystallographic planes thus the values can be connected with the microscale second- or even third-order residual stresses and the microscale mechanical properties. The influence of this problem was observed during the measurement of stresses in aluminum alloy when the determined values significantly overcame the macroscale tensile strength. The above-described differences between HDM and XRM indicate that comparisons of results should be made with caution.

The investigation was performed to analyze the influence of the heat treatment on the structure and the mechanical properties of the AA2519-Ti6Al4V laminate with interlayer of

AA1050. The main purpose of the heat treatment was to obtain an effect of precipitation hardening in aluminum alloy, which was successfully gained. Nevertheless, this process has seriously influenced the residual stresses state. The laminate was distorted because of the differences in thermophysical properties thus there were performed wide investigation on the state of residual stresses. On the basis of the obtained results the following conclusions can be stated:

1. A heat treatment affects the laminate resulting in an increase of the ultimate strength Rm about 7% to the value of 704 MPa and the yield strength $R_{0.2}$ about 14% to the value of 498 MPa, but it has also negative influence on the ductility decreasing an elongation from 6.5% to 5.8%;
2. As a result of heat treatment, the microhardness of titanium alloy remains unchanged but in AA2519 it increased from app. 95 $HV_{0.1}$ [13] to the value of 150 $HV_{0.1}$;
3. The heat treatment in AA2519 caused fine-grained structure with steady distribution of precipitates in the matrix and additionally a slight sublayer of about 4–6 μm width with some voids observed at the interface between this material and interlayer;
4. The surface residual stresses were determined by the hole drilling method and are closely related to distortion. They are compressive on the side of aluminum alloy (to −102 MPa) and tensile on the side of titanium alloy (up to +158 MPa);
5. The residual stress state in cross-section was determined by X-ray diffraction method. They are compressive reaching −1250 MPa in Ti6Al4V. The results for AA2519 are very dubious.

Author Contributions: Conceptualization, T.Ś. and I.S.; methodology, L.Ś. and I.S.; validation, T.Ś., L.Ś. and I.S.; formal analysis, I.S. and T.Ś.; investigation, I.S.; resources, I.S.; data curation, T.Ś.; writing—original draft preparation, I.S. and T.Ś.; visualization, T.Ś.; supervision, L.Ś.; All authors have read and agreed to the published version of the manuscript.

Funding: This research was supported by the Polish Ministry of National Defence [grant number PBG/13-998]. The project is carried out under Project PBS2/A5/35/2013 funded by the National Research and Development Centre.

Institutional Review Board Statement: Not applicable.

Informed Consent Statement: Informed consent was obtained from all subjects involved in the study

Data Availability Statement: Not applicable.

Acknowledgments: We would like to thank the Explomet company from Opole, which performed explosive bending of titanium and aluminum alloys.

Conflicts of Interest: The authors declare no conflict of interest.

References

1. Saravanan, S.; Raghukandan, K.; Hokamoto, K. Improved microstructure and mechanical properties of dissimilar explosive cladding by means of interlayer technique. *Arch. Civ. Mech. Eng.* **2016**, *16*, 563–568. [CrossRef]
2. Loureiro, A.; Carvalho, G.H.S.F.L.; Galvão, I.; Leal, R.M.; Mendes, R. Chapter 6–Explosive Welding. In *Advanced Joining Processes*; da Silva, L., El-Zein, M., Martins, P., Eds.; Elsevier: Amsterdam, The Netherlands, 2021; pp. 207–237. [CrossRef]
3. Sniezek, L.; Szachogluchowicz, I.; Gocman, K. The Mechanical Properties of Composites AA2519-Ti6Al4V Obtained by Detonation Method. In Proceedings of the 8th International Conference on Intelligent Technologies in Logistics and Mechatronics Systems, ITELMS 2013, Panevėžys, Lithuania, 23–24 May 2013; pp. 214–219.
4. Szachogłuchowicz, I.; Hutsaylyuk, V.; Sniezek, L. Low cycle fatigue properties of AA2519-Ti6Al4V. *Procedia Eng.* **2015**, *114*, 26–33. [CrossRef]
5. Bataev, I.A.; Bataev, A.A.; Mali, V.I.; Pavliukova, D.V. Structural and mechanical properties of metallic–intermetallic laminate composites produced by explosive welding and annealing. *Mater. Des.* **2012**, *35*, 225–234. [CrossRef]
6. Rohatgi, A.; Harach, D.J.; Vecchio, K.S.; Harvey, K.P. Resistance-curve and fracture behavior of Ti–Al_3Ti metallic–intermetallic laminate (MIL) composites. *Acta Mater.* **2003**, *51*, 2933–2957. [CrossRef]
7. Jiang, S.-Y.; Li, S.-C.; Zhang, L. Microstructure evolution of Al-Ti liquid-solid interface. *Trans. Nonferrous Met. Soc. China* **2013**, *23*, 3545–3552. [CrossRef]

8. Luo, J.G.; Acoff, V.L. Using cold roll bonding and annealing to process Ti/Al multi-layered composites from elemental foils. *Mater. Sci. Eng. A* **2004**, *379*, 164–172. [CrossRef]
9. Peng, L.M.; Li, H.; Wang, J.H. Processing and mechanical behavior of laminated titanium–titanium tri-aluminide (Ti–Al_3Ti) composites. *Mater. Sci. Eng. A* **2005**, *406*, 309–318. [CrossRef]
10. Peng, L.; Wang, J.; Li, H.; Zhao, J.; He, L. Synthesis and microstructural characterization of Ti–Al_3Ti metal–intermetallic laminate (MIL) composites. *Scr. Mater.* **2005**, *52*, 243–248. [CrossRef]
11. Walczak, W. Zgrzewanie wybuchowe metali i jego zastosowania. *Zesz. Nauk.* **2002**, *51*, 15–22.
12. *ASTM Standard E 837*; Determining Residual Stress by the Hole-Drilling Strain-Gage Method. ASTM International: West Conshohocken, PA, USA, 2020.
13. *Tech Note TN-503-6*; Measurement of Residual Stresses by the Hole-Drilling Strain Gage Method. Micro-Measurements. Vishay Precision Group: Malvern, PA, USA, 2010.
14. *ISO 6892-1:2016*; Metallic Materials—Tensile Testing—Part 1: Method of Test at Room Temperature. National Standards Authority of Ireland: Dublin, Ireland, 2016.
15. *ISO 6507-1:2007*; Metallic Materials—Vickers Hardness Test—Part 1: Test Method. European Committee for Standardization: Brussels, Belgium, 2018.
16. Szachogluchowicz, I.; Śnieżek, L.; Ślęzak, T.; Kluczyński, J.; Grzelak, K.; Torzewski, J.; Fras, T. Mechanical Properties Analysis of the AA2519-AA1050-Ti6Al4V Explosive Welded Laminate. *Materials* **2020**, *13*, 4348. [CrossRef] [PubMed]

materials

MDPI

Article

Effect of Steel-Cutting Technology on Fatigue Strength of Steel Structures: Tests and Analyses

Sławomir Rowiński

Faculty of Civil Engineering, Wrocław University of Science and Technology, 50-370 Wrocław, Poland; slawomir.rowinski@pwr.edu.pl; Tel.: +48-71-320-2905

Abstract: This paper presents the results of comparative fatigue tests carried out on steel S355J2N specimens cut out using different cutting methods, i.e., plasma cutting, water jet cutting, and oxyacetylene cutting. All the specimens were subjected to cyclic loading from which appropriate S-N curves were obtained. Furthermore, face-of-cut hardness and roughness measurements were carried out to determine the effect of the cutting method on the fatigue strength of the tested steel. The fatigue strength results were compared with the standard S-N fatigue curves. The fatigue strength of the specimens cut out with oxyacetylene was found to be higher than that of the specimens cut out with plasma even though the surface roughness after cutting with plasma was smaller than in the case of the other cutting technology. This was due to the significant effect of material hardening in the heat-affected zones. The test results indicate that, in comparison with the effect of the cutting technology, the surface condition of the specimens has a relatively small effect on their fatigue strength.

Keywords: steel structures; fatigue strength of steel; hardness; roughness; plasma cutting; water jet cutting; gas cutting; composite dowel

Citation: Rowiński, S. Effect of Steel-Cutting Technology on Fatigue Strength of Steel Structures: Tests and Analyses. *Materials* **2021**, *14*, 6097. https://doi.org/10.3390/ma14206097

Academic Editors: Jaroslaw Galkiewicz and Lucjan Śnieżek

Received: 23 August 2021
Accepted: 12 October 2021
Published: 15 October 2021

Publisher's Note: MDPI stays neutral with regard to jurisdictional claims in published maps and institutional affiliations.

1. Introduction

Most of the steel frame structures currently built and used in, i.a., the construction industry, the marine industry, and the manufacturing industry have their parts joined together and properly shaped by, e.g., welding and cutting out from larger steel elements. Currently, oxyacetylene cutting, plasma cutting, and water jet cutting are the predominant cutting technologies. In the case of oxyacetylene cutting, computer-controlled devices usually execute the cutting line with an accuracy of 0.8–1.6 mm. The width of the slit depends on the cutting parameters, i.e., the diameter and shape of the oxygen nozzle, the cutting oxygen and inflammable gas pressure, and the cutting speed. After oxygen cutting, the cut heat-affected zone (CHAZ) is relatively wide and depends on the alloying element content in the material. In the case of low-carbon steel plates, the width of CHAZ amounts to less than 0.8 mm at the thickness of 12.5 mm and to about 3 mm at the thickness of 150 mm [1]. Plasma cutting consists in melting metal and ejecting it from the slit with a strongly concentrated electric arc flowing between a nonconsumable electrode and the workpiece. The width of CHAZ is inversely proportional to the cutting speed and depends on the composition (conductivity) of the material being cut. In the case of 25 mm thick 18-8-type austenitic steels, CHAZ is 0.08–0.13 mm wide at the cutting speed of 1.2 m/min. Water jet cutting consists in using a strongly compressed water jet formed by passing water through a small-diameter nozzle. The water jet removes the cut material from the cutting slit through erosion and cutting fatigue under high local stresses and, additionally, through micromachining when abrasive powder with a (gamet, olivine, or silica) grain size of 0.3–0.4 mm is used. The temperature of the cut edges does not exceed 100 °C (cold cutting). The cut material thickness depends on the water jet cutting parameters, i.a., the cutting speed, the water pressure, the kind and grain size of the powder, and the powder feed rate [1].

Thanks to the significant differences between the above-mentioned cutting technologies, they find application in the manufacture of all kinds of steel frame structures. The choice of a cutting technology depends on the quality requirements and the manufacturer's technical and financial capabilities as well as on the requirements specified by the design engineer who takes into consideration the effect of the particular technologies of cutting out a steel frame structure on the latter's ultimate and fatigue strength.

The adverse effect of the technology of cutting out a structure on the latter's fatigue strength was observed during extensive experimental studies on innovative connectors of the composite dowel type used in composite steel and concrete bridges [2]. Such connectors are created by appropriately joining together steel structural members and concrete. The innovative composite dowel joint is based on the idea of cutting the rolled beam's steel web in two along a specifically shaped line (Figure 1) so that the dowels obtained in this way in each of the two parts when embedded in concrete will constitute a mechanical connector carrying the delamination forces between the steel and the concrete [3–5].

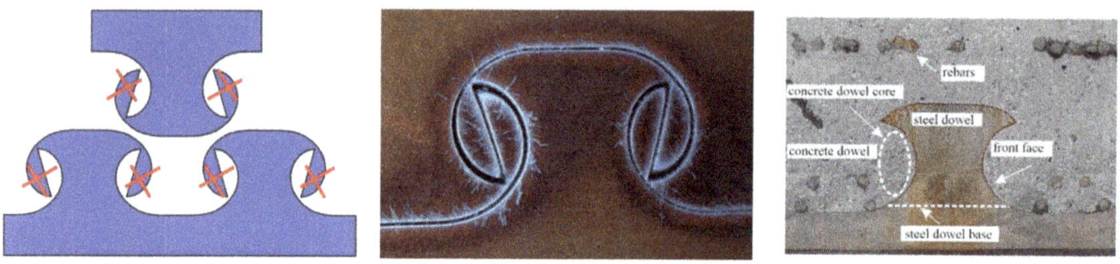

(a) (b) (c)

Figure 1. Cut-out MCL: (**a**) steel connectors with marked parts to be removed, (**b**) connector before separation of two parts of cut beam, (**c**) component parts of the innovative joint with MCL dowels.

It was only after the PreCo-Beam [2] had been completed when the effect of the cutting technology on the load-bearing capacity of the composite dowel joint was given some thought. It was found that the roughness of the dowel's front face after oxyacetylene cutting could be the cause of fatigue cracking in beams under cyclic loading [6–8].

In the literature, one can find the results of experimental studies of the effect of the technology of cutting steel on the latter's strength. However, one should bear in mind that such studies do not take into account the complexity of this problem (particularly in the construction industry) as the experiments are conducted not on full-size models of the structures but on specimens. Moreover, the results are for the element's particular shape, thickness, and steel grade. The effect of cutting technologies and the obtained specimen surface characteristics on the fatigue strength is described in the work of [9]. Specimens 6 and 12 mm thick with a steel strength of 240–900 MPa were tested. It was confirmed that the fatigue strength increases with the tensile strength of the steel and depending on the surface roughness. The fatigue strength of the specimens increases when their surface roughness is reduced by additional surface treatments, such as sandblasting [10]. The oxygen cut specimens have the highest fatigue strength, followed by the laser and plasma cut specimens. The gas-cut specimens have the highest surface roughness but also the highest compressive residual stress state. The plasma cut specimens have the lowest roughness, but their residual stresses are practically zero in comparison with the oxygen and laser-cut specimens [11]. The fatigue strength of plasma cut surfaces can be significantly improved with a post-cutting treatment applicable. The improvement is achieved by introducing compressive residual stress and reducing surface roughness height through grinding [12].

For steel S690Q and steel S355M, it is observed that when straight edges are cut with plasma and a laser, these cutting technologies improve, in comparison with gas cutting, the fatigue strength [13,14]. Laser-cut steel S890Q was found to have higher fatigue strength than when cut with gas or plasma [15].

It emerges from the tests carried out on specimens that the fatigue strength of steel frame structures is a complex problem sensitive to many factors. In construction, this problem is further compounded by the fact that it is not possible to directly observe the initiation and propagation of cracks (the connector is embedded in concrete) and also by the complicated interactions in the joint (delamination forces change from dowel to dowel and are transmitted through the direct pressure of the concrete against the front faces of the dowels and via the adhesive forces between the steel beam's flat surfaces and the concrete). Therefore in order to assess the effect of the cutting technology on the fatigue strength of the material, comparative tests were carried out on dumbbell-shaped specimens cut out using different cutting technologies.

2. Experiment

2.1. Test Specimens and Test Plan

The specific specimen shape and dimensions (thickness and fillet radius) (Figure 2) were adopted so that the test results could apply to connectors of the composite dowel type (Figure 1). The specimens were cut out of 10 mm thick plate S355J2N along the plate rolling direction. Three series of specimens were cut out using different cutting technologies designated as: A—water jet, B—oxyacetylene, and C—plasma.

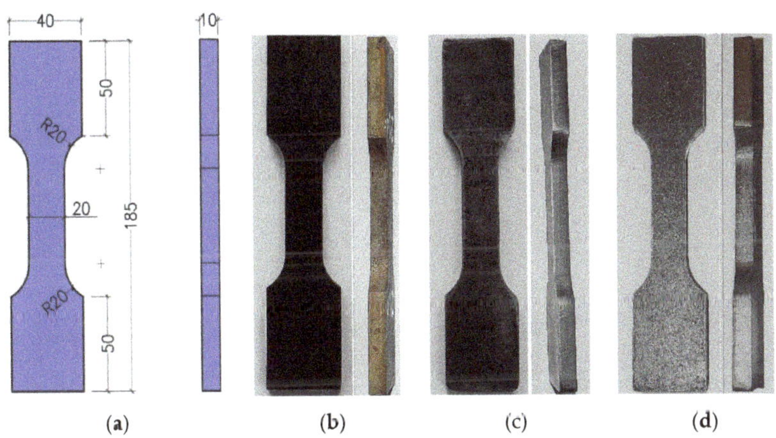

Figure 2. Specimen for comparative cyclic tests: (**a**) geometry (mm), (**b**) series A (water jet cut) specimen, (**c**) series B (oxyacetylene cut) specimen, (**d**) series C (plasma cut) specimen.

The tests were carried out on a 100 kN testing machine. The specimens were subjected to uniaxial tension-compression (R = −1) in the high cycle fatigue range until fracture. The load spectrum was sinusoidal with a frequency f = 10 Hz. Through trials, such a frequency was selected that the temperature of the specimens during tests would not exceed 60 °C. The tests were conducted in a cyclic testing machine in set stress ranges $\Delta\sigma_i$ (Table 1) on one to four specimens for each level $\Delta\sigma_i$. The number of cycles N = 5 m was set as the lifespan limit. The force signal and the total deformation signal were registered during the tests. Deformation was measured along the gauge length of 25 mm by means of an extensometer.

Table 1. Test stress ranges $\Delta\sigma_i$ for cutting technologies.

Specimen Type	Δσ (MPa)	Specimen Type	Δσ (MPa)	Specimen Type	Δσ (MPa)
A Water jet cutting	150 175 200 300 325 350	B Oxyacetylene cutting	200 225 250 300 - -	C Plasma cutting	125 150 175 200 225 -

The specimens would most often fail due to rupture in the fillet (geometric notch) area. Photographs of selected specimens after failure for each of the cutting technologies are shown below (Figure 3).

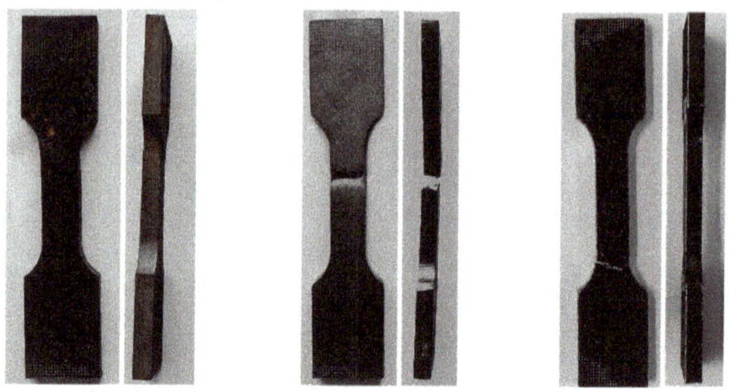

(a) (b) (c)

Figure 3. Failed specimens cut out with: (**a**) water, (**b**) oxyacetylene, (**c**) plasma.

2.2. Results

The test results are presented in Tables 2–4.

Table 2. Test results for specimens A.

Specimen No. (-)	*F* (kN)	Δσ (MPa)	N_f (Cycles)
12A 13A 18A	30	150	5,402,341 5,467,028 5,034,480
11A 16A 17A	35	175	779,609 1,209,227 2,744,983
14A 15A	40	200	536,918 336,773
1A 2A 3A	60	300	7619 9174 8339
7A 8A 9A	65	325	3722 3330 3339
5A 6A 4A 22A	70 70	350 350	2175 2225 2129 1225

Table 3. Test results for specimens B.

Specimen No. (-)	*F* (kN)	$\Delta\sigma$ (MPa)	N_f (Cycles)
13B 14B	40	200	5,476,832 3,463,974
8B 9B 10B 11B 12B	45	225	5,249,668 3,475,725 3,568,974 5,882,781 2,870,595
3B 4B	50	250	338,745 1,030,958
5B 6B 7B	60	300	13,463 32,353 15,718

Table 4. Test results for specimens C.

Specimen No. (-)	*F* (kN)	$\Delta\sigma$ (MPa)	N_f (Cycles)
12C 13C 14C	25	125	5,000,000 5,000,000 5,000,000
7C 10C 11C	30	150	5,000,000 1,714,866 430,863
3C 5C 6C	35	175	1,143,120 444,094 716,537
9C	40	200	196,487
2C 4C 8C	45	225	141,694 70,782 108,665

Figure 4 shows the test results as logarithmic stress amplitude versus the logarithmic number of cycles for the three cutting technologies. Test results regression curves for each of the technologies were calculated. The standard curves for the fatigue categories of 80 MPa and 125 MPa according to standard [16] were included for comparison.

From the regression curves for the respective cutting technologies, $\Delta\sigma_c$ stress values at N = 2 million cycles, i.e., the fatigue categories, and values of fatigue curve slope cotangent m were calculated. The results are presented in Table 5.

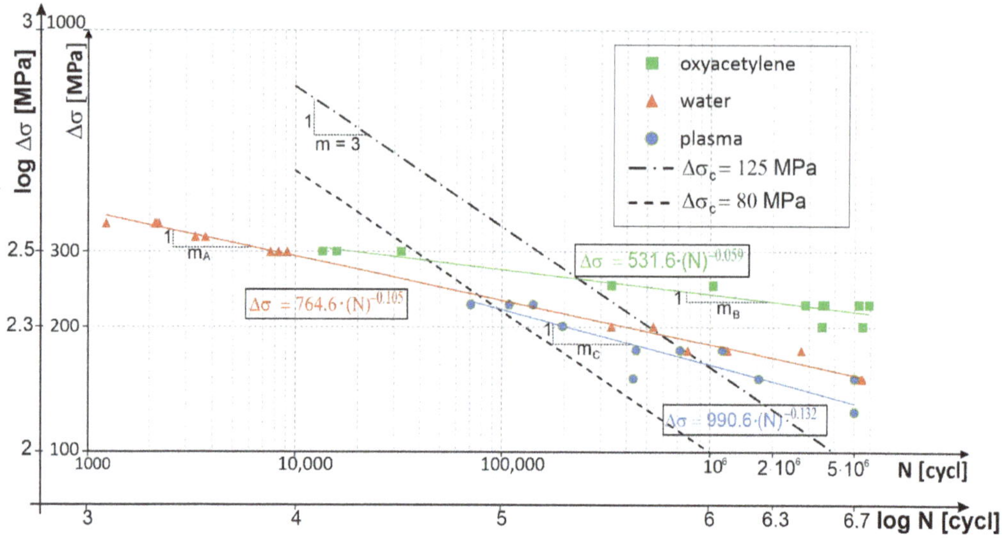

Figure 4. Results of tests carried out on specimens. (oxyacetylene, water, plasma).

Table 5. Values of $\Delta\sigma_c$ and m and regression curves for considered cutting technologies.

Cutting Technology	Regression Curve Equation	Fatigue Category $\Delta\sigma_c$ (MPa)	Fatigue Curve Slope m
water A	$\Delta\sigma = 764.6 \cdot (N)^{-0.105}$	167	10
oxyacetylene B	$\Delta\sigma = 531.6 \cdot (N)^{-0.059}$	226	17
plasma C	$\Delta\sigma = 990.6 \cdot (N)^{-0.132}$	146	8

3. Cut Edge Conditions

3.1. Macroscopic and Microscopic Examinations of Fatigue Fractures

Detailed macroscopic and microscopic analyses of selected specimens cut out with: water (2A, 14A, 17A), oxyacetylene (4B, 10B, 14B), and plasma (6C, 8C, 10C) were carried out. The specimens' faces of cut and fatigue fracture and crack surfaces (Figure 5) were examined.

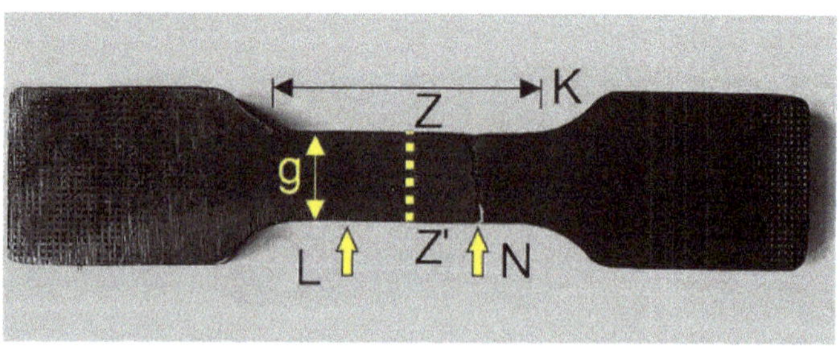

Figure 5. Test specimen: L—face of cut, N—fatigue fracture, K—area from which samples were taken to make metallurgical polished sections, Z-Z′—cross-sections on which polished sections were made, g—metallurgical polished section hardness measuring length.

The macroscopic examinations were performed under a stereoscopic light microscope, while the microscopic examinations were carried out using a confocal laser scanning microscope and a scanning electron microscope. Samples for preparing metallurgical polished sections were cut out using a precision cutter and mounted in conductive resin. The mounted samples were ground and polished on a polishing machine and subjected to etching with 5% HNO_3 solution.

The tested specimens' faces of cut showed numerous furrows resulting from cutting, along which fatigue cracks propagated (Figures 6 and 7). This is particularly visible for water jet cutting, in which case the face of the cut is distinctly varied, showing an area of the entry of the water jet with an abrasive and an exit area (Figure 8).

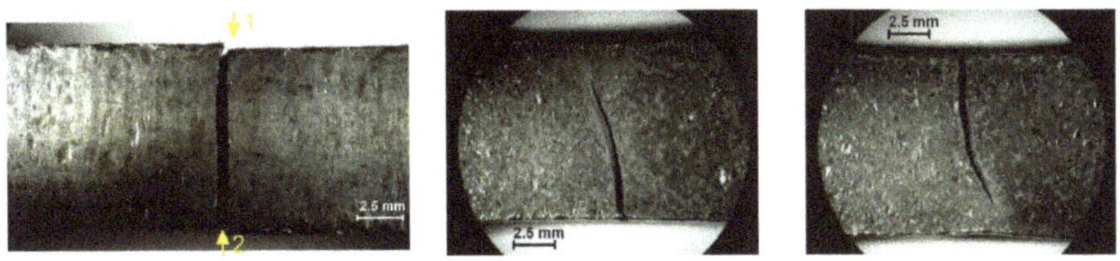

(**a**) (**b**) (**c**)

Figure 6. Fracture of specimen 6C: (**a**) face of cut made with plasma, (**b**) numerically denoted fracture sides no. 1, (**c**) numerically denoted fracture sides no. 2.

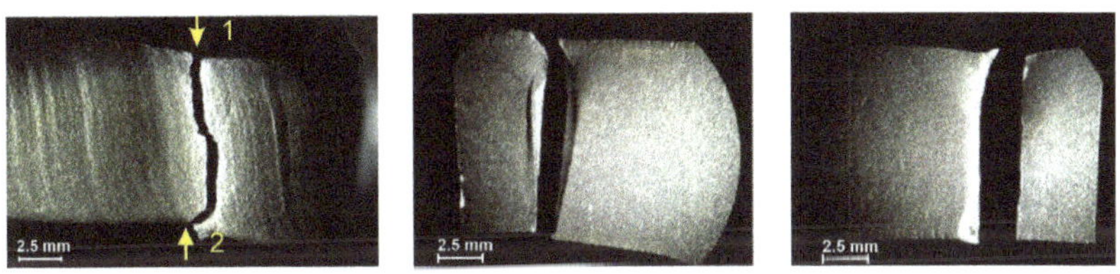

(**a**) (**b**) (**c**)

Figure 7. Fracture of specimen 14B: (**a**) face of cut made with gas, (**b**) numerically denoted fracture sides no. 1, (**c**) numerically denoted fracture sides no. 2.

(**a**) (**b**) (**c**)

Figure 8. Fracture of specimen 17A: (**a**) face of cut made with water jet, (**b**) numerically denoted fracture sides no. 1, (**c**) numerically denoted fracture sides no. 2.

The examinations of the fractured surfaces of the specimens clearly corroborated the fatigued character of the fractures, as visible in Figure 9.

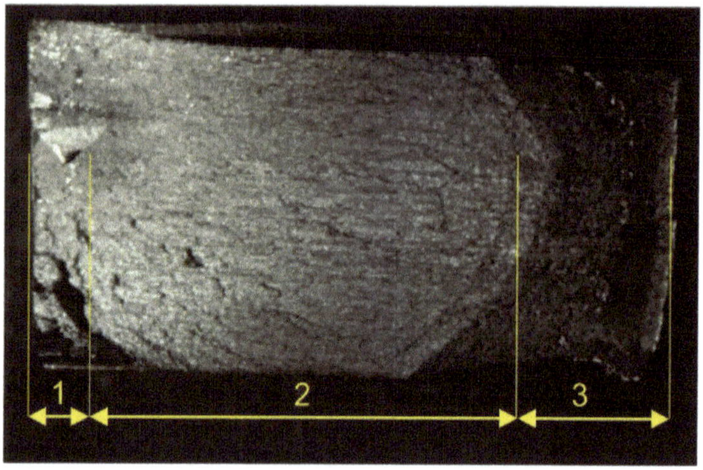

Figure 9. Fractured surface of specimen 14B: 1—brittle fracture area, 2—ductile fracture area, 3—granular area (end phase of fracture).

3.2. *Investigations of Face-of-Cut Roughness*

Roughness, i.e., the arithmetic mean deviation of the profile from the mean line, on the two faces of cut denoted in Figure 10 was investigated for each specimen. The results are presented in Table 6.

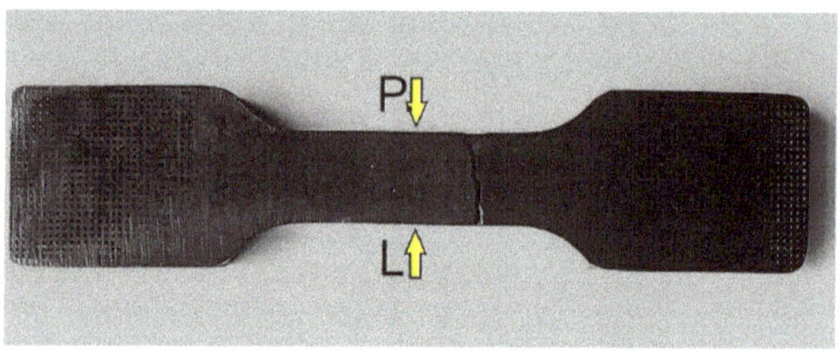

Figure 10. Denotations of surfaces subjected to roughness measurements.

Table 6. Results of roughness measurements: R_a—roughness, $R_{a,mean}$—mean roughness.

Specimen No. (-)	R_a (μm)		$R_{a,mean}$ (μm)
	L	P	
2A	3.358	3.529	3.496
14A	3.889	3.311	
17A	3.715	3.172	
4B	1.531	1.719	1.444
10B	1.158	1.166	
14CB	1.530	1.561	
6C	0.296	0.369	0.208
8C	0.314	0.269	
10C	0.355	0.329	

3.3. Microscopic Examinations of Metallurgical Polished Sections

For microscopic examinations, a sample was cut out from the measurement area of the specimens consistently with the Z-Z′ plane, parallel to the fracture surface (Figure 5). After etching with 5% HNO_3 solution, a nonequilibrium ferritic-pearlitic structure became visible (Figure 11).

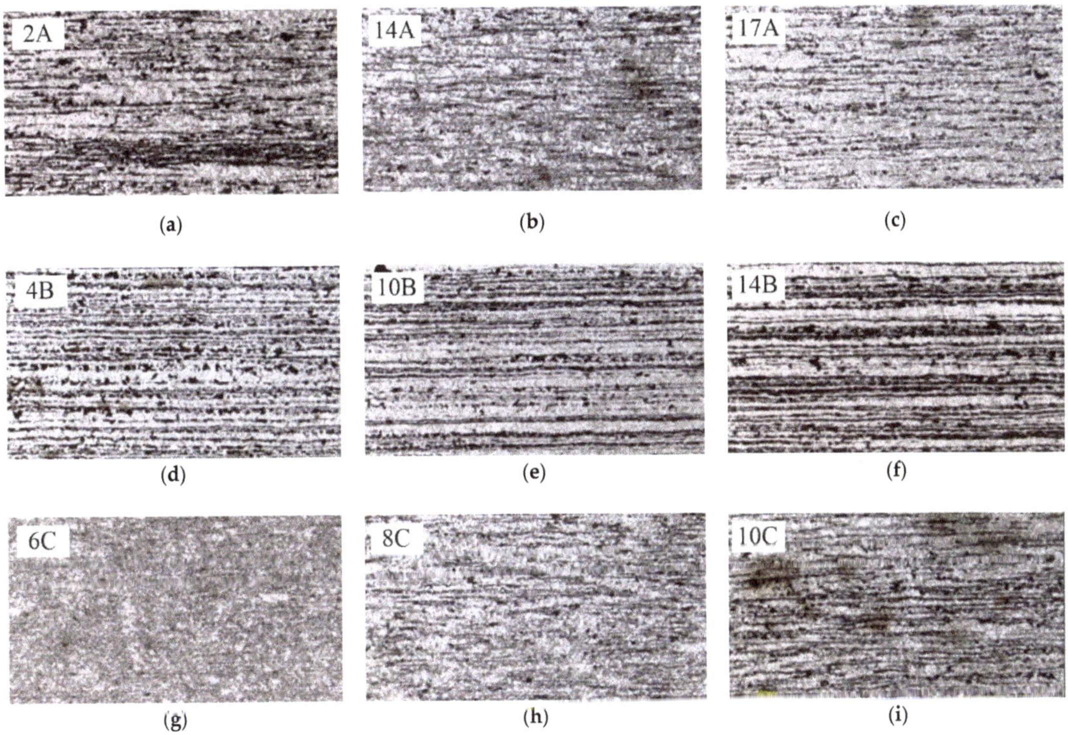

Figure 11. Microstructures of specimens: (**a**–**c**) cut out with water, (**d**–**f**) cut out with oxyacetylene, (**g**–**i**) cut out with plasma.

3.4. Investigations of Cut Heat-Affected Zone

The specimens cut out using the gas technology, and the plasma technology showed a distinctly changed structure at the cut edge (Figures 12 and 13) due to the local heating of the material. The cut heat-affected zone (CHAZ presented in Table 7) was characterized by a martensitic structure with varied carbon content. This structure, unlike the ferritic-pearlitic structure, was more brittle and susceptible to cracking (Figure 14). No significant changes in microstructure were observed in the case of the water cut specimens.

3.5. Hardness Tests

Hardness was measured using the Vickers method in accordance with the standard [17]. The measurements were performed under the load of 10 kg acting over the time of 10 s.

The hardness measurements were carried out on metallurgical polished sections along segment g (Figure 5) consistent with the pearlite-ferrite banding. The distance between the cut edge and the first impression amounted to 0.3 mm. The next impressions were spaced at every 1.5 mm. In total, 10 impressions were made for each specimen.

The hardness of the tested specimen materials ranged from 146 to 352 HV10 (Figure 15). Hardness values above 150 HV10 were measured exclusively at the edges of the specimens cut with oxyacetylene and plasma. The average hardness in these places amounted to 244 HV10 and 310 HV10 for the type B specimens and the type C specimens, respectively.

(**a**) (**b**) (**c**)

Figure 12. Structures in cut heat-affected zone in specimens: (**a**) 4B, (**b**) 10B, (**c**) 14B—oxyacetylene cutting technology.

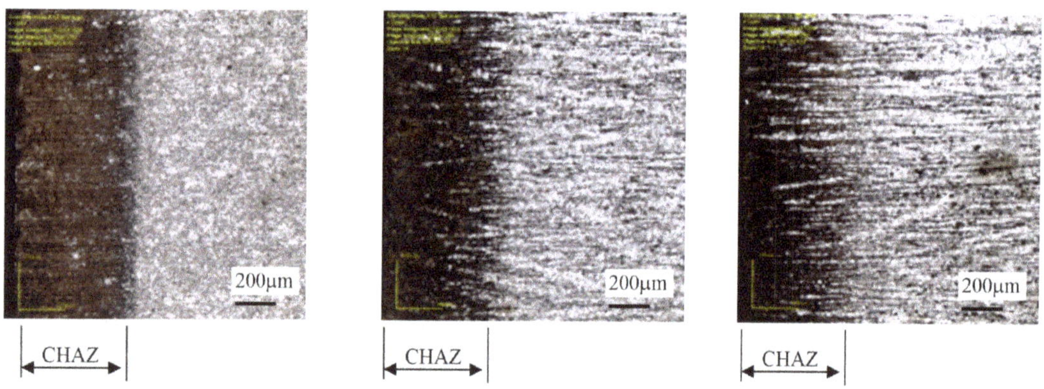

(**a**) (**b**) (**c**)

Figure 13. Structures in cut heat-affected zone in specimens: (**a**) 6C, (**b**) 8C, (**c**) 10C—plasma cutting technology.

Table 7. Depth of cut heat-affected zone (CHAZ).

Specimen No. (-)	CHAZ (µm)
4B	450
10B	550
14B	550
6C	505
8C	517
10C	475

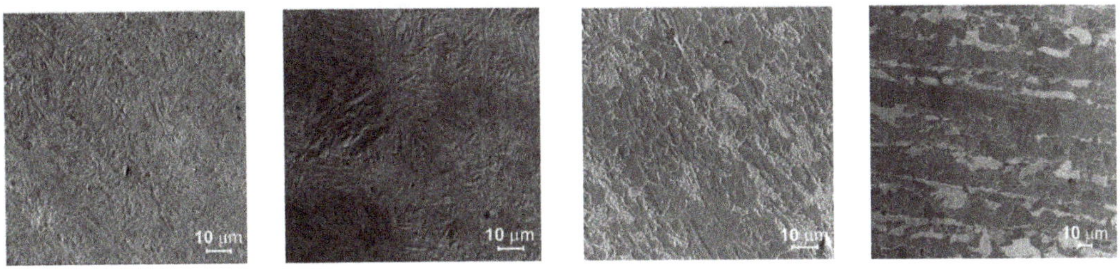

(**a**) (**b**) (**c**) (**d**)

Figure 14. Structures: (**a**) acicular structure in CHAZ in specimens cut with plasma, (**b**) acicular structure in CHAZ in specimens cut with oxyacetylene, (**c**) pearlitic-ferritic structure in specimens cut with plasma, (**d**) pearlitic-ferritic structure in specimens cut with oxyacetylene.

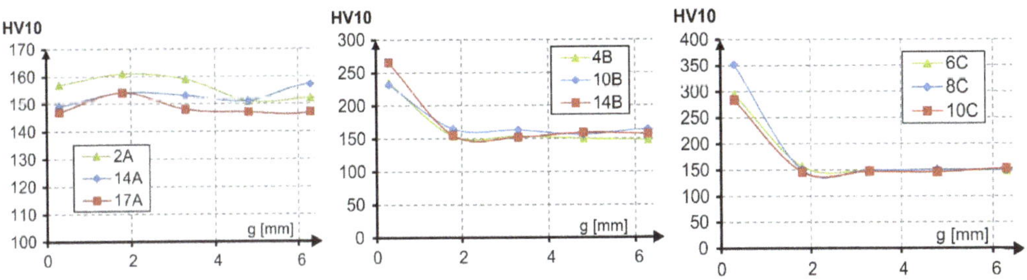

(**a**) (**b**) (**c**)

Figure 15. Hardness of type: (**a**) A specimens (water), (**b**) B specimens (oxyacetylene), (**c**) C specimens (plasma).

4. Analysis of Results

The specimens cut out using the oxyacetylene technology showed a longer fatigue lifespan than the other specimens cut out with plasma and water, as presented in Figure 6. Fatigue categories were calculated from the regression curves of the cutting technologies. The fatigue categories amounted to: $\Delta\sigma_{c,B}$ = 226 MPa for oxyacetylene cutting, $\Delta\sigma_{c,A}$ = 167 MPa for water jet cutting and $\Delta\sigma_{c,C}$ = 146 MPa for plasma cutting. The results are above fatigue category $\Delta\sigma_c$ = 125 MPa according to the work of [16], which is safely and most frequently adopted by building designers and applies to gas-cut metal plates with removed edge discontinuities. Unfortunately, in the case of series-produced connectors of the composite dowel type, each of the production operations, including the machining of the face of cut, is thought to add to the cost, and so efforts are made to reduce the latter by, i.a., limiting the additional treatments of the face of the cut. The current design guidelines according to the work of [16] do not take into account the effect of the cutting technology and the quality of the cut surface on the fatigue strength of the structure. Therefore it is necessary to clarify and specify more precisely the fatigue category for other cutting technologies, including water jet cutting and plasma cutting, which should have a beneficial effect on the design of steel frame structures.

In all the considered cases, the fatigue curve slope cotangents m were larger (m_A = 10, m_B = 17, m_C = 8) in comparison with the standard curves according to the work of [16], for which m = 3. The slope values provide information about the speed of fracture of the specimens under variable load. In the case of oxyacetylene cutting, the specimens would fracture slowest (m_B = 17), as opposed to the specimens cut out with plasma, which would fracture fastest (m_C = 8). It should be noted that the results of this test depend on, i.a., the number and shape of the specimens, the character of the fatigue load, and the yield point of the material. Therefore it is difficult to directly compare the obtained results. A convergence between the results is considered to be a satisfactory outcome. In the case of steel S355, one gets curve slope $m = 7$ for oxygen cut specimens and $m = 13$ for plasma cut specimens [11], or $m = 5.2$ for plasma, $m = 5.8$ for oxygen and $m = 16.8$ for water, as described in the work of [18]. The ongoing research confirms the conservative standard recommendations for which $m = 3$ [16]. It should be added that the obtained moderate conservatism can be proper considering that small-scale specimens, in general, ensure greater fatigue reliability than large-scale beam specimens [19].

The material tests corroborated the relatively smaller surface roughness for plasma cutting ($R_{a,mean} = 0.208$ μm) and oxyacetylene cutting ($R_{a,mean} = 1.444$ μm) than for water jet cutting ($R_{a,mean} = 3.496$ μm). The numerous furrows in the water jet entry zones in the water cut surfaces were micronotches in which the initiation of fatigue fractures would take place. It is noteworthy that the choice of cutting parameters and the thickness of the metal plate being cut have a bearing on the quality of the cut-out specimen's surface [18].

The effect of the kind of machining and the surface layer condition on the fatigue strength is expressed by surface condition coefficient β_p as a ratio of the fatigue strength of an unnotched (polished) specimen to the latter's strength after machining. The higher the surface condition coefficient, the lower the specimen's fatigue strength due to surface irregularities. Figure 16 shows the results of experiments [20] in which the effect of the kind of machining (grinding, fine rolling, coarse rolling) on the value of coefficient β_p, depending on the tensile strength, was studied. As one can see, the surface condition coefficient increases with surface roughness. Additionally, the mean surface roughness values of the faces of cut of in-house specimens of type A, B, and C for steel S355J2N with tensile strength $f_u = 510$ MPa were included in the figure.

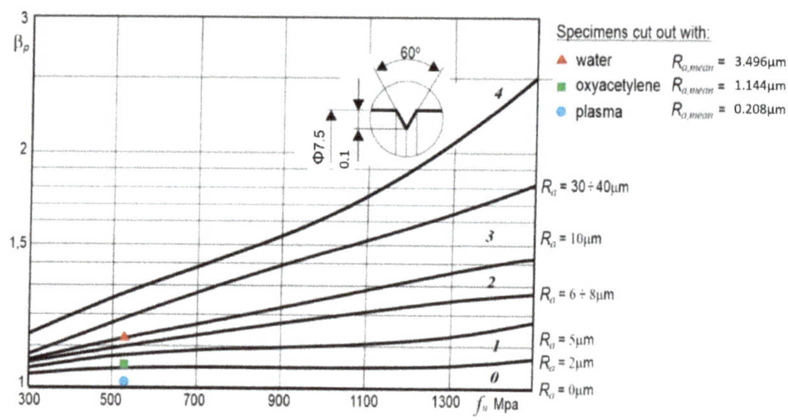

Figure 16. Effect of kind of machining on the value of surface condition coefficient β_p for tension or bending, depending on tensile strength of steel and kind of machining for: 0—polished, 1—ground, 2—fine rolled, 3—coarse rolled, and 4—sharply ring-notched (for comparison) specimens. Adapted with permission from Ref. [20], 2021, Wydawnictwo Naukowe PWN.

Table 8 contains the measured values of β_p (according to Figure 16) and the calculated fatigue categories $\Delta\sigma_c$ for each of the cutting technologies and the percentage differences relative to the results for the specimens cut out with plasma. Judging by the differences, the condition of the surface of the specimens has a relatively small (up to 12%) effect on their fatigue strength in comparison with the technologies used to cut them (55%).

Table 8. Values of β_p and $\Delta\sigma_c$ and differences between results for tested specimens.

Cutting Technology	β_p (-)	Difference between β_p Results (%)	$\Delta\sigma_c$ (MPa)	Difference between $\Delta\sigma_c$ Results (%)
Water (A)	1.13	12	167	14
Oxyacetylene (B)	1.05	4	226	55
Plasma (C)	1.01	0	146	0

The fatigue strength of the specimens cut out with oxyacetylene ($\Delta\sigma_c$ = 226 MPa) is higher than that of the specimens cut out with plasma ($\Delta\sigma_c$ = 146 MPa) even though the surface roughness after cutting with plasma is smaller than in the case of the other cutting technology. This is due to the significant effect of material hardening in the heat-affected zones. In both cases, acicular structures (Figure 12) with comparable heat-affected zone depths measured from the surface of the cut were obtained. However, in the case of plasma cutting, the hardness measured at the cut edge (310 HV10) was 27% greater than for the specimens cut out using the gas cutting technology (244 HV10). A similar agreement between the results was obtained in the tests described in the work of [18], where the specimens cut out with plasma were characterized by the highest hardness (280 HV10) and lower fatigue strength ($\Delta\sigma_c$ = 239 MPa) in comparison with the oxygen cut specimens for which hardness amounted to 190 HV10 and fatigue strength to 264 MPa.

The results of the comparative tests indicate that the gas cutting technology used so far to cut out connectors for the innovative composite dowel joint is more advantageous than the plasma cutting technology or the water cutting technology. Furthermore, oxyacetylene cutting is the cheapest and most available cutting technology in prefabrication plants.

5. Conclusions

From the results of the fatigue tests carried out on steel S355J2N specimens cut out using different cutting methods, i.e., plasma cutting, water jet cutting, and oxyacetylene cutting, the following conclusions, providing a basis for further analyses leading to the development of design guidelines for steel connectors of the composite dowel type, can be drawn:

1. The technology of cutting out dowels of the composite dowel type has a bearing on their fatigue strength. Connectors cut out using oxyacetylene cutting can have higher fatigue strength than the ones cut out using plasma cutting or water jet cutting;
2. The effect of the technology used to cut out steel connectors of the composite dowel type can be greater than that of the condition of the face of the cut;
3. The slopes of the fatigue strength curves determined for the cut-out specimens: m_A = 10 for water jet cutting, m_B = 17 for oxygen cutting, and m_C = 8 for plasma cutting, corroborate the conservative standard recommendation m = 3 according to the work of [16];
4. The FAT125 fatigue curve according to the work of [16] can be appropriate for the design of composite dowel connectors to be cut out using oxygen cutting, plasma cutting, and water jet cutting. Nevertheless, further experimental studies (the S-N curve method) need to be carried out on beam specimens of composite structures in order to verify the fatigue curve for the composite dowel connector.

Funding: This research received no external funding.

Institutional Review Board Statement: Not applicable.

Informed Consent Statement: Not applicable.

Data Availability Statement: Data sharing not applicable.

Conflicts of Interest: The authors declare no conflict of interest.

Abbreviations

The following symbols and notations, in the order in which they appear in the text, are used in this paper:

CHAZ	Cut heat-affected zone
R	Stress ratio
$\Delta\sigma$	Nominal stress range
F	Force
N_f	Number of cycles to failure
logA	Intercept of mean S-N curve
m	Slope of fatigue strength curve
R_a	Surface roughness
β_p	Coefficient of surface condition

The other symbols used in this paper are explained when they appear in the text for the first time.

References

1. Klimpel, A. *Welding, Autogenous Welding and Cutting of Metals*; Wydawnictwo Naukowo–Techniczne: Warszawa, Poland, 1999; pp. 633–654. (In Polish)
2. Seidl, G.; Popa, N.; Zanon, R.; Lorenc, W.; Kożuch, M.; Rowiński, S.; Franssen, J.-M.; Fohn, T.; Hermosilla, C.; Farhang, A.; et al. RFCS Dissemination Knowledge Project PRECO+: Prefabricated Enduring Composite Beams Based on Innovative Shear Transmission, RFS2-CT-2011-00026 "PRECO+" Design Guide. Available online: http://www.stb.rwth-aachen.de/projekte/2005/INTAB/downloadPreco.php (accessed on 21 August 2021).
3. Lorenc, W.; Kożuch, M.; Rowiński, S. The behaviour of puzzle-shaped composite dowels. Pt. 1, experimental study. *J. Constr. Steel Res.* **2014**, *101*, 482–499. [CrossRef]
4. Lorenc, W.; Kożuch, M.; Rowiński, S. The behaviour of puzzle-shaped composite dowels. Pt. 2, Theoretical investigations. *J. Constr. Steel Res.* **2014**, *101*, 500–518. [CrossRef]
5. Kożuch, M.; Rowinski, S. Elastic behaviour of the steel part of a shear connection with MCL composite dowels. *Steel Constr.* **2016**, *9*, 107–114. [CrossRef]
6. Dudziński, W.; Pękalski, G.; Harnatkiewicz, P.; Kopczyński, A.; Lorenc, W.; Kożuch, M.; Rowiński, S. Study on fatigue cracks in steel-concrete shear connection with composite dowels. *Arch. Civ. Mech. Eng.* **2011**, *11*, 839–858. [CrossRef]
7. Rowiński, S. Fatigue Strength of Steel Dowels in Innovative Shear Connection of Steel–Concrete Composite Beam. Ph.D. Thesis, Wrocław University of Technology, Wrocław, Poland, 2012; pp. 11–72. (In Polish).
8. Rykaluk, K.; Marcinczak, K.; Rowiński, S. Fatigue hazards in welded plate crane runway girders–Locations, zauses and calculation. *Arch. Civ. Mech. Eng.* **2017**, *18*, 69–82. [CrossRef]
9. Sperle, J.-O. Influence of parent metal strength on the fatigue strength of parent material with machined and thermally cut edges. *Weld. World* **2008**, *52*, 79–92. [CrossRef]
10. Remes, H.; Korhonen, E.; Lehto, P.; Romanoff, J.; Niemelä, A.; Hiltunen, P.; Kontkanen, T. Influence of surface integrity on the fatigue strength of high-strength steels. *J. Constr. Steel Res.* **2013**, *89*, 21–29. [CrossRef]
11. Stenberg, T.; Lindgren, E.; Barsoum, Z.; Barmicho, I. Fatigue assessment of cut edges in high strength steel-influence of surface quality. *Mater. Sci. Eng. Technol.* **2017**, *48*, 556–569. [CrossRef]
12. Lillemäe, I.; Liinalampi, S.; Lehtimäki, E.; Remes, H.; Lehto, P.; Romanoff, J.; Ehlers, S.; Niemelä, A. Fatigue strength of high strength steel after shipyard production process of plasma cutting, grinding and sandblasting. *Weld. World* **2018**, *62*, 1273–1284. [CrossRef]
13. García, T.; Álvarez, J.A.; Cicero, S.; Carrascal, I.; Martin-Meizoso, A. Effect of thermal cutting methods on the fatigue life of high strength structural steel S690Q. In Proceedings of the ASME Proceeding 2015 Pressure Vessel and Piping Conference, Anaheim, CA, USA, 19–23 July 2015.
14. Garcia, T.; Cicero, S.; Carrascal, I.; Madrazo, V.; Alvarez, J.A. Effect of cutting method on fatigue crack initiation and fatigue life of structural steel S355M. In Proceedings of the ASME Proceeding 2014 Pressure Vessel and Piping Conference, Anaheim, CA, USA, 20–24 July 2014; Volume 3.
15. Cicero, S.; García, T.; Álvarez, J.A.; Bannister, A.; Klimpel, A.; Martín-Meizoso, A.; Aldazabal, J. Fatigue behavior of high strength steel S890Q containing thermally cut straight edges. *Procedia Eng.* **2016**, *160*, 246–253. [CrossRef]
16. Eurocode 3: Design of Steel Structures. Part 1–9: Fatigue. 2005. Available online: https://www.phd.eng.br/wp-content/uploads/2015/12/en.1993.1.9.2005-1.pdf (accessed on 20 September 2021).
17. PN-EN ISO 6507-1:1999. Metals. Vickers hardness measurement. In *Test Method*; ISO: Geneva, Switzerland, 1999. (In Polish)
18. Diekhoff, P.; Hensel, J.; Nitschke-Pagel, T.; Dilger, K. Investigation on fatigue strength of cut edges produced by various cutting methods for high-strength steel. *Weld. World* **2020**, *64*, 545–561. [CrossRef]

19. Keating, P.B.; Fisher, J.W. Evaluation of fatigue tests and design criteria on welded details. In *National Cooperative Highway Research Program (NCHRP), Report 286*; Transportation Research Board: Washington, DC, USA, 1986.
20. Dietrich, M. *Fundamentals of Machine Construction*; Wydawnictwo Naukowo–Techniczne: Warszawa, Poland, 1999; Volume 1, pp. 319–485. (In Polish)

materials

MDPI

Article

The Influence of Heat Treatment on the Mechanical Properties and Corrosion Resistance of the Ultrafine-Grained AA7075 Obtained by Hydrostatic Extrusion

Marta Orłowska [1,*], Ewa Ura-Bińczyk [2], Lucjan Śnieżek [1], Paweł Skudniewski [2], Mariusz Kulczyk [3], Bogusława Adamczyk-Cieślak [2] and Kamil Majchrowicz [2]

1 Faculty of Mechanical Engineering, Military University of Technology, Gen. S. Kaliskiego 2, 00-908 Warsaw, Poland; lucjan.sniezek@wat.edu.pl

2 Faculty of Materials Science and Engineering, Warsaw University of Technology, Wołoska 141, 02-507 Warsaw, Poland; ewa.ura@pw.edu.pl (E.U.-B.); pskudniewski@gmail.com (P.S.); boguslawa.cieslak@pw.edu.pl (B.A.-C.); kamil.majchrowicz@pw.edu.pl (K.M.)

3 Institute of High Pressure Physics, Polish Academy of Sciences, Sokolowska 29/37 St., 01-142 Warsaw, Poland; mariusz.kulczyk@unipress.waw.pl

* Correspondence: marta.orlowska@wat.edu.pl; Tel.: +48-261-839-245

Abstract: In this paper, the corrosion resistance and mechanical properties of the 7075 aluminum alloy are studied. The alloy was deformed by hydrostatic extrusion and then aged both naturally and artificially. Results are compared with those of coarse-grained material subjected to T6 heat treatment. The aim of the research is to find the optimal correlation between the mechanical properties and the corrosion resistance of the alloy. To this end, static tensile tests with subsequent fractography, open circuit potential, and potentiodynamic polarization tests in 0.05 M NaCl were conducted. Obtained results show that a combination of precipitate hardening and a deformed microstructure leads to increased mechanical strength with high anisotropy due to the presence of fibrous grains. Plastic deformation increases susceptibility to corrosion due to the increased number of grain boundaries, which act as paths along that corrosion propagates. However, further artificial aging incurs a positive effect on corrosion resistance due to changes in the chemical composition of the matrix as a result of the precipitation process.

Keywords: 7075 aluminum alloy; hydrostatic extrusion; heat treatment; microstructure; mechanical properties; corrosion resistance

Citation: Orłowska, M.; Ura-Bińczyk, E.; Śnieżek, L.; Skudniewski, P.; Kulczyk, M.; Adamczyk-Cieślak, B.; Majchrowicz, K. The Influence of Heat Treatment on the Mechanical Properties and Corrosion Resistance of the Ultrafine-Grained AA7075 Obtained by Hydrostatic Extrusion. *Materials* **2022**, *15*, 4343. https://doi.org/10.3390/ma15124343

Academic Editor: Young Gun Ko

Received: 2 June 2022
Accepted: 18 June 2022
Published: 20 June 2022

1. Introduction

The 7XXX series of Al–Zn–Mg(–Cu) aluminum alloys (AAs) are highly important industrial materials, primarily due to their light weight, high strength, and good corrosion resistance. Such materials are commonly used in the construction of aircraft, among other uses. As a group of materials, the 7XXX alloys are particularly interesting because of the process of precipitation hardening, which causes a complex microstructure. Such a microstructure is composed of fine hardening precipitates and coarse intermetallic phases within the Al matrix [1]. For the 7XXX family, this includes Al_3Fe, Al_2CuMg, Al_2Cu, $Al_3(FeCu)$, Al_7Cu_2Fe, or Mg_2Si particles. The hardening precipitates are formed and distributed throughout the matrix via proper heat treatment: supersaturated solid solutioning followed by aging. The principal precipitation sequence that dominates hardening in most commercially used 7XXX alloys is presented in Equation (1):

$$\mathrm{SSSS}\alpha \rightarrow \mathrm{GP\ zones} \rightarrow \eta' \rightarrow \eta \quad (1)$$

where SSSSα represents a supersaturated solid solution, GP zones are Guinier–Preston zones, η' is a metastable phase (with a Mg:Zn ratio in the range from 1:1 to 1:1.15), and η is a stable $MgZn_2$ phase. $MgZn_2$ plays the most significant role in the precipitate hardening process [2].

The mechanical properties of Al–Zn–Mg–Cu quaternary alloys are determined primarily by precipitation hardening, where the nanometer-scale precipitates act as pinning centers to prevent dislocation motion [3]. Hardening is achieved via the presence of microstructural obstacles that inhibit dislocation movement and improve the overall mechanical properties of the alloy [4]. Such obstacles are also grain boundaries. Reducing the grain size increases the fraction of grain boundaries within the material and correspondingly increases the mechanical strength of the material. The application of severe plastic deformation (SPD) [5,6] leads to grain formation and achieves an ultrafine-grained (UFG) microstructure, which substantially increases mechanical strength. Following SPD, 7XXX alloys exhibited an increase in tensile strength from 228 to 434 MPa [7]. However, the greatest increases in strength are obtained when SPD is combined with precipitate strengthening. The application of hydrostatic extrusion followed by aging increases the tensile strength of AA 7475 to 700 MPa [8]. As such, the optimal microstructure to improve the mechanical strength of 7XXX alloys has a fine grain size with a uniform dispersion of small hard particles to inhibit dislocation motion.

Along with mechanical properties, corrosion resistance is also affected by both precipitates and grain boundaries. Nevertheless, there is no clear correlation, as corrosion resistance is influenced by various microstructural factors and also the environment in which the examination takes place. Therefore, determining the influence of one microstructural component on the corrosion resistance is challenging. Nevertheless, there are works devoted to these topics. In the case of grain size, susceptibility to corrosion may decrease [9]; in the case of AAs, it reduces cathodic kinetics, improves the results of mass-loss testing, and in some cases improves resistance to stress corrosion cracking [10]. Studies that suggest that corrosion resistance increases with decreasing grain size generally find that the refined microstructures are attributed to an increased readiness to passivate compared to coarse-grained (CG) materials due to a higher grain boundary density [11]. Further to grain refinement, SPD physically breaks down second-phase or intermetallic particles below a critical size, which subsequently prevents these second-phase particles from operating as efficient local cathodes and sites for the initiation of localized attack [9,12]. However, some works present an opposing view: corrosion rate may also increase as AA grain size increases [9,13]. In [14], friction stir processing was used to refine the microstructure of AA 6063. The smaller grains produced by such an approach were more susceptible to corrosion in 1 M HCl. Furthermore, Osorio et al. suggested that, when exposed to NaCl electrolytes, CG AAs showed better corrosion resistance than that of fine-grained AAs, as they had fewer corrosion initiation sites [15].

The purpose of this work is to correlate the changes in the microstructure caused by plastic deformation and subsequent heat treatment with the mechanical properties and corrosion resistance of the AA7075. However, in opposition to the previous works devoted to this topic, the approach of maintaining the size and distribution of intermetallic particles was undertaken. This is crucial in terms of corrosion resistance, as such particles are the places where corrosion attack initiates due to their electrochemical properties; therefore, their influence is the most significant. In the present study, due to preserving the size of the intermetallic particles, other structural components were investigated, i.e., grain boundaries and second-phase particles. As 7XXX AAs have a complex microstructure that is determined by thermomechanical processing, the primary goal of this study is to determine the optimal heat treatment parameters of ultrafine-grained AA 7075, obtained by hydrostatic extrusion. The aim of the study is to increase the mechanical strength of AA 7075 while maintaining its high corrosion resistance.

2. Materials and Methods

The research subject is commercially available AA 7075, with its chemical composition given in Table 1, delivered in the form of rods with a diameter of ∅20. The manufacturer of the alloy was Kamensk Uralsky Metallurgical Works. For the purpose of this study, the samples underwent different combinations of heat treatment and SPD. The SPD method used was hydrostatic extrusion [16]. In this approach, the billet is located in the container and surrounded with a pressure transmitting medium. The piston compresses the medium until the billet starts to extrude through the die. The negligible friction between the billet and the die allows for the use of small die angles that ensure that the high deformation is homogeneous. The unique features of hydrostatic extrusion are three-axial compressive stresses within the billet and high strain rates greater than 10^4 s^{-1}. In comparison to standard SPD methods, as a means of grain refinement, this process requires significantly smaller total strain. In this study, hydrostatic extrusion was performed in a single step with a reduction in the sample diameter from ∅20 to ∅10. The equivalent strain of hydrostatic extrusion is given by Equation (2):

$$\varepsilon = 2ln\left(\frac{d'}{d''}\right) \tag{2}$$

where d' is the initial diameter of the rod, and d'' is the exit diameter. The equivalent strain following the single-step hydrostatic extrusion process that we used was approximately 1.4.

The following list of samples were examined:

- CG, a coarse-grained precipitation-strengthened sample used as a reference material, solution heat-treated with T6 aging following hydrostatic extrusion to maintain the size and location of intermetallic inclusions.
- HE, an ultrafine-grained sample naturally aged for 180 days following hydrostatic extrusion.
- HT1, an ultrafine-grained sample precipitation strengthened following hydrostatic extrusion (artificially aged at 100 °C for 24 h).
- HT2, an ultrafine-grained sample precipitation strengthened following hydrostatic extrusion (artificially aged at 120 °C for 24 h).

For the aging process, an SUP-30G furnace from WAMED Co. was used, which has very high certainty of maintaining the temperature, i.e., ±0.2 °C.

Table 1. Chemical composition of AA 7075.

Element	Zn	Mg	Cu	Cr	Ti	Si	Fe	Mn	Al
Content (wt. %)	5.70	2.40	1.50	0.19	0.04	0.09	0.23	0.06	balanced

The microstructure and postcorrosion morphology of the samples were investigated using a Hitachi Su70 scanning electron microscope (SEM). The surface corrosion attack was observed with samples polarized up to 0.1 mA/cm^2, and the postcorrosion morphology (the number and shape of pits) of the sample cross-sections was examined. The depth and morphology of the pits were evaluated on the longitudinal cross-section of the sample, oriented along the grains in the direction of hydrostatic extrusion). Energy-dispersive X-ray spectroscopy (EDS) was used to perform the elemental analysis and chemical characterization of the intermetallic particles present in the microstructure. Electron backscatter diffraction (EBSD) was used to examine the crystallographic orientation of grain and grain boundary characteristics. We determined the crystallographic orientations of the grains, the grain size, the fraction of high angle grain boundaries (HAGBs), and the fraction of low angle grain boundaries (LAGBs). The data collection step size was 250 µm.

For detailed microstructure characterization, a JEOL JEM-1200 transmission electron microscope (TEM) operating at 120 kV was used. TEM samples were sectioned along the transverse and longitudinal axes of the rods. Thin TEM foils were prepared by grinding slices down from a thickness of 1 mm into a thickness of 150 µm, and then electropolishing them using a solution of Struers electrolyte A2 at 5 °C and 25 V. To complement the TEM investigations, we also used X-ray diffraction (XRD). To correctly identify the visible precipitates on the TEM samples, XRD measurements were taken using a Bruker D8 Advance diffractometer with Cu Kα radiation. Specimens were rotated at 10 rpm during the diffraction recording. Diffraction lines were matched to the pattern using EVA V3.0 software.

The mechanical properties of the processed samples were determined by tensile tests. For each sample, five measurements were taken. Static tensile tests were carried out on flat minisamples [17,18] with cross sections of 0.6 $\times$ 0.8 mm and gauge lengths of 5 mm. The tests were performed at room temperature at an initial strain rate of 1×10^{-3} s^{-1}, using digital image correlation for noncontact strain measurements. We determined quantitative average values for ultimate tensile strength (UTS), 0.2% offset yield strength (YS), and elongation at break (E_b). Due to the size of the examined tensile test samples, measurements were taken from two directions, longitudinal and transverse, with respect to the extrusion direction. This approach allowed for the anisotropy of the mechanical properties to be examined. After the tensile tests, fractography was performed on an SEM.

Corrosion resistance was investigated via electrochemical experiments. Samples were subjected to cyclic polarization in potentiodynamic (PP) mode. Before each test, each sample was immersed in the prepared electrolytes for 10 min to stabilize the rest potential. Additionally, the change in open circuit potential (E_{OCP}) was observed for 48 h to determine the behavior of the material in the corrosive environment over an extended period of time. To produce homogeneous, comparable surfaces and ensure the presence of a repeatable passivation layer, all samples were ground by silicon carbide water grinding abrasive paper, with grades of up to #4000, and then placed within a desiccator for 24 h prior to corrosion testing. All tests were performed at room temperature, with a corrosive medium of Cl– ions within an aqueous environment. To this end, a solution of 0.05 M NaCl using distilled water was prepared. A NOVA AutoLab PGSTAT302N potentiostat was used to control the potential between the working electrode (sample) and the Ag/AgCl reference electrode while measuring the current between the working electrode and the platinum sheet counter electrode. The surface of each sample was sealed in plastic, with a 20 mm^2 area left exposed. Corrosion tests were conducted on the surfaces perpendicular to the hydrostatic extrusion direction. The cyclic PP with a 1 mV/s scan rate started from -0.05 V relative to the E_{OCP}, and this was reversed when the current had reached a value of 0.3 mA. These measurements were repeated three times to ensure the reproducibility of the results. Average corrosion parameters, including corrosion potential (E_{corr}), corrosion current density (i_{corr}), and repassivation potential (E_{rep}), were obtained from the curves.

3. Results

3.1. *Microstructure*

3.1.1. SEM/EBSD

Maps of the grain boundaries of CG and HE samples are presented in Figure 1. Both the transverse and longitudinal planes are shown with respect to the extrusion direction. For the latter, the extrusion direction is indicated by black arrows. There are noticeable differences between the samples. In the transverse plane, the CG sample contained a large amount of HAGBs, with a much smaller number of LAGBs distributed throughout. The grains were equiaxial with an average grain size of 17.1 μm. The largest grain exceeded 40 μm in size. The hydrostatic extrusion process caused significant grain refinement, resulting in an average grain size of 1.3 μm (including subgrains) for the HE sample. This difference was primarily caused by a considerable increase in the density of LAGBs: the CG sample contained 5.56% LAGBs, while the HE sample contained 76%. LAGBs are predominantly located in the grain interiors, creating a network of subgrains surrounded by the HAGBs which delineate and separate each grain.

In the longitudinal plane, HAGBs lay parallel with one another in accordance with the extrusion direction, creating strongly elongated grains. The heat treatment of the CG sample resulted in a larger average grain size of 17.6 μm. Nevertheless, the elongated shape was preserved. Within the grains, a small number of LAGBs were observed. For the HE sample, a network of LAGBs existed within the interiors of significantly elongated grains, which were separated by the HAGBs. LAGBs propagated both perpendicular and parallel to the hydrostatic extrusion direction. The average grain size of the HE sample within the longitudinal plane was 1.7 µm. The length of the lamellas defined by the HAGBs exceeded 75 µm, while their maximal thickness did not exceed 10 µm.

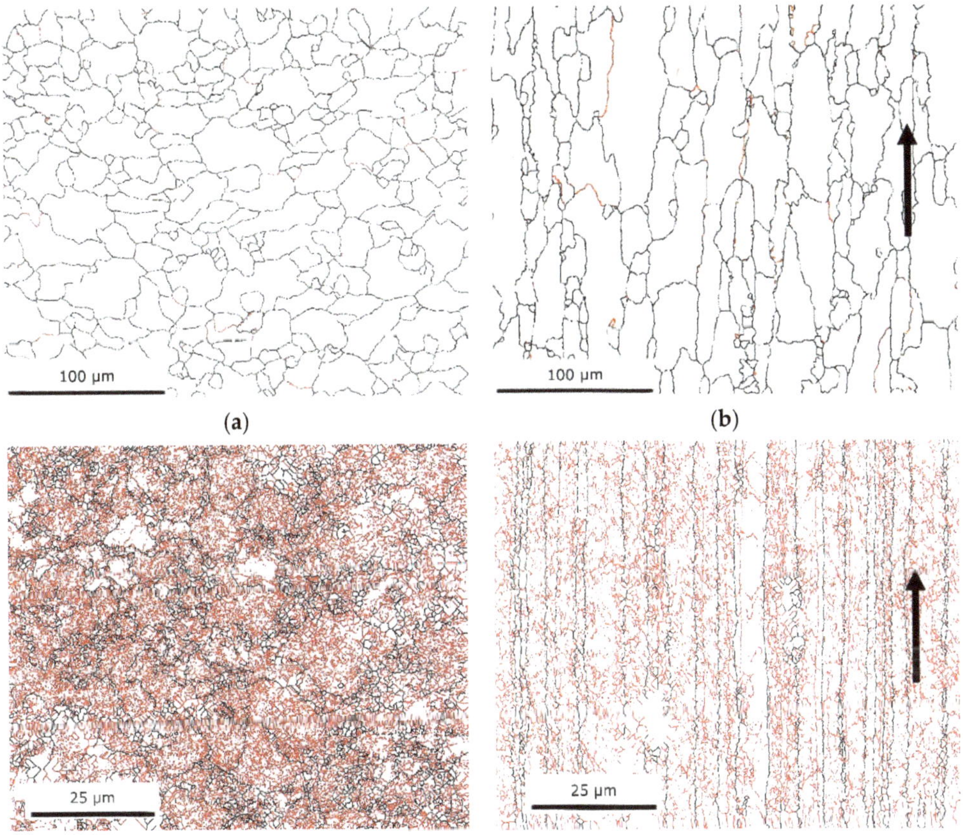

(a) (b) (c) (d)

Figure 1. Grain boundary maps, with LAGBs shown in red, and HAGBs shown in black: (**a**) CG transverse plane, (**b**) CG longitudinal plane, (**c**) HE transverse plane, and (**d**) HE longitudinal plane; black arrows indicate the direction of the hydrostatic extrusion.

The crystallographic orientation of the grains was also analyzed. Figure 2 shows grain orientation maps of the transverse sample planes. The inverse pole figures for each sample are shown as insets. A comparison of these maps shows that a strong texture developed following hydrostatic extrusion. For the CG sample, the grain orientation was scattered, with preferred orientations of <111>, <112>, and <012>. However, the maximal measured intensity was only 1.68. For the HE sample, two distinct dominant crystallographic orientations could be identified: <001> and <111>. Moreover, the maximal intensity of 5.48 was larger than that of the CG sample.

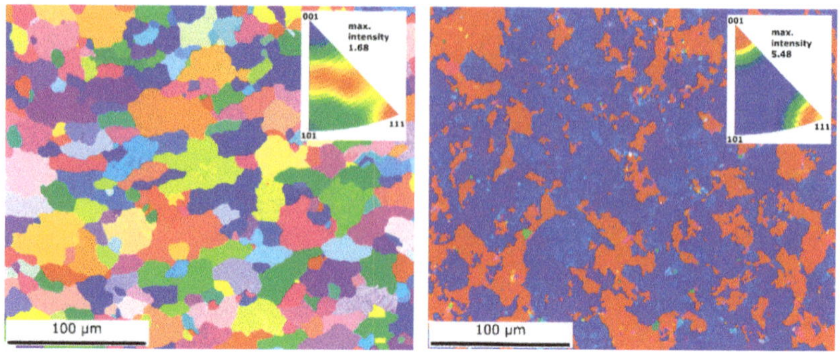

(**a**) (**b**)

Figure 2. Orientation maps of (**a**) CG and (**b**) HE samples.

3.1.2. XRD

Samples were analyzed using XRD to identify the phases. The corresponding XRD patterns are presented in Figure 3. XRD peaks could be identified mainly as ∝ Al solid solution (labeled as Al), η-precipitates (labeled as $MgZn_2$), and also some peaks as $MgCuAl_2$ and Al_3Cu_2. In general, the XRD peaks were slightly stronger for the CG sample than those for the other samples. However, the XRD diffraction peaks originating from the η precipitates were weak for all samples. Following hydrostatic extrusion, a strong texture caused the incomplete detection of peaks originating from the Al solid solution. The nonequilibrium phase (η′) was also detected, but the extremely small size of the η′-phase precipitates resulted in diminished XRD peaks and significantly complicated their proper description. Furthermore, for samples that had undergone hydrostatic extrusion, XRD graphs were very similar, with the same identified constituents.

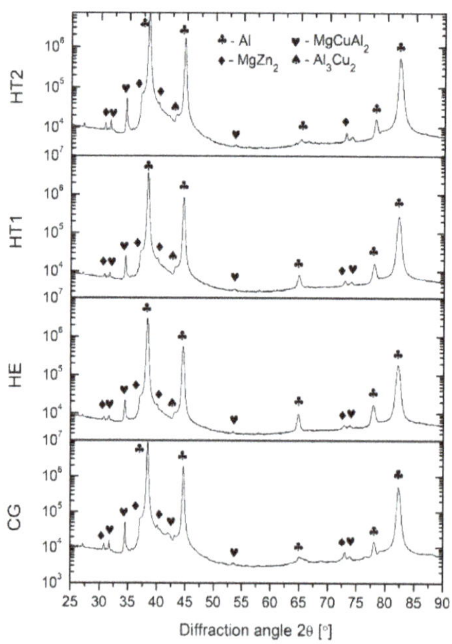

Figure 3. XRD patterns of examined samples (PDF no. of identified phases: PDF 00-004-0787—Al, PDF 01-073-5874—$MgCuAl_2$, PDF 00-034-0457—$MgZn_2$, PDF 01-071-5716—Al_3Cu_2).

3.1.3. TEM

TEM bright-field images of the CG sample are presented in Figure 4a,b. The microstructure of the CG sample consisted of equiaxed grains with finely dispersed η-precipitates ($MgZn_2$). The majority of such precipitates were located within the grains; a small fraction were located at grain boundaries. Much smaller nonequilibrium precipitates (η′) were homogeneously distributed in the grain interiors. These precipitates are the blurry disc- or rod-shaped dots in the background of the images. Figure 4c,d show that the microstructure of the naturally aged HE sample was significantly deformed. The dark grain areas likely indicate the presence of a high density of dislocations. This suggests that the microstructure obtained via the hydrostatic extrusion process was highly deformed, with significant stress applied to the material that correspondingly resulted in the generation of dislocations. Compared to the CG sample, the number and size of η precipitates visible within the grains were significantly reduced for the HE sample. However, those that were observed were homogeneously distributed within the microstructure. Moreover, the precipitates were also present at the grain boundaries. Fine metastable precipitates (η′) could also be observed within the grains.

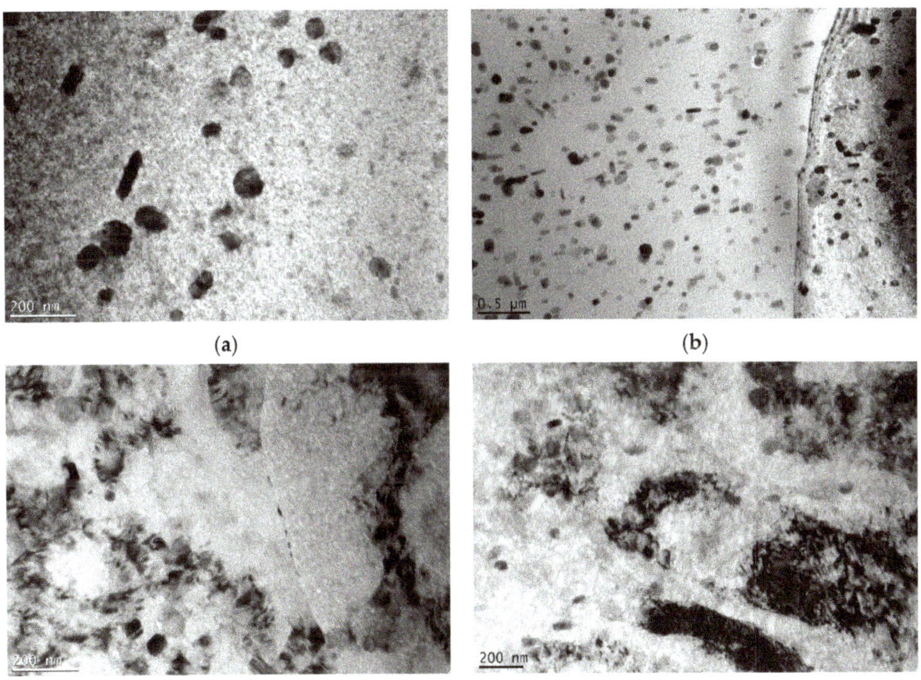

(c) (d)

Figure 4. TEM micrographs of the samples: (**a**) transverse plane of CG, (**b**) longitudinal plane of CG, (**c**) transverse plane of HE, and (**d**) longitudinal plane of HE.

TEM micrographs of samples HT1 and HT2 are presented in Figure 5. Regarding the HE sample, a deformed microstructure with a high density of dislocations could be observed. Elongated grains were visible within the longitudinal plane, with precipitates formed both within the grains and on the grain boundaries. The grains were elongated along the extrusion direction, which is marked by white arrows. The η precipitates of the HT1 and HT2 samples were much smaller than those of the CG sample, but larger than those of the HE sample, indicating that an increased aging temperature increases the size of η precipitates within the microstructure following hydrostatic extrusion. Moreover, the numbers of precipitates in the HT1 and HT2 samples were increased in comparison to those in the HE sample. As the aging temperature increased, the size of the precipitates at the grain boundaries increased. The HT1 sample had slightly larger η precipitates than those of the HT2 sample. For both samples, η′ precipitates were also visible.

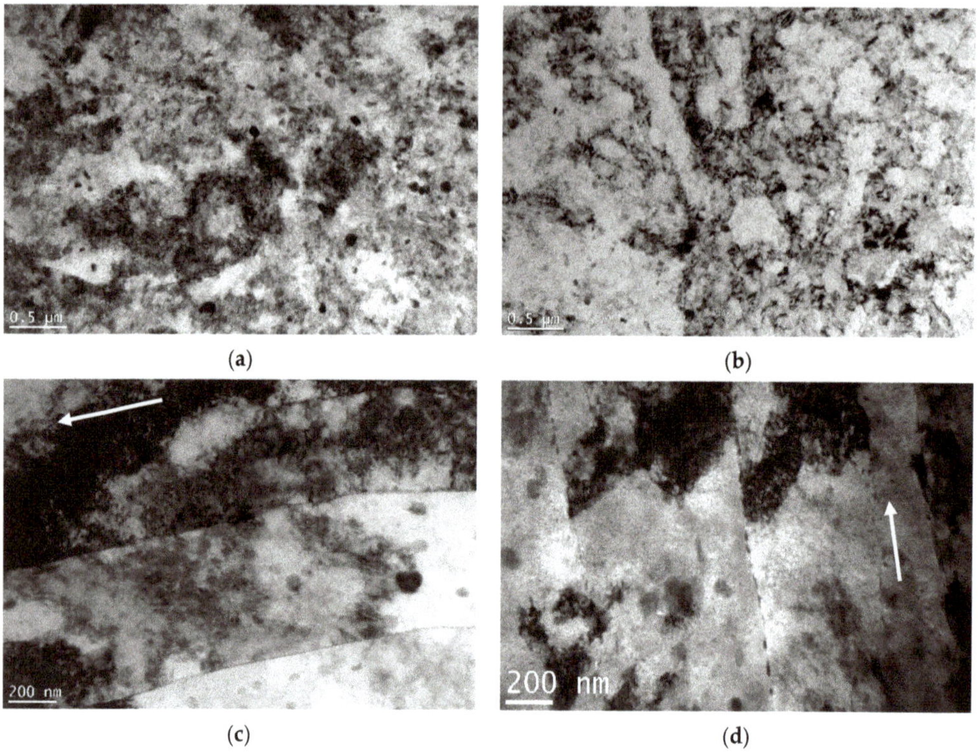

Figure 5. TEM micrographs of the samples: (**a**) HT1 transverse plane, (**b**) HT2 transverse plane, (**c**) HT1 longitudinal plane, and (**d**) HT2 longitudinal plane; white arrows indicate the direction of the hydrostatic extrusion.

3.1.4. SEM/EDS

The intermetallic particles were examined using SEM/EDS. Figure 6 presents the EDS mapping results, showing the chemical composition of the intermetallic particles (inclusions) within the microstructure. As the CG sample was heat-treated following hydrostatic extrusion to obtain a coarse-grained microstructure, there was no noticeable difference between each sample in the shape and size of the particles. Results demonstrate that the majority of the analyzed particles consisted of Al, Cu, and Fe, with a smaller amount of Mn and a very small amount of Cr. The particles were comparatively large and irregularly shaped, with sizes of 1–10 μm. The particles were formed during alloy solidification and did not dissolve during subsequent thermomechanical processing [19]. From the literature, the particles that occur most commonly in AA 7075 and correspond with the compounds that we identified are Al_7Cu_2Fe, Al_2CuMg, Al_2Cu, and Al_3Fe [20,21]. Compounds that contain Cu and Fe are nobler than pure Al, resulting in the dissolution of more active matrix and leading to the appearance of rings around a nearly intact particle or particle colony [22]. This phenomenon was observed later, during the postcorrosion damage evaluation.

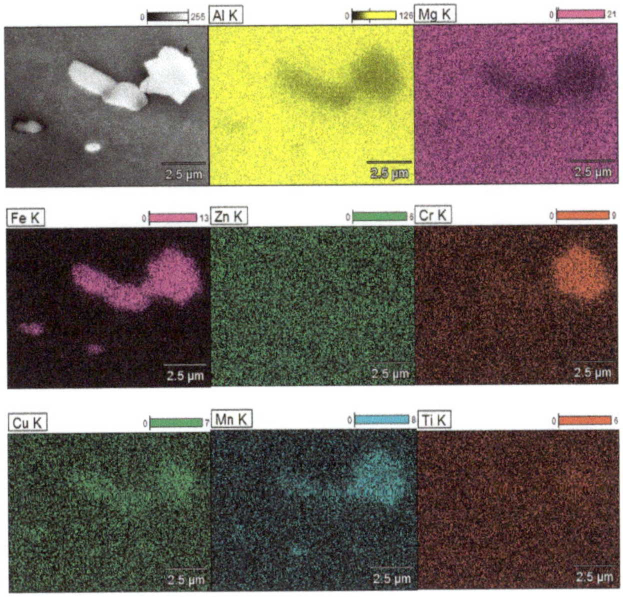

Figure 6. EDS mapping results of inclusions within AA 7075.

3.2. Mechanical Properties

The representative stress–strain curves are shown in Figure 7, while the average results of YS, UTS, and E_b tests are presented in Figure 8. The values of UTS, YS, and E_b are shown for the tests performed in the longitudinal and transverse directions with respect to the extrusion direction. The CG sample had a UTS of approximately 530 MPa and a YS of 460 MPa. Slightly higher values were obtained in the longitudinal direction. Moreover, E_b was approximately 4% larger in the longitudinal direction than it was in the transverse direction. Following both hydrostatic extrusion and heat treatment, mechanical strength increases and elongation decreases. Noticeable differences were observed on the two measured directions. The highest value of UTS was obtained for the HE sample in the longitudinal direction; in the transverse direction, HE, HT1, and HT2 all displayed similar values. The value of YS gradually increased as the aging temperature increased. The opposite relationship existed between E_b and aging temperature. A noticeable difference in results obtained for the two measured directions could be observed. The values of all three parameters were larger in the longitudinal direction. For the CG sample, the differences were relatively small—approximately 10 MPa. However, the remaining samples displayed much larger differences, in the range of 80–140 MPa for YS and 60–100 MPa for UTS. The highest anisotropy of mechanical strength was observed for HE, while it was reduced after subsequent heat treatment. The work hardening capacity, which is defined as the ratio of UTS to YS, was the largest for the HE sample and gradually decreased for a higher aging temperature.

The fracture surfaces of the samples are shown in Figure 9. Images were taken from the tensile specimens cut from the transverse direction in relation to the extrusion direction. In each case, the fracture was ductile and with a transcrystalline character. The main observed feature was the presence of numerous equiaxial dimples. The CG sample was larger than the samples after hydrostatic extrusion. In the case of the aging process, there was no correlation. This indicates that the size and subsequently the number of dimples were connected with grain size and dislocation density because these were the sites of the nucleation of the cavities, which may result in the formation of dimples. In [23], a decrease in grain size also led to a decrease in dimple size.

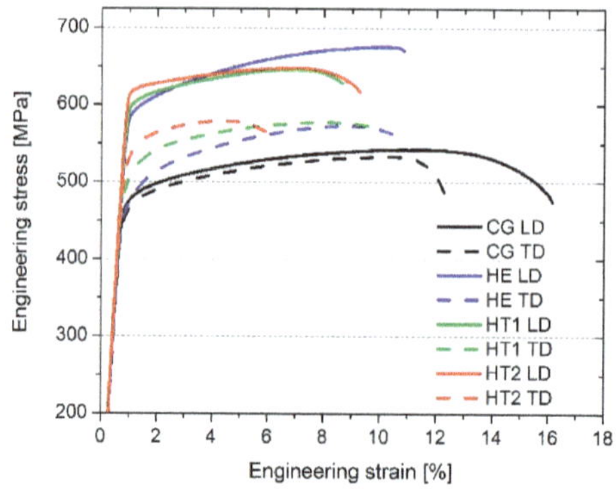

Figure 7. Representative stress–strain curves of the samples.

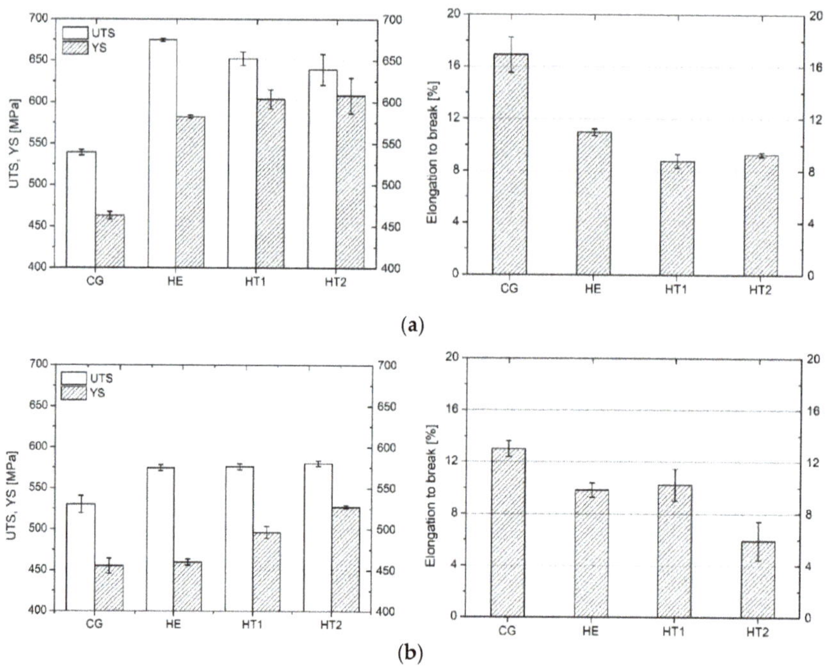

Figure 8. Values of UTS, YS, and E_b for each sample, measured in the (**a**) longitudinal and (**b**) transverse directions with respect to the extrusion direction.

3.3. Corrosion

3.3.1. Electrochemical Properties

Figure 10 shows the E_{OCP} over time for each sample during their 48 h immersion in 0.05 M NaCl. The values of E_{OCP} sharply increased, followed by a steady decrease over time. The only exception was the HE sample, which maintained a very unstable E_{OCP} level that oscillated between −0.70 and −0.67 V/Vref, producing an irregular line on the graph. For the HT2 sample, the E_{OCP} value stabilized at approximately −0.75 V/Vref, following 14 h of a sharp decrease. For the HT1 sample, the E_{OCP} value decreased steadily over almost 35 h before stabilizing at −0.72 V/Vref. Similarly, the E_{OCP} of the CG sample stabilized at −0.70 V/Vref following a gradual decrease. After 48 h, the HT2 sample displayed the lowest E_{OCP} value of approximately −0.75 V/Vref, and the HE sample displayed the highest value of approximately −0.67 V/Vref.

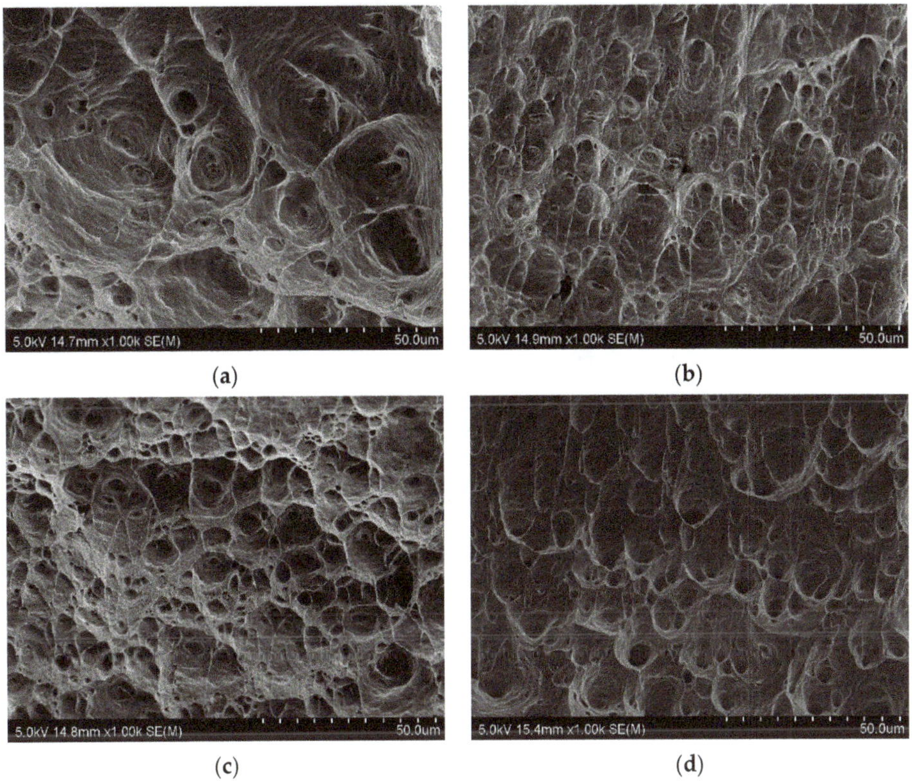

Figure 9. Fractures of the tensile samples cut from the transverse direction: (a) CG, (b) HE, (c) HT1 and (d) HT2.

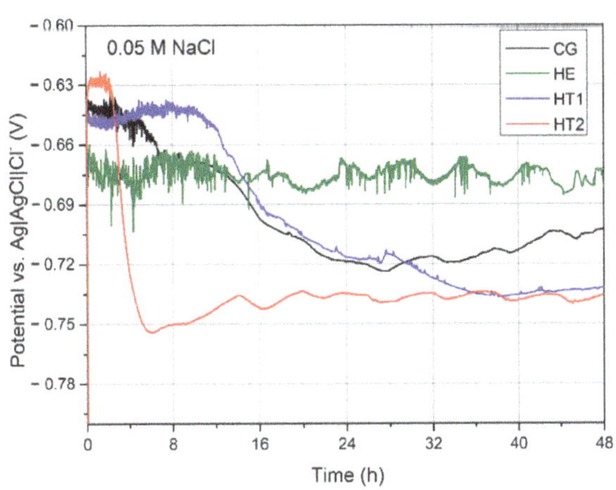

Figure 10. E_{OCP} over time for the CG, HE, HT1, and HT2 AA 7075 samples during 48 h immersion in 0.05 M NaCl.

Figure 11 shows representative cyclic PP curves during 600 s of E_{OCP} stabilization in 0.05 M NaCl solution. The electrochemical data obtained from the curves are listed in Table 2. During the forward scan, the shape of each curve was similar: a rapid increase in current density once corrosion potential (E_{corr}) had been reached. This indicates that the anodic behavior was dominated by active dissolution. The cathodic current density during the forward scan was the lowest for the HE sample, higher for the aged HT1 and HT2 samples, and highest for the CG sample. This may indicate the acceleration of the oxygen reduction reaction. The b factor, which describes the inclination of the cathodic branch, was the highest for the CG sample and decreased with increasing aging temperature. The lowest E_{corr} was observed for the HE sample. This shifted to more anodic values following aging. Moreover, artificial aging lowers the corrosion current density (i_{corr}), which was the lowest for the HT2 sample. This indicates that artificial aging at elevated temperatures reduces susceptibility to localized attacks. During the reverse scan, the HE curve significantly differed in shape. The hysteresis loops and repassivation potentials were somewhat similar for the aged and recrystallized samples. A large hysteresis and much lower repassivation potential (−780 mV/Vref) were recorded for the HE sample. This suggests differences in repassivation, and consequently in the morphology of the corrosion attack [24].

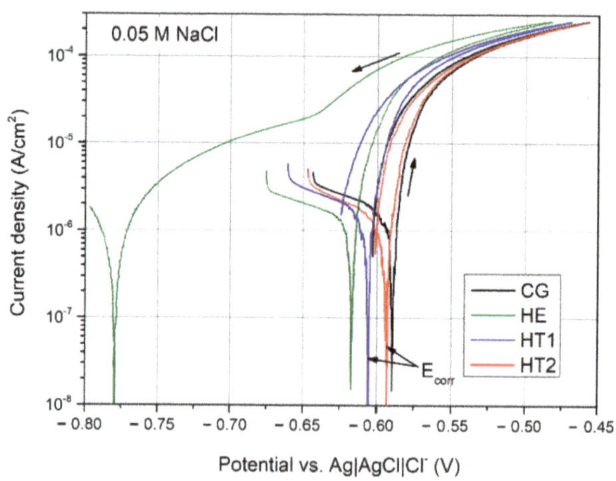

Figure 11. PP curves of the CG, HE, HT1, and HT2 AA 7075 samples following 600 s of E_{OCP} stabilization in 0.05 M NaCl solution.

Table 2. Average values of corrosion potential (E_{corr}), corrosion current density (i_{corr}), and repassivation potential (E_{rep}), b factor calculated from the potentiodynamic polarization curves.

Sample	E_{corr}, mV/Ref	i_{corr}, μA/cm^2	E_{rep}, mV	b Factor, V/dec
CG	−590 ± 5	0.72 ± 0.08	−599 ± 3	0.048
HE	−615 ± 4	0.78 ± 0.13	−780 ± 3	0.039
HT1	−604 ± 4	0.66± 0.21	−617 ± 6	0.024
HT2	−592 ± 3	0.50 ± 0.05	−600 ± 1	0.024

A comparison of the results of PP to the literature data for AA7075 shows that a similar course of the curves was observed as that in [25]; however, due to the lower pH of the 3.5% NaCl solution, the observed PP curves were shifted to less noble values, and a breakage potential was also observed. Similar findings in the case of E_{corr} were observed for AA7075 examined in 3.5% NaCl [26], but the values of i_{corr} were in the same range as that in the present study. An abrupt increase in a current density was observed here after reaching E_{corr}, which indicated rapid dissolution. AA7075 was also investigated in [27], where 0.001 M NaCl was used for corrosion tests. The obtained average values of E_{corr} were higher for about 50–100 mV, but the scattering of the results was more pronounced in comparison to the results obtained in the present study. Similar values of i_{corr} were also observed.

3.3.2. Postcorrosion Morphology

Figure 12 presents the postcorrosion morphology following the PP tests. Corrosion products were detected on the surface of several samples. The most substantial differences between the samples were the depths and shapes of the pits rather than the number of pits present. Images of the cross-sections are shown in Figure 13. CG and HE samples had the deepest pits of approximately 100 and 130–150 µm, respectively. The HT1 and HT2 samples had the shallowest pits, of depth 50–70 µm. Although the pits were relatively narrow, for samples that had undergone hydrostatic extrusion were located in close proximity to one another, creating a network of pits and creating a larger region of degradation. The pit locations appeared to be correlated with grain shape and elongation in addition to the locations of the intermetallic particles. The hydrostatic extrusion process tends to align constituent particles into bands within the alloy, but no significant difference in their sizes was noted. At multiple locations, the pitting and initial corrosion damage was located close to the intermetallic inclusions, and propagated along the intermetallic particles. Moreover, the intermetallic particles were frequently observed to be intact, with corrosion occurring at the particle–matrix interface, resulting in the dissolution of the adjacent material (matrix).

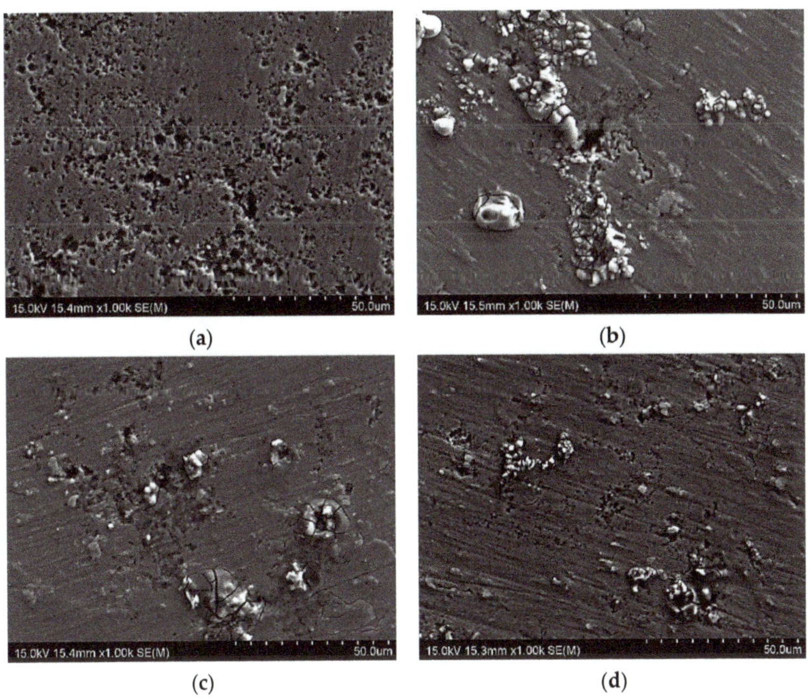

Figure 12. Sample surfaces following potentiodynamic polarization in 0.05 M NaCl: (**a**) CG, (**b**) HE, (**c**) HT1, and (**d**) HT2.

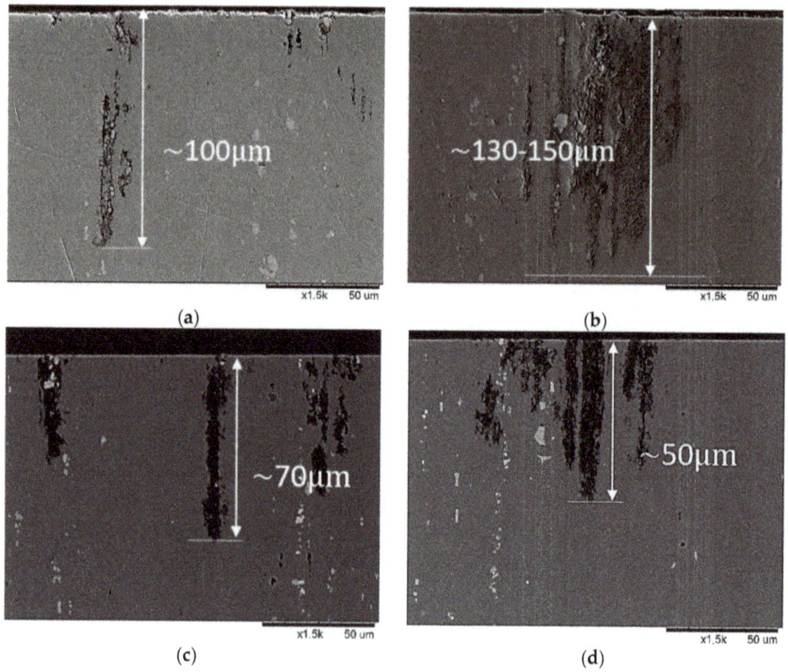

Figure 13. Sample cross-sections following potentiodynamic polarization in 0.05 M NaCl: (**a**) CG, (**b**) HE, (**c**) HT1, and (**d**) HT2.

4. Discussion

4.1. Microstructural Evolution through Hydrostatic Extrusion with Subsequent Aging

Microscopic observations of the material following hydrostatic extrusion reveal that the microstructure was severely deformed, with the grains significantly elongated (fibrous) along the extrusion direction. Microstructural evolution in Al during hydrostatic extrusion had been demonstrated [28] with three presented mechanisms of HAGB formation. The first mechanism is based on the formation of new grains as a result of dynamic recrystallization, the second describes continuous grain rotation as a result of mobile dislocation absorption into grain boundaries, and the third concerns neighbor switching due to slip banding. In general, the fraction of HAGBs increases with increasing strain. In this study, an applied equivalent strain was estimated at $\varepsilon = 1.4$. This relatively small strain resulted in the formation of a high density of dislocations. Higher strains would reduce the dislocation density and increase the misorientation angles of the grain boundaries. EBSD analysis showed that the samples had reached an average grain size of approximately 1.3 µm on the transverse plane following hydrostatic extrusion. In comparison, the reference sample displayed an average grain size of 17 µm. This shows that hydrostatic extrusion is an efficient means of grain refinement and, at this strain level, the formation of subgrain structures. The EBSD investigation further revealed that hydrostatic extrusion significantly changed the characteristics of the grain boundaries, increasing the fraction of LAGBs within the microstructure to 76%. The high fraction of LAGBs was caused by relatively low accumulated deformation. Previous work showed that applying hydrostatic extrusion to an Al–Mg–Si alloy at a similar level as that in the present study could further accommodate plastic strain via the dislocation of cells [29]. Further deformation would increase the misorientation angle between cells to above 15°, causing the formation of HAGBs.

Grain orientation analysis provided information about the favorable crystallographic orientations within the microstructure following hydrostatic extrusion. The grains aggregated into two dominant orientations: <001> and <111>. The <111> orientation is typical for a fiber texture, whereas the <001> orientation may result from dynamic recrystallization caused by the temperature increase during hydrostatic extrusion [28]. As was shown, the grain orientation stimulates the deformation substructure: for grains with the <111> orientation, dense dislocation walls are formed; for grains with the <001> orientation, stable cell dislocation structures can be observed [30]. In contrast, within the CG sample, no clear texture or preferred crystallographic orientation was identified. This may have been caused by the solution annealing performed after the hydrostatic extrusion, leading to grain recrystallization following plastic deformation.

In addition to grain size and texture, precipitates were also investigated. XRD studies showed that $MgZn_2$ precipitates formed within the structure as a result of the aging process. In addition, η' phases were observed by the TEM. An increase in aging temperature increased the size of the $MgZn_2$ precipitates within the microstructure. Hydrostatic extrusion also influenced the precipitation phenomena. The work investigating an Al–Mg–Si alloy showed that a heterogeneous microstructure following hydrostatic extrusion results in differences in the precipitate sequence [30]. This is caused by thermal shocks during the deformation process, which may induce the formation of clusters before the precipitate forms. For pure Al, adiabatic heating during hydrostatic extrusion with a strain of $\varepsilon = 1.4$ should be approximately 120–130 °C, and may influence or initiate the precipitation process in the case of age-hardenable alloys [31]. For AA 7475, the hydrostatic extrusion process changed the precipitation kinetics and phase stability induced by the grain boundaries [32]. Precipitation occurs along fast diffusion paths, such as grain boundaries and dislocations. Therefore, a larger number of precipitates at grain boundaries was observed within a deformed microstructure, as in this study. However, due to the higher number of nucleation sites, the precipitates were smaller than those in coarse-grained material. This effect was most noticeable within the grain interiors. Nevertheless, the increase in temperature of artificial aging led to an increase in the size of the precipitates.

4.2. Influence of Microstructural Evolution on Mechanical Properties

The evaluation of the mechanical properties revealed that hydrostatic extrusion caused an increase in mechanical strength. The highest UTS (approximately 675 MPa) was obtained for the HE sample in the longitudinal direction. In the transverse direction, the HE, HT1, and HT2 samples demonstrated similar values (approximately 575 MPa) and greater than those the CG sample. The increase in aging temperature caused an increase in YS. The average tensile strength of the commercially available coarse-grained AA 7075 in a T6 state is approximately 570 MPa, with a YS of 500 MPa and an elongation at break of approximately 9% [33]. This indicates that the combination of hydrostatic extrusion and T6 thermal treatment increased tensile strength by up to 18% when compared with the heat treatment of CG AA 7075.

Hydrostatic extrusion significantly increases mechanical strength, which is caused by the introduction of a considerable number of structural defects that inhibit the motion of dislocations during deformation. In the case of pure metals, when these defects are the only strengthening mechanism, the increase in mechanical strength is dependent on the value of true strain [34]. Nevertheless, in the case of age-hardened alloys such as AA 7075, strengthening from precipitates must also be considered. Following hydrostatic extrusion, and with dislocation slip in the dominant deformation mechanism, the particle strengthening of AA 7475 is limited by both enhanced grain boundary precipitation that does not contribute to an increase in mechanical strength, and the smaller size of precipitates within grain interiors, which are stronger barriers for moving dislocations [35]. Precipitates within grain interiors were also larger for the CG sample (see Figure 4) than for samples that had undergone hydrostatic extrusion. Therefore, precipitation strengthening is less effective for UFG material, with a larger contribution by structural defects such as grain boundaries and dislocations.

The anisotropy of mechanical properties was also investigated, as tensile tests were performed on samples cut from two directions, transverse and longitudinal, with respect to the extrusion direction. The microstructural anisotropy was significant as grains form fibrous shapes along the extrusion direction. This caused long HAGBs with interior LAGBs to create a network of subgrains. Regardless of the deformation ratio during hydrostatic extrusion, the AA 6060 alloy was characterized by weaker strength in the cross-section perpendicular to the extrusion direction [36], as was found in the present study for AA 7075. The anisotropy of AA 6060 increased with an increase in ε UTS and YS values indicated approximately 10% anisotropy, with higher values in the longitudinal direction for $\varepsilon = 2.28$. The anisotropy of mechanical properties was caused by the strong crystallographic macrotexture that develops in the material during hydrostatic extrusion. As a result, the spacing between HAGBs was smaller in the longitudinal plane than that in the transverse plane (see Figure 1). This caused a higher volume of HAGBs, which are stronger barriers for moving dislocations [34,37]. As a result, the samples cut in the longitudinal direction possessed higher mechanical strength.

4.3. *Influence of Microstructural Evolution on Corrosion Resistance*

The obtained results of the CG and HE samples show that the plastic deformation increased the corrosion susceptibility of the material. This may have resulted from the grain boundary characteristics (the fraction of LAGBs and HAGBs) of the samples being modified by the HE process. EBSD examination confirmed that HE contained a much larger fraction of LAGBs in comparison to CG. However, the absolute number and length of HAGBs were also greater. The dissolution of the material preferably occurred along HAGBs, since the HAGBs indicate regions with elevated levels of excess energy [38]. This phenomenon may have been the principal reason for the significantly lower corrosion resistance of the HE sample.

Postcorrosion analysis of the pit morphologies showed that, for CG, the pits were relatively narrow; for the samples that had undergone hydrostatic extrusion, the pits tended to develop inside the material, creating larger networks of dissolved material. Simultaneously, the depth of the corrosion pits decreased for samples that had been subjected to artificial aging. As the aging conditions and hydrostatic extrusion did not change the composition and size of the Al–Cu–Fe particles, the obtained results must be connected with differences in the matrix composition due to the growth of precipitates. In particular, the naturally aged HE sample exhibited more active breakdown potential than that of the peak-aged T6 HT2 sample, because the former had a higher potential difference between the intermetallic particles and the matrix and thus exhibited a stronger galvanic relationship [37]. The improvement of corrosion resistance following artificial aging can be attributed to the number of precipitates, as artificial aging significantly accelerates the precipitation effect. That is, the matrix composition of the material changes with aging due to precipitation, rendering it nobler and improving the corrosion resistance of the material. This may explain why the naturally aged HE was much more prone to dissolution and more susceptible to corrosion when compared to the artificially aged samples. Furthermore, artificial aging at higher temperatures leads to the coarsening of large precipitates at the expense of smaller precipitates, which in turn reduces the number of the precipitates within the microstructure. This effect may influence the overall corrosion behavior of the material, since corrosion is centered on specific sites in the heterogeneous microstructure, such as precipitates and intermetallic particles. Therefore, their limited number reduced the number of preferential nucleation sites of a corrosion attack [39].

5. Conclusions

In this paper, the effect of combined hydrostatic extrusion processing and various thermal treatments on the corrosion and mechanical properties of AA 7075 was investigated. Due to preserving the size of intermetallic particles that significantly influence the corrosion resistance, other microstructural components and their influence could be examined. The main conclusions from the present study are as follows:

- Hydrostatic extrusion with an equivalent strain of $\varepsilon = 1.4$ resulted in grain refinement from 17 to 1.3 μm. A microstructure consisting of fibrous grains oriented in the <001> and <111> directions with a fraction of LAGBs of 76% was obtained.
- As a result of thermomechanical treatment, an increase in tensile strength of up to 100 MPa was obtained when compared to the CG T6 state. Hydrostatic extrusion caused a noticeable anisotropy of the mechanical properties, as differences in tensile strength between longitudinal and transverse directions were in the range of 60–100 MPa.
- Artificial aging caused an increase in the size of $MgZn_2$ precipitates within the microstructure, and led to improved corrosion resistance due to changes in the chemical composition of the matrix. The differences in mechanical strength in relation to the aging temperature were not significant due to the location of precipitates at the grain boundaries, which were not effective in strengthening the material.
- The optimal correlation of high mechanical strength and high corrosion resistance was obtained for the sample subjected to hydrostatic extrusion and subsequently aged at 120 °C for 24 h, corresponding to a T6 state.

Author Contributions: Conceptualization, M.O. and E.U.-B.; methodology, M.O., E.U.-B. and M.K.; validation, E.U.-B., Ł.Ś. and M.K.; formal analysis, M.O.; investigation, M.O., E.U.-B., P.S., K.M., M.K. and B.A.-C.; resources, E.U.-B.; data curation, M.O., P.S., K.M. and B.A.-C.; writing—original draft preparation, M.O., E.U.-B. and Ł.Ś.; writing—review and editing, M.O. and E.U.-B.; visualization, M.O.; supervision, E.U.-B. and Ł.Ś.; project administration, E.U.-B.; funding acquisition, E.U.-B. and Ł.Ś. All authors have read and agreed to the published version of the manuscript.

Funding: This research was funded by POB "Technologie Materiałowe" of Warsaw University of Technology (Poland) within the Excellence Initiative: Research University (IDUB) program under the project titled "Optimization of the corrosion resistance and mechanical properties of precipitation strengthened 7075 aluminum alloy subjected to hydrostatic extrusion".

Institutional Review Board Statement: Not applicable.

Informed Consent Statement: Not applicable.

Data Availability Statement: The data presented in this study are available on request from the corresponding author.

Conflicts of Interest: The authors declare no conflict of interest.

References

1. Meng, Q.; Frankel, G.S. Effect of Cu Content on Corrosion Behavior of 7xxx Series Aluminum Alloys. *J. Electrochem. Soc.* **2004**, *151*, B271. [CrossRef]
2. Polmear, I.P. *Light Alloys—From Traditional Alloys to Nanocrystals*; Butterworth-Heinemann: Oxford, UK, 2006; Volume 2.
3. Osamura, K.; Kubota, O.; Protit, P.; Okuda, H.; Ochiai, S.; Fujii, K.; Kusui, J.; Yokote, T.; Kubo, K. Development of high-strength aluminum alloys by mesoscopic structure control. *Metall. Mater. Trans. A* **1995**, *26*, 1597–1599. [CrossRef]
4. Ralston, K.D.; Birbilis, N.; Weyland, M.; Hutchinson, C.R. The effect of precipitate size on the yield strength-pitting corrosion correlation in Al-Cu-Mg alloys. *Acta Mater.* **2010**, *58*, 5941–5948. [CrossRef]
5. Valiev, R. Nanostructuring of metals by severe plastic deformation for advanced properties. *Nat. Mater.* **2004**, *3*, 511–516. [CrossRef]
6. Majchrowicz, K.; Pakieła, Z.; Giżyński, M.; Karny, M.; Kulczyk, M. High-cycle fatigue strength of ultrafine-grained 5483 Al-Mg alloy at low and elevated temperature in comparison to conventional coarse-grained Al alloys. *Int. J. Fatigue* **2018**, *106*, 81–91. [CrossRef]
7. Darban, H.; Mohammadi, B.; Djavanroodi, F. Effect of equal channel angular pressing on fracture toughness of Al-7075. *Eng. Fail. Anal.* **2016**, *65*, 1–10. [CrossRef]
8. Wawer, K.; Lewandowska, M.; Kurzydłowski, K.J. Improvement of mechanical properties of a nanoaluminium alloy by precipitate strengthening. *Arch. Metall. Mater.* **2012**, *57*, 877–881. [CrossRef]
9. Miyamoto, H. Corrosion of ultrafine grained materials by severe plastic deformation, an overview. *Mater. Trans.* **2016**, *57*, 559–572. [CrossRef]

10. Miyamoto, H.; Yuasa, M.; Rifai, M.; Fujiwara, H. Corrosion behavior of severely deformed pure and single-phase materials. *Mater. Trans.* **2019**, *60*, 1243–1255. [CrossRef]
11. Song, D.; Ma, A.B.; Jiang, J.H.; Lin, P.H.; Shi, J. Improving corrosion resistance of pure Al through ECAP. *Corros. Eng. Sci. Technol.* **2011**, *46*, 505–512. [CrossRef]
12. Son, I.J.; Nakano, H.; Oue, S.; Kobayashi, S.; Fukushima, H.; Horita, Z. Pitting corrosion resistance of anodized aluminum-copper alloy processed by severe plastic deformation. *Mater. Trans.* **2008**, *49*, 2648–2655. [CrossRef]
13. Nakano, H.; Yamaguchi, H.; Yamada, Y.; Oue, S.; Son, I.J.; Horita, Z.; Koga, H. Effects of high-pressure torsion on the pitting corrosion resistance of aluminum-iron alloys. *Nippon Kinzoku Gakkaishi/J. Jpn. Inst. Met.* **2013**, *77*, 543–549. [CrossRef]
14. Mahmoud, T.S. Effect of friction stir processing on electrical conductivity and corrosion resistance of AA6063-T6 Al alloy. *Proc. Inst. Mech. Eng. Part C J. Mech. Eng. Sci.* **2008**, *222*, 1117–1123. [CrossRef]
15. Osório, W.R.; Freire, C.M.; Garcia, A. The role of macrostructural morphology and grain size on the corrosion resistance of Zn and Al castings. *Mater. Sci. Eng. A* **2005**, *402*, 22–32. [CrossRef]
16. Lewandowska, M.; Kurzydlowski, K.J. Recent development in grain refinement by hydrostatic extrusion. *J. Mater. Sci.* **2008**, *43*, 7299–7306. [CrossRef]
17. Molak, R.M.; Paradowski, K.; Brynk, T.; Ciupinski, L.; Pakiela, Z.; Kurzydlowski, K.J. Measurement of mechanical properties in a 316L stainless steel welded joint. *Int. J. Press. Vessel. Pip.* **2009**, *86*, 43–47. [CrossRef]
18. Majchrowicz, K.; Jóźwik, P.; Chromiński, W.; Adamczyk-Cieślak, B.; Pakieła, Z. Microstructure, texture and mechanical properties of mg-6sn alloy processed by differential speed rolling. *Materials* **2021**, *14*, 83. [CrossRef]
19. Birbilis, N.; Buchheit, R.G. Investigation and Discussion of Characteristics for Intermetallic Phases Common to Aluminum Alloys as a Function of Solution pH. *J. Electrochem. Soc.* **2008**, *155*, C117. [CrossRef]
20. Huo, W.; Hou, L.; Cui, H.; Zhuang, L.; Zhang, J. Fine-grained AA 7075 processed by different thermo-mechanical processings. *Mater. Sci. Eng. A* **2014**, *618*, 244–253. [CrossRef]
21. Málek, P.; Cieslar, M. The influence of processing route on the plastic deformation of Al-Zn-Mg-Cu alloys. *Mater. Sci. Eng. A* **2002**, *324*, 90–95. [CrossRef]
22. Birbilis, N.; Cavanaugh, M.K.; Buchheit, R.G. Electrochemical behavior and localized corrosion associated with Al7Cu2Fe particles in aluminum alloy 7075-T651. *Corros. Sci.* **2006**, *48*, 4202–4215. [CrossRef]
23. Abd El Aal, M.I.; Um, H.Y.; Yoon, E.Y.; Kim, H.S. Microstructure evolution and mechanical properties of pure aluminum deformed by equal channel angular pressing and direct extrusion in one step through an integrated die. *Mater. Sci. Eng. A* **2015**, *625*, 252–263. [CrossRef]
24. Brunner, J.G.; May, J.; Höppel, H.W.; Göken, M.; Virtanen, S. Localized corrosion of ultrafine-grained Al-Mg model alloys. *Electrochim. Acta* **2010**, *55*, 1966–1970. [CrossRef]
25. Andreatta, F.; Terryn, H.; de Wit, J.H.W. Corrosion behaviour of different tempers of AA7075 aluminium alloy. *Electrochim. Acta* **2004**, *49*, 2851–2862. [CrossRef]
26. Sun, Q.; Yang, M.; Jiang, Y.; Lei, L.; Zhang, Y. Achieving excellent corrosion resistance properties of 7075 Al alloy via ultrasonic surface rolling treatment. *J. Alloys Compd.* **2022**, *911*, 165009. [CrossRef]
27. Tian, W.; Li, S.; Wang, B.; Liu, J.; Yu, M. Pitting corrosion of naturally aged AA 7075 aluminum alloys with bimodal grain size. *Corros. Sci.* **2016**, *113*, 1–16. [CrossRef]
28. Lewandowska, M. Mechanism of grain refinement in aluminium in the process of hydrostatic extrusion. *Solid State Phenom.* **2006**, *114*, 109–116. [CrossRef]
29. Chrominski, W.; Majchrowicz, K.; Lewandowska, M. Microstructural response to compression deformation of ultrafine-grained aluminum with various microstructures. *Mater. Sci. Eng. A* **2019**, *763*, 138184. [CrossRef]
30. Chrominski, W.; Lewandowska, M. Precipitation phenomena in ultrafine grained Al-Mg-Si alloy with heterogeneous microstructure. *Acta Mater.* **2016**, *103*, 547–557. [CrossRef]
31. Pachla, W.; Kulczyk, M.; Smalc-Koziorowska, J.; Wróblewska, M.; Skiba, J.; Przybysz, S.; Przybysz, M. Mechanical properties and microstructure of ultrafine grained commercial purity aluminium prepared by cryo-hydrostatic extrusion. *Mater. Sci. Eng. A* **2017**, *695*, 178–192. [CrossRef]
32. Ura-Bińczyk, E. Improvement of Pitting-Corrosion Resistance of Ultrafine-Grained 7475 Al Alloy by Aging. *Materials* **2022**, *15*, 360. [CrossRef] [PubMed]
33. Available online: http://www.matweb.com/search/DataSheet.aspx?MatGUID=4f19a42be94546b686bbf43f79c51b7d&ckck=1 (accessed on 29 January 2022).
34. Hansen, N. Hall–Petch relation and boundary strengthening. *Scr. Mater.* **2004**, *51*, 801–806. [CrossRef]
35. Lewandowska, M.; Wawer, K.; Kozikowski, P.; Ohnuma, M.; Kurzydlowski, K.J. Precipitation in a nanograined 7475 aluminium alloy—Processing, properties and nanoanalysis. *Adv. Eng. Mater.* **2014**, *16*, 482–485. [CrossRef]
36. Przybysz, S.; Kulczyk, M.; Pachla, W.; Skiba, J.; Wróblewska, M.; Mizera, J.; Moszczynska, D. Anisotropy of mechanical and structural properties in aa 6060 aluminum alloy following hydrostatic extrusion process. *Bull. Pol. Acad. Sci. Tech. Sci.* **2019**, *67*, 709–717. [CrossRef]
37. Hansen, N. Boundary strengthening in undeformed and deformed polycrystals. *Mater. Sci. Eng. A* **2005**, *409*, 39–45. [CrossRef]

38. Orłowska, M.; Ura-Bińczyk, E.; Olejnik, L.; Lewandowska, M. The effect of grain size and grain boundary misorientation on the corrosion resistance of commercially pure aluminium. *Corros. Sci.* **2019**, *148*, 57–70. [CrossRef]
39. Murer, N.; Oltra, R.; Vuillemin, B.; Néel, O. Numerical modelling of the galvanic coupling in aluminium alloys: A discussion on the application of local probe techniques. *Corros. Sci.* **2010**, *52*, 130–139. [CrossRef]

materials

MDPI

Article

Fatigue Fracture Analysis on 2524 Aluminum Alloy with the Influence of Creep-Aging Forming Processes

Liyong Ma [1,†], Chi Liu [2,†], Minglei Ma [3], Zhanying Wang [1,4,*], Donghao Wu [1], Lijuan Liu [1] and Mingxing Song [1,*]

1 School of Mechanical Engineering, Hebei University of Architecture, Zhangjiakou 075031, China; maliyong@buaa.edu.cn (L.M.); wzx2124@hebiace.edu.cn (D.W.); zyq1292@hebiace.edu.cn (L.L.)
2 School of Mechanical and Electrical Engineering, Changsha University, Changsha 410199, China; liuchi@ccsu.edu.cn
3 Zhangjiakou Cigarette Factory Co., Ltd., Zhangjiakou 075001, China; why2048@hebiace.edu.cn
4 Zhangjiakou Special Equipment Intelligent Monitoring Operation and Maintenance Technology Innovation Center, Zhangjiakou 075031, China
* Correspondence: wzy1293@hebiace.edu.cn (Z.W.); bz1907006@buaa.edu.cn (M.S.); Tel.: +86-313-418-7755 (Z.W.); +86-313-418-7749 (M.S.)
† These authors contributed equally to this work.

Abstract: The different creep-aging forming processes of 2524 aluminum alloy were taken as the research object, and the effects of creep-aging temperature and creep stress on the fatigue-crack propagation properties of the alloy were studied. The research results showed the following under the same sintering time of 9 h, at creep-aging temperatures of 100 °C, 130 °C, 160 °C, and 180 °C, respectively, with an increase in creep-aging temperature: the fatigue-crack propagation rate was promoted, the spacing of fatigue striations increased, and the sizes of dimples decreased while the number was enlarged; this proves that the fatigue property of the alloy was weakened. Compared with the specimens with creep deformation radii of 1000 mm and 1500 mm, the creep deformation stress was the smallest when the forming radius was 1800 mm, with a higher threshold value of fatigue-crack growth in the near-threshold region of fatigue-crack propagation ($\Delta K \leq 8$ MPa·m$^{1/2}$). Under the same fatigue cycle, the specimens under the action of larger creep stress endured a longer fatigue stable-propagation time and a faster fracture speed. Comparing the effect of creep-aging temperature and creep stress, the creep-aging temperature plays a dominant role in the fatigue-crack propagation of creep-aged 2524 aluminum alloy.

Keywords: creep-aging forming; creep-aging temperature; creep stress; 2524 aluminum alloy; fatigue-crack propagation rate; fatigue fracture morphology

Citation: Ma, L.; Liu, C.; Ma, M.; Wang, Z.; Wu, D.; Liu, L.; Song, M. Fatigue Fracture Analysis on 2524 Aluminum Alloy with the Influence of Creep-Aging Forming Processes. *Materials* **2022**, *15*, 3244. https://doi.org/10.3390/ma15093244

Academic Editors: Daolun Chen, Lucjan Śnieżek and Jaroslaw Galkiewicz

Received: 31 March 2022
Accepted: 27 April 2022
Published: 30 April 2022

1. Introduction

2524 aluminum alloy is a new type of high-strength aluminum alloy for aviation [1] that is mainly used for wing skins [2]. During service, it is subjected to loads in complex environments, so its fatigue performance is particularly critical [3]. Compared with aluminum alloys such as 2024, 2124 and 2224, 2524 aluminum alloy has higher strength and better fatigue properties [4,5]. The fracture toughness is increased by 15~20%, the fatigue resistance doubles, and the fatigue life is increased by 27~45% [6,7], which is well suited to the requirements of modern aircraft design for material damage tolerance [8]. Meanwhile, with the maturity of the creep-aging forming process, combined with the good hot-working properties of 2524 aluminum alloy, the application prospects of 2524 aluminum alloy in the future aerospace industry are broad [9].

During the creep-aging forming process, the microstructure and evolution process of the material are complex, and the factors that affect the fatigue properties of the creep-aging forming material also become complicated. Pitcher et al. [10] studied creep-aging-formed 2024A, 8090, and 7449 aluminum alloys from the two aspects of springback and damage tolerance, and conducted a large number of fatigue-crack propagation rate experiments. The result showed that 2024A and 8090 had sufficient damage resistance after creep deformation, which can be useful for making lower wing skins. Rafiq A. Siddiqui et al. [11] studied the effects of age temperature and duration on the fatigue properties of 6063 aluminum alloys. The study found the vacancy diffusion mechanism played a significant role in the formation of Guinier–Preston (GP) zones, in which the solid solution is precipitated due to uniform nucleation during the dissolution process, and a fine quasi-stable phase is precipitated. The structure of this stable phase was similar to that of the main matrix, and the two were coherent with each other, thus disturbing the regularity of the lattice, leading to the increase in the fatigue defects of the alloy. Brav G.H. et al. [12] studied the effects of different aging processes on the crack propagation rate of the 2024-T351 alloy and Al–Cu–Mg–Li alloy. The results showed that artificial aging increases the strength of the alloy, but it also reduces the crack propagation resistance of the alloy. In addition, with the prolongation of artificial aging duration, the crack propagation rate increased monotonically. The authors believed that the reduction in crack propagation resistance was related to the precipitation of the T1 phase and S' phase in the alloy; however, the influence mechanism of the aging process on the precipitation phase was not discussed, nor was the effect of fine precipitation on the crack-tip slip-mode. Sarioglu F. and Orhaner F.Ö. [13] researched the fatigue property of the 2024-T3 aluminum alloy after aging at a low temperature of 130 °C for 100 h/1000 h. The results showed that long-term aging eliminated the difference in the crack propagation rate of alloys in the sampling directions (L-T, T-L, 60°). Burba et al. [14] found that the minimum fatigue life of the alloy was mainly affected by the density of the θ' precipitation phase, but had little effect on the average fatigue life. Yang [15] conducted a lot of research on the effect of the creep-aging forming process on the high-cycle fatigue properties of Al-Zn-Mg series-7075 high-strength aluminum alloys. The study found that the increase in creep-aging temperature and time has a positive effect on the improvement of alloy fatigue properties. The effect of preloading stress will reduce the fatigue resistance of the alloy. However, due to the differences in microstructure and chemical composition, the conclusions of this study were not applicable to 2524 aluminum alloys.

Xu et al. [16] studied the effect of tensile pre-strain before creep-aging forming on the mechanical properties of 2524 aluminum alloy using a constant-stress creep-aging test. they found that the magnitude of creep strain was greatly increased with the increase in pre-strain and that the pre-strain improved the formability, mechanical properties and microstructures of 2524. In addition, Xu et al. [17] also conducted research on the tension and compression creep-aging behaviors of Al-Cu-Mg alloy. They demonstrated that the creep strains under tensile stresses were larger than the creep strains under compressive stresses. Meanwhile, the formation of the S phase in the aluminum matrix was caused by the compressive stress, inhibiting the precipitates in the grain boundary, which was beneficial to the hardness of the compression creep-aged alloy. Liu et al. [18] studied the effects of creep-aging and artificial aging on the fatigue-crack propagation of 2524 aluminum alloy. They found that the prolonged aging time resulted in excessive precipitation of the needle-like S' phase, changed the dislocation slip mode, reduced the reversibility of the slip, and accelerated the accumulation of fatigue damage. In addition, creep stress accelerated the aging precipitation process of the alloy. Compared with artificial aging, under the same aging time, the size of the precipitates in the creep-aged alloy was larger and the yield strength and hardness were increased, but the fatigue resistance was decreased. In our previous work [19], we studied the effects of creep time on the microstructural and mechanical properties, as well as the fatigue-crack propagation, in 2524 aluminum alloy. TEM was employed for the observation of precipitation and dislocation.

Most scholars have focused on the relationship between the creep/stress relaxation mechanism and the material deformation. However, the relationship between creep-aging temperature, creep stress, and the fatigue properties of materials has not been comprehensively and systematically studied.

This paper will focus on analyzing the macro-microscopic characteristics of the fatigue-crack fractures of creep-formed specimens, and analyze the fracture mode, crack propagation path and fracture mechanism of creep-formed components. The fatigue property of creep-formed alloys at different creep-aging temperatures and creep stresses will be explored from the micro-morphological features.

2. Experimental Processes

2.1. Materials

In this study, a 2524 aluminum alloy sheet for aviation (the thickness was 3.5 mm) was selected for research. After the alloy was solution-treated, it was cold-machined, and then naturally aged to a stable state. The main chemical composition of 2524 aluminum alloy is shown in Table 1 [9,16].

Table 1. Chemical composition of 2524 aluminum alloy (mass fraction: wt%).

Cu	Mg	Mn	Fe	Zn	Si	Ti	Cr	Al
4.62	1.32	0.57	0.035	0.004	0.025	0.02	0.001	Bal.

2.2. Creep-Aging Forming

The creep-aging forming experiment specimen and mold are shown in Figure 1a, and the size of the experiment sheet was 360 mm × 220 mm. The 2524 aluminum alloy sheet was placed in the center of the mold, then the mold and sheet were wrapped with air felt and a vacuum bag; then, the bag was sealed with heat-resistant glue. The bag was gradually evacuated to a vacuum, and a negative pressure of 0.1 MPa was maintained. The sheet was elastically deformed under uniform load until it was completely fitted with the mold surface. The deformed sheet and mold were put into an autoclave (Figure 1b), with a heating rate of 1.5 °C/min. In addition, the creep-aging treatment was carried out according to the set creep-aging temperature, and the vacuum bag was kept in a sealed state during the creep-forming process. After the set aging time was reached, the load and temperature were removed, and the components were cooled in a furnace, obtaining the final desired shape.

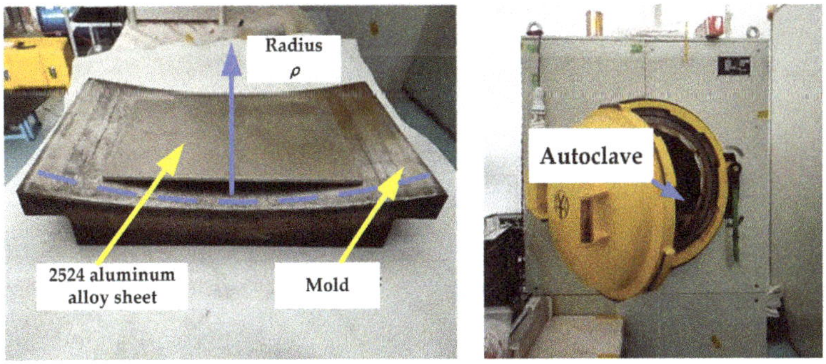

(a) (b)

Figure 1. Creep-aging forming mold and tooling: (**a**) mold and sheet; (**b**) autoclave.

(1) To study the effect of creep-aging temperature on fatigue property, creep deformation experiments were carried out at creep-aging temperatures of 100 °C, 130 °C, 160 °C and 180 °C, respectively, and the creep-aging time of all specimens was 9 h.

(2) To study the effect of creep stress, the radii of curvature ρ were 1000 mm, 1500 mm and 1800 mm, respectively. The larger the radius ρ, the smaller the creep stress. The creep-aging temperature was 160 °C, and the creep time was 9 h.

2.3. Fatigue Experiment

The fatigue-crack propagation experiment was carried out at room temperature. The sample preparation and stress loading method of the fatigue-crack propagation specimen were designed according to ASTM-E647 [20]. The experimental equipment was the MTS-810 fatigue experimenting machine made in the U.S. The fatigue-crack propagation experiment utilized standard compact tensile specimens (CT specimens), and the specimens were cut using a wire electric discharge. The size of the sample is shown in Figure 2. Before the fatigue-crack propagation experiment, a crack of 2.75 mm was prefabricated using the pre-crack module in the experiment system. The experiment adopted sine wave loading, with the maximum loading of 2400 N. The stress ratio R of 0.1~0.5 ($R = P_{min}/P_{max}$), the loading frequency f of 10 Hz, and the crack length a were detected by the compliance control method (COD).

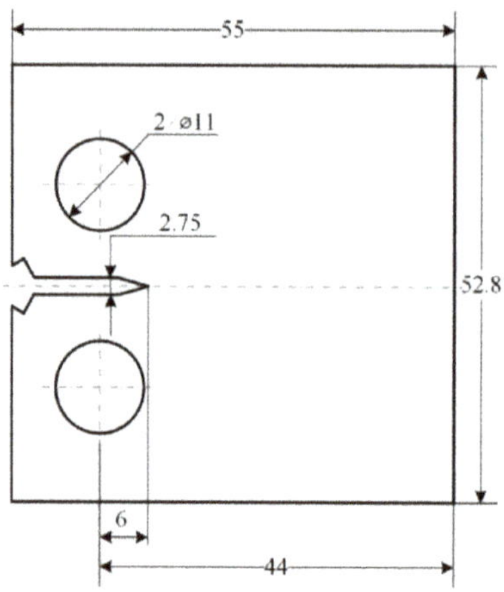

Figure 2. Schematic diagram of CT specimen for fatigue-crack propagation experiment.

The fatigue-crack propagation data obtained in the experiment were processed using the seven-point incremental polynomial method. For data point i and its front 3 points and back 3 points—a total of 7 continuous data points—the quadratic polynomial was employed to perform local fitting and derivation, and the fitting values of the fatigue-crack propagation rate were obtained using Equation (1):

$$a_i = b_0 + b_1\left[\frac{N_i - C_1}{C_2}\right] + b_2\left[\frac{N_i - C_1}{C_2}\right]^2 \tag{1}$$

where N_i is the number of cycles, a_i is the fitting crack length value; b_0, b_1 and b_2 are the regression parameters determined according to the minimum squared deviation between the observed value of the crack length and the fitting value; $C_1 = (N_{i-3} - N_{i+3})/2$; $C_2 = (N_{i+3} - N_{i-3})/2$; and $-1 \leq (N_{i-3} - C_1)/C_2 \leq 1$. By derivation of Equation (1), the crack propagation rate at N_i can be obtained:

$$\left(\frac{da}{dN}\right)_{a_i} = \frac{b_1}{C_2} + \frac{2b_2(N_i - C_1)}{C_2^2} \tag{2}$$

For CT specimens with type-I open cracks, the range of stress intensity factor ΔK at the crack tip can be calculated using Equation (3) [19,20]:

$$\Delta K = \frac{\Delta P}{B\sqrt{W}} \cdot \frac{(2+\alpha)}{(1-\alpha)^{3/2}}(0.886 + 4.64\alpha - 13.32\alpha^2 + 14.72\alpha^3 - 5.6\alpha^4) \tag{3}$$

where ΔP is the force value range; $\Delta P = P_{max} - P_{min}$, P_{max} is the maximum loading force; P_{min} is the minimum loading force; $\alpha = a/W$, a is the crack length; B is the width of the specimen (B = 5 mm); and W is the width of the specimen (W = 44 mm).

3. Result and Discussion

3.1. Effect of Creep-Aging Temperature on Fatigue-Crack Propagation

3.1.1. Effect of Creep-Aging Temperature on Fatigue-Crack Growth Rate

The creep-aging temperatures were 100 °C, 130 °C, 160 °C and 180 °C, respectively, and the aging time was the same, i.e., 9 h. The fatigue-crack propagation rate experiment was carried out. The $da/dN - \Delta K$ curve is shown in Figure 3. In Figure 3a, the crack propagation curves of the creep-formed alloy at 100 °C and 130 °C coincide with those of 2524 aluminum alloy without creep forming, indicating a fatigue property with no significant change at lower creep-aging temperatures (<130 °C), because of insufficient effective deformation of the alloy at lower temperatures. Additionally, the $da/dN - \Delta K$ curve of 2524-T3 aluminum alloy without creep forming was taken as the comparison data. 2524-T3 is a kind of alloy obtained through cold-working of 2524 aluminum alloy after solution treatment, then stabilization by natural aging.

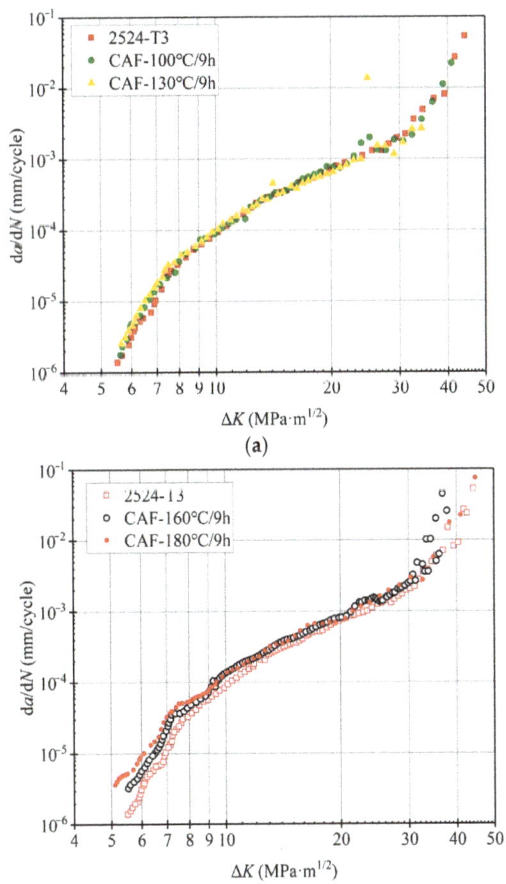

(b)

Figure 3. Fatigue-crack propagation rate da/dN-ΔK curves of 2524 aluminum alloy at different creep-aging temperatures: (**a**) 100 °C and 130 °C; (**b**) 160 °C and 180 °C.

Nevertheless, under the age temperatures of 160 °C and 180 °C, the crack propagation rate curves show significant differences (Figure 3b). In the near-threshold region ($\Delta K \leq 8$ MPa·m$^{1/2}$), the crack propagation resistance of the alloy is significantly reduced under high-temperature creep-aging forming. Under the same ΔK of 6 MPa·m$^{1/2}$, when the creep-aging temperature is 160 °C, $da/dN = 5.62 \times 10^{-6}$ mm/cycles^{-1}, but when the creep-aging temperature is 180 °C, da/dN is increased to 9.11×10^{-6} mm/cycles^{-1}, showing that the crack propagation resistance of 2524 aluminum alloy is decreased to a certain extent under high-temperature creep-aging forming. In addition, with the increase in ΔK, the difference in crack propagation rate gradually decreases. According to the curve, it is delineated in the Paris region of 10 MPa·m$^{1/2} \leq \Delta K \leq 30$ MPa·m$^{1/2}$, and the curve shows an obvious linear relationship.

The straight part of the $da/dN - \Delta K$ curve in the double logarithmic coordinate is fitted by the Paris equation, and the corresponding fitting constants C, n and the fatigue-crack propagation rate under the same ΔK are shown in Table 2. The error value is 2.16%. The values of n are close, ranging from 2.7 to 3.0, indicating that the creep-aging temperature has little effect on the crack propagation rate of 2524 aluminum alloy in the medium and high stress range. When ΔK exceeds 30 MPa·m$^{1/2}$, the curve of $da/dN - \Delta K$ has an obvious turning point, and the crack propagation rate da/dN increases rapidly from 10^{-3} mm/cycles to 0.1 mm/cycles until an instability fracture occurs.

Table 2. Paris fitting parameters C and n at different creep-aging temperatures.

Aging Status	C	n	$da/dN = C\Delta K^n$/(mm·cycle^{-1})		
			$\Delta K = 8$	$\Delta K = 12$	$\Delta K = 16$ (MPa·m$^{1/2}$)
100 °C/9 h	1.77×10^{-7}	2.99	3.51×10^{-5}	1.52×10^{-4}	4.31×10^{-4}
130 °C/9 h	1.77×10^{-7}	2.99	4.15×10^{-5}	1.75×10^{-4}	3.97×10^{-4}
160 °C/9 h	4.32×10^{-7}	2.67	4.52×10^{-5}	2.25×10^{-4}	4.78×10^{-4}
180 °C/9 h	2.81×10^{-7}	2.80	5.02×10^{-5}	2.13×10^{-4}	5.74×10^{-4}

3.1.2. Fracture Morphologies at Different Creep-Aging Temperatures

(1) Fatigue-crack propagation zone

The creep-forming temperature has a decisive effect on the solute precipitation in the supersaturated state of 2524 aluminum alloy [21]. The increase in the age temperature enhances the atomic activity in the alloy, resulting in a quick increase in the precipitation rate of the precipitation phase [22]. Therefore, the nucleation, growth and enrichment of the precipitates during the aging process will be affected by the age temperature, which will lead to changes in the fatigue properties of the material. Figure 4 shows the fracture morphology of 2524 aluminum alloy at crack length $a = 5$ mm after aging at different creep-aging temperatures (100 °C, 130 °C, 160 °C, 180 °C) for 9 h. The corresponding ΔK at this time is about 15 MPa·m$^{1/2}$, and the crack is in the stable propagating stage.

In Figure 4, the fatigue sections of the specimens are all flat, and the relative torsion of the cleavage planes in adjacent grains makes the crack propagate along many transgranular planes, forming smooth and flat sections, i.e., large fatigue platforms. These fatigue platforms are connected by the tearing edge, indicated by mark 2 in the figure. The tearing edge is deflected by an angle of 10°~40° relative to the main crack propagation direction, and the fracture shows the characteristics of a ductile transgranular fracture. There are also many micro-pores (mark 1) distributed on the cross-section, which originate from the tiny plastic deformations confined around the coarser second-phase particles during the fatigue process. Meanwhile, secondary cracks (mark 4) approximately perpendicular to the direction of the main crack-propagation plane are also observed, which propagate into the material. There are more fracture cleavage steps in the fracture morphology under high-temperature aging of 180 °C—9 h, and the fracture surface is rough, with part of the fracture morphology even showing slight brittle fracture characteristics. From Figure 4d, it can be seen that the obvious small fatigue steps are connected by the shear edges at mark 3; moreover, the heights are different, indicating that the aging precipitation and hardening rate of 2524 aluminum alloy are promoted due to high-temperature aging. The alloy enters the overaging state in advance, and the ductility decreases significantly, which adversely affects the fatigue property and toughness of the alloy.

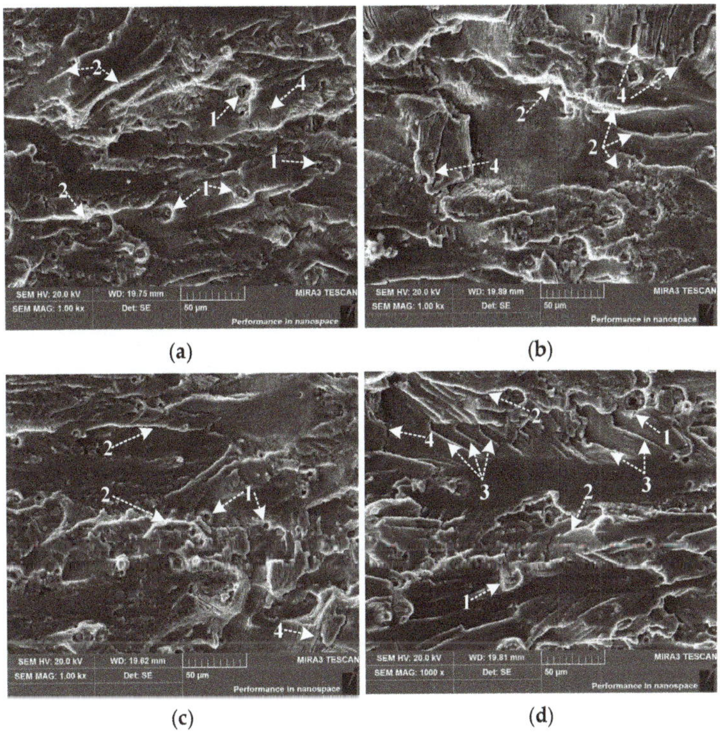

Figure 4. SEM images of stable fatigue-crack propagating zone at different creep-aging temperatures (a = 5 mm): (**a**) 100 °C—9 h; (**b**) 130 °C—9 h; (**c**) 160 °C—9 h; (**d**) 180 °C—9 h. Marks in the figure: 1—micropore; 2—tear edge; 3—shear edge; 4—secondary crack.

The micro-morphology of the corresponding position in Figure 4 is magnified to 20,000 times for observation, and fatigue striations can be observed, as shown in Figure 5. There are obvious differences in the fatigue striation spacing of the specimens under different creep-aging temperatures. Since the spacing of only one fatigue striation is too small to measure, and the measured value of only one fatigue striation often brings in error, the spacing of five fatigue striations was measured for precision in mirroring the effect of creep temperature and stress on the fatigue-crack propagation. Under the low-temperature aging of 100 °C—9 h and 130 °C—9 h, the spacing of five fatigue striations are relatively small, at 1.22 μm and 2.41 μm, respectively, and the striation morphology is very clear and regular. With the increase in creep-aging temperature, under the same crack length, the striation spacing increases, and the appearance is rough. Under the high temperature aging of 160 °C—9 h and 180 °C—9 h, the average striation spacing is 3.52 μm and 3.86 μm. The difference in fatigue striation spacing reflects the increase in the size and volume fraction of the alloy precipitates in the peak or overaging state, which increases the strength of the material, but reduces its elongation and increases its brittleness, which makes fatigue streaks less likely to occur.

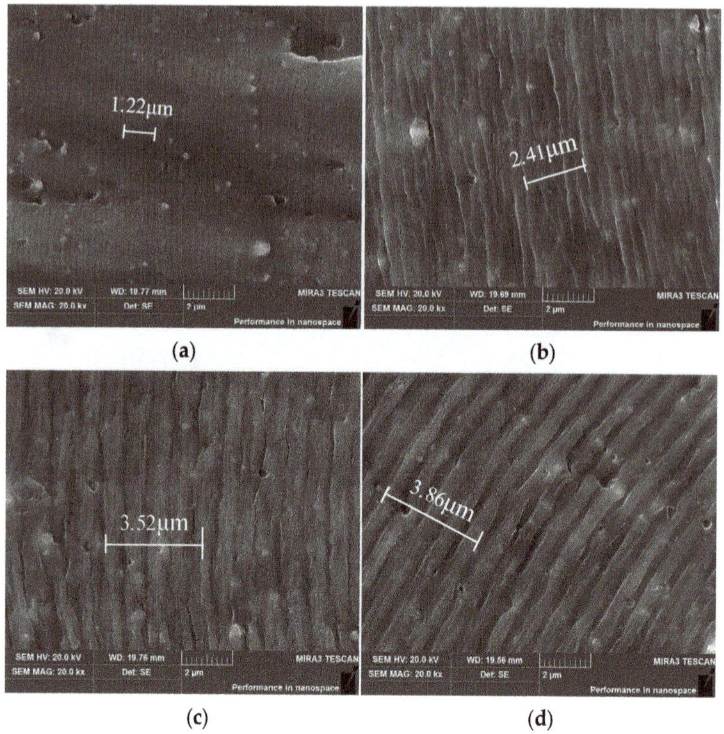

Figure 5. SEM images of fatigue striations in the stable propagation zone at different creep-aging temperatures (a = 5 mm): (**a**) 100 °C—9 h; (**b**) 130 °C—9 h; (**c**) 160 °C—9 h; (**d**) 180 °C—9 h.

At different creep-aging temperatures, when the precipitated strengthening phase maintains a coherent or semi-coherent relationship with the matrix, it is generally believed that dislocations can cut through the precipitated phase [23], and then plane slip occurs, resulting in uneven deformation in local areas. As the slip plane continues to expand, the crack propagation path may deflect, kink, and bifurcate, reducing the rate of crack propagation. When the strengthening phase is incoherent with the matrix, the dislocations bypass the precipitation phase, and the deformation in the local area is relatively uniform; this reduces the possibility of deflection, kink and bifurcation during crack propagation, thereby increasing the rate of crack propagation. Therefore, for samples with an age temperature of 100 °C and 130 °C, the precipitates maintain a coherent or semi-coherent relationship with the matrix, and the fatigue-crack propagation rate is slow. However, when the age temperature is 180 °C, the precipitation phase of the material grows up and breaks away from the semi-coherent relationship with the matrix, so the crack propagation rate of the sample is at its fastest, as shown in Figure 3b.

(2) Fatigue fracture zone

Figure 6 shows the fracture morphologies of 2524 aluminum alloy in the Fatigue fracture zone (a = 25 mm) after aging at different creep-aging temperatures (100 °C, 130 °C, 160 °C, 180 °C) for 9 h. From the fracture surface of the specimen with the creep-aging temperature of 130 °C (see Figure 6b), the large pits caused by the debonding between the precipitation phase and the interface of the aluminum matrix are surrounded by smaller pits. Meanwhile, the large pits are not connected to each other during the fracture process, and there are obvious tearing edges along the small pits, which indicates that the specimen also has good plastic deformation ability. the higher the creep-aging temperature, the larger the dimple size. However, there is no obvious tearing edge, and the fracture morphology is also flatter. The high-temperature creep accelerates the nucleation and growth of the precipitation phase at 180 °C. During the tearing fracture process, the precipitation phase will hinder the dislocation slip due to the incompatibility of the precipitation phase with the aluminum matrix [24], resulting in stress concentration. When the stress concentration exceeds the critical value that the material can withstand, the interface between the precipitate and the aluminum matrix is debonded [25], or the precipitate is fractured, which reduces the plastic deformation ability of the aluminum alloy. With the continuous increase in plastic strain, the interface debonding between the precipitation phase and the aluminum matrix causes the aggregation of large pits. When the effective bearing area is reduced to a critical value, the aluminum matrix will break rapidly.

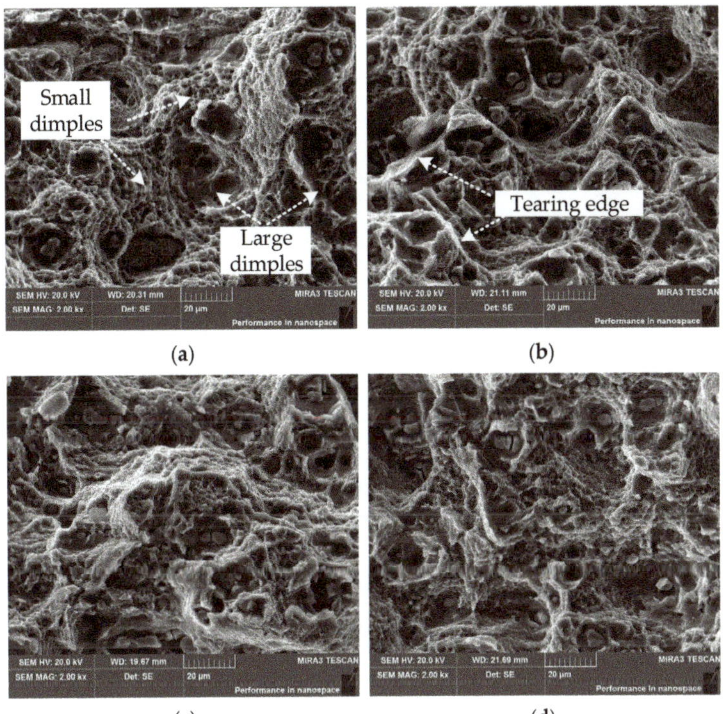

Figure 6. SEM images of fatigue-crack transient region at different creep-aging temperatures: (**a**) 100 °C—9 h; (**b**) 130 °C—9 h; (**c**) 160 °C—9 h; (**d**) 180 °C—9 h.

Under the impact of the creep-aging temperature and the critical size, the enrichment speed of the precipitates are different, and the type, density and length of the precipitates also change [19,26]. In addition, 2524 aluminum alloys under different creep-aging temperature have differences in microstructures and mechanical properties such as yield strength, hardness, and elongation [26–28]. Therefore, by controlling the creep-aging temperature, a 2524 aluminum alloy with good plasticity and fatigue properties can be obtained.

3.2. *Effect of Creep-Aging Stress on Fatigue-Crack Propagation*

3.2.1. Effect of Creep Stress on Fatigue-Crack Propagation Rate

The fatigue-crack propagation rate curves of 2524 aluminum alloy after forming under different creep stresses are shown in Figure 7. In the near-threshold region of fatigue-crack propagation ($\Delta K \leq 8$ MPa·m$^{1/2}$), the fatigue-crack propagation threshold of 2524 aluminum alloy with a forming radius of 1800 mm is higher than that at 1500 mm and 1000 mm, and its propagation rate is also lower than alloys in the other two states. However, the propagation rate with the forming radius of 1000 mm and 1500 mm are relatively close. This shows that the increase in creep stress also reduces the crack propagation resistance of the alloy, but the effect is not as significant as creep-aging temperature. When 8 MPa·m$^{1/2}$ $\leq \Delta K \leq$ 20 MPa·m$^{1/2}$, the crack propagation enters the stable expansion stage, and the curve shows an obvious linear relationship. When ΔK exceeds about 18 MPa·m$^{1/2}$, the propagation rate goes up sharply and da/dN rapidly expands until fracture occurs.

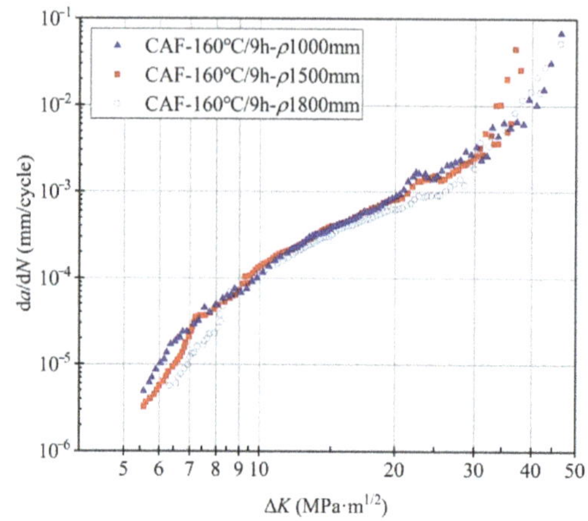

Figure 7. Fatigue-crack propagation rate curves of 2524 aluminum alloy under different creep stresses.

The linear part in Figure 7 is fitted according to the Paris equation, and the corresponding fitting constants C, n and the fatigue-crack propagation rate under the same ΔK are shown in Table 3. The error value is 4.65%. From the fitting results, the values of the exponents n are very close, ranging from 2.7 to 3.0. In the medium stress region and high stress region, the propagation rate of the formed specimens at a radius of 1000 mm is the fastest, indicating that the increase in creep stress reduces the fatigue-crack propagation resistance.

Table 3. Paris fitting parameters C and n of 2524 aluminum alloy under different creep stresses.

ρ	C	n	$da/dN = C\Delta K^n$/(mm·cycle^{-1})			
			$\Delta K = 7$	$\Delta K = 12$	$\Delta K = 16$	$\Delta K = 21$ (MPa·m$^{1/2}$)
1000 mm	1.77×10^{-7}	2.99	2.90×10^{-5}	2.25×10^{-4}	4.75×10^{-4}	1.12×10^{-3}
1500 mm	4.32×10^{-7}	2.67	2.24×10^{-5}	2.23×10^{-4}	4.78×10^{-4}	9.67×10^{-4}
1800 mm	2.81×10^{-7}	2.80	1.21×10^{-5}	2.95×10^{-4}	4.10×10^{-4}	6.83×10^{-4}

On the one hand, this is because the alloy formed under the radius of 1000 mm has the largest bending deformation and the highest dislocation density contained in the crystal [29]. The large-scale dislocation accumulation and entanglement lead to the work-hardening of the alloy itself. On the other hand, these dislocations provide a large number of nucleation sites, and meanwhile, facilitate the short-circuit expansion of solute atoms in the alloy matrix, promote the precipitation and coarsening of the precipitation phase [30], and indirectly improve the yield strength. In addition, the effect of alloy yield strength on the fatigue-crack propagation rate is mainly reflected in the size of the plastic zone at the crack tip. The plastic zone at the crack tip has the following relationship with the yield strength [31]:

$$r_p(\alpha) = \frac{\Delta K^2}{4\pi(\sigma_{0.2})^2}\left(\frac{3}{2}\sin^2\alpha + (1-2\mu)^2(1+\cos\alpha)\right) \tag{4}$$

where r_p is the radius vector; ΔK is the stress intensity factor; $\sigma_{0.2}$ is the yield stress; α is the polar angle; and μ is the Poisson's ratio of 2524 aluminum alloy. Under the same stress intensity factor amplitude ΔK, the size r_p of the plastic zone at the crack tip is inversely proportional to the yield strength $\sigma_{0.2}$. The larger the r_p, the more energy is absorbed under each cyclic load, and the better the fatigue damage resistance of the alloy (i.e., the increase in the yield strength reduces the crack propagation resistance of the alloy). Therefore, the increase in creep stress has an adverse effect on the fatigue performance of 2524 aluminum alloy.

In different creep-aging specimens, the larger the aging stress, the lower the apparent activation energy, and the lower the resistance encountered by the movement of dislocations [12]. Meanwhile, a large number of dislocations brought by pre-deformation provide a large number of mobile dislocations for creep, and also promote the nucleation of the second phase [32]. Therefore, when the creep specimen is creep formed under larger stress, the precipitation and growth rate of the precipitates are also faster, but the density of the precipitates declines. The enhancement of the pinning effect of precipitation relative to dislocations improves the plastic deformation resistance, and the damage caused by the same fatigue cyclic load is smaller.

The increase in creep stress also reduces the crack propagation resistance relative to unstressed aging. However, compared with the effect of creep-aging temperature, the effect of creep stress is not as significant as that of creep temperature.

3.2.2. Fracture Morphologies at Different Creep Stresses

(1) Fatigue-crack propagation zone

Figure 8 shows the fatigue striation spacing under the action of different creep curvature radii of 1000 mm, 1500 mm and 1800 mm, respectively. The widths of the five fatigue striations under different creep curvature radii are significantly different, at 2.27 μm, 2.66 μm and 2.93 μm, respectively. Therefore, in the stable propagation stage with the stress intensity factor ΔK ranging from 10 to 25 MPa MPa·m$^{1/2}$, the fatigue striation spacing of the specimen with a smaller creep radius of curvature (i.e., a larger creep stress) is larger than that of the specimen with a larger radius of curvature; moreover, the crack propagation rate is faster, and the fatigue resistance is lower.

For 2524 aluminum alloy under stress aging, the preferential growth orientation of the precipitates is sensitive to the applied stress. The applied stress field changes the degree of mismatch between the matrix and the precipitates, which causes the elastic distortion field and the elastic energy of the coherent precipitates to change, thus affecting the precipitation and evolution of the precipitates [33]. Larger aging stress results in lower apparent activation energy, less resistance encountered by the movement of dislocations, and a large number of dislocations caused by pre-deformation; these provide a large number of movable dislocations for creep, and promote the nucleation of the second phase [34]. Therefore, when the creep specimen creeps under the larger stress, the damage caused by the same fatigue cyclic load to the specimen is smaller; however, after the alloy enters the over-aging stage, the fatigue property declines significantly. This is because the precipitation and propagation rate of the precipitation phase goes up with the increase in stress [19], while the density of the precipitation phase goes down with an increase in stress, which makes the plasticity of the aluminum alloy stronger. While the fatigue life of the specimen decreases, the fatigue striation spacing becomes wider.

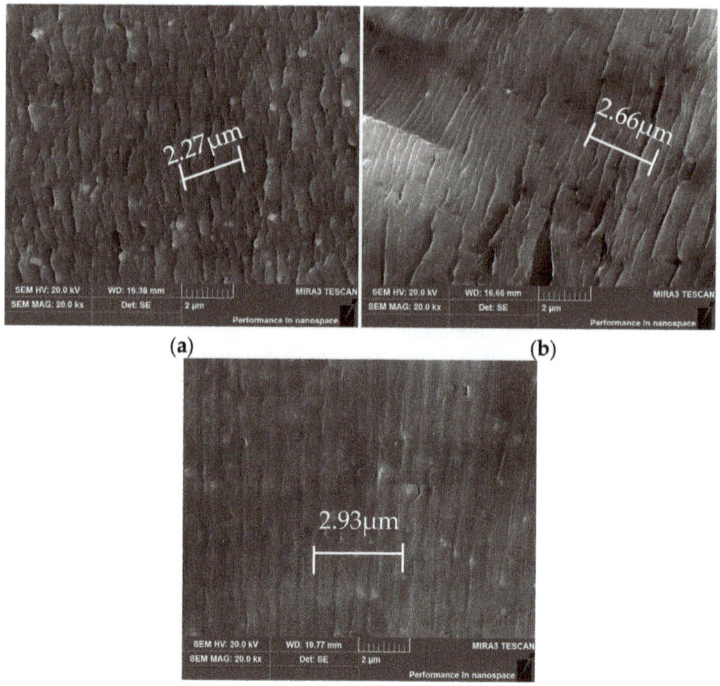

Figure 8. SEM images of fatigue striations under different curvature radii ρ: (**a**) ρ = 1000 mm; (**b**) ρ = 1500 mm; (**c**) ρ = 1800 mm.

(2) Fatigue fracture zone

Figure 9 illustrates the SEM images of the fracture zone under different creep stresses, and the curvature radii are 1000 mm, 1500 mm and 1800 mm, respectively. With the increase in creep stress, the fracture morphologies are slightly different; dimples and fractured second-phase particles are clearly seen. In the sample with a curvature radius of 1000 mm, the fracture contains dimples and cleavage fracture morphology. Comparing the fatigue properties of the specimens treated by creep-aging under different curvature radii, under the same fatigue cycle, the specimen with smaller curvature radius has a larger area of the fatigue-crack stable-growth zone, and a smaller area of fracture zone; this demonstrates that the specimen under the action of larger creep stress has longer fatigue stability propagation time and faster instantaneous fracture speed.

The applied creep stress changes the precipitation process of the precipitation phase. Under the action of high stress, the *S* phase is more likely to precipitate and grow. During creep, dislocation density increases and becomes entangled, affecting the dislocation slip. As the creep deformation increases, the dislocations in the grains are rearranged, and the entangled dislocations drive the formation of subgrains [35]. The dislocations entangled with high density form the unit cell-walls of the subgrains, and the dislocation density in the subgrains is low. In addition, the creep stress breaks the balance of the precipitates and changes the precipitation process, and the growth of the precipitates hinders the movement of grain boundaries and dislocations, improving the properties of the aluminum alloy.

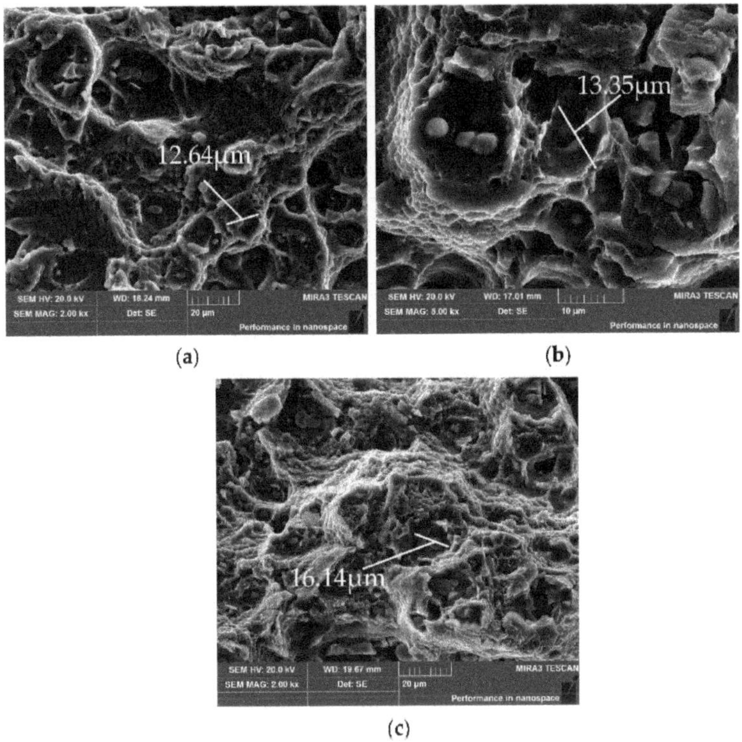

Figure 9. SEM images of the instantaneous break area under different curvature radii ρ: (**a**) ρ = 1000 mm; (**b**) ρ = 1500 mm; (**c**) ρ = 1800 mm.

4. Conclusions

(1) With the same ΔK, the crack propagation rate increases with the increase in creep-aging temperature. With the increase in ΔK, the difference in crack propagation rate gradually decreases.

(2) Under the same crack length, with the increase in creep-aging temperature, the spacing of fatigue striations increases and the size of dimples decreases, while the number of dimples increases, and the fatigue resistance of the alloy decreases.

(3) In the near-threshold region of fatigue-crack propagation ($\Delta K \leq 8$ MPa·m$^{1/2}$), the fatigue-crack propagation threshold of 2524 aluminum alloy with a forming radius of 1800 mm is higher than that of alloys at 1500 mm and 1000 mm.

(4) Under the same fatigue cycle, the specimens under the action of larger creep stress have longer fatigue stable-propagation time and a faster transient fracture speed. However, compared with the effect of creep-aging temperature, the effect of creep stress is not as significant as that of creep temperature. Hence, the creep-aging temperature plays a dominant role in the fatigue-crack propagation of creep-aged 2524 aluminum alloy.

Author Contributions: Conceptualization, L.M., C.L. and M.S.; methodology, L.M. and M.S.; software, C.L. and M.M.; validation, L.M., C.L. and M.M.; formal analysis, D.W. and L.L.; investigation, L.M., C.L., M.M., Z.W., D.W. and L.L.; resources, L.M. and C.L.; data curation, Z.W.; writing—original draft preparation, L.M., C.L., M.M. and Z.W.; writing—review and editing, M.M., Z.W., and L.L.; supervision, Z.W. and M.S.; project administration, L.M., C.L., Z.W. and L.L.; funding acquisition, L.M., C.L., Z.W., and L.L. All authors have read and agreed to the published version of the manuscript.

Funding: This research was funded by the Hebei Province Technology Innovation Guidance Program Project: Science and Technology Winter Olympics Special, grant number 20475501D; the Basic

Scientific Research Business Project of Hebei University of Architecture, grant numbers 2022QNJS02 and 2021QNJS08; and the Changsha Municipal Natural Science Foundation, grant number kq2007085.

Data Availability Statement: The data presented in this study are available on request from the corresponding authors.

Conflicts of Interest: The authors declare no conflict of interest.

References

1. Li, S.B.; Li, X.; Liang, W.; Liu, Y.L.; Yan, H.Z.; Liu, C. Effects of laser shock peening on fatigue crack growth rate and fracture properties of AA2524 aluminum alloy. *J. Cent. South Univ.* **2022**, *29*, 848–859. [CrossRef]
2. Song, M.; Wu, L.; Liu, J.; Hu, Y. Effects of laser cladding on crack resistance improvement for aluminum alloy used in aircraft skin. *Opt. Laser Technol.* **2021**, *133*, 106531. [CrossRef]
3. Mason, C.J.T.; Avery, D.Z.; Phillips, B.J.; Jordon, J.B.; Allison, P.G. Strain Rate Dependent Plasticity Model for Precipitate Hardened Aerospace Aluminum Alloy Produced with Solid-State Additive Manufacturing. *J. Dyn. Behav. Mater.* **2021**. [CrossRef]
4. Song, B.; Liu, Z.; Gu, Y.; Zhou, X.; Zeng, S. Microstructures and fatigue fracture behavior of an Al–Cu–Mg–Ag alloy with a low Cu/Mg ratio. *Mater. Sci. Eng. A* **2011**, *530*, 473–480.
5. Hu, Y.-J.; Sun, Y.-P.; Zhou, S.-P.; He, J.-M.; Yang, C.-Y. Effect of a cooling method on the structural and mechanical properties of friction stir spot welding with a 2524 aluminum alloy. *Mater. Res. Express* **2021**, *8*, 026517. [CrossRef]
6. Srivatsan, T.S.; Kolar, D.; Magnusen, P. Influence of temperature on cyclic stress response, strain resistance, and fracture behavior of aluminum alloy 2524. *Mater. Sci. Eng. A* **2001**, *314*, 118–130. [CrossRef]
7. Zheng, Z.Q.; Cai, B.; Zhai, T.; Li, S.C. The behavior of fatigue crack initiation and propagation in AA2524-T34 alloy. *Mater. Sci. Eng. A* **2011**, *528*, 2017–2022. [CrossRef]
8. Srivatsan, T.S.; Kolar, D.; Magnusen, P. The cyclic fatigue and final fracture behavior of aluminum alloy 2524. *Key Eng. Mater.* **2008**, *378–379*, 207–230. [CrossRef]
9. Liu, C.; Ma, L.; Zhang, Z.; Fu, Z.; Liu, L. Research on the Corrosion Fatigue Property of 2524-T3 Aluminum Alloy. *Metals* **2021**, *11*, 1754. [CrossRef]
10. Pitcher, P.D.; Styles, C.M. Creep Age Forming of 2024A, 8090 and 7449 Alloys. *Mater. Sci. Forum* **2000**, *331–337*, 455–460. [CrossRef]
11. Siddiqui, R.A.; Abdullah, H.A.; Al-Belushi, K.R. Influence of aging parameters on the mechanical properties of 6063 aluminium alloy. *J. Mater. Processing Technol.* **2000**, *102*, 234–240. [CrossRef]
12. Bray, G.H.; Glazov, M.; Rioj, R.J.; Lib, D.; Gangloffb, R.P. Effect of artificial aging on the fatigue crack propagation resistance of 2000 series aluminum alloys. *Int. J. Fatigue* **2001**, *23*, 265–276. [CrossRef]
13. Sarioğlu, F.; Orhaner, F.Ö. Effect of prolonged heating at 130°C on fatigue crack propagation of 2024 Al alloy in three orientations. *Mater. Sci. Eng. A* **1998**, *248*, 115–119. [CrossRef]
14. Burba, M.E.; Caton, M.J.; Jha, S.K.; Szczepanski, C.J. Effect of Aging Treatment on Fatigue Behavior of an Al-Cu-Mg-Ag Alloy. *Metall. Mater. Trans. A* **2013**, *44*, 4954–4967. [CrossRef]
15. Songbai, L.I.; Liyong, M.A.; Chi, L.; Jiuhuo, Y.I. Effects of Aging Temperature on Microstructure and High Cycle Fatigue Performance of 7075 Aluminum Alloy. *J. Wuhan Univ. Technol. (Mater. Sci.)* **2017**, *32*, 677–684.
16. Xu, Y.; Zhan, L.; Li, W. Effect of pre-strain on creep aging behavior of 2524 aluminum alloy. *J. Alloys Compd.* **2017**, *691*, 564–571. [CrossRef]
17. Xu, Y.; Zhan, L.; Xu, L.; Huang, M. Experimental research on creep aging behavior of Al-Cu-Mg alloy with tensile and compressive stresses. *Mater. Sci. Eng. A* **2017**, *682*, 54–62. [CrossRef]
18. Liu, Y.L.; Wang, Q.; Liu, C.; Song-Bai, L.I.; Jun, H.; Zhao, X.Q. Effect of creep and artificial aging on fatigue crack growth performance of 2524 aluminum alloy. *J. Jilin Univ. (Eng. Ed.)* **2019**, *49*, 1636–1643.
19. Liu, C.; Liu, Y.; Li, S.; Ma, L.; Zhao, X.; Wang, Q. Effect of creep aging forming on the fatigue crack growth of an AA2524 alloy. *Mater. Sci. Eng. A* **2018**, *725*, 375–381. [CrossRef]
20. ASTM E647-08. Standard Test Method for Measurement of Fatigue Crack Growth Rates. ASTM International: West Conshohocken, PA, USA, 2008.
21. Khan, I.N.; Starink, M.J.; Yan, J.L. A model for precipitation kinetics and strengthening in Al–Cu–Mg alloys. *Mater. Sci. Eng. A* **2008**, *472*, 66–74. [CrossRef]
22. Quan, L.W.; Zhao, G.; Tian, N.; Huang, M.L. Effect of stress on microstructures of creep-aged 2524 alloy. *Chin. J. Nonferrous Met.* **2013**, *23*, 2209–2214. [CrossRef]
23. Zhang, H.; Qiu, X.; Xu, D.; Liu, Y.; Zhao, X. Effect of precipitated phase on dislocation activity under high-frequency impacting and rolling. *Micro Nano Lett.* **2018**, *13*, 1542–1544. [CrossRef]
24. Benachour, M.; Hadjoui, A.; Benguediab, M.; Benachour, N. Stress Ratio Effect on Fatigue Behavior of Aircraft Aluminum Alloy 2024 T351. *MRS Proc.* **2011**, *1276*. [CrossRef]
25. Masoudi Nejad, R.; Berto, F.; Tohidi, M.; Jalayerian Darbandi, A.; Sina, N. An investigation on fatigue behavior of AA2024 aluminum alloy sheets in fuselage lap joints. *Eng. Fail. Anal.* **2021**, *126*, 105457. [CrossRef]
26. Xu, Y.; Yang, L.; Zhan, L.; Yu, H.; Huang, M.J.M. Creep Mechanisms of an Al–Cu–Mg Alloy at the Macro- and Micro-Scale: Effect of the S′/S Precipitate. *Materials* **2019**, *12*, 2907. [CrossRef]

27. Xu, Y.; Zhan, L.; Ma, Z.; Huang, M.; Wang, K.; Sun, Z. Effect of heating rate on creep aging behavior of Al-Cu-Mg alloy. *Mater. Sci. Eng. A* **2017**, *688*, 488–497. [CrossRef]
28. Zhan, L.H.; Tan, S.G.; Yang, Y.L.; Huang, M.H.; Shen, W.Q.; Xing, Z. A Research on the Creep Age Forming of 2524 Aluminum Alloy: Springback, Mechanical Properties, and Microstructures. *Adv. Mech. Eng.* **2014**, *6*, 707628. [CrossRef]
29. Lin, Y.C.; Jiang, Y.Q.; Zhang, X.C.; Deng, J.; Chen, X.M. Effect of creep-aging processing on corrosion resistance of an Al–Zn–Mg–Cu alloy. *Mater. Des.* **2014**, *61*, 228–238. [CrossRef]
30. Gouma, P.I.; Lloyd, D.J.; Mills, M.J. Precipitation processes in Al–Mg–Cu alloys. *Mater. Sci. Eng. A* **2001**, *319–321*, 439–442. [CrossRef]
31. Kurguzov, V.D.; Kornev, V.M.; Moskvichev, V.V.; Kozlov, A.A. Influence of periodic change in the yield strength in a plate on the development of plastic zones near a crack tip. *J. Appl. Mech. Tech. Phys.* **2014**, *55*, 1037–1044. [CrossRef]
32. Chen, Y.Q.; Pan, S.P.; Zhou, M.Z.; Yi, D.Q.; Xu, D.Z.; Xu, Y.F. Effects of inclusions, grain boundaries and grain orientations on the fatigue crack initiation and propagation behavior of 2524-T3 Al alloy. *Mater. Sci. Eng. A* **2013**, *580*, 150–158. [CrossRef]
33. Liu, C.; Liu, Y.; Ma, L.; Li, S.; Zhao, X.; Wang, Q. Precipitate Evolution and Fatigue Crack Growth in Creep and Artificially Aged Aluminum Alloy. *Metals* **2018**, *8*, 1039. [CrossRef]
34. Lin, Y.C.; Xia, Y.C.; Jiang, Y.Q.; Zhou, H.M.; Li, L.T. Precipitation hardening of 2024-T3 aluminum alloy during creep aging. *Mater. Sci. Eng. A* **2013**, *565*, 420–429. [CrossRef]
35. Ungár, T.; Victoria, M.; Marmy, P.; Hanák, P.; Szenes, G. A new procedure of X-ray line profile analysis applied to study the dislocation structure and subgrain size-distributions in fatigued MANET steel. *J. Nucl. Mater.* **2000**, *276*, 278–282. [CrossRef]

materials

Article

An Investigation of the Contact Fatigue Characteristics of an RV Reducer Crankshaft, Considering the Hardness Gradients and Initial Residual Stress

Xin Li, Wen Shao *, Jinyuan Tang, Han Ding and Weihua Zhou

State Key Laboratory of High Performance Complex Manufacturing, College of Mechanical and Electrical Engineering, Central South University, Changsha 410083, China
* Correspondence: wen.shao@csu.edu.cn or shaowen_2013@163.com

Abstract: The crankshaft is one of the core components of a Rotate Vector (RV) reducer. The fatigue life of the RV reducer is severely hindered by fatigue failure on the eccentric cylindrical surface of the crankshaft. The hardness gradients and residual stress in the crankshaft, associated with machining operations, exert an enormous impact on the rolling contact fatigue (RCF). In this work, a finite element method (FEM)-based three-dimensional elasto-plastic contact model is established to calculate the stress–strain field by taking hardness gradients and initial residual stress into account. The RCF characteristics of an RV reducer crankshaft is investigated by applying modified Fatemi–Socie (FS) multiaxial fatigue criterion. The results indicate that initial residual stress plays an influential role in the fatigue damage by altering the distribution of the maximum normal stress near the contact surface. The modified FS fatigue criterion could better consider the effect of initial residual stress and the shear stress, which significantly improves the prediction accuracy of the contact fatigue life model. The contact fatigue performance could be considerably improved by designing appropriate shot peening parameters to obtain optimized residual stress distribution. Therefore, the technique presented may serve as an important guideline for the anti fatigue design of an RV reducer crankshaft.

Keywords: RV reducer crankshaft; initial residual stress; hardness gradients; rolling contact fatigue; modified Fatemi–Socie criterion

Citation: Li, X.; Shao, W.; Tang, J.; Ding, H.; Zhou, W. An Investigation of the Contact Fatigue Characteristics of an RV Reducer Crankshaft, Considering the Hardness Gradients and Initial Residual Stress. *Materials* **2022**, *15*, 7850. https://doi.org/10.3390/ma15217850

Academic Editors: Lucjan Śnieżek, Jaroslaw Galkiewicz and Sebastian Lipiec

Received: 31 August 2022
Accepted: 3 November 2022
Published: 7 November 2022

1. Introduction

The RV (Rotate Vector) reducer is extensively used in industry robots owing to its peculiar and fascinating properties, such as compact structure, small size, light weight, high reduction ratio, high transmission accuracy and efficiency, high torsional rigidity, etc. [1]. The crankshaft and cylindrical roller bearing are the core components of the RV reducer. Due to the limited available space of the RV reducer, the cylindrical roller bearing is usually without the inner ring and outer ring. That is, the inner and outer rings of the bearing are directly composed of the eccentric cylindrical surface of the crankshaft and the bearing inner hole on the cycloid wheel. The transmission errors and fatigue life of the RV reducer is severely hindered by failure modes, such as pitting and spalling on the eccentric cylindrical surface caused by the long-term high cyclic contact stress between the eccentric cylindrical surface of the crankshaft and the roller bearing [2]. With increasingly higher performance requirements, such as high transmission accuracy, high load-carrying capacity, and long service life of the RV reducer, the RCF of the crankshaft has also become the limiting factor affecting the reliability of the RV reducer.

Several attempts have been made towards improving the transmission performance of RV reducers. Zhang et al. [3] established a mixed lubrication analysis model for RV reducers. The contact load, surface roughness, and geometry of the cylindrical roller bearings were innovatively included in their model. Xu et al. [4,5] developed a dynamic model for the transmission systems of an RV reducer that took into account the cylindrical roller bearing's radial clearance. Wang et al. [6] presented a contact force and transmission error analysis of an RV Reducer. Their results revealed that the modified model, based on contact force curves, could improve the transmission performance of the RV reducer as the load increased. Deng et al. [7] calculated the rated life of the RV reducer's angular contact ball bearing. They also discovered that bearing life had a significant impact on the total life of the RV reducer under heavy load conditions. The present investigations on RV reducers have been mostly focused on the meshing characteristics of the cycloid pinwheel, the dynamic characteristics of the RV reducer, and the life estimation of the bearing. Few studies have been reported on the RCF performances of the crankshaft of the RV reducer due to the complexity of the contact load between the RV reducer crankshaft and the cylindrical roller bearing, although it exerts an enormous impact on the reliability and fatigue life of the RV reducer.

Concerted efforts were directed towards the fatigue life and damage prediction methods for gears and rolling bearings. Li et al. [8] presented a multiaxial fatigue model considering mixed lubrication for crack initiation life prediction of spur gears. Liu et al. [9] proposed an improved multiaxial fatigue life model with higher life prediction accuracy compared with classical models. Vijay et al. [10] presented a novel model to simulate the crack initiation and propagation in bearing steels, considering the anisotropy of crystal. Their results indicated that the Fatemi–Socie (FS) criterion could be used to estimate the RCF life. Continuum damage mechanics (CDM) and the elasto-plastic model with damage-coupling were widely used to investigate the spalling initiation and propagation behaviors of cylindrical rolling bearings [11,12]. Despite the efforts of these earlier studies, prediction of the fatigue life and damage from these investigations may inevitably suffer from the disadvantage of neglecting the effect of mechanical property that is introduced in the manufacturing processes, such as heat treatment, grinding, shot peening, etc.

In recent years, considerable progress has been achieved for the evaluation of the contact fatigue performance by taking surface integrity such as surface topography, residual stress, hardness gradient, and microstructure into account. Numerous researchers have shown that surface integrity significantly affects the fatigue life in rolling contact [13]. Choi et al. [14] pointed out that the prediction accuracy of fatigue life increased more than 40% when the residual stress was taken into account. The results also showed that increments of more than 12 and 8 times could be reached for the crack initiation propagation lives if residual stress was considered. The critical plane approach was recognized as an effective method to solve multiaxial fatigue problems [9]. The Dang–Van fatigue criterion, the FS criterion, and Zaretsky fatigue life model were applied to study the contact fatigue performance of a carburized gear under heavy loading by incorporating residual stress [15,16]. Mahdavi et al. [17] investigated the effect of superposed residual stresses on micro plasticity around inclusions in bearing steel. Ooi et al. [18] experimentally studied the impact of restrained austenite and residual stress on the fatigue life of carburized AISI 8620 steel. Although residual stress was considered in [17,18], the distribution of residual stress was not well represented in these studies for the estimation of RCF. Guan et al. [19] investigated the influence of compressive residual stress (CRS) induced by shot peening on fatigue risks and found that appropriate CRS distribution could decrease the rate of damage accumulation in bearing steel containing carbide. Walvekar et al. [20] studied the combined impact of hardness gradient and residual stress curves on the RCF lives of bearing steel materials. It was found that optimized carburizing depth could prolong the fatigue life of bearing steel materials to a large extent. Furthermore, the relationship between associated parameters such as effective case hardening depth (CHD), surface hardness, and hardness curve shape and fatigue performance was also demonstrated [21,22].

The above studies confirm that it is of theoretical value and engineering significance to consider the mechanical property gradient and initial residual stress when evaluating the contact fatigue performance of carburized steel materials or components. The researchers above also provided a theoretical basis for the contact fatigue characteristic analysis of the crankshaft. Moreover, the successful prediction of the fatigue life of the crankshaft incorporating residual stress and hardness gradient through numerical modelling provides the superiorities of yielding numerous results in a short period and sometimes offers beneficial insight into different states of RCF processes, which can be also used to optimize the machining parameters. Therefore, it is of great importance and necessity to consider the initial state of the crankshaft in the contact fatigue life prediction model.

In this work, focusing on fatigue life assessment of an RV reducer crankshaft, an FEM-based three-dimensional elasto-plastic contact model is established. The hardness gradients and initial residual stress are obtained by Vickers hardness tests and X-ray diffraction method. The collected hardness and residual stress data are incorporated into the elasto-plastic contact model and then the stress–strain histories are obtained. Contact fatigue life assessment is performed by the critical plane method and multiaxial fatigue criterion. The influence of friction factor and initial residual stress on fatigue damage and fatigue life is investigated. Moreover, the effect of residual stress induced by shot peening on improving fatigue life is also analyzed quantitatively.

2. Failure Analysis of Crankshaft

The reliability and fatigue life are important performance indicators of the RV reducer, so it is necessary to carry out fatigue life tests of the RV reducer. In addition, the fatigue life test is not only conducive to analyzing the performance degradation law and failure mechanism of the reducer, but also provides guidance for the design and manufacture of key parts of the RV reducer.

Figure 1 shows fatigue life test system of an RV reducer. Because the RV reducer is a piece of high-precision transmission equipment, transmission efficiency can be selected as the failure judgment criterion, and comprehensive judgment can be made in combination with vibration, temperature rise, and noise. One RV reducer testing machine is selected for fatigue test. The test is stopped when the transmission efficiency is detected to be lower than 85% of the threshold, and a service life of approximately 2150 h was obtained.

Figure 1. Fatigue life test system of RV Reducer.

The reducer was disassembled after the fatigue test. Comparing the new crankshaft parts, it was found that the cylindrical surface of the needle roller bearing in contact with the crankshaft was seriously worn, as shown in Figure 2.

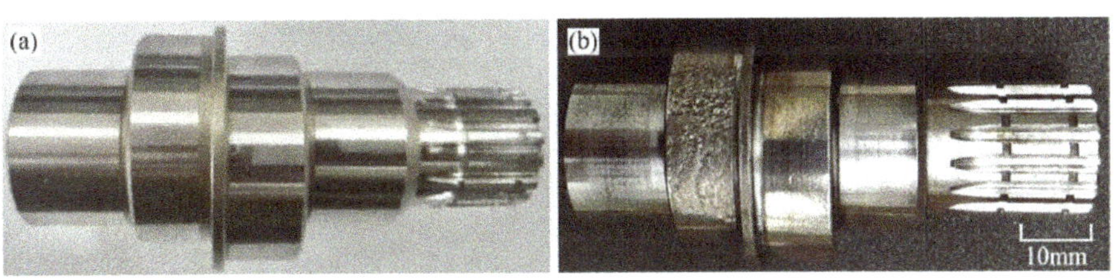

Figure 2. Macroscopic morphology of crankshaft: (**a**) the new part; (**b**) the failed part.

The SEM micrographs of the eccentric cylindrical surface of the crankshaft before and after failure are shown in Figure 3. Regular grinding grooves are left on the new crankshaft surface (as shown in Figure 3a). There are thin strip-like scratches along the rolling direction, and numerous pit-like cracks appeared on the surface of the failed crankshaft (Figure 3b,c).

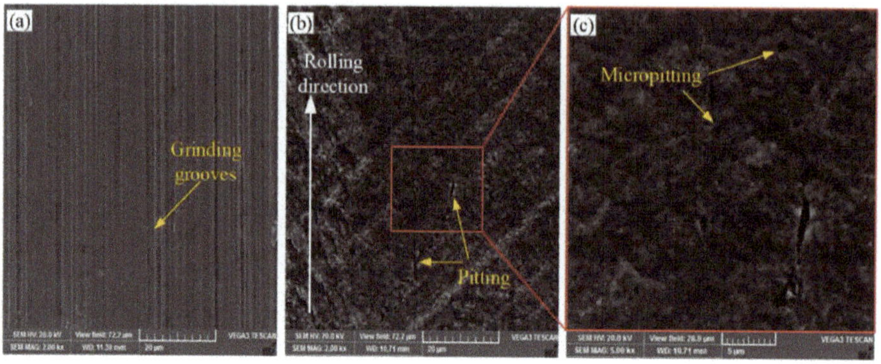

Figure 3. The SEM micrographs of the eccentric cylindrical surface of the crankshaft: (**a**) surface of new parts; (**b**,**c**) surface after failure.

Because the crank shaft is the core part of the power input and output of the reducer, it bears periodic radial load and dynamic load. Moreover, the reducer often experiences impact load, continuous start and stop, and other working conditions, which leads to complex load changes on the crankshaft. Therefore, the eccentric cylindrical surface of the crankshaft is also the most prone to failure in engineering practice. Fatigue pitting occurs on the eccentric cylindrical surface of the crankshaft under the long-term high cyclic contact stress, and a series of pits are formed. The crankshaft is one of the weakest links of the RV reducer, which severely restricts the fatigue life of the RV reducer.

3. Methodology

3.1. Stress Analysis of RV Reducer

The crankshaft used in this study is taken from the RV reducer of an industrial robot. The structure of the RV reducer is shown in Figure 4. The critical parameters of the rolling contact pair between the crankshaft and roller are given in Table 1. The crankshaft material is 20CrNi2MoA. The composition of it is listed in Table 2. The crankshaft has undergone several manufacturing processes, such as carburizing, quenching, tempering, and finally precision grinding. The detailed thermal treatment process is shown in Figure 5. The crankshaft sample is carburized and diffused for 6 h after the temperature soared to 930 °C, then the sample is quickly quenched in oil, followed by low temperature tempering at 230 °C for 2 h. After heat treatment, the lath martensite structure is finally obtained. Lath martensite can better resist impact and crack propagation, so that the material has higher hardness, good wear resistance, and higher contact fatigue properties.

Through the motion and force analysis of the cycloidal gear and cylindrical roller bearing (Figure 6), the resultant force acting on the cylindrical roller bearing is obtained. The cylindrical roller bearing bears the force from the cycloidal gear and the crankshaft. F_{j1}, F_{j2}, and F_{j3} are the components of force, F, which acts on the cycloidal gear via the cylindrical roller bearing (as shown in Figure 6a). F can also be decomposed into the normal force, F_r, and tangential force, F_t (as shown in Figure 6b).

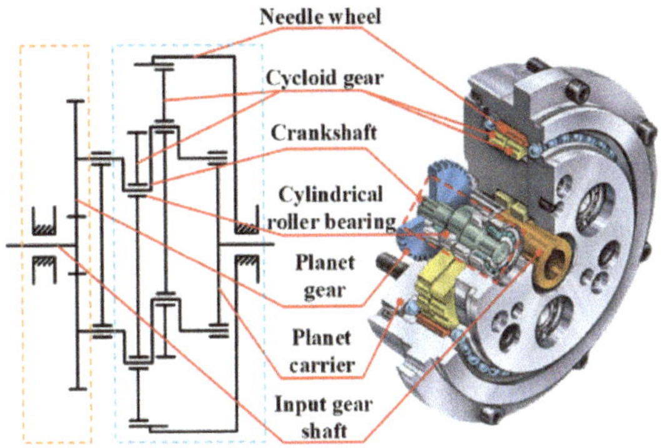

Figure 4. Transmission schematic diagram of RV Reducer.

Table 1. Crankshaft-bearing rolling pair parameters.

Parameters	Value	Parameters	Value
Crankshaft rotation speed (r/min)	$n_s = 585$	Rated output torque (N·m)	$T_{out} = 800$
Radius of the rolling element (mm)	$R_1 = 4.0$	Rolling element material	GCr15
Radius of the eccentric cylindrical surface (mm)	$R_2 = 16.6$	Crankshaft material	20CrNi2MoA
Length of the rolling element (mm)	$l_1 = 12.0$	Young's modulus (GPa)	$E_1 = 219$, $E_2 = 210$
Radius of needle tooth distribution circle (mm)	$R_z = 82$	Poisson ratio	$v_1 = 0.3$, $v_2 = 0.275$
Radius of crankshaft distribution circle (mm)	$r_o = 46.77$	Eccentricity (mm)	$e = 1.5$
Number of teeth of needle wheel	$z_4 = 39$	Number of crankshafts	$n = 3$
Number of teeth of cycloid gear	$z_5 = 40$	Short amplitude coefficient	$k = 0.7317$

Table 2. The composition of 20CrNi2MoA.

Element	C	Mn	Cr	Ni	Si	Mo	Cu	S	P	Fe
Wt.%	0.21	0.63	0.57	1.8	0.33	0.25	0.30	0.015	0.02	Bal.

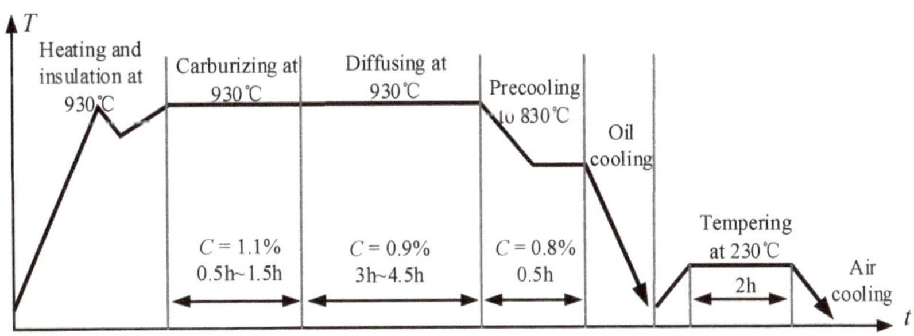

Figure 5. The curve of the thermal treatment process.

Based on the structural parameters of the RV reducer and the force equilibrium, F can be expressed as follows, and the detailed calculation process can be seen in [23].

$$F = \sqrt{F_t^2 + F_r^2} = \frac{T_{out}}{2nez_5r_o}\sqrt{(ez_5)^2 + r_o^2 + k_y^2r_o^2 + 2ez_5r_o\cos\varphi - 2k_yez_5r_o\sin\varphi} \quad (1)$$

Combining Table 1 and Equation (1), the resultant force distribution of the cycloid gear acting on the cylindrical roller is obtained, as depicted in Figure 7. Periodic sinusoids are found for the forces acting on the cylindrical rollers. The load fluctuation of three cylindrical roller bearings in the same cycloid gear are found to have a phase difference of 120°. According to Harris et al. [24], the equivalent load of cylindrical roller bearings can be calculated as [23]:

$$F_m = \frac{T_{out}}{2}\left(\frac{(ez_5)^4 + 4\left(1+k_y^2\right)(ez_5)^2 r_o^2 + \left(1+k_y^2\right)^2 r_o^4}{n^4 (ez_5)^4 r_o^4}\right)^{1/4} \tag{2}$$

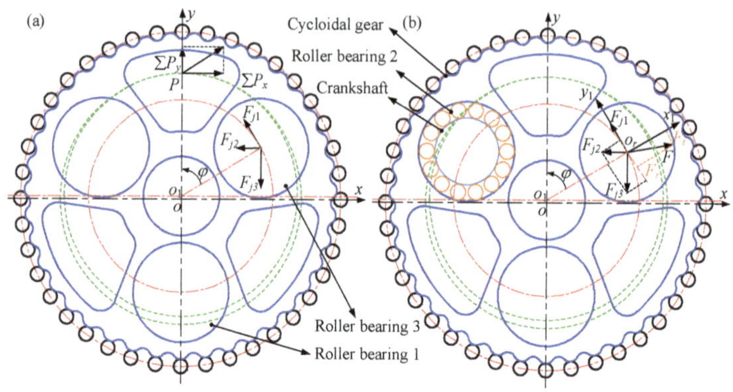

Figure 6. Schematic diagram: (**a**) force on cycloid gear; (**b**) force on cylindrical roller bearing.

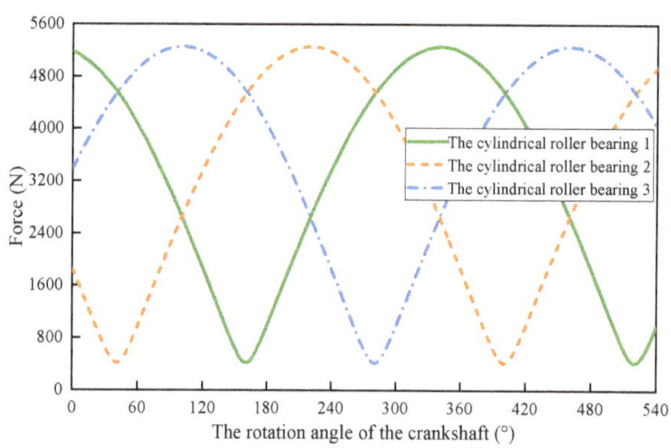

Figure 7. The variation of forces acting on the cylindrical roller bearings.

Power is transmitted between the crankshaft and the cylindrical rollers. Dynamic loads and relatively large radial forces act on the cylindrical rollers. The force analysis of each roller on the cylindrical roller bearing is carried out after obtaining the resultant force on the cylindrical roller bearing. The Newton–Raphson algorithm is employed to calculate the load distribution of each rolling element acting on the crankshaft. The crankshaft is subjected to periodic load, hence the equivalent contact load acting on the crankshaft by the cylindrical roller bearing will be used as input in the following section.

3.2. Modeling of Residual Stress and Hardness Gradient

The x, y, and z directions in the crankshaft coordinates act along the rolling, depth, and axial directions, respectively. Several empirical methods for fitting the hardness distribution profile introduced by carburizing have been proposed by Lang and Kernen [25] and Thomas [26]. Among them, the Thomas method is the closest to the measured results and has been widely used in ISO standards. The empirical formula can be expressed as [26]:

$$HV(y) = \begin{cases} a_a \cdot y^2 + b_a \cdot y + c_a (0 \le y < CHD) \\ a_b \cdot y^2 + b_b \cdot y + c_b (CHD \le y < y_{cor}) \\ HV_{cor} (y_{cor} \le y) \end{cases} \tag{3}$$

$$\begin{cases} a_a = \frac{550 - HV_{sur}}{CHD^2 - 2 \cdot y_{HV,\max} \cdot CHD}; b_a = -2 \cdot a_a \cdot y_{HV,\max}; c_a = HV_{sur}; \\ a_b = \frac{H'(CHD)}{2 \cdot (CHD - y_{cor})}; b_b = -2 \cdot a_b \cdot y_{cor}; \\ c_b = 550 - a_b \cdot CHD^2 - b_b \cdot CHD; H'(CHD) = 2 \cdot a_a \cdot CHD + b_a; \end{cases} \tag{4}$$

where HV_{sur} is the hardness on the surface, HV_{cor} is the hardness in the core, and y_{cor} and $y_{HV,max}$ denote the depths with hardness equal to HV_{cor} and the maximum hardness, respectively. In this study, $y_{HV,max}$ is equal to zero. The depth with a hardness of 550 HV is defined as CHD (the case hardening depth). The CHD, HV_{sur}, and HV_{cor} are designed as 1.0 mm, 670 HV, and 450 HV, respectively, based on the engineering practice of the crankshaft.

The crankshaft sample is cut along the cross-section by a wire cutter, inlaid and polished, and then the cross-sectional microhardness is measured by a Vickers hardness tester. A pyramid diamond indenter is selected for the hardness test, with a load of 0.5 kgf and duration of 15 s. Different positions at the same depth were measured at least two times, and the average hardness was obtained. Figure 8 shows the hardness data (the green circle) measured by the hardness tester machine. The red and black dashed lines are used for determine the value of CHD. The empirical hardness gradient curve (the blue solid line) based on the Thomas method is also shown in Figure 8. It can be seen that the empirical curve correlates well with the experimental data.

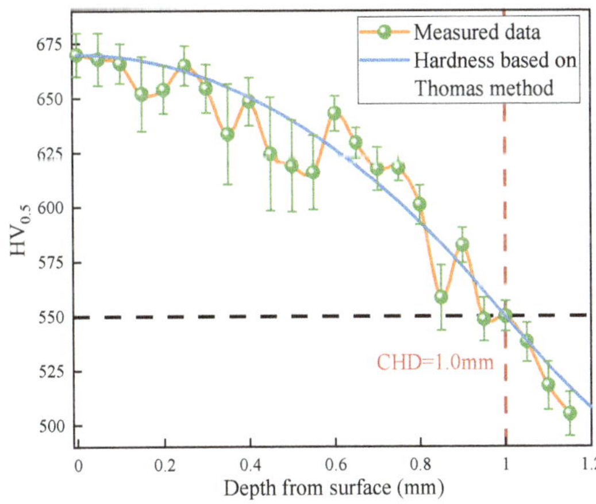

Figure 8. The measured hardness data and the empirical hardness curve.

The tensile strength and yield strength and residual stress distribution will be altered by the variation of martensite, retained austenite, and other structural components during the carburizing-quenching process. A linear relationship existed between Vickers hardness and tensile strength and yield strength, which was given by [27,28].

$$\sigma_{ys}(y) = \frac{HV(y)}{3} \cdot (0.1)^{m-2} \tag{5}$$

$$\sigma_b(y) = \frac{HV(y)}{3} [1 - (m-2)] \left[\frac{12.5(m-2)}{1-(m-2)} \right]^{(m-2)} \tag{6}$$

where $\sigma_{ys}(y)$, $HV(y)$, and $\sigma_b(y)$ are the yield strength, Vickers hardness, and the tensile strength, respectively, and m is the Meyer hardness coefficient, which can be taken as 2.19 for high-strength steel materials.

The generally recognized X-ray diffraction test with electrolytic polisher and empirical formula can be used to determine the residual stress at the contact surface or near-surface areas of the crankshaft. In the present work, the electropolishing method (as shown in Figure 9b) is employed to remove the material layer by layer, and the corresponding residual stress distribution is examined through a Proto iXRD system with Cr-Kα radiation (as shown in Figure 9a). The magnitudes of the residual stress along the x and z axis ($\sigma_{r,x}$ and $\sigma_{r,z}$) are found to be equivalent at the same depth, thus the residual stress is expressed as symbol σ_r in the subsequent section. The residual stress along the depth direction, $\sigma_{r,y}$, is always negligible compared with the other two components, and is thus ignored in this study.

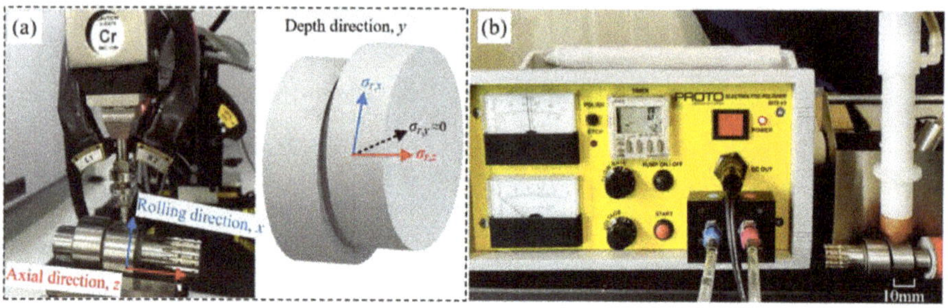

Figure 9. Residual stress measurement: (**a**) X-ray diffractometer; (**b**) Electrolytic polisher.

Empirical methods have also been provided to characterize the linear relationship between the hardness curve and the residual stress distribution [29,30]. Hertter's empirical formula employed in this study is given by [30]:

$$\sigma_r(y) = \begin{cases} -1.25\cdot(HV(y) - HV_{core}) & (HV(y) - HV_{core} \leq 300) \\ 0.2857\cdot(HV(y) - HV_{core}) - 460 & (HV(y) - HV_{core} > 300) \end{cases} \tag{7}$$

Figure 10 shows the residual stress data measured by X-ray diffraction (the green circle), and the empirical curve (the yellow solid line) is obtained by the Hertter's method. It can be observed that the measured residual stresses are, in general, consistent with those fitted by the empirical formula. Therefore, the empirical residual stress curve is employed in the numerical model.

Moreover, the influence of tensile residual stress caused by improper carburizing or grinding burns during the machining process on fatigue damage will also be analyzed in this study [31]. Ultrasonic vibration assisted grinding (UAG) has many advantages, such as reducing grinding force and improving surface quality [32]. A hypothetic tensile residual stress curve (the blue dashed line) with the same amplitude as compressive residual compressive stress is depicted in Figure 10.

The residual stress generated by the plastic deformation near the surface is also caused by surface strengthening processes such as shot peening and ultrasonic rolling. Shot peening produces spoon-shaped residual compressive stress distribution, which is conducive to improving the fatigue performance [33]. Zhao at al. [34] proposed a model for calculating the residual stress distribution after shot peening. The maximum error between the measured and simulated residual stress was 15.8%, which verified the accuracy of the method. A residual stress curve distributed along the depth is designed based on the residual stress data of shot peening in [34], as shown in Figure 11. The maximum plastic deformation near the surface layer leads to the maximum residual compressive stress. The maximum residual stress depth is 50 μm. The compressive stress decreases gradually with the increase of depth and then tends to be stable.

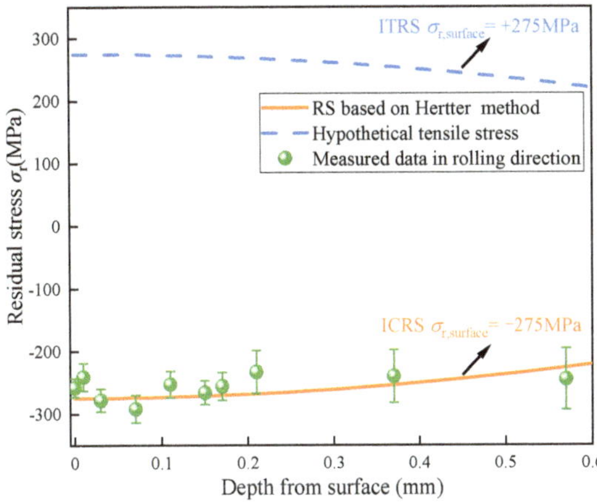

Figure 10. The measured residual stress data and the empirical stress curve.

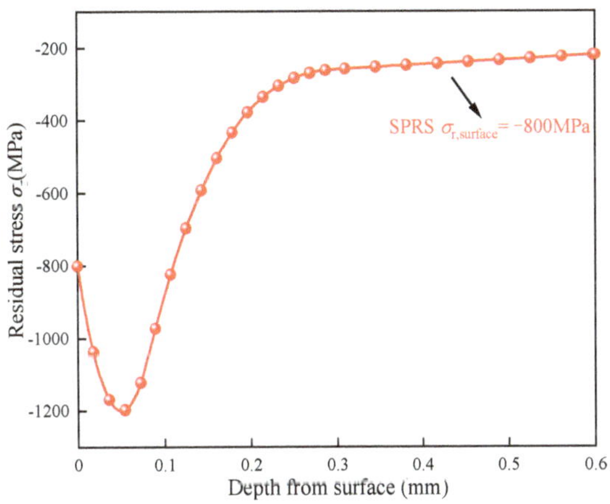

Figure 11. The residual stress distributions induced by shot peening.

3.3. *FEM-Based Elasto-Plastic Contact Analysis*

The rolling contact between the eccentric cylindrical surface of the crankshaft and the cylindrical roller bearing can be simplified as rigid cylindrical surface and a semi-infinite space according to the contact mechanics theory [35]. The diagram of the contact model of the crankshaft and the cylindrical roller is depicted in Figure 12.

In order to better reflect the three-dimensional stress–strain field response, a three-dimensional elasto-plastic finite element contact model is established in the commercial finite element software ABAQUS (Figure 13). The equivalent curvature radius of the rigid cylindrical surface is R = 3.22 mm. The normal load, F, applied on the rigid cylindrical surface, is determined by the output torque of the RV reducer during numerical simulation. The bottom nodes of the mesh model are fully fixed, and symmetric boundary conditions are set around the mesh model. That is, the nodes on the left and right sides of the mesh model are constrained in x and y directions. A static implicit solver is selected for simulation calculation. The computational domain is determined as -500 μm $\leq x \leq$ 500 μm,

0 μm ≤ y ≤ 600 μm, and 0 μm ≤ z ≤ 200 μm. The rolling direction, x, is long enough to minimize the influence of the boundary on the stress calculation in this direction.

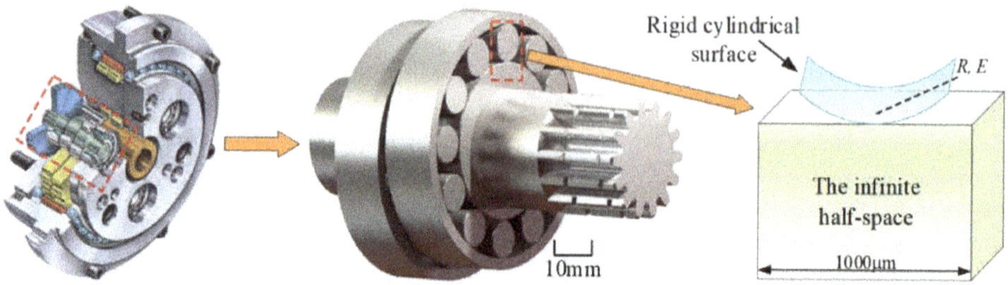

Figure 12. The contact model of the crankshaft and the cylindrical roller bearing.

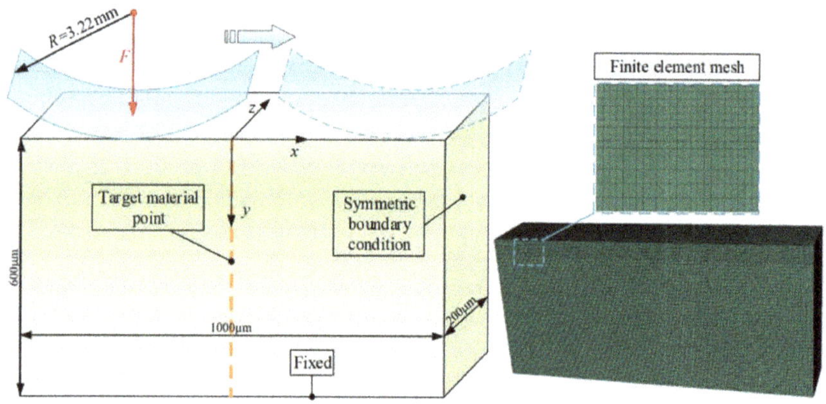

Figure 13. The numerical elasto-plastic contact model.

Considering the efficiency and accuracy of finite element calculation, the grid size is gradually expanded from the contact surface to the bottom. The grid size in the x and z directions is 5 μm. The grid size of the near surface layer in the y direction is 5 μm, and the grid size is set to 10 μm as the depth increases. The C3D8R element is selected because it can well withstand distortion, and its stress–strain calculation is also more accurate. The tangential contact property is defined as frictional contact with a friction coefficient of 0.05. The friction coefficient of 0.05, which was also used in [36,37], representing excellent lubrication in rolling contact, is employed in this study. The rigid cylindrical surface rolls from $x = -400$ μm to $x = 400$ μm, and multiple analysis steps are set to realize the reciprocating circular motion of the cylindrical surface. The material points lying on the yellow dashed line ($x = 0$ and $y \in [0, 600]$ μm) will be analyzed in the present study because the material at the same depth experiences the same cyclical stress.

Ultra-high stress exists in the kinematic pair (crankshaft and the cylindrical roller bearing) during extreme operation conditions, which inevitably produces plastic deformation. To accurately characterize the elastoplastic response, the varying yield stress with depth is incorporated into the FEM model, according to the linear positive correlation between the hardness and yield strength (Equation (5)). The efficiency for assigning material properties can be improved through the secondary development by Python. Therefore, the isotropic hardening model is applied according to the uniaxial tensile experiment, and the true stress-plastic strain fitting curve is shown in Figure 14. The uniaxial tensile test of material 20CrNi2Mo specimens was completed to investigate the effect of quenching temperature on the properties of strength and toughness [38]. Moreover, in [39], quasi-static tensile/compressive experiments were carried out to obtain the mechanical properties of high strength steel. These experimental results play a great role in this study. Because the yield strength of the bearing rolling element (1617 MPa) is much higher than that of the crankshaft (1413 MPa), contact fatigue failure is more likely to occur on the crankshaft. Therefore, the following mainly analyzes the contact fatigue characteristics of the crankshaft.

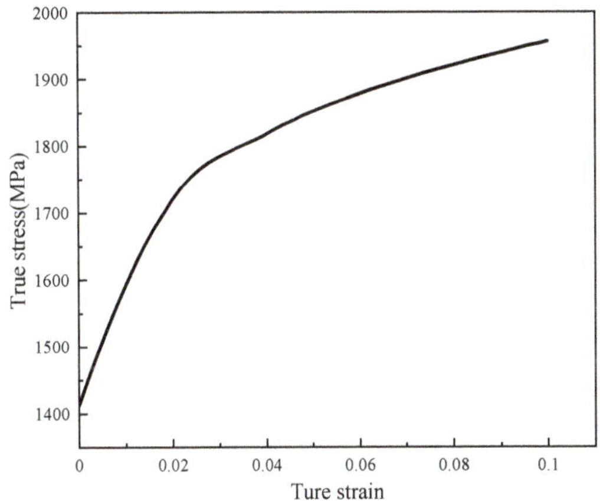

Figure 14. The true stress-plastic strain curve.

The initial residual stress field is defined before starting the response history. The hexahedral element has six stress components (σ_{11}, σ_{22}, σ_{33}, σ_{12}, σ_{13}, σ_{23}). The shear stress and the stress in the depth direction are negligible according to the above residual stress measurement. Therefore, normal stress components σ_{11} and σ_{33}, confirmed at residual stress σ_r (Figures 10 and 11), are applied to the element integration points of the model (Figure 13). The repetitive work of initial residual stress application can be reduced through the secondary development of Python. After the initial stress field is applied in the finite element model, the equivalent node load formed by the stress field need to be in equilibrium with the specified boundary conditions. Therefore, a static analysis step without external load is set to obtain the actual residual stress field. The actual residual stress obtained by adding different types of residual stress curves (Figures 10 and 11) is shown in Figure 15. The variation of initial residual stress after equilibrium is negligible, which verifies that the method of applying the initial residual stress field to this model is reasonable.

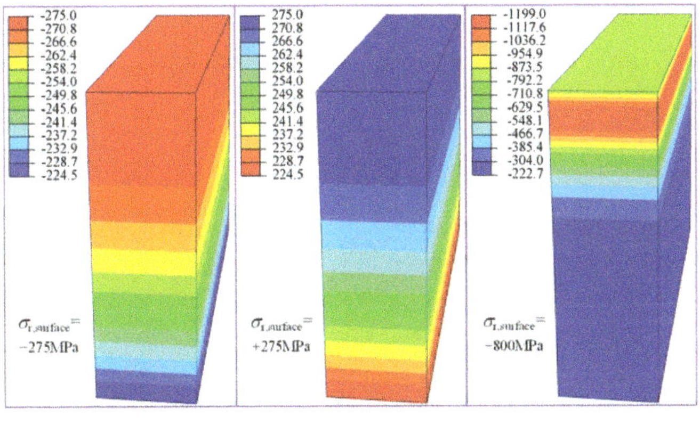

Figure 15. The actual residual stress after equilibrium (without external load).

3.4. Contact Fatigue Life Assessment Model

The stress in the contact region varies non-proportionally in the loading cycle due to the multiaxial characteristics of the stress state. Thus, it is necessary to apply a preferable multiaxial fatigue criterion to capture stress–strain response and then estimate the fatigue life of the crankshaft in such a complicated time-varying stress state. The FS multiaxial fatigue criterion dominated by shear-type fatigue failure is employed in this work. The normal stress and shear strain with maximum amplitudes are used as fatigue damage parameters. The fatigue life and fatigue damage (*FD*) in this criterion are respectively expressed as [40,41]:

$$\frac{\Delta\gamma_{\max}}{2}(1+k\frac{\sigma_{n,\max}}{\sigma_{ys}}) = \frac{\tau'_f}{G}(2N_f)^b + \gamma'_f(2N_f)^c \tag{8}$$

$$FD_{FS} = \frac{\Delta\gamma_{\max}}{2}[1+k\frac{\sigma_{n,\max}}{\sigma_{ys}}] \tag{9}$$

where $\Delta\gamma_{\max}/2 = \gamma_a$ denotes the maximum shear strain amplitude, $\sigma_{n,\max}$ denotes the maximum normal stress perpendicular to the critical plane, which is assumed to be the plane experiencing maximum shear strain amplitude; b and c denote the shear fatigue strength and the shear fatigue ductility indexes, set as -0.087 and -0.58 [42]; G and N_f are the shear elastic modulus and the crack initiation life, respectively; k is the material constant (set as 1 in this study) [43]; and τ'_f and γ'_f denote the shear fatigue strength and shear fatigue ductility coefficients, which are defined as follows [44]:

$$\tau'_f = \sigma'_f/\sqrt{3}, \gamma'_f = \sqrt{3}\varepsilon'_f \tag{10}$$

where σ'_f and ε'_f denote the axial fatigue strength and axial fatigue ductility coefficients, which are calculated by Baumel and Seeger's method [42]:

$$\sigma'_f = 1.5\sigma_b, \varepsilon'_f = 0.59\psi \tag{11}$$

$$\psi = \begin{cases} 1.0 & (\sigma_b/E < 0.003) \\ 1.375 - 125(\sigma_b/E) & (\sigma_b/E > 0.003) \end{cases} \tag{12}$$

where E is the Young's modulus and σ_b is the tensile strength.

When analyzing the fatigue data, including combined axial-torsion load paths, it is found that if the maximum normal stress is normalized by the shear stress range, the prediction accuracy of fatigue life can be improved. That is, replacing yield stress, σ_{ys}, with the shear stress range, $G\Delta\gamma$, the modified fatigue damage (FD_{mod}) is calculated as follows [45]:

$$FD_{\text{mod}} = \frac{\Delta\gamma_{\max}}{2}(1+k\frac{\sigma_{n,\max}}{G\Delta\gamma}) = \frac{\tau'_f}{G}(2N_f)^b + \gamma'_f(2N_f)^c \tag{13}$$

The multi-axial stress–strain histories are calculated by using the model outlined in Section 3.3. The stress tensor time history and the strain tensor time history at each element is expressed as $\sigma(t)$ and $\varepsilon(t)$, respectively.

$$\sigma(t) = \begin{bmatrix} \sigma_{xx}(t) & \tau_{xy}(t) & \tau_{xz}(t) \\ \tau_{xy}(t) & \sigma_{yy}(t) & \tau_{yz}(t) \\ \tau_{xz}(t) & \tau_{yz}(t) & \sigma_{zz}(t) \end{bmatrix}, \varepsilon(t) = \begin{bmatrix} \varepsilon_{xx}(t) & \frac{1}{2}\gamma_{xy}(t) & \frac{1}{2}\gamma_{xz}(t) \\ \frac{1}{2}\gamma_{xy}(t) & \varepsilon_{yy}(t) & \frac{1}{2}\gamma_{yz}(t) \\ \frac{1}{2}\gamma_{xz}(t) & \frac{1}{2}\gamma_{yz}(t) & \varepsilon_{zz}(t) \end{bmatrix} \tag{14}$$

Once the multi-axial stress–strain histories are obtained, Euler angle-based axis transformation is employed to search for the critical planes that cover all the possible directions in the 3D space [46], as shown in Figure 16. The transformation of stress vector and strain vector is similar, so the former will be explained in detail below. Considering each element integration point, O is in the center of the absolute reference frame, $\{O; x, y, z\}$. The orientation of a material plane, Ω, with unit normal vector, n (n_x, n_y, n_z), can be located by using spherical coordinate parameters α_p and β_p. Parameter α is the angle between the projection of unit normal vector n on plane x–y and axis x, while β is the angle between unit vector n and axis z [47].

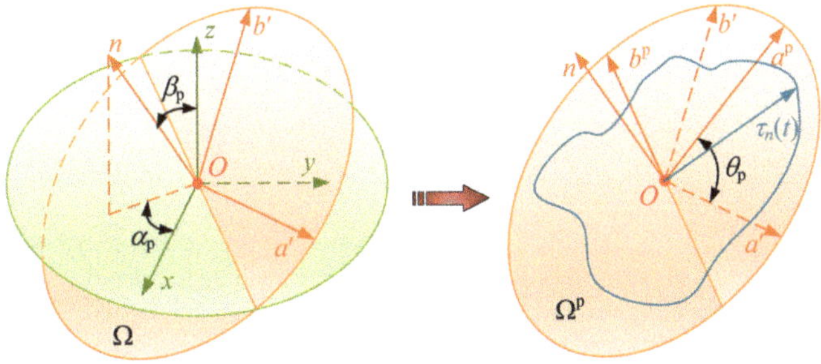

Figure 16. Definition of the Euler transformations and critical plane search method.

The transformation coordinate system $\{O; a', b', n\}$ is defined by searching across two angular ranges ($\alpha_p \in [0, 2\pi]$, $\beta_p \in [0, \pi]$), as shown in Figure 16. Axis n is parallel to unit vector n, whereas axes a' and b' are parallel to material plane Ω, and the corresponding unit vectors are a and b, respectively.

$$n = \begin{bmatrix} n_x \\ n_y \\ n_z \end{bmatrix} = \begin{bmatrix} \sin\beta_p \cdot \cos\alpha_p \\ \sin\beta_p \cdot \sin\alpha_p \\ \cos\beta_p \end{bmatrix}, \; a = \begin{bmatrix} a_x \\ a_y \\ a_z \end{bmatrix} = \begin{bmatrix} -\sin\alpha_p \\ \cos\alpha_p \\ 0 \end{bmatrix}, \; b = \begin{bmatrix} b_x \\ b_y \\ b_z \end{bmatrix} = \begin{bmatrix} -\cos\beta_p \cdot \cos\alpha_p \\ -\cos\beta_p \cdot \sin\alpha_p \\ \sin\beta_p \end{bmatrix} \tag{15}$$

In order to easily calculate the normal stress, $\sigma_n(t)$, and shear stress, $\tau_n(t)$, associated with the material plane, Ω, the total stress vector, $t(t)$, depending on the stress tensor, $\sigma(t)$, needs to be calculated as:

$$t(t) = \begin{bmatrix} t_x(t) \\ t_y(t) \\ t_z(t) \end{bmatrix} = \sigma(t) \cdot n = \begin{bmatrix} \sigma_{xx}(t) & \tau_{xy}(t) & \tau_{xz}(t) \\ \tau_{xy}(t) & \sigma_{yy}(t) & \tau_{yz}(t) \\ \tau_{xz}(t) & \tau_{yz}(t) & \sigma_{zz}(t) \end{bmatrix} \begin{bmatrix} n_x \\ n_y \\ n_z \end{bmatrix} \tag{16}$$

The normal stress and component of shear stress on two axes can be obtained from the above formula, as follows:

$$\{O; a', b', n\} \Rightarrow \begin{cases} \sigma_n(t) = t_x(t) \cdot n_x + t_y(t) \cdot n_y + t_z(t) \cdot n_z \\ \tau_{na}(t) = t_x(t) \cdot a_x + t_y(t) \cdot a_y + t_z(t) \cdot a_z \\ \tau_{nb}(t) = t_x(t) \cdot b_x + t_y(t) \cdot b_y + t_z(t) \cdot b_z \end{cases} \tag{17}$$

The shear stress, $\tau_n(t)$, is obtained as follows:

$$\tau_n(t) = \sqrt{\tau_{na}^2(t) + \tau_{nb}^2(t)} \tag{18}$$

During the load cycle, normal stress, $\sigma_n(t)$, varies its magnitude, but the direction remains parallel to vector n. Therefore, its maximum value can simply be calculated as follows:

$$\sigma_{n,\max} = \max[\sigma_n(t)] \tag{19}$$

Meanwhile, shear stress, $\tau_n(t)$, varies its magnitude and direction with time, which makes the solution of the shear stress amplitude more complicated. This coordinate system, $\{O; a', b', n\}$, is then rotated about axis n until the amplitude of the shear stress component along axis a^{P} is maximized. The new coordinate system is recorded as $\{O; \boldsymbol{a}^{\mathrm{P}}, \boldsymbol{b}^{\mathrm{P}}, \boldsymbol{n}\}$, and $\boldsymbol{\Omega}_{\mathbf{p}}$ is used to represent the critical plane, as shown in Figure 16, assuming that the rotation angle is θ_p ($\theta_p \in [0, \pi]$) and the transformation matrix is $\mathbf{T}(\theta_p)$, defined by:

$$T(\theta_p) = \begin{bmatrix} 1 & 0 & 0 \\ 0 & \cos\theta_p & \sin\theta_p \\ 0 & -\sin\theta_p & \cos\theta_p \end{bmatrix} \tag{20}$$

The stress tensors around this coordinate system are given as:

$$\{O; \boldsymbol{a}^{\mathrm{P}}, \boldsymbol{b}^{\mathrm{P}}, \boldsymbol{n}\} \Rightarrow \begin{cases} \sigma_n^P(t) = T(\theta_p)\cdot\sigma_{\mathrm{n}}(t) \\ \tau_{na}^P(t) = T(\theta_p)\cdot\tau_{\mathrm{na}}(t) \\ \tau_{nb}^P(t) = T(\theta_p)\cdot\tau_{\mathrm{nb}}(t) \end{cases} \tag{21}$$

That is, the θ_p that maximizes the in-plane shear stress amplitude along axes a^{P} is searched for, and at this moment the amplitude of $\tau_{nb}^p(t)$ is the smallest [48]. Thus, the stress vector satisfying the following relationship can be obtained:

$$\tau_{n,a} = \frac{1}{2}[\max\tau_n(t) - \min\tau_n(t)] \tag{22}$$

The overall computational methodology of estimating the fatigue life demonstrated above is summarized in the flow chart of Figure 17. Firstly, the stable stress–strain field is calculated by the elasto-plastic contact model, and the stress–strain components of the target material pointing along the depth is extracted through the secondary development of Python. Then, the stress, strain ($\sigma_{n,\max}$ and $\Delta\gamma_{\max}$), and FD_{mod} are calculated. Finally, the fatigue initiation life is evaluated via the Newton–Raphson method.

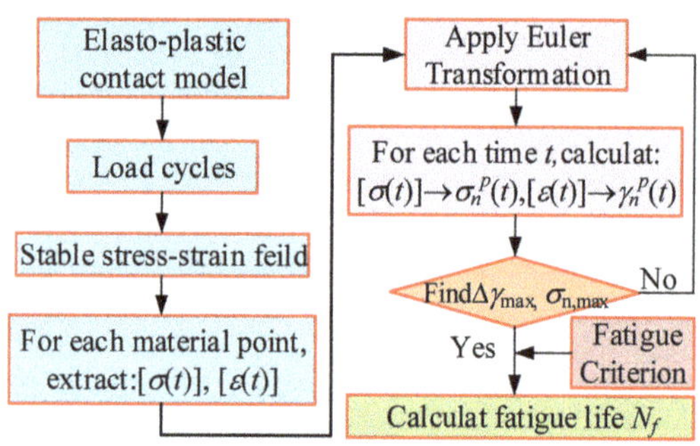

Figure 17. Computational methodology of predicting fatigue life.

4. Results and Discussion

4.1. Effect of Friction Coefficient and Normal Stress on Fatigue Damage

The friction coefficients of 0, 0.05, and 0.2 correspond to ideal smooth contact without friction, the reasonable friction coefficient for electrohydrodynamic lubricated bearing contact, and the state where the excellent lubrication environment of the reducer is destroyed. These three friction coefficients are selected firstly to investigate the effect of lubrication on the shear strain, γ_{a}, and the modified fatigue damage, FD_{mod}, on the material plane ($\theta_p \in [0, \pi]$) with the initial compressive residual stress (ICRS) case under the framework of modified FS criterion. The output torque is set as the rated output torque of 800 N·m and the Maximum contact stress P_h = 1.56 GPa. As can be seen from Figure 18, the contour of shear strain amplitude, γ_{a}, is symmetrical, and $\gamma_{\mathrm{a,max}}$ lies at the plane orientation angles of 90° and 0° (180°) when the friction coefficient, μ, is equal to 0. The distribution of γ_{a} and FD_{mod} over θ_p is asymmetric when $\mu \neq 0$. Moreover, the worse lubrication, with a friction coefficient of 0.2, leads to a much more obvious asymmetry of the FD_{mod} contour.

According to the definition of the critical plane, the plane orientations are along $\theta_p = 97°$ and $\theta_p = 101°$ when the friction coefficients are 0.05 and 0.2. The location of the maximum damage gradually moves to the contact surface, and the fatigue damage increases as the friction coefficient increases; similar results are also found in Ref [49].

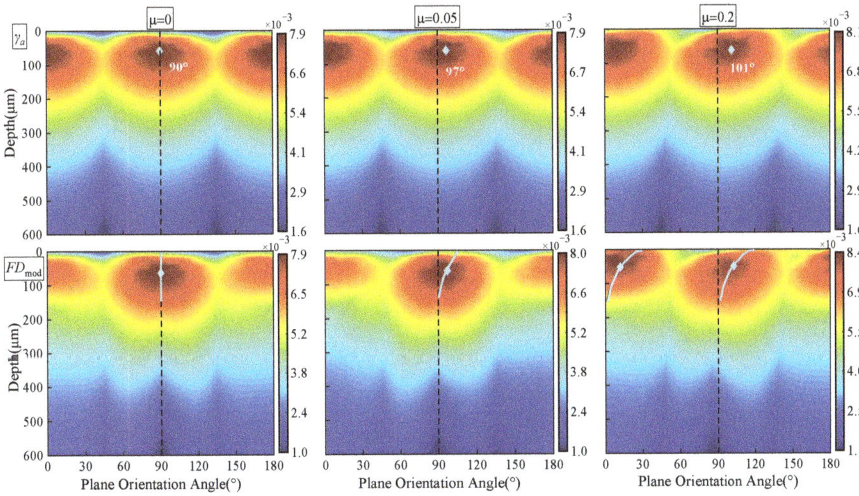

Figure 18. Variation of γ_a and FD_{mod} under different lubrication conditions (T_{out} = 800 N·m, P_h = 1.56 GPa).

Three residual stress conditions corresponding to the residual stress curves shown in Figure 10, such as initial compressive residual stress (ICRS), without initial residual stress (IRS), and initial tensile residual stress (ITRS), are chosen to explore the influence of initial residual stress on the fatigue damage (Figure 19). The output torque and friction coefficient are set as 800 N·m and 0.05, respectively. It can be observed from Figure 19 that the distributions of the maximum normal stress under three initial residual stress states are completely different at the subsurface, and the initial residual stress states appreciably influences the maximum normal stress, σ_{max}, near the $\theta_{\text{p}} = 0°$ (180°) plane. Compared with the without the IRS case, ICRS reduces σ_{max} by 48.4% while ITRS increases the value of σ_{max} by 99.8%.

The variations of FS damage (FD_{FS}) and modified FS damage (FD_{mod}) are also plotted versus plane orientation in Figure 19, which can provide additional insight into the damage mechanisms. Figure 19 shows that the variations of damage value against different plane angles under the modified FS criterion is more obvious than that under the FS criterion. The maximum fatigue damage, $FD_{\text{mod,max}}$, occurs at the $\theta_p = 97°$ plane with the ICRS case, while the maximum damage, $FD_{\text{mod,max}}$, shifts from $\theta_p = 97°$ to $\theta_p = 5°$ as the initial residual stress changes from compressive stress to tensile stress. This might be attributed to the fact that the increase of maximum normal stress leads to the increase of the damage near the $\theta_p = 0°$ (180°) plane. Therefore, the initial residual stress affects the position of $FD_{\text{mod,max}}$ by influencing the distribution of the maximum normal stress under the FS criterion; similar results are also found in [50]. Moreover, the tensile normal stress with the ITRS case results in higher damage predictions near the 0° (180°) plane for modified FS damage. Tensile stress help to accelerate shear crack growth, while compressive normal stress with the ICRS case may serve to increase friction between the cracks and reduce the crack driving force. The above also demonstrates that the ratio of maximum normal stress to shear stress in modified FS damage (FD_{mod}) can better consider the interaction effect between the two stress [45]. Therefore, the ICRS would lead to a decrease in fatigue damage and increase the crack initiation life, N_f.

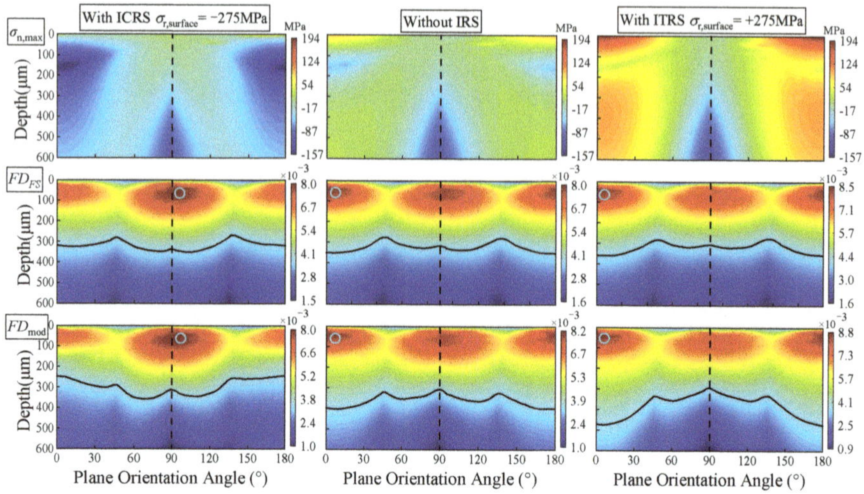

Figure 19. Effect of initial residual stress on $\sigma_{n,\max}$ and the fatigue damage (T_{out} = 800 N·m).

4.2. *Effect of Initial Residual Stress on Plastic Strain*

The equivalent plastic strain is chosen as the evaluation index for the analysis of contact fatigue damage. Figure 20 illustrates the variation of the maximum equivalent plastic strain (Max PEEQ) corresponding to different output torques in the RCF cycles. The Max PEEQ is close to 0 when the output torque is small, indicating that a pure elastic response exists under this circumstance. It is evident that the Max PEEQ increases remarkably with the increase of output torque. It can be concluded that heavy load conditions would aggravate the rate of plastic strain accumulation, resulting in the redistribution of stress.

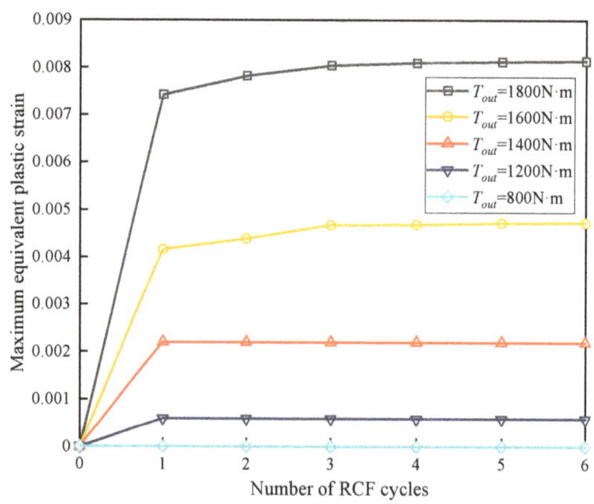

Figure 20. Effect of output torque on the maximum equivalent plastic strain (Max PEEQ).

Figure 21 demonstrates the variation of the maximum equivalent plastic strain (Max PEEQ) corresponding to different residual stress cases in the RCF cycles when the output torque is set as T_{out} = 1800 N·m and the maximum contact stress P_h = 2.34 GPa. Compared with the case without IRS, ICRS reduces PEEQ by 6.3%, while ITRS increases PEEQ by 37.5%. The increase rate of PEEQ for ITRS is obviously greater than the decrease rate of PEEQ for ICRS. RCF behaviors such as pitting and spalling more likely appear near the surface owing to the position of Max PEEQ near to the contact surface with ITRS.

Meanwhile, the presence of ICRS makes the position where the maximum equivalent plastic strain appear deeper. It is also found in [14] that the compressive residual stresses make the crack initiation depth deeper. Because the crack initiation depth determines the crack propagation length required to reach the surface, the ICRS enhanced the fatigue life by prolonging the number of cycles of cracks reaching the surface.

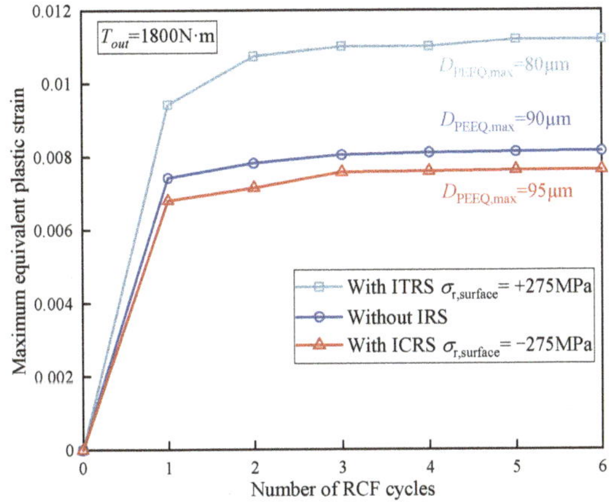

Figure 21. The effect of initial residual stresses on the maximum equivalent plastic strain (Max PEEQ) in RCF cycle (T_{out} = 1800 N·m, P_h = 2.34 GPa).

The initial residual stress also significantly affects the scope of the plastic region. Figure 22 depicts the variation of the maximum normal stress corresponding to different residual stress cases when the output torque is set as T_{out} = 2000 N·m (P_h = 2.47 GPa). It can be clearly seen from the figure that the plastic region forms at a certain depth of material rather than initiates on the surface. Under the same loading condition, the presence of ICRS shrinks the plastic region, while ITRS expands the plastic region compared with that without IRS. Almost the same maximum normal stress in the plastic region (around −750 MPa) are found for all the three scenarios. However, the distribution and magnitude of the maximum normal stress at a depth of 40 μm beneath the surface are totally different. The maximum normal stress in this region for ITRS is the largest, while the increase of the maximum normal stress is inhibited by the ICRS.

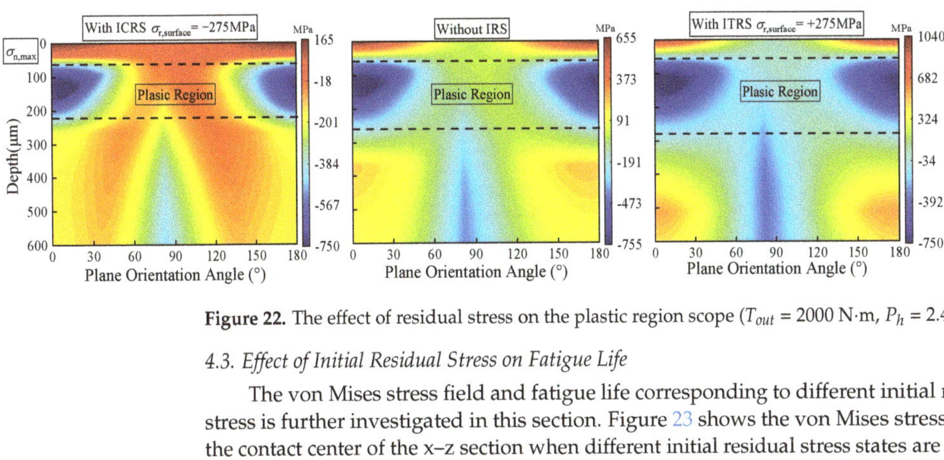

Figure 22. The effect of residual stress on the plastic region scope (T_{out} = 2000 N·m, P_h = 2.47 GPa).

4.3. Effect of Initial Residual Stress on Fatigue Life

The von Mises stress field and fatigue life corresponding to different initial residual stress is further investigated in this section. Figure 23 shows the von Mises stress field at the contact center of the x–z section when different initial residual stress states are applied. It is worth noting that the SPRS represents the residual stress introduced by shot peening (as shown in Figure 11). The maximum von Mises stress under the contact center in IRS state is 615.9 MPa, which is 30.4% lower than that without IRS, while the von Mises stress increase amplitude in ITRS state is 30.7% higher than that without IRS. What is more noteworthy is that there is an obvious low stress region under the contact center with the SPRS state. The maximum von Mises stress is an effective criterion to completely avoid the occurrence of plastic deformation. The significant reduction of the maximum von Mises stress means that the occurrence of plastic deformation can be suppressed or delayed [17]. The application of appropriate initial compressive residual stress can significantly reduce the plastic deformation and thus improve the contact fatigue performance.

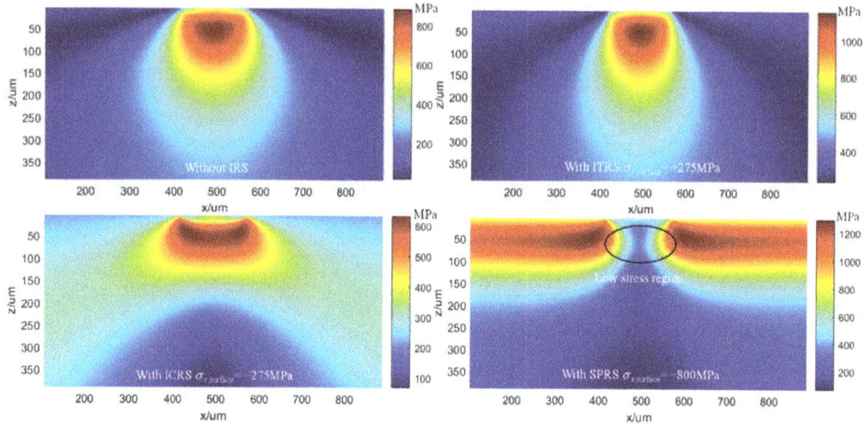

Figure 23. The effect of initial residual stress on the von Mises stress field (P_h = 1.56 GPa).

Figure 24 illustrates the minimum fatigue life, namely the fatigue limit on the basis of the modified FS criterion when the output torque is set as 800 N·m. Under the rated load condition, the minimum contact fatigue life with the ICRS case is 9.08×10^7, which is 32.7% larger than that without IRS, while the fatigue life with ITRS is 55.3% lower than that without IRS. It is apparent that the fatigue life after applying residual stress introduced by shot peening is increased by 125.7% compared with the state with ICRS. Therefore, the fatigue life can be considerably improved by designing an appropriate shot peening process and optimizing residual stress distribution.

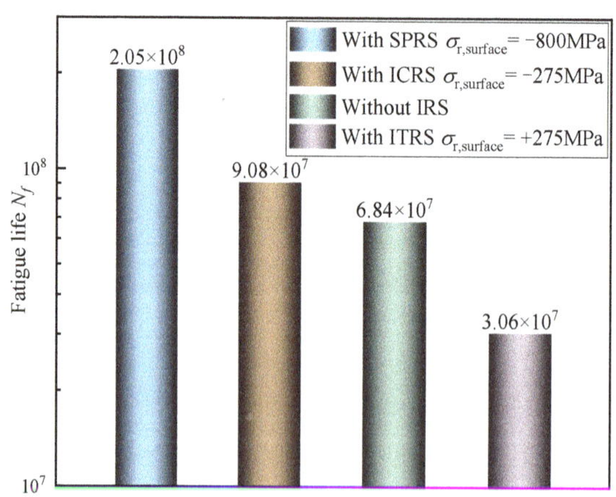

Figure 24. Variation of contact fatigue life with different initial residual stress states (P_h = 1.56 GPa).

Based on the modified Fatemi–Socie (FS) multiaxial fatigue criterion, the fatigue life forecasting has considerably improved by taken hardness gradients and initial residual stress into account especially under multiaxial loading conditions. Under the rated output torque of 800 N·m, the experimental fatigue life of the RV reducer is 2150 h (7.55×10^7). The error between the predicted fatigue life (9.08×10^7)and the experimental value is 20.3%. Considering the complex environmental conditions and the fatigue crack growth path in the actual work process of the RV reducer, the calculation results under this error are acceptable. However, there are still many shortcomings in the proposed method, such as the wear evolution behavior of morphology during service, contact pressure distribution under lubrication conditions, cumulative calculation of fatigue damage, etc. Fatigue life assessment would be more challenging considering the coupling effect of these factors.

5. Conclusions

In this work, focusing on fatigue life assessment of an RV reducer crankshaft, an innovative FEM-based three-dimensional elasto-plastic contact model is proposed. The RCF characteristics of the crankshaft are investigated by applying the modified Fatemi–Socie (FS) multiaxial fatigue criterion. Some distinct features can be highlighted, as follows:

(1) The location of the maximum shear strain depends on the friction coefficient. When the friction coefficient is low, the position of the maximum shear strain is still on the subsurface. Meanwhile, the fatigue initial location moves from the subsurface to the surface, and the fatigue damage increases as the friction coefficient μ increases.

(2) The initial residual stress plays an influential role in fatigue damage and crack initiation depth by altering the distribution of the maximum normal stress, $\sigma_{n,\max}$, near the contact surface. The compressive residual stress can reduce $\sigma_{n,\max}$ by 48.4% compared with that without residual stress. Therefore, the ratio of maximum normal stress to shear stress in the modified FS fatigue criterion can better consider the interaction effect between the residual stress and the shear stress, which significantly improves the prediction accuracy of the contact fatigue life model.

(3) The ICRS makes the plastic region shrink and improves the contact fatigue performance by delaying the time of cracks propagating to the surface. Under the rated load condition, the minimum contact fatigue life with ICRS is 9.03×10^7, which is 29.6% larger than that without IRS, while the minimum contact fatigue life with SPRS is 125.7% larger than that with ICRS. Residual stress distribution introduced by shot peening significantly enhances the fatigue life of the crankshaft.

(4) Moreover, the fatigue life could be maximized by designing appropriate shot peening parameters to obtain optimized residual stress distribution. The experimental verification of the proposed fatigue life assessment method can also be conducted in future study though it is costly and time-consuming.

Author Contributions: Conceptualization, X.L. and W.S.; methodology, X.L.; software, X.L.; formal analysis, W.S. and J.T.; investigation, J.T. and H.D.; data curation, W.Z.; writing—original draft preparation, X.L.; writing—review and editing, X.L., W.S., J.T., H.D. and W.Z.; visualization, X.L. and W.S.; supervision, W.S.; project administration, J.T. and W.S.; funding acquisition, J.T. and W.S. All authors have read and agreed to the published version of the manuscript.

Funding: This research was funded by the National Key R&D Program of China (No. 2019YFB2004700), the State Key Laboratory of High Performance Complex Manufacturing (No. ZZYJKT2019-08); and the China Postdoctoral International Exchange Program (No. 140050004).

Institutional Review Board Statement: Not applicable.

Informed Consent Statement: Not applicable.

Data Availability Statement: Not applicable.

Conflicts of Interest: The authors declare no conflict of interest.

References

1. Pham, A.-D.; Ahn, H.-J. High Precision Reducers for Industrial Robots Driving 4th Industrial Revolution: State of Arts, Analysis, Design, Performance Evaluation and Perspective. *Int. J. Precis. Eng. Manuf.-Green Technol.* **2018**, *5*, 519–533. [CrossRef]
2. Han, L.; Guo, F. Global sensitivity analysis of transmission accuracy for RV-type cycloid-pin drive. *J. Mech. Sci. Technol.* **2016**, *30*, 1225–1231. [CrossRef]
3. Zhang, Z.; Wang, J.; Zhou, G.; Pei, X. Analysis of mixed lubrication of RV reducer turning arm roller bearing. *Ind. Lubr. Tribol.* **2018**, *70*, 161–171. [CrossRef]
4. Xu, L.; Yang, Y. Dynamic modeling and contact analysis of a cycloid-pin gear mechanism with a turning arm cylindrical roller bearing. *Mech. Mach. Theory* **2016**, *104*, 327–349. [CrossRef]
5. Xu, L.X.; Chen, B.K.; Li, C.Y. Dynamic modelling and contact analysis of bearing-cycloid-pinwheel transmission mechanisms used in joint rotate vector reducers. *Mech. Mach. Theory* **2019**, *137*, 432–458. [CrossRef]
6. Wang, H.; Shi, Z.-Y.; Yu, B.; Xu, H. Transmission Performance Analysis of RV Reducers Influenced by Profile Modification and Load. *Appl. Sci.* **2019**, *9*, 4099. [CrossRef]
7. Deng, F.; Li, K.; Hu, X.; Jiang, H.; Huang, F. Life calculation of angular contact ball bearings for industrial robot RV reducer. *Ind. Lubr. Tribol.* **2019**, *71*, 826–831. [CrossRef]
8. Li, S.; Kahraman, A.; Klein, M. A fatigue model for spur gear contacts operating under mixed elastohydrodynamic lubrication conditions. *J. Mech. Design* **2012**, *134*, 041007. [CrossRef]
9. Liu, J.; Lv, X.; Wei, Y.; Pan, X.; Wang, Y. A novel model for low-cycle multiaxial fatigue life prediction based on the critical plane-damage parameter. *Sci. Prog.* **2020**, *103*, 003685042093622. [CrossRef]
10. Vijay, A.; Sadeghi, F. A continuum damage mechanics framework for modeling the effect of crystalline anisotropy on rolling contact fatigue. *Tribol. Int.* **2019**, *140*, 105845. [CrossRef]
11. Li, F.; Hu, W.; Meng, Q.; Zhan, Z.; Shen, F. A new damage-mechanics-based model for rolling contact fatigue analysis of cylindrical roller bearing. *Tribol. Int.* **2018**, *120*, 105–114. [CrossRef]
12. Shen, F.; Zhou, K. An elasto-plastic-damage model for initiation and propagation of spalling in rolling bearings. *Int. J. Mesh. Sci.* **2019**, *161–162*, 105058. [CrossRef]
13. Guo, Y.B.; Warren, A.W. The impact of surface integrity by hard turning vs. grinding on fatigue damage mechanisms in rolling contact. *Surf. Coat. Technol.* **2008**, *203*, 291–299. [CrossRef]
14. Choi, Y. A study on the effects of machining-induced residual stress on rolling contact fatigue. *Int. J. Fatigue* **2009**, *31*, 1517–1523. [CrossRef]
15. Liu, H.; Liu, H.; Zhu, C.; He, H.; Wei, P. Evaluation of Contact Fatigue Life of a Wind Turbine Gear Pair Considering Residual Stress. *J. Tribol.* **2018**, *140*, 041102. [CrossRef]
16. Wang, W.; Liu, H.; Zhu, C.; Bocher, P.; Liu, H.; Sun, Z. Evaluation of Rolling Contact Fatigue of a Carburized Wind Turbine Gear Considering the Residual Stress and Hardness Gradient. *J. Tribol.* **2018**, *140*, 061401. [CrossRef]
17. Mahdavi, H.; Poulios, K.; Kadin, Y.; Niordson, C.F. Finite element study of cyclic plasticity near a subsurface inclusion under rolling contact and macro-residual stresses. *Int. J. Fatigue* **2021**, *143*, 105981. [CrossRef]
18. Ooi, G.T.C.; Roy, S.; Sundararajan, S. Investigating the effect of retained austenite and residual stress on rolling contact fatigue of carburized steel with XFEM and experimental approaches. *Mater. Sci. Eng. A* **2018**, *732*, 311–319. [CrossRef]
19. Guan, J.; Wang, L.; Mao, Y.; Shi, X.; Ma, X.; Hu, B. A continuum damage mechanics based approach to damage evolution of M50 bearing steel considering residual stress induced by shot peening. *Tribol. Int.* **2018**, *126*, 218–228. [CrossRef]
20. Walvekar, A.A.; Sadeghi, F. Rolling contact fatigue of case carburized steels. *Int. J. Fatigue* **2017**, *95*, 264–281. [CrossRef]
21. Zhang, S.; Wang, W.; Zhang, H.; Zhao, Z. The effect of hardness distribution by carburizing on the elastic–plastic contact performance. *Tribol. Int.* **2016**, *100*, 24–34. [CrossRef]
22. Wang, W.; Liu, H.; Zhu, C.; Tang, J.; Jiang, C. Evaluation of contact fatigue risk of a carburized gear considering gradients of mechanical properties. *Friction* **2019**, *8*, 1039–1050. [CrossRef]
23. Huang, J.; Li, C.; Chen, B. Optimization Design of RV Reducer Crankshaft Bearing. *Appl. Sci.* **2020**, *10*, 6520. [CrossRef]
24. Harris, T.A.; Kotzalas, M.N. *Advanced Concepts of Bearing Technology: Rolling Bearing Analysis*; CRC Press: Boca Raton, FL, USA, 2006.
25. Lang, O.R.; Kernen i, R. Dimensionierung komplizierter Bauteile aus Stahl im Bereich der Zeit-und Dauerfestigkeit. *Materialwiss. Werkst.* **1979**, *10*, 24–29. [CrossRef]
26. Thomas, J. Flankentragfähigkeit und Laufverhalten von Hartfeinbearbeiteten Kegelrädern. Ph.D. Thesis, Technische Universität München, Munich, Germany, 1998.
27. Cahoon, J.; Broughton, W.; Kutzak, A. The determination of yield strength from hardness measurements. *Metall. Trans.* **1971**, *2*, 1979–1983. [CrossRef]
28. Pavlina, E.J.; Van Tyne, C.J. Correlation of Yield Strength and Tensile Strength with Hardness for Steels. *J. Mater. Eng. Perform.* **2008**, *17*, 888–893. [CrossRef]
29. MackAldener, M.; Olsson, M. Tooth interior fatigue fracture—Computational and material aspects. *Int. J. Fatigue* **2001**, *23*, 329–340. [CrossRef]
30. Hertter, T. Rechnerischer Festigkeitsnachweis der Ermüdungstragfähigkeit Vergüteter und Einsatzgehärteter Stirnräder. Ph.D. Thesis, Technische Universität München, Munich, Germany, 2003.

31. Réti, T. Residual stresses in carburized, carbonitrided, and case-hardened components. In *Handbook of Residual Stress and Deformation of Steel*; ASM International: Almere, The Netherlands, 2002; pp. 189–208.
32. Zhou, W.; Tang, J.; Shao, W.; Wen, J. Towards understanding the ploughing friction mechanism in ultrasonic assisted grinding with single grain. *Int. J. Mesh. Sci.* **2022**, *222*, 107248. [CrossRef]
33. You, S.; Tang, J.; Zhou, W.; Zhou, W.; Zhao, J.; Chen, H. Research on calculation of contact fatigue life of rough tooth surface considering residual stress. *Eng. Fail. Anal.* **2022**, *140*, 106459. [CrossRef]
34. Zhao, J.; Tang, J.; Zhou, W.; Jiang, T.; Liu, H.; Xing, B. Numerical modeling and experimental verification of residual stress distribution evolution of 12Cr2Ni4A steel generated by shot peening. *Surf. Coat. Technol.* **2022**, *430*, 127993. [CrossRef]
35. Johnson, K.L. *Contact Mechanics*; Cambridge University Press: Cambridge, UK, 1987.
36. Slack, T.; Sadeghi, F. Explicit finite element modeling of subsurface initiated spalling in rolling contacts. *Tribol. Int.* **2010**, *43*, 1693–1702. [CrossRef]
37. Bomidi, J.A.R.; Sadeghi, F. Three-Dimensional Finite Element Elastic–Plastic Model for Subsurface Initiated Spalling in Rolling Contacts. *J. Tribol.* **2014**, *136*, 011402. [CrossRef]
38. Long, S.-L.; Liang, Y.-L.; Jiang, Y.; Liang, Y.; Yang, M.; Yi, Y.-L. Effect of quenching temperature on martensite multi-level microstructures and properties of strength and toughness in 20CrNi2Mo steel. *Mater. Sci. Eng. A* **2016**, *676*, 38–47. [CrossRef]
39. Jiang, T.; Zhou, W.; Tang, J.; Zhao, X.; Zhao, J.; Liu, H. Constitutive modelling of AISI 9310 alloy steel and numerical calculation of residual stress after shot peening. *Int. J. Impact Eng.* **2022**, *166*, 104235. [CrossRef]
40. Fatemi, A.; Socie, D.F. A critical plane approach to multiaxial fatigue damage including out-of-phase loading. *Fatigue Fract. Eng. Mater. Struct.* **1988**, *11*, 149–165. [CrossRef]
41. Bannantine, A.; Socie, D. A Multiaxial Fatigue Life Estimation. *Adv. Fatigue Lifetime Predict. Tech.* **1992**, *1122*, 249.
42. Bäumel, A.J.; Seeger, T. Materials data for cyclic loading. *Mater. Sci. Monogr.* **1990**, *61*, 1076.
43. Shamsaei, N.; Fatemi, A. Effect of hardness on multiaxial fatigue behaviour and some simple approximations for steels. *Fatigue Fract. Eng. Mater. Struct.* **2009**, *32*, 631–646. [CrossRef]
44. Dowling, N.E. *Mechanical Behavior of Materials: Engineering Methods for Deformation, Fracture, and Fatigue*; Pearson: London, UK, 2012.
45. Gates, N.R.; Fatemi, A. On the consideration of normal and shear stress interaction in multiaxial fatigue damage analysis. *Int. J. Fatigue* **2017**, *100*, 322–336. [CrossRef]
46. Hotait, M.A.; Kahraman, A. Estimation of Bending Fatigue Life of Hypoid Gears Using a Multiaxial Fatigue Criterion. *J. Mech. Design* **2013**, *135*, 101005. [CrossRef]
47. Susmel, L. A simple and efficient numerical algorithm to determine the orientation of the critical plane in multiaxial fatigue problems. *Int. J. Fatigue* **2010**, *32*, 1875–1883. [CrossRef]
48. Ding, H.; Zhang, Y.; Li, H.; Rong, K.; Tang, J.; Chen, S. Bending fatigue life oriented tooth flank dry-grinding tool modification for cleaner manufacturing of spiral bevel gear product. *J. Clean. Prod.* **2021**, *328*, 129566. [CrossRef]
49. Chen, Z.; Jiang, Y.; Tong, Z.; Tong, S. Residual Stress Distribution Design for Gear Surfaces Based on Genetic Algorithm Optimization. *Materials* **2021**, *14*, 366. [CrossRef] [PubMed]
50. Wang, W.; Liu, H.; Zhu, C.; Du, X.; Tang, J. Effect of the residual stress on contact fatigue of a wind turbine carburized gear with multiaxial fatigue criteria. *Int. J. Mesh. Sci.* **2019**, *151*, 263–273. [CrossRef]

materials

MDPI

Article

Analysis of the Failure Process of Elements Subjected to Monotonic and Cyclic Loading Using the Wierzbicki–Bai Model

Urszula Janus-Galkiewicz and Jaroslaw Galkiewicz *

Faculty of Mechatronics and Mechanical Engineering, Kielce University of Technology, 25-314 Kielce, Poland; janus.urszula@gmail.com

* Correspondence: jgalka@tu.kielce.pl; Tel.: +48-41-342-4711

Abstract: This article presents the results of a simulation in which smooth cylindrical and ring-notched samples were subjected to monotonic and fatigue loads in an ultra-short-life range, made of Inconel 718 super alloy. The samples displayed different behaviors as a result of different geometries that introduced varying levels of stress triaxiality and loading methods. The simulations used the Wierzbicki–Bai model, which took into account the influence of stress tensors and stress-deviator invariants on the behavior of the material. The difference in the behaviors of the smoothed and notched specimens subjected to tensile and fatigue loads were identified and described. The numerical results were qualitatively supported by the results of the experiments presented in the literature.

Keywords: Wierzbicki–Bai model; fatigue; tensile test; stress triaxiality; Lode parameter

Citation: Janus-Galkiewicz, U.; Galkiewicz, J. Analysis of the Failure Process of Elements Subjected to Monotonic and Cyclic Loading Using the Wierzbicki–Bai Model. *Materials* **2021**, *14*, 6265. https://doi.org/10.3390/ma14216265

Academic Editor: Andrea Spagnoli

Received: 24 September 2021
Accepted: 18 October 2021
Published: 21 October 2021

1. Introduction

Understanding the processes that occur in structural elements allows one to design more durable parts. The finite element method is currently the basic tool for analyzing the influence of various factors on material behavior. In the case of slowly changing monotonic loads, it seems that we obtain reliable results, but there are still problems with both the calculations and the interpretation of the results for other types of loads. Methods to speed up the calculation time [1] and to take into account various aspects of the loading process, such as dynamic loading [2] and the appearance of cracks [3], which are the final feature of the critical cross-section, are constantly being developed.

A fundamental problem with these methods is the evaluation of the strength of the material. The static tensile test, in many cases, does not give sufficient answers to the question of when the failure will occur. When assessing the strength of elastic–plastic materials, the basic problem is with how to determine the moment of transition into a plastic state. Many models can be used for this; the most popular models are Mises–Huber, Tresca, Drucker–Prager, and Mohr–Coulomb. They can be used with greater or lesser success for various types of materials, although they have their limitations. For example, the Mises–Huber model is not sensitive to stress triaxiality and the Lode parameter, whereas the Drucker–Prager model considers the fact that the material behaves differently under tension and compression, which means that it is sensitive to stress triaxiality. The influence of the Lode parameter can be found in the Tresca and Mohr–Coulomb models. The experiments that were initiated by Wierzbicki and Bai [4], and later conducted by many others, have shown [5,6] that the level of plastic deformation at the moment of failure of the tested element depends on the stress triaxiality and the Lode parameter. On this basis, a new model of plasticity was proposed. Numerous examples of the use of this model have shown very good compliance between the numerical simulations and the experimental results. The model has previously been used for monotonic loads, but recently articles in the field of fatigue (so far, of ultra-low cycles) have started to appear, presenting very reliable results. In [7], the Wierzbicki–Bai model was used to reproduce measurable displacement and force parameters in the range of monotonic tensile or compression tests of samples with different geometries, which can be characterized by the triaxial level and the Lode parameter. The above-mentioned model was used in [8,9] to analyze the fatigue loading process in the range of up to 100 cycles.

The model itself has been gradually developed for several years [10–13]. The yield criterion was extended to take into account the influence of triaxiality and the Lode parameter, as well as the numerical application and the parameters supporting the calculation process. In the present work, our attention was focused on analyzing the load process of smooth cylindrical specimens and specimens with a ring notch, and tracing the development of the damage in the material, as the accumulated effective plastic strain properly normalized. As a result, the influence of the notch-root radius on the behavior of the specimens that were subjected to a tensile load was identified and described, and the behavior of specimens subjected to a fatigue load was compared with those under a tensile load. All results were verified by experimental data.

The structure of the article is as follows: in the first section, the theoretical background is provided; then, the materials and calculation methods are discussed; in the third section, the achieved results are presented; and in the last section, our results and conclusions are discussed.

2. Theoretical Background

Stress triaxiality is understood to mean the ratio of hydrostatic stress to effective stress, that is, the ratio of the first stress tensor invariant by a function that depends only on the second invariant of the deviatoric stress tensor [14].

$$\eta = \frac{\sigma_m}{\sigma_e} \tag{1}$$

The Lode parameter is slightly more complex to explain [15]. If the greatest tangential stress expressed by principal stresses is $\tau = (\sigma_I - \sigma_{III})/2$ and the normal stress in the plane of maximum tangential stress is equal to $\sigma_N = (\sigma_I + \sigma_{III})/2$, the Lode parameter can be written by the equation $L = (\sigma_{II} - \sigma_N)/\tau$ [16]. Although it is not visible at first glance, this parameter is related to the third invariant of the deviatoric stress tensor. Let us introduce the parameter denoting the normalized third invariant of the deviatoric stress tensor [17]:

$$\xi = \frac{27}{2}\frac{J_3}{\sigma_e^3} \tag{2}$$

This quantity is related to the Lode angle, determined on the deviatoric plane according to the equation:

$$\xi = \cos(3\theta) \tag{3}$$

and with the Lode parameter L:

$$\xi = L\left(9 - L^2\right)/\sqrt{\left(L^2 + 3\right)^3} \tag{4}$$

Knowing the value of the effective stress, the Lode angle, and the stress triaxiality, it is possible to uniquely describe the stress state, as shown in Figure 1.

On the deviatoric plane, the yield surface is reduced to a circle (Figure 1a). However, the triaxiality of the stress affects its radius, and the Lode angle affects the position of the point on the circle that defines the load state. Moreover, the Lode angle may influence the shape of the yield surface [12]. It does not necessarily have to be a circle; this shape should be determined and verified experimentally. Wierzbicki et al., therefore, proposed universal functions based on both parameters. This function consists of two terms separated by variables.

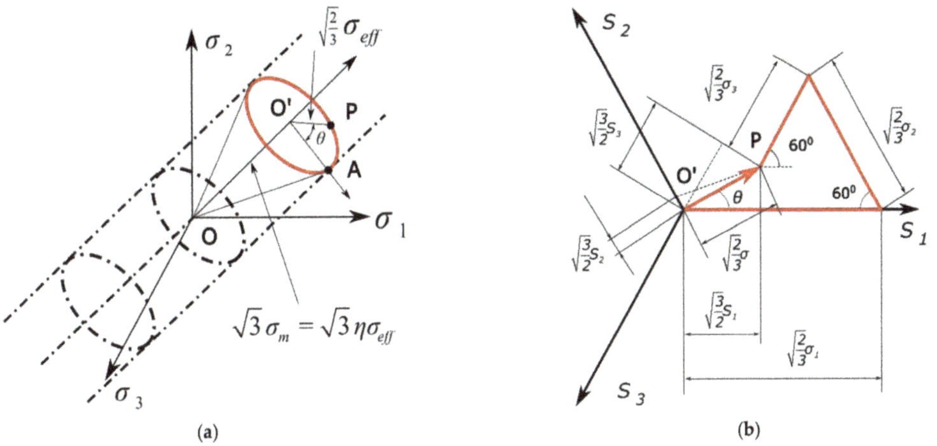

Figure 1. Components of the stress tensor in the principal stress space (**a**). Components of the stress tensor on the deviatoric plane (**b**).

The impact of stress triaxiality is expressed by the function:

$$f(\eta) = 1 - c_\eta(\eta - \eta_0) \tag{5}$$

where c_η is the coefficient of stress triaxiality depending on plasticity, η_0 is the reference value of the stress triaxiality, and η is the current value of the stress triaxiality.

The influence of the Lode angle on the shape of the yield surface is more challenging. The function that describes this impact has evolved as new, more complete experimental results have appeared, and finally reached the form:

$$f(\theta) = c_\theta^s + (c_\theta^{ax} - c_\theta^s)\left(\frac{m+1}{m}\right)\left(\gamma - \frac{\gamma^{m+1}}{m+1}\right) \tag{6}$$

where $\gamma = \frac{\cos(\pi/6)}{1-\cos(\pi/6)}\left(\frac{1}{\cos(\theta-\pi/6)} - 1\right)$. The values of "c" can be treated as material constants, but in the fullest version they are described by functions:

$$c_\theta^{ax} = \begin{cases} c_\theta^t & \bar{\theta} \geq 0 \\ c_\theta^c & \bar{\theta} < 0 \end{cases} \tag{7}$$

$$c_\theta^s = \sqrt{3}/2 + \left(B_1 e^{-B_2 \varepsilon_{pl}}\right) f(\bar{\theta}) \tag{8}$$

$$f(\bar{\theta}) = \left(1 - |\bar{\theta}|^{B_3}\right)^{B_4} \tag{9}$$

where B parameters are the quantities selected, so that the simulation results are as close as possible to the experimental results. Their task is to consider the influence of large deformations and to correct the influence of the Lode angle. The constants c_θ^t and c_θ^c allow the distinction between compression and tension.

It should be emphasized that Equation (6), proposed by Wierzbicki, was not the only function used. In the papers inspired by the works of Wierzbicki et al. [16,18], other functions were used based on the parameters (σ_{eff}, $\bar{\theta}$, η), creating the so-called Haigh–Westergaard space.

Taking into account Equations (5) and (6), the plasticity function proposed by Wierzbicki et al. can be written as [7]:

$$\sigma\left(\varepsilon_{pl}, \eta, \theta\right) = \sigma(\bar{\varepsilon}_p)\left[1 - c_\eta(\eta - \eta_0)\right]\left[c_\theta^s + (c_\theta^{ax} - c_\theta^s)\left(\frac{m+1}{m}\right)\left(\gamma - \frac{\gamma^{m+1}}{m+1}\right)\right] \quad (10)$$

where $\sigma(\bar{\varepsilon}_p)$ is the effective-stress value read from the tensile diagram presented in the logarithmic strain-true stress system. The terms $(m+1)/m$ and $\gamma^{m+1}/(m+1)$ have been added to facilitate numerical calculations.

An example of the results of plasticity surface modification through Equations (5) and (6) is presented in Figure 2.

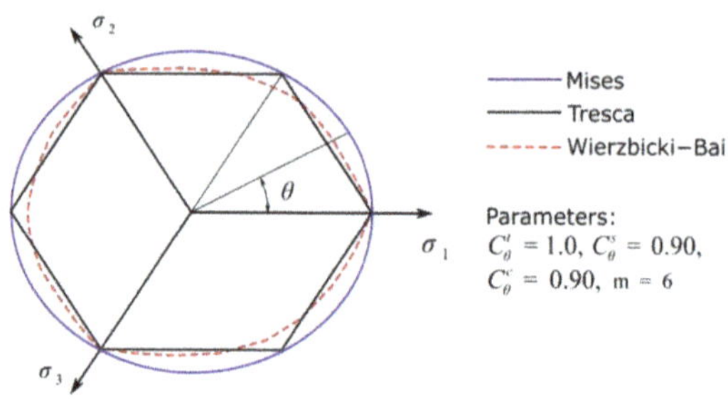

Figure 2. Yield surfaces for various yield criteria.

The fully developed Wierzbicki–Bai model contains many material constants, which can be considered a weakness; however, these constants are easily determinable, and the plasticity function itself can be used in a simplified form.

Research shows that in the case of elastic–plastic metals, both quantities, that is, the Lode parameter and the stress triaxiality, play essential roles when the yield conditions are analyzed. Triaxiality controls the void growth [19–21], whereas the Lode parameter is associated with a change in the shape of the growing voids [22–24]. As a result, they influence the critical strain, creating a fracture locus. An example diagram showing the dependence of the strain at the critical moment of triaxiality and the Lode angle is shown in Figure 3.

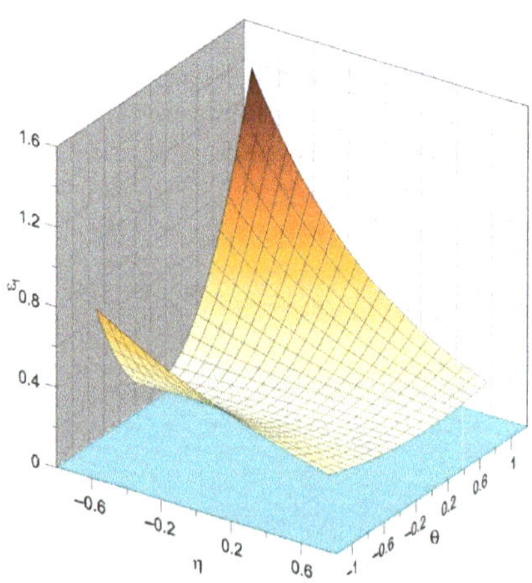

Figure 3. Exemplary fracture locus (source: own simulations).

3. Material

The material used in the work was Inconel 718 alloy, which is a nickel–chromium alloy characterized by its high resistance to corrosion and creep. The chemical composition of the alloy is presented in are given in Table 1 in [7].

This material was selected since the parameters required to fully characterize it in the Wierzbicki–Bai model are readily available in the literature, and the process of obtaining them is described in detail in [7]. The tensile curve was reconstructed with the help of Equation (4), and is presented in Figure 4.

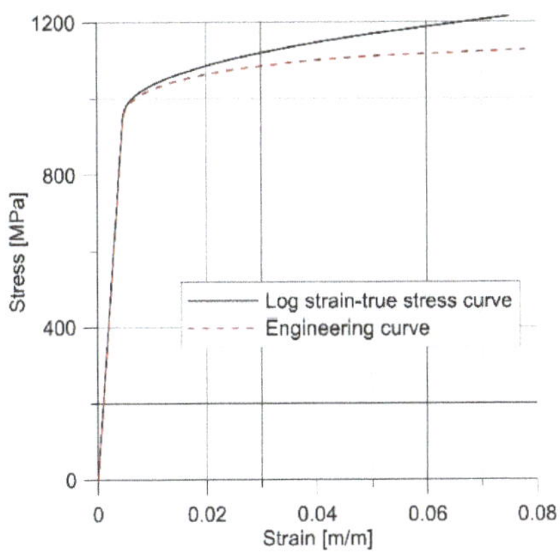

Figure 4. Comparison of engineering and true stress–strain curves (source: own computations).

After reaching the yield point, the material was described with the Ludwik curve in the form:

$$\sigma\left(\bar{\varepsilon}_{pl}\right) = \sigma_0 + K\bar{\varepsilon}_{pl}^{-n} \tag{11}$$

The entire true stress–strain curve was represented by a set of material constants: Young's modulus $E = 200{,}000$ MPa, Poisson's ratio $\nu = 0.284$, yield stress $\sigma_0 = 945.1$ MPa, and parameters of the Ludwik curve $K = 835.4$ MPa and $n = 0.425$.

However, the numerical calculations required a much wider set of necessary data to determine the yield surface. These constants are given in Table 5 in [7].

It is worth emphasizing at this point that the symmetry of the yield locus concerning tension and compression was adopted.

Another problem was determining the moment of material failure. The model should follow the experimental data; however, the damage can be modeled in various ways [25]. The parameter D was used for this, and calculated according to the equation:

$$D = \int_0^{\bar{\varepsilon}_{pl}} \frac{d\bar{\varepsilon}_{pl}}{\bar{\varepsilon}_f\left(\eta, \bar{\theta}\right)} \tag{12}$$

where $\bar{\varepsilon}_{pl}$ is the equivalent plastic strain. The critical strain $\bar{\varepsilon}_f$ depended on the stress triaxiality and the Lode parameter. Its full analytical form was described in [7]. The easiest way to create it was to fit the experimental data as described in article [26]. The 3D fracture locus in this case was constructed as follows:

$$\begin{aligned} \varepsilon f = {} & (N_{1,1}\eta^2 + N_{1,2}\eta + N_{1,3})\bar{\theta}^3 + \\ & (N_{2,1}\eta^2 + N_{2,2}\eta + N_{2,3})\bar{\theta}^2 + \\ & (N_{3,1}\eta^2 + N_{3,2}\eta + N_{3,3})\bar{\theta} + \\ & (N_{4,1}\eta^2 + N_{4,2}\eta + N_{4,3}) \end{aligned} \tag{13}$$

where the table of coefficients had the form of Table 1. The fracture locus is shown in Figure 5. To decrease the dynamic effects after crack initiation, an additional "softening" function was introduced:

$$\sigma_o = \begin{cases} \sigma_0 & 0 \leq D < 1 \\ \beta\sigma_0 & 1 \leq D \leq D_c \end{cases} \tag{14}$$

where the softening factor β is:

$$\beta = \left(\frac{D_c - D}{D_c - 1} \right)^w \tag{15}$$

Table 1. Coefficients of the fracture locus (source: own computations).

N=	−0.4773	0.319	−0.7304
	0.6683	−0.5705	1.5615
	−0.2423	0.514	−1.8897
	−0.0526	−0.3344	1.4992

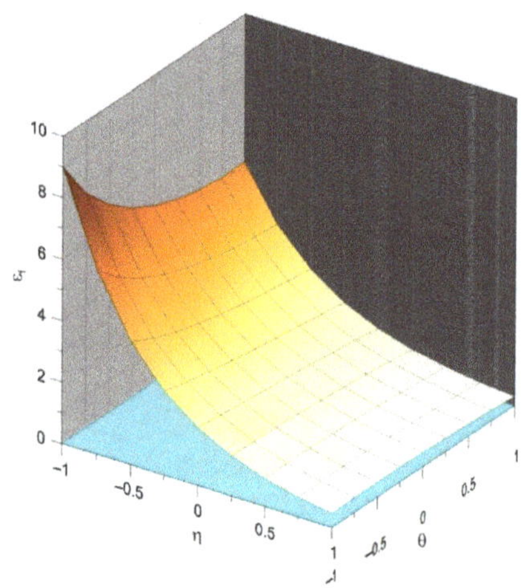

Figure 5. Simplified fracture locus.

The parameters D_c and "w" after from [16] were assumed to be equal to 1.2 and 6, respectively. The fracture process onset when D = 1, and when $D = D_c$, complete split occurred. The shape of the dependence of β parameter on the damage indicator D described by (15) is linear for w = 1, but in general it is nonlinear.

4. Numerical Model

The geometries shown in Figure 6 were adopted for the calculations.

The specimens were modeled in such a way that the minimum diameter (critical section) was always a circle with a diameter of 4 mm.

The calculations were performed in Abaqus/Explicit version 6.12-2, in the Linux environment. The VUMAT procedure was used for modeling the material, as it allows the user to program the effect of the stress triaxiality on the development of plasticity and failure of the element.

The geometry was modeled using CAX4R linear axisymmetric elements. The critical cross-section for each geometry was filled with elements with a size of 2 mm/32 = 0.0625 mm. The step-in time was determined automatically by the Abaqus routine (version 6.12-2), based on the size of the smallest element.

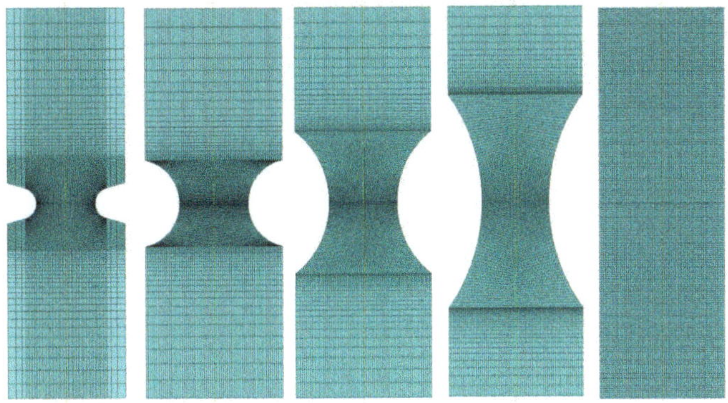

Figure 6. Geometry of the specimens tested in the program. R = 1.0 (**a**), R = 2.5 (**b**), R = 5.0 (**c**), R = 10.0 (**d**), smooth round (**e**).

As the plasticity theory used in the paper is valid under several assumptions, that is, the homogeneity and material isotropy, and the material is taken to be elastic–plastic with isotropic hardening, we applied isotropic hardening.

Using the existing symmetries, a quarter of the geometry shown in Figure 6 was always modeled.

The uniform displacement was applied to the upper edge of each model (Figure 7).

Figure 7. Boundary conditions.

5. Tensile Tests

The reference state for simulating the tensile test was a cylindrical specimen with a diameter of 4 mm (Figure 6e). The result of the simulation was the force-displacement plot (Figure 8), and the distribution of the effective stress and the parameter *D*.

From the distribution of the effective stress and the *D* parameter, it can be seen (Figure 9) that in a smooth round bar the damage process began in the specimen axis. The result of this behavior on the specimen surface was the appearance of a cup-and-cone. Figure 10 shows how the parameter *D* changed at the edge and in the center of the specimen. It shows that at low loads the failure parameter had a constant value in the cross-section, whereas increasing the load caused a slightly faster increase in *D* in the specimen center.

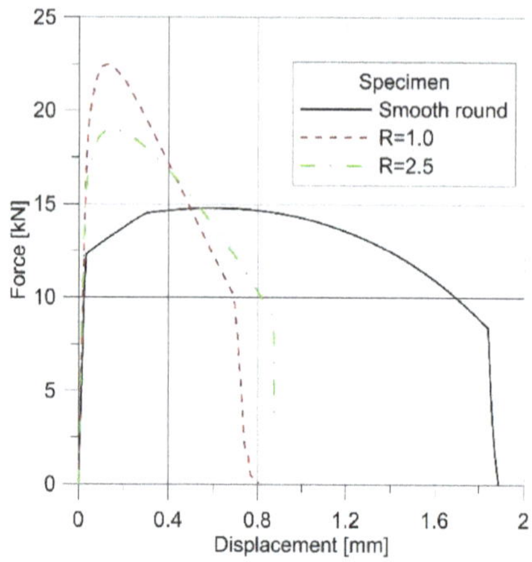

Figure 8. Force–displacement plots for different geometries.

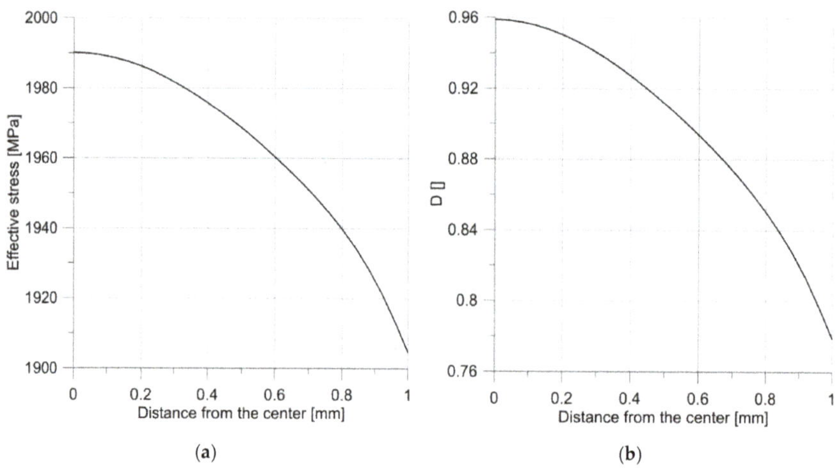

Figure 9. Distribution of effective stress (**a**) and parameter *D* (**b**) just before the fracture initiation of the smooth round specimen.

Introducing a small-radius notch influenced the behavior of the sample during the test. Loading the specimen with a notch of the radius of 1 mm allowed us to obtain a much higher maximum force (Figure 8), and the specimen fractured with much less displacement (strain). This was associated with the change in the level of triaxiality in the specimen.

The changes in effective stress and the *D* parameter are shown in Figure 11. The presented results show that the failure process started from the notch root. The changes in the *D* parameter in the specimen center, and the notch root during loading, as shown in Figure 12, proved that the increase in the parameter describing the level of damage was much higher compared with the smooth round specimen. Moreover, from the beginning of loading, this process developed more intensively in the notch root.

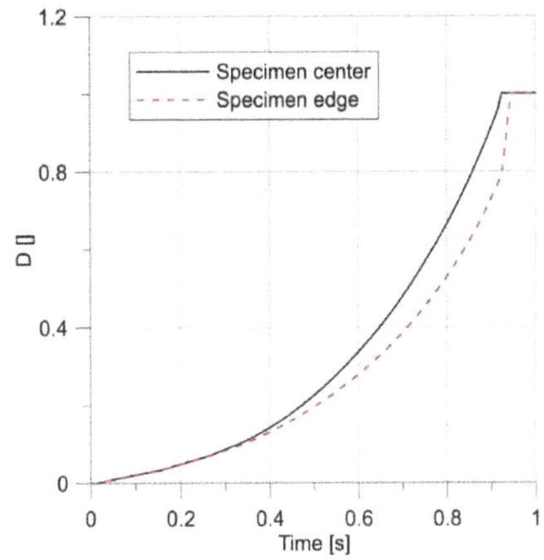

Figure 10. Changes in the *D* parameter over time for a smooth round specimen.

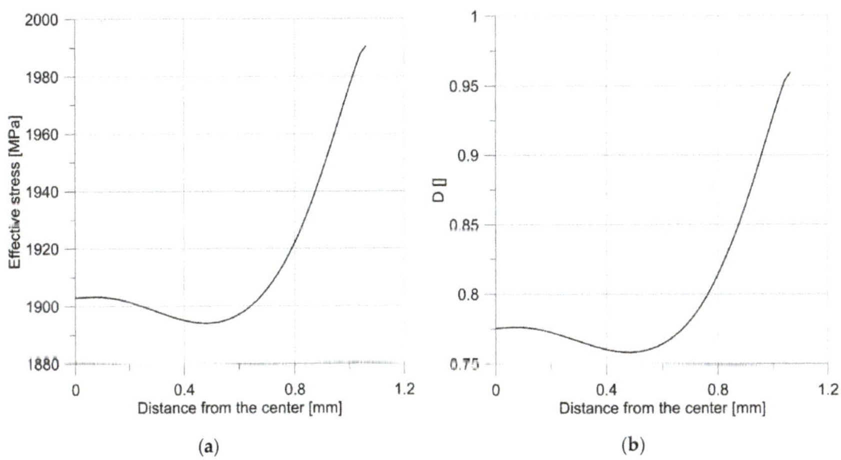

Figure 11. Distribution of effective stress (**a**) and parameter *D* (**b**) just before the moment of failure of the specimen with the notch R = 1 mm.

It may seem that such a situation will be true for every notch; however, increasing the radius to only 2.5 mm resulted in a change in the behavior of the specimen. As it is clear to see in Figure 6, the maximum force obtained during the tensile test was much lower, but the critical displacement was much larger. The diagrams of effective stress and the *D* parameter were much more interesting (Figure 13).

It can be seen that the courses of the mentioned parameters were irregular, and the maximum was not at the notch root, but in the specimen center.

The plot comparing the *D* parameter changes in the center of the specimen and at the notch root (Figure 14) resembled that of a smooth sample (Figure 10).

It was interesting that these changes in the behavior of the specimens subjected to monotonic loading translated into the behavior of cyclic loading.

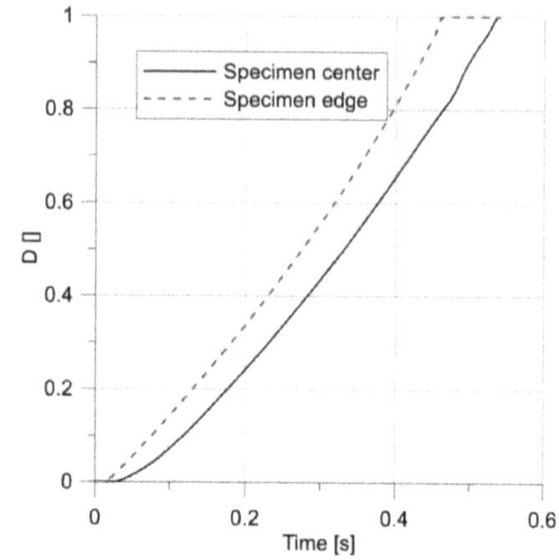

Figure 12. Changes in parameter D over time for a specimen with a notch R = 1.0 mm.

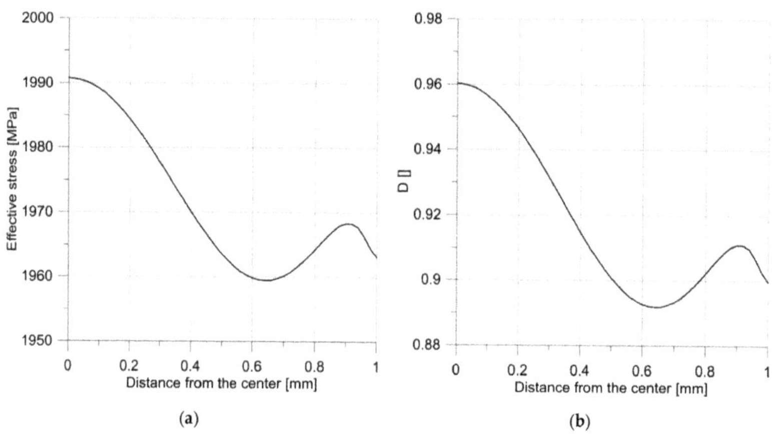

Figure 13. The distribution of effective stress (**a**) and the parameter D (**b**) just before the moment of failure of the sample, with the notch R = 2.5 mm.

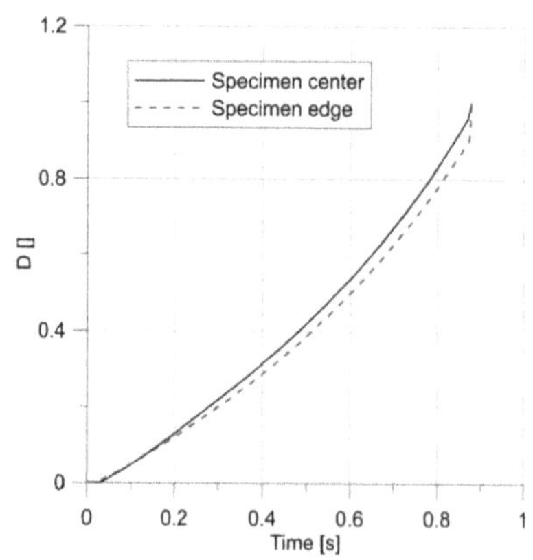

Figure 14. Changes in parameter D over time for a specimen with a notch R = 2.5 mm.

6. Fatigue Tests

All specimens were loaded with repeated stress cycles. The first sample for fatigue testing was a specimen with a notch radius of 2.5 mm, loaded with a sinusoidal variable load with an amplitude of 0.2 mm and a mean displacement in a cycle of 0.2 mm. The material response for cyclic loading is shown in Figure 15.

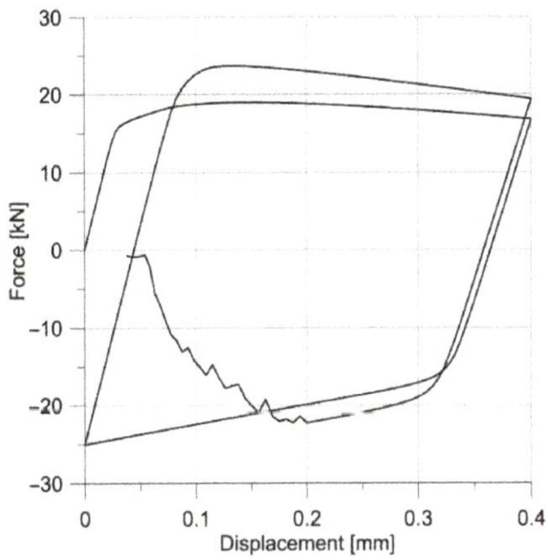

Figure 15. Changes in the stress–strain hysteresis loop for a specimen with a notch R = 2.5 mm.

The changes in the *D* parameter during loading are shown in Figure 16. It can be seen that from the initial stage of loading until the first maximum, the specimen behaved in accordance with the results obtained for the monotonic loading, but the first unloading caused a change in the behavior of the specimen, and the maximum shifted towards the notch root.

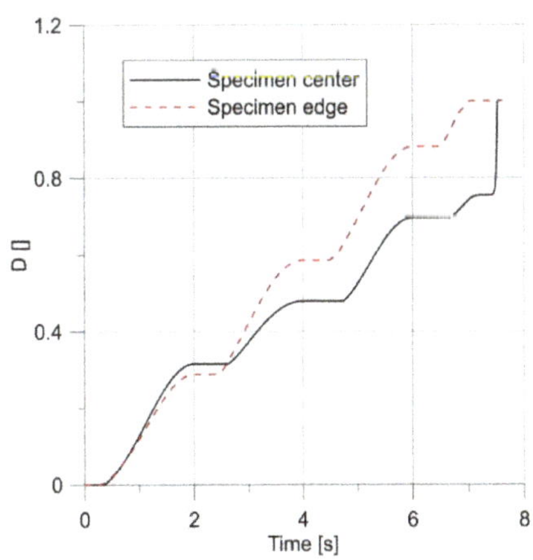

Figure 16. Changes in parameter *D* under load.

The tests were repeated for specimens with much larger notch radii, i.e., 5 and 10 mm; the load amplitude remained the same. The material response is shown in Figure 17.

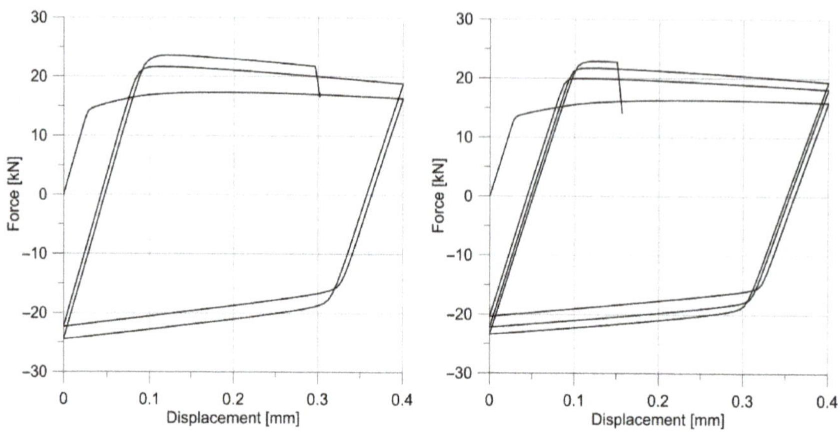

(**a**) (**b**)

Figure 17. Changes in the stress–strain hysteresis loops for specimens with notches R = 5.0 mm (**a**) and R = 10 mm (**b**).

For these specimens, the loading process ended with a fracture initiated in the notch root (Figure 18).

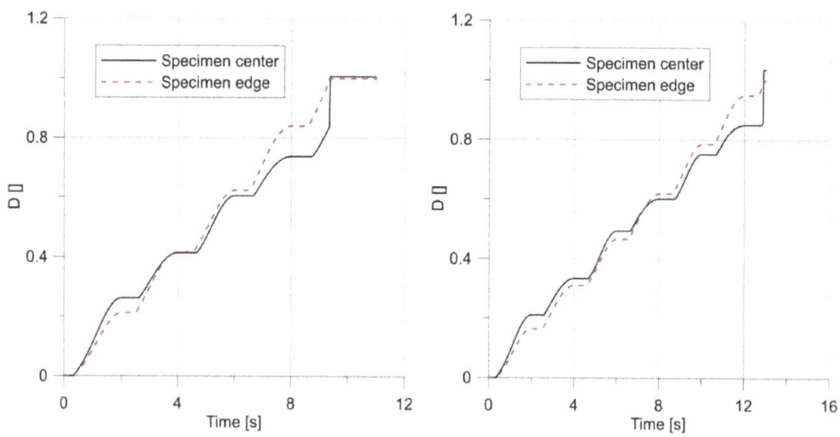

(**a**) (**b**)

Figure 18. Changes in parameter D under a load, for specimens with notch radius R = 5.0 mm (**a**) and R = 10 mm (**b**).

The presented results showed that increasing the radius of the notch root improved the fatigue strength, with a greater number of load cycles. During the fatigue test, at some point the damage of the material, initially developing faster in the specimen center, began to dominate at the notch root [27,28]. The change in the place of damage dominance depended on the notch root radius; the later it occurred, the greater the notch radius was. Unfortunately, the behavior of the specimen similar to this under a under a monotonic load was not obtained. In an attempt to obtain such behavior during the fatigue loading, tests were carried out on a smooth round specimen.

Two levels of loading were used: the standard load used for other specimens, and a load decreased by 25%. Reducing the load increased the number of cycles from 5 to 9 (Figure 19).

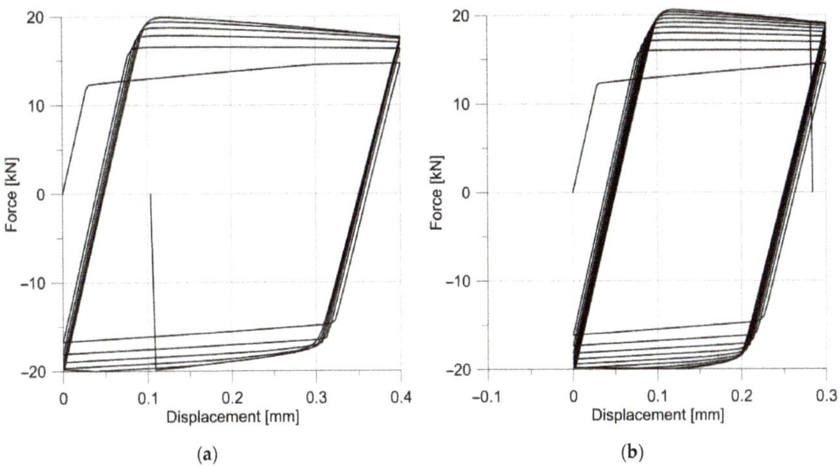

Figure 19. Change in the stress–strain hysteresis loop for smooth round specimens with a regular load (**a**) and a load decreased by 25% (**b**).

Changes in the damage level at two characteristic points are shown in Figure 20.

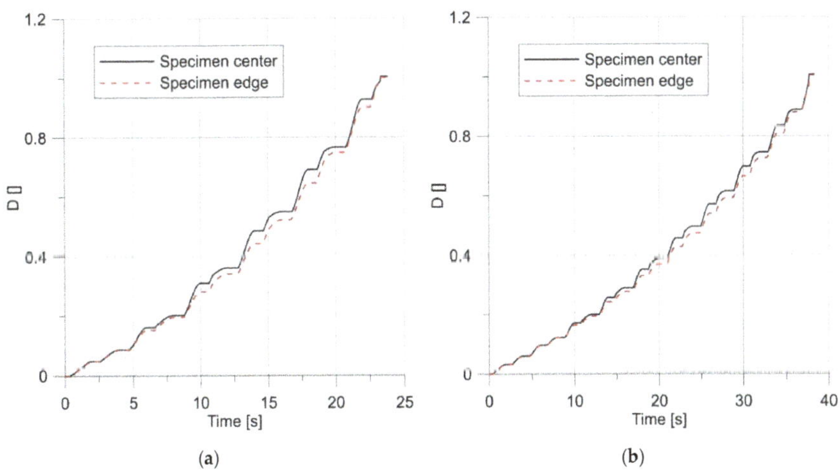

Figure 20. Changes in parameter *D* for a smooth round specimen with a regular load (**a**) and a reduced load (**b**).

Parameter *D*, in both cases, had a higher value at the specimen center during almost the entire loading time, but in the final phase the maximum damage position changed, probably due to the influence of the neck curvature. Determining the location of the *D* maximum required us to trace the distribution of the *D* parameter at the last moment before the crack initiation (Figure 21). In the case of a smooth round specimen subjected to ultra-low-cycle fatigue, the crack initiation occurred near the outer surface, but not at the root of the notch produced by the neck. This was confirmed by numerous examples of fracture surfaces obtained during the research [29–31].

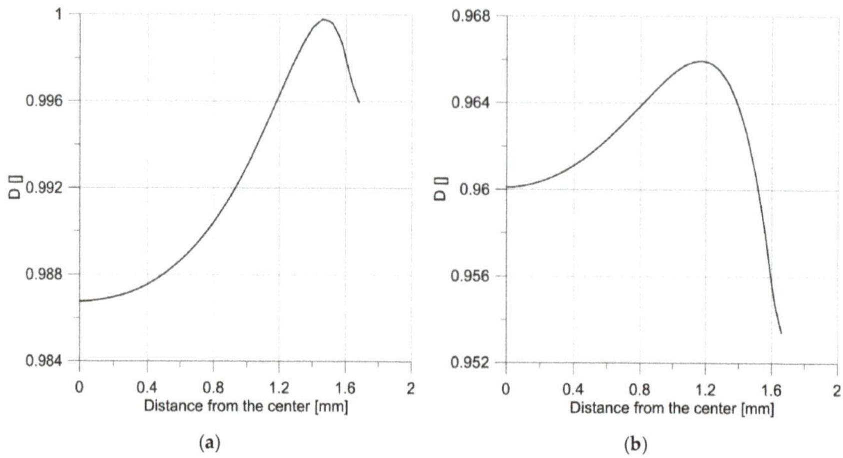

Figure 21. Distribution of the *D* parameter just before the crack initiation for the regular load (**a**) and the decreased load (**b**).

Figure 22 shows how the rate of material damage accelerated with the cycle. This diagram also reveals the asymmetry of the damage accumulation. Despite the assumption of the symmetry of behavior concerning tension and compression, a greater increase in damage was recorded in the case of tension.

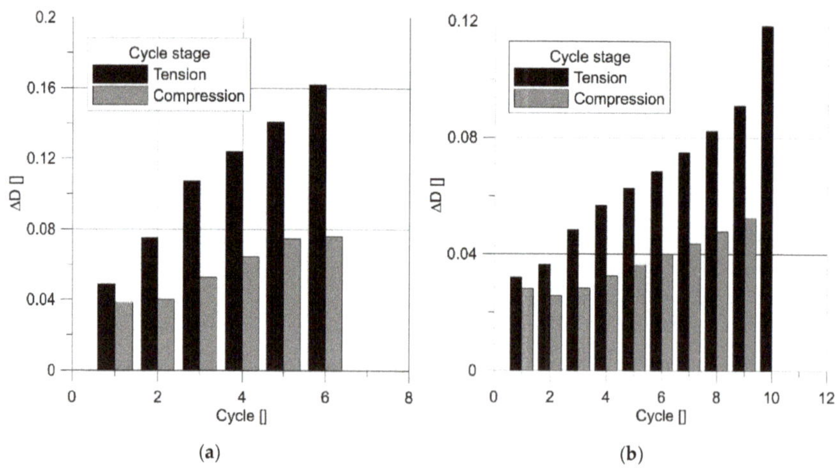

Figure 22. Increment of parameter *D* for subsequent cycles.

Similar behavior was also observed with the notched specimens.

7. Discussion and Conclusions

This paper used the Wierzbicki–Bai model to analyze the damage process of elements subjected to monotonic and fatigue loads. Smooth round and notched specimens were analyzed. For the material of the specimens, Inconel 718 was selected with the assumption of isotropic hardening. As this assumption could be too rough, further investigations are necessary on the combined-material hardening rule. Nevertheless, the results showed that the Wierzbicki–Bai model captured the differences in the behavior of the specimens with different geometries, and the development of the damage in the analyzed elements. In a smooth round specimen subjected to tension, the greatest increase in damage occurred in the center of the sample, demonstrating that they are harder to fracture than notched specimens. The introduction of the small radius notch to the specimen geometry in the tensile specimens caused a crack initiation in the root of the notch; however, a sufficiently large increase in the notch radius caused the fracture process to initiate in the specimen center, as in the case of a smooth round specimen. As a result, the smooth specimens fractured more dynamically than the notched ones that we could observe during the simulation. The results proved that it was possible to locate the hotspot of the specimen: the place where a combination of high-effective stress, stress triaxiality, and the Lode parameter was favorable for the initiation of cracks. This information is of great importance for engineers, with regard to how they can change the geometry of machine members to improve their strength.

The situation was completely different for fatigue loads. In the case of the specimens with a notch, a fracture began at the root of the notch, regardless of the size of the radius. In the case of smooth round specimens, the crack initiation did not occur at the specimen center, as in the case of monotonic loading. The initiation site was close to the outer surface, that is, near the root of the forming neck. Such a state is confirmed by numerous experimental studies. The position of the crack initiation site in the smooth round specimen was to some extent affected by the load level; unfortunately, it was not possible to obtain a state in which the fatigue crack initiation would occur in the center of the specimen. The probable cause of this was the notch influence. Even in the smooth specimens at the final stage, one can observe a notch, as is presented in Figure 23.

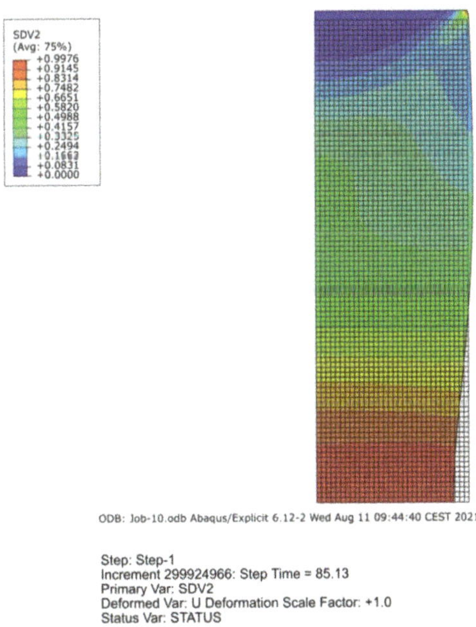

Figure 23. Smooth specimen shape at the final stage.

The increase in the damage of the specimens subjected to fatigue load was much more intense during a tensile stage in the cycle, even though the symmetry of the yield locus was adopted, but only when the Lode parameter was considered. This proved that the main source of the phenomenon was a different level of stress triaxiality for tension and compression. This result shows that ultra-low-cycle fatigue can be utilized to calibrate the value of the coefficient of stress triaxiality dependency on plasticity c_η. The most popular calibration method so far has been to compare the results of the tensile test and the upsetting test; however, the ultra-low-cycle fatigue test would be faster and less expensive, as it uses only a single specimen.

Author Contributions: Conceptualization, J.G.; methodology, J.G.; software, J.G.; validation, J.G., U.J.-G.; formal analysis, J.G., U.J.-G.; investigation, J.G., U.J.-G.; data curation, U.J.-G.; writing—original draft preparation, J.G.; writing—review and editing, U.J.-G.; visualization, U.J.-G. All authors have read and agreed to the published version of the manuscript.

Funding: This research received no external funding.

Institutional Review Board Statement: Not applicable.

Informed Consent Statement: Not applicable.

Data Availability Statement: No new data were created or analyzed in this study. Data sharing is not applicable to this article.

Conflicts of Interest: The authors declare no conflict of interest.

References

1. Funari, M.F.; Lonetti, P.; Spadea, S. A crack growth strategy based on moving mesh method and fracture mechanics. *Theor. Appl. Fract. Mech.* **2019**, *102*, 103–115. [CrossRef]
2. Amini, M.R.; Shahani, A.R. Finite element simulation of dynamic crack propagation process using an arbitrary Lagrangian Eulerian formulation. *Fatigue Fract. Eng. Mater. Struct.* **2013**, *36*, 533–547. [CrossRef]
3. Xu, D.; Liu, Z.; Liu, X.; Zeng, Q.; Zhuang, Z. Modeling of dynamic crack branching by enhanced extended finite element method. *Comput. Mech.* **2014**, *54*, 489–502. [CrossRef]
4. Bai, Y.; Wierzbicki, T. Application of extended Mohr-Coulomb criterion to ductile fracture. *Int. J. Fract.* **2010**, *161*, 1–20. [CrossRef]
5. Khan, A.S.; Liu, H. A new approach for ductile fracture prediction on Al 2024-T351 alloy. *Int. J. Plast.* **2012**, *35*, 1–12. [CrossRef]
6. Danas, K.; Ponte Castañeda, P. Influence of the Lode parameter and the stress triaxiality on the failure of elasto-plastic porous materials. *Int. J. Solids Struct.* **2012**, *49*, 1325–1342. [CrossRef]
7. Algarni, M.; Bai, Y.; Choi, Y. A study of Inconel 718 dependency on stress triaxiality and Lode angle in plastic deformation and ductile fracture. *Eng. Fract. Mech.* **2015**, *147*, 140–157. [CrossRef]
8. Algarni, M.; Choi, Y.; Bai, Y. A unified material model for multiaxial ductile fracture and extremely low cycle fatigue of Inconel 718. *Int. J. Fatigue* **2017**, *96*, 162–177. [CrossRef]
9. Algarni, M.; Bai, Y.; Zwawi, M.; Ghazali, S. Damage evolution due to extremely low-cycle fatigue for inconel 718 alloy. *Metals* **2019**, *9*, 1109. [CrossRef]
10. Bao, Y.; Wierzbicki, T. On fracture locus in the equivalent strain and stress triaxiality space. *Int. J. Mech. Sci.* **2004**, *46*, 81–98. [CrossRef]
11. Bao, Y.; Wierzbicki, T. On the cut-off value of negative triaxiality for fracture. *Eng. Fract. Mech.* **2005**, *72*, 1049–1069. [CrossRef]
12. Xue, L.; Wierzbicki, T. Ductile fracture initiation and propagation modeling using damage plasticity theory. *Eng. Fract. Mech.* **2008**, *75*, 3276–3293. [CrossRef]
13. Kofiani, K.; Nonn, A.; Wierzbicki, T. New calibration method for high and low triaxiality and validation onSENT specimens of API X70. *Int. J. Press. Vessel. Pip.* **2013**, *111–112*, 187–201. [CrossRef]
14. Brocks, W.; Kunecke, G. On the Influence of Triaxiality of the Stress State on Ductile Tearing Resistance. In *Defect Assessment in Components—Fundamentals and Applications (ESIS/EGF9)*; Blauel, J.G., Schwalbe, K.-H., Eds.; Mechanical Engineering Publications Limited: London, UK, 1991.
15. Gao, X.; Zhang, G.; Roe, C. A study on the effect of the stress state on ductile fracture. *Int. J. Damage Mech.* **2010**, *19*, 75–94. [CrossRef]
16. Mohr, D.; Marcadet, S.J. Micromechanically-motivated phenomenological Hosford-Coulomb model for predicting ductile fracture initiation at low stress triaxialities. *Int. J. Solids Struct.* **2015**, *67–68*, 40–55. [CrossRef]
17. Malvern, L.E. *Introduction to the Mechanics of a Continuous Medium*; Prentice-Hall, Inc.: Englewood Cliffs, NJ, USA, 1969.
18. Erice, B.; Gálvez, F. A coupled elastoplastic-damage constitutive model with Lode angle dependent failure criterion. *Int. J. Solids Struct.* **2014**, *51*, 93–110. [CrossRef]
19. Rice, J.R.; Tracey, D.M. On the ductile enlargement of voids in triaxial stress fields*. *J. Mech. Phys. Solids* **1969**, *17*, 201–217. [CrossRef]
20. McClintock, F.A. A criterion for ductile fracture by the growth of holes. *J. Appl. Mech.* **1968**, *35*, 363–371. [CrossRef]
21. Besson, J. Continuum models of ductile fracture: A review. *Int. J. Damage Mech.* **2010**, *19*, 3–52. [CrossRef]
22. Neimitz, A.; Galkiewicz, J.; Lipiec, S.; Dzioba, I. Estimation of the onset of crack growth in ductile materials. *Materials* **2018**, *11*, 2026. [CrossRef]
23. Barsoum, I.; Faleskog, J. Rupture mechanisms in combined tension and shear-Experiments. *Int. J. Solids Struct.* **2007**, *44*, 1768–1786. [CrossRef]

24. Barsoum, I.; Faleskog, J. Rupture mechanisms in combined tension and shear-Micromechanics. *Int. J. Solids Struct.* **2007**, *44*, 5481–5498. [CrossRef]
25. Hamdia, K.M.; Msekh, M.A.; Silani, M.; Thai, T.Q.; Budarapu, P.R.; Rabczuk, T. Assessment of computational fracture models using Bayesian method. *Eng. Fract. Mech.* **2019**, *205*, 387–398. [CrossRef]
26. Bai, Y.; Wierzbicki, T. A new model of metal plasticity and fracture with pressure and Lode dependence. *Int. J. Plast.* **2008**, *24*, 1071–1096. [CrossRef]
27. Pereira, F.G.L.; Lourenço, J.M.; Nascimento, R.M.D.; Castro, N.A. Fracture Behavior and Fatigue Performance of Inconel 625. *Mater. Res.* **2018**, *21*. [CrossRef]
28. Sun, C.; Song, Q. A method for predicting the effects of specimen geometry and loading condition on fatigue strength. *Metals* **2018**, *8*, 811. [CrossRef]
29. Sharma, A.; Oh, M.C.; Ahn, B. Recent advances in very high cycle fatigue behavior of metals and alloys—A review. *Metals* **2020**, *10*, 1200. [CrossRef]
30. Mahtabi, M.J.; Shamsaei, N.; Rutherford, B. Mean Strain Effects on the Fatigue Behavior of Superelastic Nitinol Alloys: An Experimental Investigation. *Procedia Eng.* **2015**, *133*, 646–654. [CrossRef]
31. Li, X.; Xiong, S.M.; Guo, Z. Failure behavior of high pressure die casting AZ91D magnesium alloy. *Mater. Sci. Eng. A* **2016**, *672*, 216–225. [CrossRef]

materials

MDPI

Article

Robust Determination of Fatigue Crack Propagation Thresholds from Crack Growth Data

Josef Arthur Schönherr [1,*], Larissa Duarte [2], Mauro Madia [2], Uwe Zerbst [2], Max Benedikt Geilen [1], Marcus Klein [1] and Matthias Oechsner [1]

1 Center for Structural Materials (MPA-IfW), Technical University of Darmstadt, 64283 Darmstadt, Germany; max.geilen@tu-darmstadt.de (M.B.G.); marcus.klein@tu-darmstadt.de (M.K.); matthias.oechsner@tu-darmstadt.de (M.O.)
2 Bundesanstalt für Materialforschung und -Prüfung (BAM), Division 9.4, 12205 Berlin, Germany; larissa.duarte@bam.de (L.D.); mauro.madia@bam.de (M.M.); uwe.zerbst@bam.de (U.Z.)
* Correspondence: josef.schoenherr@tu-darmstadt.de; Tel.: +49-6151-16-20348

Abstract: The robust determination of the threshold against fatigue crack propagation ΔK_{th} is of paramount importance in fracture mechanics based fatigue assessment procedures. The standards ASTM E647 and ISO 12108 introduce operational definitions of ΔK_{th} based on the crack propagation rate $\mathrm{d}a/\mathrm{d}N$ and suggest linear fits of logarithmic ΔK– $\mathrm{d}a/\mathrm{d}N$ test data to calculate ΔK_{th}. Since these fits typically suffer from a poor representation of the actual curvature of the crack propagation curve, a method for evaluating ΔK_{th} using a nonlinear function is proposed. It is shown that the proposed method reduces the artificial conservativeness induced by the evaluation method as well as the susceptibility to scatter in test data and the influence of test data density.

Keywords: fatigue crack propagation threshold; ISO 12108; ASTM E647; data evaluation methods; experimental determination

Citation: Schönherr, J.A.; Duarte, L.; Madia, M.; Zerbst, U.; Geilen, M.B.; Klein, M.; Oechsner, M. Robust Determination of Fatigue Crack Propagation Thresholds from Crack Growth Data. *Materials* **2022**, *15*, 4737. https://doi.org/10.3390/ma15144737

Academic Editors: Lucjan Śnieżek, Jaroslaw Galkiewicz and Sebastian Lipiec

Received: 1 June 2022
Accepted: 1 July 2022
Published: 6 July 2022

1. Procedures for the Determination of the Fatigue Crack Propagation Threshold from Crack Propagation Data

Typically, the outcome of fatigue crack growth (FCG) tests for the determination of the fatigue crack propagation threshold ΔK_{th} are the crack length a and load history data (e.g., minimum and maximum force) on dependence of the number of cycles N. Considering linear elastic fracture mechanics, the stress intensity factor range ΔK can be calculated from the load and crack length history [1–3]. By using, e.g., the numerical differentiation technique like the secant method or the incremental polynomial method [4], fatigue crack propagation rate $\mathrm{d}a/\mathrm{d}N$ can be computed from the crack length readings and the cycles count. Due to measurement inaccuracy, influence of the testing environment, material inhomogeneities and other effects, test data are always affected by scatter. Furthermore, in most cases, there is no distinct reading at $\mathrm{d}a/\mathrm{d}N = \mathrm{d}a/\mathrm{d}N_{th,ASTM}$ and $\mathrm{d}a/\mathrm{d}N = \mathrm{d}a/\mathrm{d}N_{th,ISO}$, respectively. Hence, the direct determination of the corresponding stress intensity ranges $\Delta K_{th,ASTM}$ and $\Delta K_{th,ISO}$ is not possible. Therefore, data fitting including inter- or extrapolation techniques to determine $\Delta K_{th,ASTM}$ and $\Delta K_{th,ISO}$ are needed, which will be discussed in the following.

Especially in the presence of extrinsic effects, i.e., e.g., crack closure effects [5], the accurate determination of the fatigue crack propagation threshold ΔK_{th} is not trivial. The goal of this contribution is to investigate different methods for the evaluation of the fatigue crack propagation threshold and discuss their robustness and conservativeness.

Following the two most well-known standards regarding FCG tests, namely ASTM E647 [4] and ISO 12108 [6], ΔK_{th} is defined as the asymptotic value of stress intensity factor range ΔK at which the fatigue crack propagation rate $\mathrm{d}a/\mathrm{d}N$ approaches zero. The technical (or also operational) definition of ΔK_{th} for most materials is given at finite crack growth rates

$da/dN_{th,ASTM} = 10^{-7}$ mm/cycle according to ASTM and $da/dN_{th,ISO} = 10^{-8}$ mm/cycle according to ISO. Since neither the fatigue crack growth rate nor the fatigue crack growth threshold stress intensity factor range can be measured directly, the data evaluation is of paramount importance.

1.1. Procedure Suggested by Both ASTM and ISO Standards

According to ASTM and ISO standards, the threshold stress intensity factor ranges are evaluated by determining 'the best fit straight line' [4,6] to $\log \Delta K$–$\log da/dN$ data,

$$\log_{10} \Delta K = P_1 \cdot \log_{10} da/dN + P_0, \tag{1}$$

where P_0, P_1 are fitting parameters, and then calculating the stress intensity ranges corresponding to $da/dN_{th,ASTM}$ and $da/dN_{th,ISO}$, respectively. Both standards define a minimum number of five data points, approximately equally spaced in da/dN. The fitting interval includes data pairs between 10^{-7} mm/cycle and 10^{-6} mm/cycle for ASTM and between 10^{-8} mm/cycle and 10^{-7} mm/cycle for ISO, even though both standards allow for using additional data with lower fatigue crack propagation rates, but require documenting the modified range within the test protocol. Since nowadays FCG tests typically yield far more than five data points within one decade of da/dN data, there is plenty of room for interpretation of the suggested methods. Probably, the most straightforward interpretation (named "interpretation one" in the following) is to just take all data points within the specified ranges (as long as they are approximately equally distributed in da/dN direction) and then identify the best fit straight line for example by utilizing the least-squares method; see Figure 1a. Thereby, Equation (1) is fitted to the (logarithmic) test data, using ΔK as the dependent variable (i.e., the direction of the estimated error). The optimal set of parameters obtained by the least-squares parameter optimization returns the "best fit". Another interpretation (named "interpretation two" in the following) might be that one may freely select $n \geq 5$ approximately equally spaced points that lie within the defined boundaries, for example starting with the point next to the desired threshold fatigue crack propagation rate. Then, the fit showing the maximum Pearson correlation coefficient is selected; see Figure 1b. The results differ and, by adding data generated at lower decades of da/dN, even a non-conservative (= higher) FCG threshold stress intensity range might be calculated, see Section 3.1.3. Furthermore, neither of both interpretations (nor any other straight line) is able to reflect the curvature of the depicted test data.

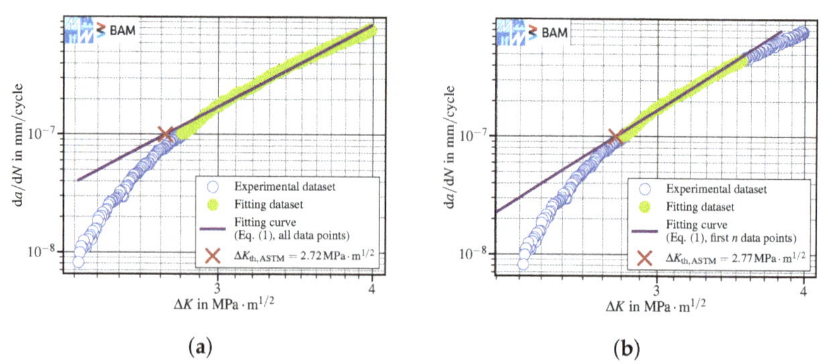

Figure 1. FCG test conducted with conventional K-decreasing at $R = 0.8$ in lab air, see corresponding paragraph in Section 3.1.1: (**a**) linear fit over all data points with 10^{-7} mm/cycle $\leq da/dN \leq 10^{-6}$mm/cycle; (**b**) best linear fit over the first n data points with $da/dN \geq 10^{-7}$ mm/cycle, where $n = 112$ (out of 177 within the interval) is selected yielding the highest Pearson correlation coefficient.

The results obtained following interpretation are strongly affected by the test data curvature because of its averaging character. The higher the curvature, the more conservative is the calculated ΔK_{th}. In contrast, interpretation two is less affected by data curvature because of its definition using the correlation coefficient of the straight line. In case of a more pronounced scatter in the vicinity of da/dN_{th}, interpretation two would have accounted for a higher number of data points and therefore be more conservative. It is trivial to state that a straight line is unable to characterize the curvature of the crack growth curve and therefore leads to conservative ΔK_{th} results for strictly monotonic increasing FCG data fitted within the recommended ranges. In turn, the evaluation error depends on the degree of the curvature, and it is therefore sensitive to the number (and location) of data pairs taken into account as well as to the curve's gradient, see also [7]. Therefore—as also mentioned in the ASTM standard [4]—a nonlinear relationship between ΔK and da/dN might be beneficial to obtain a good fit.

1.2. Procedures Suggested in the Literature

Bucci [7] considered a four-parameter Weibull function (see also [8,9]), fitted to the entire data set, see Figure 2a. He stated that the Weibull approach addresses nonlinearity in a better way, but if there are enough data points in the near-threshold region, straight line fits are performing quite well and are easier to use. In Figure 2a, the test data have been fitted using the four-parameter Weibull function

$$1 - \frac{\Delta K}{K_b} = \exp\left(-\left(\frac{da/dN - e}{v - e}\right)^k\right) \tag{2}$$

and $\Delta K_{th,ASTM}$ has been determined at $da/dN_{th,ASTM}$. Smith and Hoeppner [9] observed that the least-squares method is not suitable to fit the instability parameter K_b, the threshold parameter e, the characteristic value v and the shape parameter k to da/dN–ΔK data accurately within the threshold region due to the differences in orders of magnitude of the non-logarithmic test data (da/dN ranges between 10^{-7} mm/cycle and 10^{-6} mm/cycle vs. ΔK ranges between 10^0 MPa·m$^{1/2}$ and 10^1 MPa·m$^{1/2}$, approximately). Therefore, they proposed to use a least-squares optimization to calculate a preliminary optimized parameter set and afterwards improve the fitting results by optimizing only e and v in an orthogonal distance regression (ODR) [10]. Since an ODR optimization is connected to a quite high computational cost and compared to the 1990s the computation power increased dramatically until nowadays, this might have been the reason why in [9] only the two parameters showing the major contribution with regard to the mentioned errors were optimized using an ODR. In fact, it has been observed in this work that the curve fit may be improved by optimizing all four parameters instead of just two in an ODR after least-squares minimization. As one can clearly see, the overall agreement to the test data is quite good but locally diverges slightly. This can be observed in Figure 2a as well as in Figure 2b, where the four-parameter Weibull fit predicts a too steep curvature towards the ISO operational threshold definition ($da/dN_{th} = 10^{-8}$ mm/cycle) and therefore leads to non-conservative results for $\Delta K_{th,ISO}$. Since non-conservative values are not acceptable, the Weibull fit has not been considered any further.

Another well established method to evaluate the threshold stress intensity range was proposed by Döker [11]. The method uses a straight line fit applied to the (non-logarithmic) da/dN–ΔK data in the range 5×10^{-8}mm/cycle $\leq da/dN \leq 10^{-6}$ mm/cycle, see Figure 3a. The threshold definition is independent from the ASTM or ISO standards, and ΔK_{th} is evaluated at $da/dN_{th} = 0$. The resulting threshold stress intensity factor range is located somewhere between the values one would expect using the ISO or ASTM operational definitions. Considering that only the ASTM E647 standard [12] is cited within the original publication, this method shows quite conservative results using the ASTM operational definition of ΔK_{th}. The straightforward extension to the ISO threshold by shifting the fit range by one decade of da/dN would require test results as low as $da/dN = 5 \times 10^{-9}$ mm/cycle, which would be very time consuming to obtain and therefore not practicable. The curve behavior is poorly described, both for linear, Figure 3a, and double-logarithmic, Figure 3b, scaled da/dN–ΔK data. Therefore, further analyses with this approach have also been discarded.

Furthermore, there is a multitude of different crack propagation laws aiming at describing the whole FCG data starting from the threshold regime over the range where the FCG curve grows linear in a double-logarithmic scaled plot (also known as Paris regime) and some even include the region of instable crack growth; see [13–19]. Several of these models contain ΔK_{th} as a model parameter, which should not be confused with the operational definitions of the threshold stress intensity factor range included in the standards.

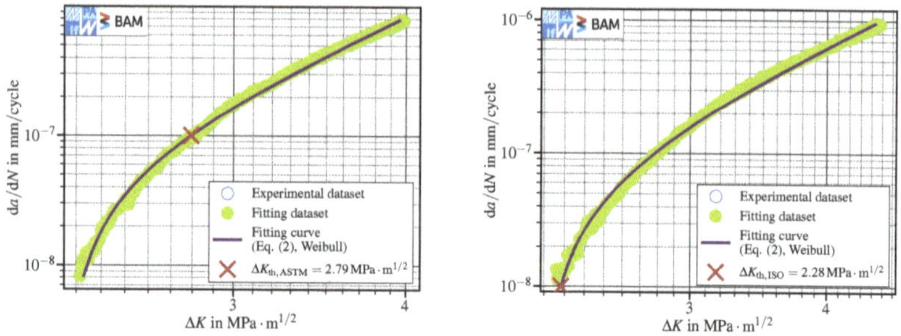

(a) (b)

Figure 2. Application of a four-parameter Weibull function on all test data, referring to: (**a**) a threshold definition at $da/dN_{th,ASTM} = 10^{-7}$ mm/cycle. K-decreasing test at $R = 0.8$ in lab air, see corresponding paragraph in Section 3.1.1; (**b**) a slightly non-conservative FCG threshold determination at $da/dN_{th,ISO} = 10^{-8}$ mm/cycle. K_{max} = const., $R_{max} \approx 0.8$, lab air, see corresponding paragraph in Section 3.1.1.

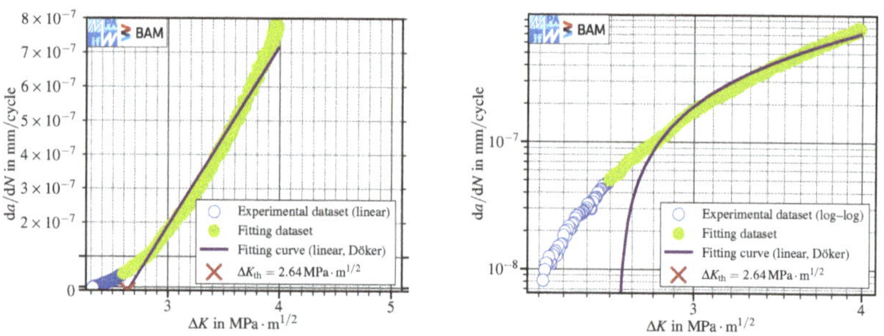

(a) (b)

Figure 3. FCG threshold determination using a linear fit, applied to regular (non-logarithmic) test data, following [11]. K-decreasing test at $R = 0.8$ in lab air, see corresponding paragraph in Section 3.1.1. (**a**) linear-scaled data; (**b**) double-logarithmic scaled data. The method clearly provides a poor data fit and determination of ΔK_{th} at $da/dN_{th} = 0$.

2. Experimental Procedure

Since none of the methods shown in Sections 1.1 and 1.2 allowed universal application for both standards and simultaneously led to robust and not overly conservative results, further fitting functions have been investigated. A comprehensive set of fatigue crack growth data recorded at BAM Berlin and MPA-IfW Darmstadt has been used to calibrate and validate the evaluation procedure. The fitting has been performed using least squares minimization, where the threshold stress intensity factor range has been used as dependent variable, if not otherwise mentioned.

2.1. Investigated Fitting Functions

The most straightforward approach was to extend Equation (1) to a more general polynomial form as

$$\log_{10} \Delta K = \sum_{i=0}^{I} P_i \cdot (\log_{10} \mathrm{d}a/\mathrm{d}N)^i, \quad (3)$$

where I denotes the polynomial degree and P_i are fit parameters. For $I = 1$, Equation (3) equals Equation (1). Regarding $I \geq 2$, in case there are only very few (or even no) data points available in the vicinity of $\mathrm{d}a/\mathrm{d}N_{\mathrm{th}}$, the fit may yield a curve having an inflection point. Hence, the fit does not represent the data in a satisfactory manner, and (for $I = 2$) it is even possible that the fit does not intersect the $\mathrm{d}a/\mathrm{d}N_{\mathrm{th}}$ line. Therefore, it is not recommended to use polynomial fits of the type given in Equation (3) with $I \neq 1$. The use of non-logarithmic data, like in the approach of Döker [11], leads to the same problems. Hence, further approaches have been investigated.

The $\log - \log$ data depicted in Figure 1 considered swapping the axes, suggesting a hyperbolic trend, leading to the fit function

$$\log_{10} \Delta K = P_1 \cdot (-\log_{10} \mathrm{d}a/\mathrm{d}N)^{-1} + P_2, \quad (4)$$

with the fitting parameters P_1 and P_2. Fitting Equation (4) to the dataset leads to results comparable to the linear fit using all data points in accordance with the standards, since the exponent -1 does not represent the curvature of the crack growth curve, see Figure 4a. By extending Equation (4) to a variable exponent as

$$\log_{10} \Delta K = P_1 \cdot (-\log_{10} \mathrm{d}a/\mathrm{d}N)^{-P_3} + P_2, \quad (5)$$

with $P_3 \geq 1$, a much better fitting to the test data is possible, see Figure 4b. Rounding the fit result of $P_3 \approx 4.60$ to the previous and next integer number, namely $P_3 = 4$,

$$\log_{10} \Delta K = P_1 \cdot (-\log_{10} \mathrm{d}a/\mathrm{d}N)^{-4} + P_2, \quad (6)$$

see Figure 4c and $P_3 = 5$,

$$\log_{10} \Delta K = P_1 \cdot (-\log_{10} \mathrm{d}a/\mathrm{d}N)^{-5} + P_2, \quad (7)$$

see Figure 4d provides two straightforward variants of Equation (5) with only two free parameters. The curvature of the fitting curve increases with increasing P_3. It is worth noting that, although $P_3 = 5$ leads to non-conservative estimation of the FCG threshold according to ISO, it is still conservative following the ASTM operational definition.

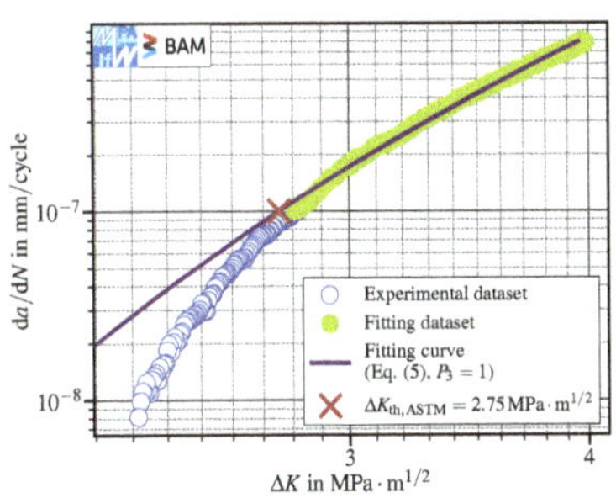

(a)

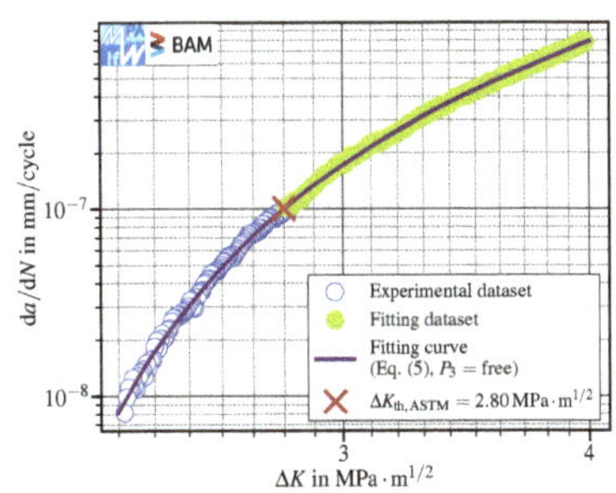

(b)

Figure 4. *Cont.*

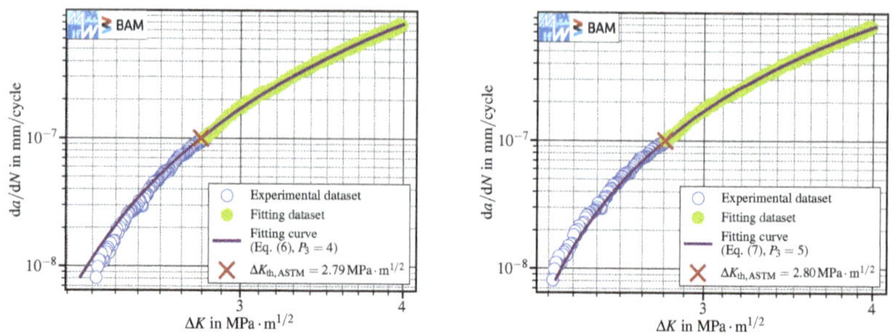

Figure 4. FCG threshold evaluation according to ASTM ($da/dN_{th,ASTM} = 10^{-7}$ mm/cycle) for the data set presented in Figure 1 using Equation (5) with fixed or free parameter P_3: (**a**) fixed, $P_3 = 1$; (**b**) free, optimized value $P_3 \approx 4.60$; (**c**) fixed, $P_3 = 4$; (**d**) fixed, $P_3 = 5$.

2.2. Quantitative Data Analysis

The main datasets investigated stemmed from a total of 47 specimens manufactured from 12 mm thick S690QL hot-rolled plates, tested at BAM Berlin and MPA-IfW Darmstadt. The materials chemical composition and mechanical properties are given in Tables 1 and 2. The microscopic analysis on etched samples showed a fine grained quenched and tempered martensitic-bainitic microstructure, see Figure 5.

Table 1. Chemical composition (in weight percent) obtained by means of spark optical emission spectrometry.

C	Si	Mn	P	S	Cr	Mo	Ni	Al	Cu	Nb	Fe
0.16	0.23	1.15	0.01	< 0.01	0.41	0.18	0.04	0.08	0.02	0.04	97.65

Table 2. Mechanical properties.

σ_y in MPa	σ_u in MPa	E in GPa	A in %	KV_2 in J (Orientation: T–L [20])
810	825	207	16	126

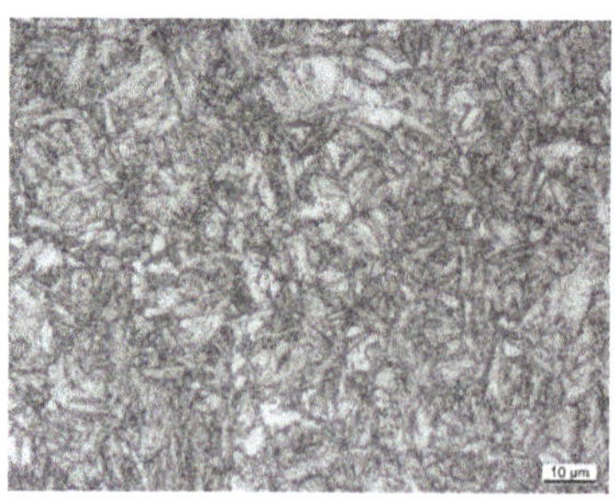

Figure 5. S690QL microstructure (T–L plane).

All tests were performed using SENB specimens [6] with a cross section of 19 mm × 6 mm. The specimens were oriented, such that the direction of crack propagation is parallel to the rolling direction, i.e., the orientation T–L according to [20]. The test data have been obtained on three resonance testing machines equipped with an eight-point bending fixture, see [6]. These were a RUMUL MIKROTRON 654 with a maximum load capacity of 20 kN and an average testing frequency of about 108 Hz, a RUMUL TESTRONIC with a maximum load capacity of 100 kN and an average testing frequency of about 60 Hz, both at BAM Berlin, and a RUMUL TESTRONIC with a maximum load capacity of 250 kN and an average testing frequency of about 90 Hz at MPA-IfW Darmstadt. The crack length was monitored using direct current potential drop techniques with current reversal and active temperature compensation (current source: HP 6033A, nanovolt meter: Keithley 2182A) at BAM and specimen compliance techniques (clip-gage: Sandner EXR10-0.5o) at MPA-IfW. The crack length was corrected a posteriori by means of optical measurements on the broken open fracture surfaces. Then, the crack propagation rates have been calculated using the slope of piecewise straight line fits performed on filtered test data. Each segment of the piecewise function referred to a crack extension of 0.02 mm. The corresponding stress intensity factors have been calculated using the formulations reported in [6].

Since in these tests the same specimen types, manufactured from the same material batch in the same specimen orientation, have been used, a high repeatability was expected. Consequently, the standard deviation in ΔK_{th} calculated for each method was the result of the data scatter within the test (stemming from small variations in environment conditions, material inhomogeneities, specimen misalignment, errors in the calculation of da/dN, etc.) and an error induced by the fit used to evaluate ΔK_{th}. Since the first part is independent of the fitting procedure, the differences in standard deviations are a measure for the fit robustness, whereas the corresponding mean value gives information on the fit quality and therefore the inter- and extrapolation error, respectively.

3. Results and Discussion

3.1. Application to Data Obtained at $R \approx 0.8$

First, data obtained at a load ratio of approximately $R \approx 0.8$, which produce only a negligible influence of crack closure effects have been investigated.

3.1.1. Evaluation for the Intervals Suggested by the Standards

In order to assess the performance of the polynomial functions with negative exponent Equation (5) in comparison to the fit suggested by the standards, see Equation (1), test data obtained at either a fixed load ratio $R = 0.8$ (K-decreasing procedures) or at $R_{max} \approx 0.8$ (K_{max} tests) have been considered, see ([4], Section 8.6). To minimize influences of the extrapolation method, the smallest recorded crack propagation rate has been required to be smaller than $1.1 \cdot da/dN_{th}$, i.e., $\min(da/dN) \leq 1.1 \times 10^{-8}$ mm/cycle for the ISO and $\min(da/dN) \leq 1.1 \times 10^{-7}$ mm/cycle for the ASTM operational definition of ΔK_{th}. It shall be noted that these boundaries have been used only for comparability between tests within this work and neither define the actual application boundaries of the method regarding data extrapolation nor represent a general recommendation.

Since the fixed parameters $P_3 = 4$ in Equation (6) and $P_3 = 5$ in Equation (7) provided a better description of the data curvature and a less conservative determination of ΔK_{th} in addition to the more general function with variable P_3, these have been compared with the two interpretations of the linear fit method suggested by the standards, see Figure 1a,b. The range used to fit the data was in all cases fixed to one decade of da/dN data, starting from da/dN_{th} and therefore equal to the ranges suggested by the standards in order to ensure comparability between the methods.

Conventional K-Decreasing at $R = 0.8$ (ΔK_{LR})

The first datasets investigated stemmed from a total of nine SENB specimens made of S690QL. These tests have been conducted using the standard K-decreasing (or load shedding) procedure suggested by the standards at constant $R = 0.8$, tested in lab air. Further information on the experimental procedure may be found in [21]. All nine specimens contained data points below $da/dN = 1.1 \times 10^{-7}$ mm/cycle and are valid for ASTM operational threshold evaluation, and four of them were also valid according to the ISO definition. The threshold stress intensity factor range results with corresponding standard deviations are presented in Table 3.

Table 3. Comparison of effective threshold values for S690QL obtained with various fit methods. Tests conducted using the K-decreasing procedure at $R = 0.8$ in lab air.

Method	$\Delta K_{th,ASTM}$ in MPa·m$^{1/2}$	$\Delta K_{th,ISO}$ in MPa·m$^{1/2}$
Linear, all data points (Equation (1))	2.71 ± 0.06	2.27 ± 0.04
Linear, first n data points (Equation (1))	2.76 ± 0.05	2.29 ± 0.04
Polynomial, neg. exp. (Equation (5))	2.78 ± 0.04	2.34 ± 0.04
Polynomial, neg. exp. $P_3 = 4$ (Equation (6))	2.77 ± 0.05	2.32 ± 0.04
Polynomial, neg. exp. $P_3 = 5$ (Equation (7))	2.78 ± 0.05	2.32 ± 0.03

The comparison between the determined threshold stress intensity factor ranges (on single specimen basis) and the value read out from test data proved conservative for every single method and dataset. Considering the mean value of the threshold stress intensity ranges, both linear fits showed a more pronounced underestimation of ΔK_{th}, inducing an artificial conservativeness, as already observed in Section 2.1. This is proven true especially for the linear fit Equation (1) performed on all data points. For the linear fit over the first n points, the artificial conservativeness of $\Delta K_{th,ASTM}$ was found to be comparatively higher than for $\Delta K_{th,ISO}$. The results obtained using Equation (5) agreed fairly well, whereas the fixed exponents $P_3 = 4$ and $P_3 = 5$ showed a slightly higher conservativeness for $\Delta K_{th,ISO}$ compared to the three-parameter version of Equation (5). The standard deviation is very low and comparable for all tests.

Load Shedding at Constant K_{max} ($R_{max} \approx 0.8$)

A set of nine SENB prepared from the same material batch has been investigated using a load shedding scheme at constant K_{max} with a final load ratio $R_{max} \approx 0.8$ at about $\Delta K_{th,ISO}$ and $R \approx 0.72 \ldots 0.76$ at $\Delta K_{th,ASTM}$. All nine specimens provided data for evaluating $\Delta K_{th,ASTM}$ and among them four were also valid for $\Delta K_{th,ISO}$ evaluation, see Table 4.

Table 4. Comparison of effective threshold values for S690QL obtained with various fit methods. Tests conducted using the constant K_{max} procedure ($R_{max} \approx 0.8$) in lab air.

Method	$\Delta K_{th,ASTM}$ in MPa·m$^{1/2}$	$\Delta K_{th,ISO}$ in MPa·m$^{1/2}$
Linear, all data points (Equation (1))	2.67 ± 0.03	2.22 ± 0.01
Linear, first n data points (Equation (1))	2.77 ± 0.03	2.24 ± 0.01
Polynomial, neg. exp. (Equation (5))	2.79 ± 0.03	2.29 ± 0.03
Polynomial, neg. exp. $P_3 = 4$ (Equation (6))	2.77 ± 0.03	2.27 ± 0.02
Polynomial, neg. exp. $P_3 = 5$ (Equation (7))	2.79 ± 0.03	2.27 ± 0.02

Since the crack propagation data obtained from different procedures at $R \approx 0.8$ nearly coincide and exhibit a very low scatter, the same conclusion as in the case of conventional K-decreasing tests can be drawn.

Compression Precracking Load Reduction (CPLR) at $R = 0.8$

A further set of four SENB extracted from the same batch has been investigated using compression precracking followed by a K-decreasing test with a constant load ratio $R = 0.8$. Applying the same criteria for selecting valid data sets for comparison returned four specimens for evaluating $\Delta K_{th,ASTM}$ and two for $\Delta K_{th,ISO}$, see Table 5. Here, the standard deviation for ISO is omitted due to the insufficient number of available data sets that include points below $da/dN = 1.1 \times 10^{-8}$ mm/cycle. The results confirm those previously shown in Table 4.

Table 5. Comparison of threshold values for S690QL obtained with various fit methods. Tests conducted using Compression Precracking Load Reduction at $R = 0.8$ in lab air.

Method	$\Delta K_{th,ASTM}$ in MPa·m$^{1/2}$	$\Delta K_{th,ISO}$ in MPa·m$^{1/2}$
Linear, all data points (Equation (1))	2.67 ± 0.02	2.22
Linear, first n data points (Equation (1))	2.74 ± 0.03	2.23
Polynomial, neg. exp. (Equation (5))	2.77 ± 0.04	2.29
Polynomial, neg. exp. $P_3 = 4$ (Equation (6))	2.73 ± 0.02	2.26
Polynomial, neg. exp. $P_3 = 5$ (Equation (7))	2.74 ± 0.03	2.27

Constant Force Range (ΔF-Constant) at $R = 0.8$

The fourth and last test to determine the threshold at $R \approx 0.8$ in ambient air has been based on another set of four SENB specimens produced from the same material batch, using conventional precracking followed by a test at constant force amplitude (ΔF-constant) at a load ratio $R = 0.8$. All four specimens have been considered valid for evaluating $\Delta K_{th,ASTM}$ and three for $\Delta K_{th,ISO}$, see Table 6. Here, the same observations as above apply.

Table 6. Comparison of threshold values for S690QL obtained with various fit methods. Tests conducted using ΔF-constant at $R = 0.8$ in lab air.

Method	$\Delta K_{th,ASTM}$ in MPa·m$^{1/2}$	$\Delta K_{th,ISO}$ in MPa·m$^{1/2}$
Linear, all data points (Equation (1))	2.58 ± 0.03	2.18 ± 0.06
Linear, first n data points (Equation (1))	2.64 ± 0.03	2.21 ± 0.06
Polynomial, neg. exp. (Equation (5))	2.69 ± 0.02	2.31 ± 0.02
Polynomial, neg. exp. $P_3 = 4$ (Equation (6))	2.67 ± 0.02	2.25 ± 0.03
Polynomial, neg. exp. $P_3 = 5$ (Equation (7))	2.69 ± 0.02	2.25 ± 0.02

Summary

The linear fits induce artificial conservativeness in the evaluation of fatigue crack propagation thresholds obtained at $R \approx 0.8$. Nevertheless, this behavior is observed less pronounced for the linear fit incorporating only the first n points. In all cases, the three-parameter polynomial Equation (5) provides less conservative results. Fixing its parameter P_3 to $P_3 = 4$ or $P_3 = 5$ sometimes induces conservativeness, but in most of the cases less pronounced than the linear fits.

Figure 6 summarizes the evaluation of the four datasets presented earlier in Section 3.1.1. The threshold stress intensity factor ranges determined according to the ASTM operational definition (see Figure 6a) as well as those following the ISO operational definition (see Figure 6b) for the four test methods (ΔK_{LR}, K_{max}, CPLR and ΔF-constant) agree fairly well within each data evaluation method.

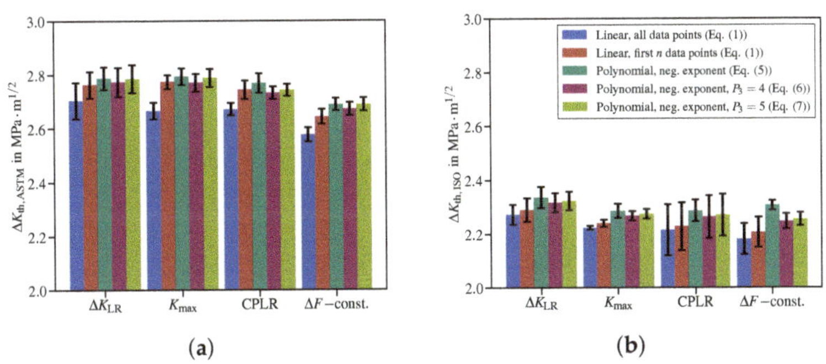

Figure 6. Fatigue crack propagation thresholds obtained at $R \approx 0.8$ applying the fit methods to the four datasets presented in Section 3.1.1: (**a**) according to the ASTM operational definition; (**b**) according to the ISO operational definition.

3.1.2. Robustness of the Fitting Methods in Handling Data Subjected to Augmented Artificial Scatter

In order to assess the ability and robustness of the fitting methods to handle scattered data, artificial scatter has been added to the test data presented in Figure 1. The additional scatter has been generated by sampling random values from a normal distribution with a mean $\mu = 1$ and a standard deviation SD $= 0.02$ and multiplying them with ΔK data, whereas da/dN remained unchanged.

The comparison of the linear fit using all scattered data points within the defined interval ($\Delta K_{th,ASTM} = 2.69\,\text{MPa}\cdot\text{m}^{1/2}$, Figure 7a) with the original dataset ($\Delta K_{th,ASTM} = 2.72\,\text{MPa}\cdot\text{m}^{1/2}$, Figure 1a) did not reveal a notable difference. The polynomials with negative exponents were also almost insensitive to scatter. The three-parameter polynomial provided $\Delta K_{th,ASTM} = 2.78\,\text{MPa}\cdot\text{m}^{1/2}$ for the scattered data (Figure 7d) compared to $\Delta K_{th,ASTM} = 2.80\,\text{MPa}\cdot\text{m}^{1/2}$ for the original dataset (Figure 4b). The two-parameter polynomial with $P_3 = 4$ showed a similar trend with $\Delta K_{th,ASTM} = 2.79\,\text{MPa}\cdot\text{m}^{1/2}$ for scattered data (Figure 7c) in comparison to $\Delta K_{th,ASTM} = 2.79\,\text{MPa}\cdot\text{m}^{1/2}$ for the original dataset (Figure 4c).

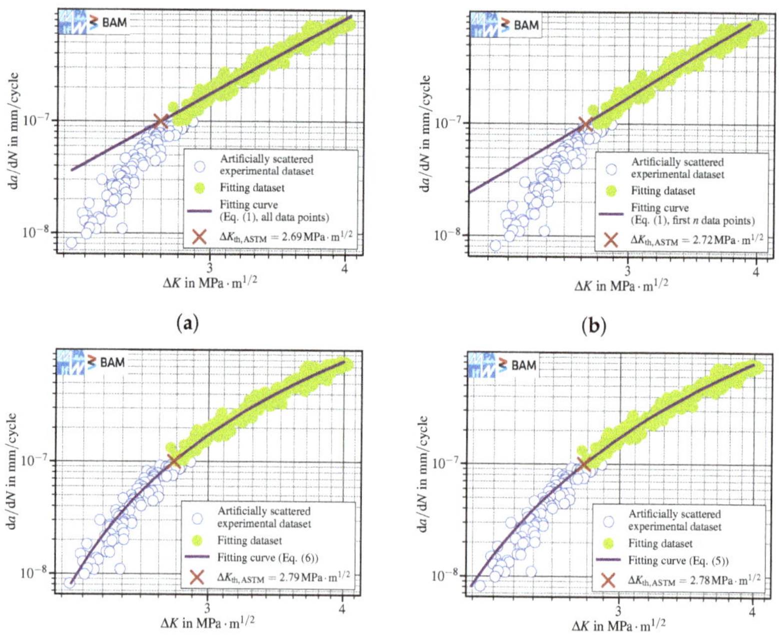

Figure 7. FCG data reported in Figure 1 with additional artificial scatter. The $da/dN_{th,ASTM}$ has been evaluated using four different methods: (**a**) linear, all data points Equation (1); (**b**) linear, first $n = 176$ data points (out of 177 within the interval) Equation (1); (**c**) polynomial, negative exponent, $P_3 = 4$ Equation (6); (**d**) polynomial, negative exponent Equation (5).

In contrast, the best linear fit over the first n points showed a pronounced sensitivity to scattered data. The best fit interval coincided almost with all data points (see Figures 7b and 1b). Consequently, $\Delta K_{th,ASTM} = 2.72\,\text{MPa}\cdot\text{m}^{1/2}$ calculated for scattered data was more conservative than $\Delta K_{th,ASTM} = 2.77\,\text{MPa}\cdot\text{m}^{1/2}$ calculated in case of the original dataset.

3.1.3. Influence of an Augmented Fit Interval

Both ASTM E647 and ISO 12108 suggest a fit interval of one decade of da/dN data, starting from $da/dN_{th,ASTM}$ and $da/dN_{th,ISO}$, respectively. Nevertheless, both allow for use data obtained at lower fatigue crack propagation rates for determining the threshold stress intensity factor range. Therefore, the impact of an augmented fit interval on the determination of fatigue crack propagation thresholds has been investigated. Since no datasets with crack propagation rates momentously below $da/dN_{th,ISO} = 1 \times 10^{-8}$ mm/cycle were available, only the threshold following the ASTM operational definition has been considered. Therefore, the data shown in Table 3, which have been obtained using a fit interval of 10^{-7} mm/cycle $\leq da/dN \leq 10^{-6}$ mm/cycle, have been compared with threshold stress intensity factor ranges obtained with augmented intervals. The upper bound has been held constant, whereas the lower bound has been varied from 2.5×10^{-8} mm/cycle up to 10^{-7} mm/cycle. The respective threshold stress intensity factor ranges have been named as $\Delta K_{th,ASTM,2.5}$, $\Delta K_{th,ASTM,5}$, $\Delta K_{th,ASTM,7.5}$ and $\Delta K_{th,ASTM,10}$, according to the lower FCG propagation rate bounds (2.5, 5, 7.5 and 10×10^{-8} mm/cycle, see Figure 8).

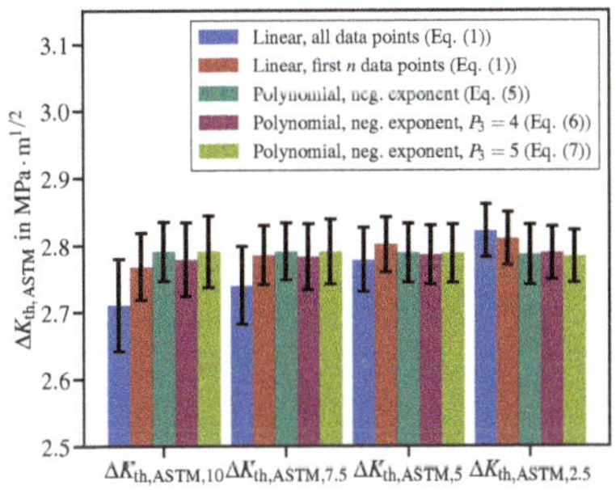

Figure 8. Comparison of threshold values obtained with various fit methods using varying fit intervals from 2.5×10^{-8} mm/cycle $\leq da/dN \leq 10^{-6}$ mm/cycle ($\Delta K_{th,ASTM,2.5}$) up to 10×10^{-8} mm/cycle $\leq da/dN \leq 10^{-6}$ mm/cycle ($\Delta K_{th,ASTM,10}$). The results refer to eight specimens made of S690QL, tested using a K-decreasing procedure at $R = 0.8$ in lab air.

Regarding both linear fits, there is a clear tendency that, with augmenting the fit interval towards lower minimal crack propagation rates (displayed in Figure 8 from left to right), there is an increase in $\Delta K_{th,ASTM}$ and therefore a decrease in conservativeness. In contrast, the results for the polynomial with negative exponent, Equations (5)–(7), are almost insensitive to the interval augmentation, but for $P_3 = 4$ and $P_3 = 5$, a reduction in standard deviation can be observed with increasing the interval. For Equation (6), the optimal lower bound was found at 2.5×10^{-8} mm/cycle and 5×10^{-8} mm/cycle with equal magnitudes in mean and standard deviation.

Using the linear functions with augmented intervals increases the risk of non-conservative extrapolation, as one can see comparing the values for $\Delta K_{th,ASTM,2.5}$, where the linear functions provided the highest threshold stress intensity factor ranges among all five methods under comparison. This issue can be clearly understood looking at the evaluation depicted in Figure 9. In particular, Figure 9a shows that the $\Delta K_{th,ASTM}$ calculated using the linear fit over all data points is on the right-hand side of the dataset, i.e., in the non-conservative region. In contrast, the polynomial with $P_3 = 4$ does not show this issue (Figure 9b).

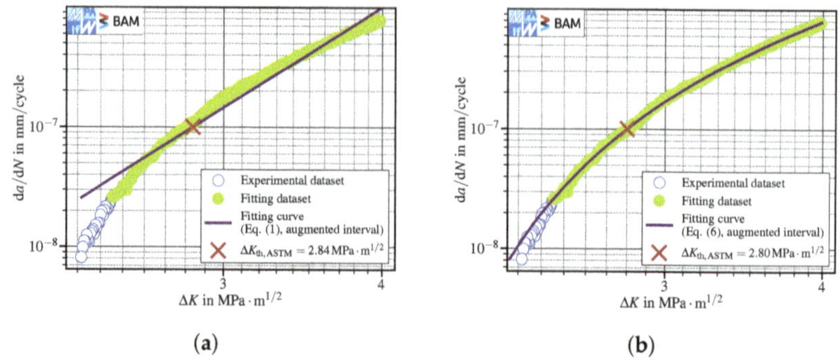

Figure 9. Influence of augmented fit interval on various fits, using an interval 2.5×10^{-8} mm/cycle $\leq$ $\mathrm{d}a/\mathrm{d}N \leq 10^{-6}$ mm/cycle: (**a**) the linear fit using all data within the specified range leads to a non-conservative result; (**b**) the polynomial fit with negative exponent $P_3 = 4$ gives a conservative result. The same dataset as reported in Figure 1 has been used.

3.1.4. Data Extrapolation

The investigations presented in Section 3.1.1 included only data at crack propagation rates as low as $1.1 \cdot \mathrm{d}a/\mathrm{d}N_{\mathrm{th}}$. Nevertheless, it shall be noted that no data might be available at low crack propagation rates, especially for the ISO operational definition with a threshold crack propagation rate as low as $\mathrm{d}a/\mathrm{d}N_{\mathrm{th,ISO}} = 10^{-8}$ mm/cycle. Therefore, to ensure a reliable and conservative evaluation of the fatigue crack propagation thresholds, a robust extrapolation technique is needed. To assess the goodness of the extrapolation, the test results given in Section 3.1.1 have been compared to artificially censored datasets, using only data with $\mathrm{d}a/\mathrm{d}N \geq 3 \cdot \mathrm{d}a/\mathrm{d}N_{\mathrm{th}}$. Hence, the resulting fit intervals after censoring were 3×10^{-8} mm/cycle $\leq \mathrm{d}a/\mathrm{d}N \leq 10^{-7}$ mm/cycle for ISO operational definition and 3×10^{-7}mm/cycle $\leq \mathrm{d}a/\mathrm{d}N \leq 10^{-6}$ mm/cycle for ASTM operational definition. Since extrapolation is very sensitive to the data range available, in order to have a reliable comparison, the investigations have been restricted to datasets that had data within the whole censored interval (including the upper bound). Hence, the number of tests with valid data for ASTM threshold determination is reduced in comparison to Section 3.1.1.

The comparison has been based on the change in ΔK_{th} induced by censoring the FCG data. $\Delta K_{\mathrm{th,cens}}$ denotes the fatigue crack propagation threshold obtained for the censored version of the data set used to evaluate ΔK_{th}. It follows that $\Delta K_{\mathrm{th}} - \Delta K_{\mathrm{th,cens}}$ values greater than or equal to zero are considered conservative, whereas values lower than zero are non-conservative. The minimum difference throughout all specimens shows whether all tests are extrapolated conservatively, whereas the mean value can be regarded as an index of the goodness of the extrapolation. The results are given in Table 7. Conservative extrapolation has been obtained for both linear fits and for the polynomial with negative exponent fixed to $P_3 = 4$, whilst the versions with $P_3 = 5$ or free P_3 returned a very limited number of negative results, meaning non-conservative extrapolation results. Even though the latter are just slightly non-conservative and rare, the occurrence of a non-conservative extrapolation should be avoided whenever possible. Regarding the "mean(·)" columns, denoting the extrapolation error, the linear functions performed far worse than the polynomial Equation (5) with $P_3 = 4$, especially with regard to the $\Delta K_{\mathrm{th,ASTM}} - \Delta K_{\mathrm{th,cens,ASTM}}$ values, where the mean and minimum extrapolation error were 0.20 MPa·m$^{1/2}$ and 0.17 MPa·m$^{1/2}$, regarding the fit over all data points and 0.23 MPa·m$^{1/2}$ and 0.18 MPa·m$^{1/2}$ for the fit over the first n points, respectively, compared to 0.03 MPa·m$^{1/2}$ and 0.01 MPa·m$^{1/2}$ for the polynomial with $P_3 = 4$. Hence, in case extrapolation is inevitable, the polynomial with $P_3 = 4$ brings a notable improvement over the linear fits, suggested by the standards. With regard to both standards, neither a statement on the minimum required crack propagation rate for a valid evaluation of the FCG threshold nor any comments on the legitimacy of a potentially necessary data extrapolation is given.

Table 7. Comparison of extrapolation errors obtained with various fit methods. The data refer to K-decreasing tests at $R = 0.8$ and K_{max} tests conducted on S690QL in lab air. The evaluation comprised ten tests for ASTM threshold and thirteen specimens for ISO threshold. All values are given in MPa·m$^{1/2}$.

Method	$\Delta K_{th,ASTM} - \Delta K_{th,cens,ASTM}$ mean(·)	min(·)	$\Delta K_{th,ISO} - \Delta K_{th,cens,ISO}$ mean(·)	min(·)
Linear, all data points (Equation (1))	0.20	0.17	0.07	0.02
Linear, first n data points (Equation (1))	0.23	0.18	0.07	0.02
Polynomial, neg. exp. (Equation (5))	0.04	−0.02	0.03	−0.05
Polynomial, neg. exp. $P_3 = 4$ (Equation (6))	0.03	0.01	0.03	0.00
Polynomial, neg. exp. $P_3 = 5$ (Equation (7))	0.01	−0.02	0.02	−0.02

3.1.5. Application to the Full Dataset

Based on the conclusions drawn in Section 3.1.4, where a robust conservative extrapolation could be obtained for the linear fits as well as for the polynomial with negative exponent fixed to $P_3 = 4$, the full dataset, including all datasets that may be extrapolated, has been reevaluated. Because an augmented fit interval may lead to non-conservative results in case of linear fits, the intervals suggested in the standards have been used (see Section 1.1). In contrast, a beneficial effect of an augmented interval has been observed when using Equation (6), see Section 3.1.3. Hence, an augmented interval for calculating $\Delta K_{th,ASTM}$ has been used in this case. Consequently, the fitting intervals have been defined as $da/dN = 5 \times 10^{-8}$ mm/cycle $\leq da/dN \leq 10^{-6}$ mm/cycle for ASTM and $da/dN = 10^{-8}$ mm/cycle $\leq da/dN \leq 10^{-7}$ mm/cycle for ISO. The results are presented in Table 8. The linear fit over the full interval resulted in the highest conservativeness in combination with a small standard deviation. Using only the first n points of the interval to generate the linear fit reduced the conservativeness with the drawback of increasing the standard deviation, especially for ASTM. By using an appropriate nonlinear function like Equation (6), the conservativeness as well as the standard deviation can be reduced.

Table 8. Comparison of threshold values obtained at $R \approx 0.8$ with various fit methods using all 29 specimen data sets for $\Delta K_{th,ASTM}$ and $\Delta K_{th,ISO}$; results for S690QL in lab air.

Method	$\Delta K_{th,ASTM}$ in MPa·m$^{1/2}$	$\Delta K_{th,ISO}$ in MPa·m$^{1/2}$
Linear, all data points (Equation (1))	2.67 ± 0.06	2.22 ± 0.04
Linear, first n data points (Equation (1))	2.73 ± 0.08	2.24 ± 0.05
Polynomial, neg. exp. $P_3 = 4$ (Equation (6))	2.76 ± 0.05	2.28 ± 0.04

In order to assess the methods capabilities when dealing with an augmented scatter in ΔK, artificial scatter, as described in Section 3.1.2, is added to each specimen data set. The results are given in Table 9. Like already observed in Figure 7, the two-parameter polynomial Equation (6) is almost insensitive to scatter in test data, whereas both linear fits partially suffer from pronounced susceptibility to scattered data.

Table 9. Comparison of threshold values determined from artificial scattered test data, obtained at $R \approx 0.8$ with various fit methods using all 29 specimen data sets for $\Delta K_{th,ASTM}$ and $\Delta K_{th,ISO}$; results for S690QL in lab air.

Method	$\Delta K_{th,ASTM}$ in MPa·m$^{1/2}$	$\Delta K_{th,ISO}$ in MPa·m$^{1/2}$
Linear, all data points (Equation (1))	2.65 ± 0.06	2.16 ± 0.08
Linear, first n data points (Equation (1))	2.68 ± 0.07	2.23 ± 0.05
Polynomial, neg. exp. $P_3 = 4$ (Equation (6))	2.76 ± 0.07	2.27 ± 0.04

3.2. Definition of the Fitting Function and Interval for the Determination of Thresholds Obtained at $R \approx 0.8$

Based on the results presented in Section 3.1, we suggested to use Equation (6) for fitting the test data obtained at $R \approx 0.8$. This polynomial exhibits only a minimal dependency on scatter in test data and returns robust results that simultaneously show only a small, but persistent conservativeness. Furthermore, data extrapolation has been shown to be valid if the lowest available crack propagation rate fulfills $da/dN_{min} \leq 3 \cdot da/dN_{th}$.

Since an augmented fit interval shows an additional reduction in standard deviation, we suggested to use an augmented interval of 5×10^{-8} mm/cycle $\leq da/dN \leq 10^{-6}$ mm/cycle compared to the one proposed in the ASTM standard (10^{-7} mm/cycle $\leq da/dN \leq 10^{-6}$ mm/cycle). Regarding the ISO operational definition, we recommended to use the suggested interval of 10^{-8} mm/cycle $\leq da/dN \leq 10^{-7}$ mm/cycle, since crack propagation rates much lower than 10^{-8} mm/cycle are time consuming using conventional FCG testing.

Using the herein proposed method for evaluating ΔK_{th}, the artificial conservativeness can be reduced and the fits' robustness improved compared to the linear functions, as shown in the previous paragraphs. Since the evaluation following the proposed method involves a very low effort compared to the conducted experiments, the application thereof is generally preferable over the linear fits.

3.3. Evaluation of the Fatigue Crack Propagation Threshold at $R = -1$

In contrast to tests carried out at $R = 0.8$, specimens tested at lower load ratios like $R = -1$ may exhibit a distinct influence of extrinsic effects such as crack closure. This can be observed in the example reported in Figure 10: due to the progressive development of crack closure during the load shedding test, the crack propagation rate decreases rapidly in the near-threshold regime, leading to a steep crack propagation curve towards the threshold. This has major consequences on the evaluation of the fatigue crack propagation threshold due to the fact that much fewer experimental points are available in the selected fitting intervals.

The linear fit over all points within the interval (Figure 10a) is not capable of handling the pronounced curvature. Hence, $\Delta K_{th,ASTM}$ is calculated overly conservatively. This holds true also for the polynomial with $P_3 = 4$, Equation (6) (Figure 10c). In contrast, the linear fit using the first n points (Figure 10b) shows a fairly good threshold approximation. The best results are obtained using the three-parameter polynomial Equation (5) (Figure 10d), which adapts to the curvature displayed by the data fairly well and results in the least (but still) conservative $\Delta K_{th,ASTM}$ value.

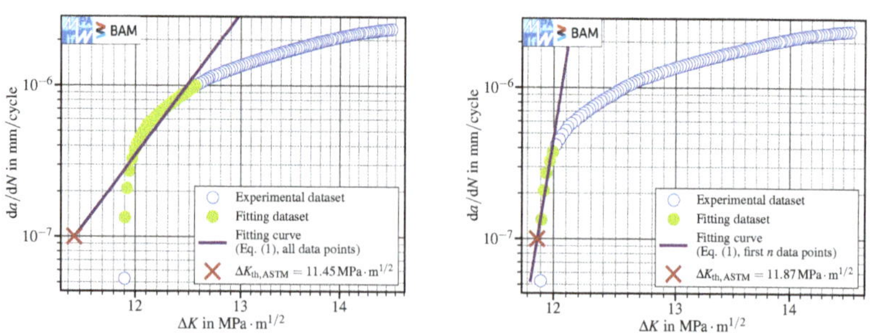

(a) (b)

Figure 10. *Cont.*

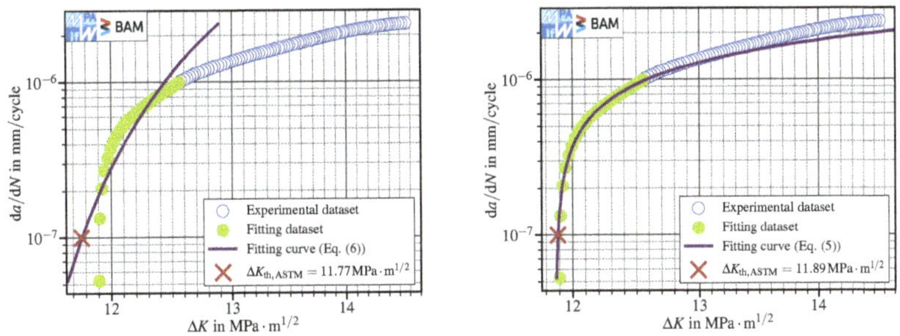

(c) (d)

Figure 10. FCG data showing a distinct effect of fatigue crack closure leading to crack arrest. The tests have been conducted following a compression precracking load reduction procedure at $R = -1$ in lab air. Different fitting strategies have been used: (**a**) the linear fit using all data in the interval 10^{-7} mm/cycle $\leq da/dN \leq 10^{-6}$ mm/cycle provided an overly conservative $\Delta K_{th,ASTM}$ value; (**b**) the linear fit using the first $n = 5$ data points (out of 29 within the interval) showed a fairly good $\Delta K_{th,ASTM}$ approximation; (**c**) Equation (6), using a fixed exponent $P_3 = 4$, was not capable of reproducing the curvature and therefore gave a very conservative result; (**d**) a very good result could be achieved using the three-parameter variant of the polynomial ($P_3 \approx 24.1$).

3.3.1. Handling of Data Affected by Extrinsic Mechanisms

Extrinsic mechanisms affect the cyclic deformation in the crack wake and therefore influence the crack growth rate [22]. The results about tests carried out at $R = 0.1$ and $R = -1$ showed a kink of the crack propagation curve in the near-threshold regime. For an in detail discussion on these findings, see [21]. The results depicted in Figure 11 about two specimens tested in lab air, using a conventional K-decreasing procedure at $R = -1$, pose the question of how to analyze the data to evaluate the fatigue crack propagation threshold. In fact, both specimens show a distinct kink in the FCG data at about $da/dN \approx 10^{-7}$ mm/cycle. In such cases, the calculation of a threshold stress intensity factor range is—regardless of the standard used—questionable, since no asymptotic behavior of the da/dN-ΔK curve towards ΔK_{th} is observable (see Section 1).

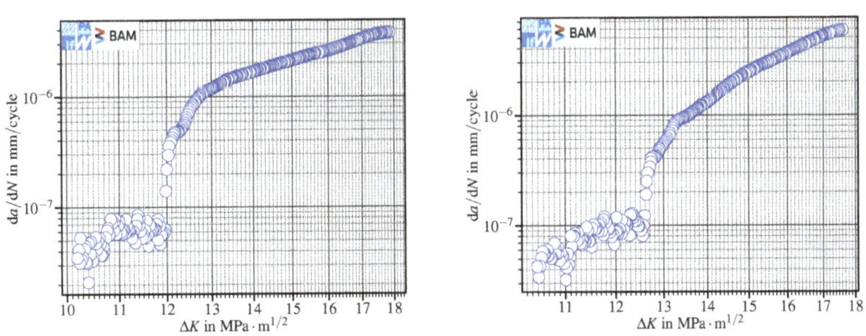

(a) (b)

Figure 11. FCG data showing a kink at about the ASTM threshold ($da/dN_{th,ASTM} = 10^{-7}$ mm/cycle): (**a**) kinking starts below $da/dN = 10^{-7}$ mm/cycle; (**b**) kinking starts slightly above 10^{-7} mm/cycle. The data refer to conventional K-decreasing at $R = -1$ in lab air. Both tests have been interrupted, since neither crack arrest has been observed nor $da/dN_{th,ISO}$ has been reached.

No general rule is applicable to the evaluation of the fatigue crack propagation thresholds, but each dataset shall be analysed separately. For instance, Figure 11a shows that the determination of $\Delta K_{th,ASTM}$ would have been possible, since the kink started below $da/dN_{th,ASTM} =$

10^{-7} mm/cycle. If the aim of this K-decreasing test would have been to determine $\Delta K_{th,ASTM}$, one might have stopped the test after reaching $da/dN < da/dN_{th,ASTM} = 10^{-7}$ mm/cycle for the first time. Hence, one would have never observed the effect of corrosion on FCG data starting just below $\Delta K \approx 12\,\text{MPa·m}^{1/2}$. Nevertheless, as the crack propagation data did not display a pronounced threshold behavior, a value for $\Delta K_{th,ASTM}$ should not be provided. This holds true also for the dataset depicted in Figure 11b, in which corrosion effects started above $da/dN_{th,ASTM} = 10^{-7}$ mm/cycle. In these cases, we recommend providing the last da/dN–ΔK reading recorded within the test as a pure indication of the lowest stress intensity factor range obtained in the load reduction test, which, nevertheless, must not be taken as a fatigue crack propagation threshold. Hence, an automated evaluation of FCG data should only be performed after checking the crack propagation data in the near-threshold regime.

3.3.2. Influence of a Lower Data Density

The majority of test results presented herein show very low noise in conjunction with a fairly high data density. Both properties depend on the quality of raw data and the methodology used to calculate da/dN and ΔK. In Figure 12, the test data presented in Figure 10 are reduced by a factor of two by skipping every second data point. Since still 14 data points distributed over the whole interval of 10^{-6} mm/cycle $\leq da/dN \leq 10^{-7}$ mm/cycle are left, this data set clearly fulfills the requirement of providing at least five points, defined in [4].

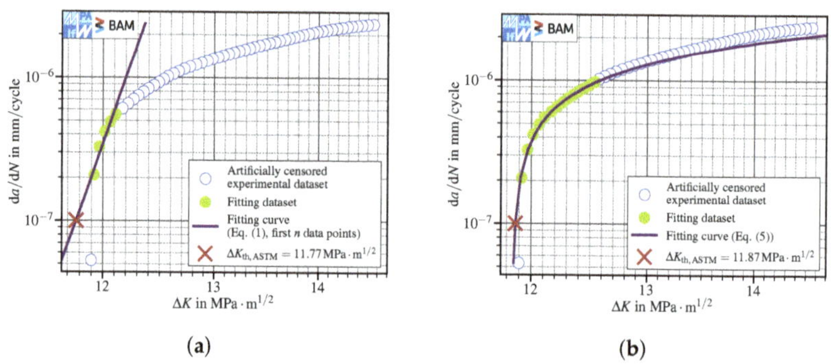

Figure 12. Test data presented in Figure 10 reduced by 50% to assess the influence of a lower data density next to $da/dN_{th,ASTM}$. (**a**) the linear fit using the best first n data points gives an overly conservative $\Delta K_{th,ASTM}$ value; (**b**) using the three-parameter polynomial Equation (5) shows a very good agreement with test data.

Here, the advantages of an appropriate nonlinear fit functions apply. The linear fit using the first n points shows a pronounced underestimation of the threshold stress intensity range, whereas the three-parameter polynomial's sensitivity to the number of data points is very limited.

3.3.3. Validation of the Proposed Method

After examining all the possible issues which might influence the robust determination of the fatigue crack propagation threshold, the methods have been validated against various datasets.

Conventional K-Decreasing at $R = -1$ in Lab Air

The first datasets investigated stemmed from a total of eight SENB specimens made of S690QL, whereof three showed a kink (see Figure 11) and therefore have not been considered in the analysis. These tests have been conducted using the K-decreasing procedure included in the standards [4,6] at constant load ratio $R = -1$ in lab air. All five remaining datasets contained data points below $da/dN = 1.1 \times 10^{-7}$ mm/cycle and therefore have been considered valid for ASTM operational threshold evaluation, while two of them were also valid regarding the ISO definition. The evaluated fatigue crack propagation thresholds, including the corresponding standard deviations, are presented in Table 10. Note that, in case of ISO, no standard deviation has been calculated due to insufficient data available.

Table 10. Comparison of threshold values obtained with various fit methods in case of K-decreasing tests conducted on S690QL at $R = -1$ in lab air.

Method	$\Delta K_{th,ASTM}$ in MPa·m$^{1/2}$	$\Delta K_{th,ISO}$ in MPa·m$^{1/2}$
Linear, all data points (Equation (1))	13.43 ± 0.91	12.78
Linear, first n data points (Equation (1))	13.67 ± 0.80	12.89
Polynomial, neg. exp. (Equation (5))	13.69 ± 0.82	12.92
Polynomial, neg. exp. $P_3 = 4$ (Equation (6))	13.56 ± 0.85	12.90
Polynomial, neg. exp. $P_3 = 5$ (Equation (7))	13.57 ± 0.85	12.90

The conservativeness of each method has been analyzed: the two-parameter polynomials Equation (6) and Equation (7), and especially the linear fit over all points, showed a pronounced underestimation of $\Delta K_{th,ASTM}$ in conjunction with a higher standard deviation compared to the linear fit over the first n points or the three parameter polynomial Equation (5). The observations with regard to artificial conservativeness also apply to the results for $\Delta K_{th,ISO}$. Comparing the latter two methods, both show comparable results, both regarding the mean ΔK_{th} value and the standard deviation.

Compression Precracking Load Reduction at $R = -1$ in Lab Air

The datasets investigated stemmed from a total of ten SENB specimens made of S690QL. All specimens showed a distinct threshold behavior and therefore all have been considered in the analysis. These tests have been conducted using compression precracking followed by a load reduction procedure at constant $R = -1$ in lab air. All ten datasets contained data points below $da/dN = 1.1 \times 10^{-7}$ mm/cycle and therefore have been considered valid for ASTM operational threshold evaluation, whereas none of them contained points in order to calculate a threshold value according to ISO. The threshold stress intensity factor ranges with corresponding standard deviations are presented in Table 11.

Table 11. Comparison of threshold values obtained with various fit methods in case of compression precracking load reduction tests carried out at $R = -1$ in lab air.

Method	$\Delta K_{th,ASTM}$ in MPa·m$^{1/2}$
Linear, all data points (Equation (1))	9.88 ± 0.99
Linear, first n data points (Equation (1))	10.52 ± 0.89
Polynomial, neg. exp. (Equation (5))	10.54 ± 0.91
Polynomial, neg. exp. $P_3 = 4$ (Equation (6))	10.20 ± 0.94
Polynomial, neg. exp. $P_3 = 5$ (Equation (7))	10.22 ± 0.93

The same conclusions as in the previous paragraph can be drawn: the linear fit over the first n points and the three-parameter polynomial Equation (5) provided the best results.

3.4. *Definition of the Fitting Function and Interval for Tests Conducted at $R \ll 0.8$*

Data obtained at load ratios momentously lower than $R = 0.8$ might be affected by phenomena like crack-closure or corrosion, which make the determination of the fatigue crack propagation threshold difficult. When a steep gradient or a kink in the near-threshold data are observed, no general or automatic extrapolation of these test datasets without further investigation is advisable. Furthermore, the augmentation of fit intervals to crack propagation rates smaller than da/dN_{th} may lead to non-conservative results and therefore it is not recommended.

If a valid threshold behavior is observed, the evaluation using the three-parameter polynomial Equation (5) provided constantly conservative, but not overly conservative, results that are neither sensitive to the data density nor to scatter. Therefore, we recommend this function over the linear fit over the first n points, which indeed performed well on regular shaped datasets. The fitting intervals shall follow the recommendations provided in the standards (10^{-7} mm/cycle $\leq da/dN \leq 10^{-6}$ mm/cycle for ASTM and 10^{-8} mm/cycle $\leq da/dN \leq 10^{-7}$ mm/cycle for ISO). In case no definite threshold is reached, i.e., no crack arrest is observed (due for instance to anti-shielding effects, see Figure 11a), ΔK_{th} cannot be determined. Therefore, we recommended providing the last da/dN- ΔK data pair for pure orientation, which shall not be intended as substitute for ΔK_{th}.

3.5. Application of the Fitting Methods to the IBESS Dataset

The proposed method has been validated further against the data from the IBESS project [23,24].Within the IBESS project, fatigue crack propagation tests have been performed on two different structural steels, S355NL and S960QL, using both standard K-decreasing procedures and CPLR tests at constant load ratios varying between $R = -1$ and $R = 0.7$. The number of specimens tested at each stress ratio was limited; therefore, the present validation considered just those tests for which a meaningful data analysis could be performed. In particular, three tests at $R = 0$ for the S355NL and three tests at $R = 0.5$ for the S960QL have been considered. Furthermore, according to the recommendations on the fitting intervals given in this work, the data have been further narrowed. In case of the S355NL, all three datasets have been considered valid with respect to the ASTM fitting interval, whereas just two among them could be used for the determination of the fatigue crack propagation threshold according to ISO. For the S960QL, only two tests for ASTM and one for ISO have been included in the analysis. The results displayed in Tables 12 and 13 confirm the conclusions drawn for the S690QL: the linear fit using the first n points and the polynomial with three parameters reduce the conservativeness in the evaluation of the fatigue crack propagation thresholds. The method proposed in the standards always provides the most conservative results. Furthermore, for the datasets with enough valid data in the fitting interval ($\Delta K_{\text{th,ASTM}}$), the polynomial with three parameters provided the smallest standard deviation.

Table 12. Comparison of the threshold values for the S355NL tested at $R = 0$ obtained with various fitting methods—a total of three sets are eligible for $\Delta K_{\text{th,ASTM}}$ evaluation and two for $\Delta K_{\text{th,ISO}}$.

Method	$\Delta K_{\text{th,ASTM}}$ in MPa·m$^{1/2}$	$\Delta K_{\text{th,ISO}}$ in MPa·m$^{1/2}$
Linear, all data points (Equation (1))	5.96 ± 0.40	5.75
Linear, first n data points (Equation (1))	6.02 ± 0.38	5.86
Polynomial, neg. exp. (Equation (5))	6.08 ± 0.33	5.88

Table 13. Comparison of the threshold values for the S960QL tested at $R = 0.5$ obtained with various fitting methods—a total of two sets eligible for $\Delta K_{\text{th,ASTM}}$ evaluation and one for $\Delta K_{\text{th,ISO}}$.

Method	$\Delta K_{\text{th,ASTM}}$ in MPa·m$^{1/2}$	$\Delta K_{\text{th,ISO}}$ in MPa·m$^{1/2}$
Linear, all data points (Equation (1))	3.30	3.04
Linear, first n data points (Equation (1))	3.31	3.10
Polynomial, neg. exp. (Equation (5))	3.31	3.08

4. Conclusions

The present paper compared several methods for the evaluation of the fatigue crack propagation thresholds. New fitting strategies have been introduced and calibrated on a large dataset of crack growth data for the S690QL. The goodness of the fitting methods has been validated further against a dataset for S355NL and S960QL.

The following conclusions can be drawn:

- The ASTM E647 and ISO 12108 standards suggest to fit $\log \Delta K$ over $\log da/dN$ data using a linear fit, but leave plenty of room for interpretation with respect to the choice of the points in the fitting interval;

- When using all data points within the suggested fitting intervals, the most conservative values of ΔK_{th} are obtained. However, the fit is not very subjected to scattered data;
- To use only the first n data points starting from the threshold crack propagation rate in order to ensure the best linear fit reduces the conservativeness at the cost of a more pronounced susceptibility to scatter and lower density of the data;
- The proposed fitting polynomials provided an improvement with respect to the goodness of the fit and susceptibility to scatter;
- An extrapolation of data was possible within given bounds for the structural steel S690QL, tested in lab air at room temperature at $R \approx 0.8$. Further tests comprising changes in materials, temperatures and the test environment should be conducted to assess the validity ranges;
- Tests subjected to crack closure phenomenon cannot be assessed in a fully automatic manner and require a manual dataset evaluation.

Author Contributions: Conceptualization, J.A.S., L.D., M.B.G., M.M., U.Z. and M.K.; methodology, J.A.S., L.D., M.B.G., M.M., U.Z. and M.K.; software, J.A.S. and L.D.; validation, J.A.S. and L.D.; formal analysis, J.A.S., L.D., M.B.G. and M.M.; investigation, J.A.S., L.D., M.B.G., M.M., U.Z. and M.K.; resources, M.O.; data curation, J.A.S. and L.D.; writing—original draft preparation, J.A.S.; writing—review and editing, L.D., M.B.G., M.M., U.Z., M.K. and M.O.; visualization, J.A.S.; supervision, M.M., U.Z., M.K. and M.O.; project administration, M.M., U.Z., M.K. and M.O.; funding acquisition, U.Z. and M.O. All authors have read and agreed to the published version of the manuscript.

Funding: This work is part of the research project IGF 20530 N / 1263 "Ermittlung des intrinsischen Schwellenwerts und dessen Validierung als Werkstoffparameter" from the Research Association for Steel Application (FOSTA), Düsseldorf, which is supported by the Federal Ministry of Economic Affairs and Climate Action through the German Federation of Industrial Research Associations (AiF) as part of the program for promoting industrial cooperative research (IGF) on the basis of a decision by the German Bundestag. The project is carried out at BAM Berlin and MPA-IfW Darmstadt.

Institutional Review Board Statement: Not applicable.

Informed Consent Statement: Not applicable.

Data Availability Statement: Not applicable.

Acknowledgments: We acknowledge support by the Deutsche Forschungsgemeinschaft (DFG—German Research Foundation) and the Open Access Publishing Fund of the Technical University of Darmstadt.

Conflicts of Interest: The authors declare no conflict of interest.

Symbols and Abbreviations

The following symbols and abbreviations are used in this manuscript:

ΔF	applied force range
ΔK	applied stress intensity factor range
ΔK_{LR}	K-decreasing FCG test procedure at constant load ratio
ΔK_{th}	fatigue crack propagation threshold
$\Delta K_{th,ASTM}$	ΔK_{th} referring to the ASTM operational definition
$\Delta K_{th,ISO}$	ΔK_{th} referring to the ISO operational definition
$\Delta K_{th,cens}$	ΔK_{th} obtained for the censoring of the data set
μ	mean value of the distribution
σ_u	ultimate tensile strength
σ_y	upper yield strength
a	crack size

A	elongation at break
da/dN	fatigue crack propagation rate
da/dN_{th}	da/dN referring to an operational threshold definition
$da/dN_{th,ASTM}$	ASTM operational threshold definition of da/dN_{th}
$da/dN_{th,ISO}$	ISO operational threshold definition of da/dN_{th}
e	Weibull threshold parameter
E	Young's modulus
I	degree of the polynomial
k	Weibull shape parameter
K_b	Weibull instability parameter
K_{max}	maximum stress intensity factor in a loading cycle
KV_2	impact energy
n	number of data points
N	number of loading cycles
P_i	fitting parameters, $i \in \mathbb{N}$
R	stress ratio
R_{max}	maximum stress ratio within K_{max}-FCG test
v	Weibull characteristic value
ASTM	American Society for Testing and Materials
BAM	Bundesanstalt für Materialforschung und -prüfung
CPLR	compression precracking load reduction
FCG	fatigue crack growth
ISO	International Organization for Standardization
MPA-IfW	Materialprüfungsanstalt Darmstadt, Institut für Werkstoffkunde
ODR	orthogonal distance regression
SD	standard deviation
SENB	single edge notch bending

References

1. Paris, P.C.; Sih, G.C. *Stress Analysis of Cracks*; Technical Report; NASA: Washington, DC, USA, 1964.
2. Knott, J.F. *Fundamentals of Fracture Mechanics*; Butterworth: London, UK, 1973.
3. Tada, H.; Paris, P.C.; Irwin, G.R. *The Stress Analysis of Cracks Handbook*, 3rd ed.; ASME Press: New York, NY, USA, 2000.
4. *ASTM E647-15e1*; Standard Test Method for Measurement of Fatigue Crack Growth Rates. ASTM International: West Conshohocken, PA, USA, 2015. [CrossRef]
5. Ritchie, R. Mechanisms of Fatigue-Crack Propagation in Ductile and Brittle Solids. *Int. J. Fract.* **1999**, *100*, 55–83. [CrossRef]
6. *ISO 12108:2018*; Metallic Materials—Fatigue Testing—Fatigue Crack Growth Method. ISO: Geneva, Switzerland, 2018.
7. Bucci, R.J. Development of a proposed ASTM standard test method for near-threshold fatigue crack growth rate measurement. In *Fatigue Crack Growth Measurement and Data Analysis. ASTM STP 738*; Hudak, S.J., Bucci, R.J., Eds.; ASTM International: West Conshohocken, PA, USA, 1981; pp. 5–28. [CrossRef]
8. Salivar, G.; Hoeppner, D. A Weibull Analysis of Fatigue-Crack Propagation Data from a Nuclear Pressure Vessel Steel. *Eng. Fract. Mech.* **1979**, *12*, 181–184. [CrossRef]
9. Smith, F.; Hoeppner, D.W. Use of the Four Parameter Weibull Function for Fitting Fatigue and Compliance Calibration Data. *Eng. Fract. Mech.* **1990**, *36*, 173–178. [CrossRef]
10. Boggs, P.T.; Rogers, J.E. Orthogonal distance regression. In *Statistical Analysis of Measurement Error Models and Applications*; American Mathematical Society: Providence, RI, USA, 1990; Volume 112, pp. 183–194. [CrossRef]
11. Döker, H. Fatigue Crack Growth Threshold: Implications, Determination and Data Evaluation. *Int. J. Fatigue* **1997**, *19*, 145–149. [CrossRef]
12. ASTM E 647. *Annual Book of ASTM Standards*; Section 4; American Society of Testing and Materials: Philadelphia, PA, USA, 1995; Volume 03.01, pp. 578–614.
13. Klesnil, M.; Lukáš, P. Influence of Strength and Stress History on Growth and Stabilisation of Fatigue Cracks. *Eng. Fract. Mech.* **1972**, *4*, 77–92. [CrossRef]
14. Romvari, P.; Tot, L.; Nad', D. Analysis of Irregularities in the Distribution of Fatigue Cracks in Metals. *Strength Mater.* **1980**, *12*, 1481–1492. [CrossRef]
15. Zheng, X. A Simple Formula for Fatigue Crack Propagation and a New Method for the Determination of ΔKth. *Eng. Fract. Mech.* **1987**, *27*, 465–475. [CrossRef]
16. Ramsamooj, D.; Shugar, T. Model Prediction of Fatigue Crack Propagation in Metal Alloys in Laboratory Air. *Int. J. Fatigue* **2001**, *23*, 287–300. [CrossRef]
17. Day, B.; Goswami, T. Weibull Model Development for Fatigue Crack Growth. *J. Mech. Behav. Mater.* **2002**, *13*, 283–296. [CrossRef]

18. Paolino, D.S.; Cavatorta, M.P. Sigmoidal Crack Growth Rate Curve: Statistical Modelling and Applications: SIGMOIDAL CRACK-GROWTH-RATE CURVE. *Fatigue Fract. Eng. Mater. Struct.* **2013**, *36*, 316–326. [CrossRef]
19. Maierhofer, J.; Pippan, R.; Gänser, H.P. Modified NASGRO Equation for Physically Short Cracks. *Int. J. Fatigue* **2014**, *59*, 200–207. [CrossRef]
20. *E1823-20*; Standard Terminology Relating to Fatigue and Fracture Testing. ASTM International: West Conshohocken, PA, USA, 2020. [CrossRef]
21. Duarte, L.; Schönherr, J.A.; Madia, M.; Zerbst, U.; Geilen, M.B.; Klein, M.; Oechsner, M. Recent Developments in Fatigue Crack Propagation Threshold Determination. *Int. J. Fatigue* 2022. *submitted*.
22. Pippan, R.; Hohenwarter, A. Fatigue crack closure: A review of the physical phenomena. *Fatigue Fract. Eng. Mater. Struct.* **2017**, *40*, 471–495. [CrossRef] [PubMed]
23. Zerbst, U. Analytische bruchmechanische Ermittlung der Schwingfestigkeit von Schweißverbindungen (IBESS-A3); Technical Report; Bundesanstalt für Materialforschung und -prüfung (BAM): Berlin, Germany, 2016.
24. Kucharczyk, P.; Madia, M.; Zerbst, U.; Schork, B.; Gerwin, P.; Münstermann, S. Fracture-mechanics based prediction of the fatigue strength of weldments. Material aspects. *Eng. Fract. Mech.* **2018**, *198*, 79–102. [CrossRef]

materials

MDPI

Article

Influence of Creep Damage on the Fatigue Life of P91 Steel

Stanisław Mroziński [1], Zbigniew Lis [1] and Halina Egner [2,*]

[1] Faculty of Mechanical Engineering, Bydgosz University of Science and Technology, Al. Prof. S. Kaliskiego 7, 85-796 Bydgoszcz, Poland; stmpkm@pbs.edu.pl (S.M.); zbigniew.lis@pbs.edu.pl (Z.L.)

[2] Faculty of Mechanical Engineering, Cracow University of Technology, Al. Jana Pawła II 37, 31-864 Kraków, Poland

* Correspondence: halina.egner@pk.edu.pl

Abstract: The following paper presents the results of tests on samples made of P91 steel under the conditions of simultaneously occurring fatigue and creep at a temperature of 600 °C. The load program consisted of symmetrical fatigue cycles with tensile dwell times to introduce creep. Static load (creep) was carried out by stopping the alternating load at the maximum value of the alternating stress. The tests were carried out for two load dwell times, 5 s and 30 s. A comparative analysis of the test results of fatigue load with a dwell time on each cycle confirmed that creep accompanying the variable load causes a significant reduction in sample durability. It was shown in the paper that regarding the creep influence in the linear fatigue damage summation approach, it is possible to improve the compliance of the fatigue life predictions with the experimental results.

Keywords: low-cycle fatigue; creep; damage; strain energy

Citation: Mroziński, S.; Lis, Z.; Egner, H. Influence of Creep Damage on the Fatigue Life of P91 Steel. *Materials* **2022**, *15*, 4917. https://doi.org/10.3390/ma15144917

Academic Editors: Jaroslaw Galkiewicz, Lucjan Śnieżek and Sebastian Lipiec

Received: 9 June 2022
Accepted: 12 July 2022
Published: 14 July 2022

1. Introduction

Analysis of the operational loads of many structural elements indicates that they are often subject to mechanical loads where the independent quantity is force (e.g., pressure) and the dependent quantity is the deformation of the element. This applies, inter alia, to facilities operating at elevated temperatures. This type of load may be additionally accompanied by material creep, which changes the nature of the load and the durability of the designed elements. Creep is much more pronounced at high temperatures, for example, in pipelines that contain a pressurized hot medium or gas turbine components that are statically loaded but operate at high and variable temperatures. In such cases, when predicting the fatigue life of a technical object, only the fatigue characteristics are taken into account, obtained for example, under the conditions of controlled deformation (ε_{ac} = const) or stresses (σ_a = const), while in real applications, the conditions are different, which may lead to divergent results obtained from calculations and tests. In papers [1–4], the results of low-cycle fatigue tests are presented for samples made from the same material, obtained in the conditions of σ_a = const and ε_{ac} = const. Based on the comparative analysis of the fatigue diagrams in the $2N_f - \varepsilon_{ac}$ coordinate system, it was found that at the same deformation levels, the fatigue life under the conditions σ_a = const is lower than that obtained in the conditions of controlled deformation (ε_{ac} = const). The deformation asymmetry during the half-cycles of tension and compression, which caused the cyclic creep of the material, was given as the reason for the reduction in durability under the conditions σ_a = const. Creep-fatigue experiments were conducted in [5] under stress-controlled conditions to understand dwell time's effects on the durability of a DS superalloy. It was shown that the most important damage source is creep-fatigue interaction damage which is controlled by load amplitude, stress ratio, and dwell time per cycle. In paper [6], multiaxial creep-fatigue life and its variations with the increasing holding time under different loading programs were investigated for 304 stainless steel. The introductions of holding periods under creep-fatigue loadings resulted in lifetime reductions as compared with pure fatigue loadings. The problem of creep-fatigue interaction for various engineering materials was also recently investigated in [7–9]. In papers [10,11], tests were carried out to determine the effect of the sequence of fatigue cycles and static loads in the load program. It was found that the sequence of both loads in the load program affects the durability. Variable loads preceding static loads result in obtaining higher durability than the variant in which the static load precedes the variable loads. In light of the obtained results, it was found that creep damage and fatigue damage are not independent. These results confirm the observations included, inter alia, in [12], where the authors found that short cracks appear during the variable load, and their density is larger if the material is initially subjected to a static load. In paper [10], experimental verification of the linear hypothesis of damage summation [13] was also carried out. Since the linear summation hypothesis is insensitive to the sequence of events in the load program, it may provide erroneous results of the durability assessments in comparison to the experimental data.

The problem of fatigue life predictions under the conditions of simultaneous occurrence of the variable and the static load was analyzed with the use of various fatigue descriptions, i.e., deformation, stress, or energy approach. In the works [11,14–20], the problem of material fatigue was attempted to be explained with the use of energy description. The results indicate significant advantages of the energy approach, even if a relatively simple combination of the static and variable load effects was applied. The main advantage of the energy description is the ability to accumulate damage using plastic strain energy as a criterion parameter.

In papers [21–23], the authors investigated the influence of creep on fatigue life and verified currently used calculation models. Several modifications of the classic Palmgren-Miner's linear fatigue damage summation hypothesis [13] were proposed. The authors concluded that disregarding creep damage in the calculations may lead to a significant differentiation in comparison with the test results. The creep duration influences the results diversion from the experiment.

The present work is a continuation of research on the improvement of constitutive modeling of low cycle fatigue (cf. [1,4,11]), as well as on the explanation of the phenomena accompanying the load program containing constant and variable loads.

2. Materials and Methods

Experimental Testing

Samples for fatigue tests were prepared from P91 steel (R_m = 716 MPa, R_e = 564 MPa, A_5 = 35% at room temperature). The chemical composition of P91 steel is shown in Table 1.

Table 1. Chemical composition of P91 steel.

C	Si	Mn	P	S	Cr	Mo	Ni
0.197	0.442	0.489	0.017	0.005	8.82	0.971	0.307
Al	Co	Cu	Nb	Ti	V	W	
0.012	0.017	0.036	0.074	0.004	0.201	0.02	

The samples were cut out from a thick-walled pipe for power industry applications, with an external diameter of 200 mm and a wall thickness of 20 mm. The samples were shaped under the guidelines specified in the standard [24]. The dimensions of the sample are shown in Figure 1.

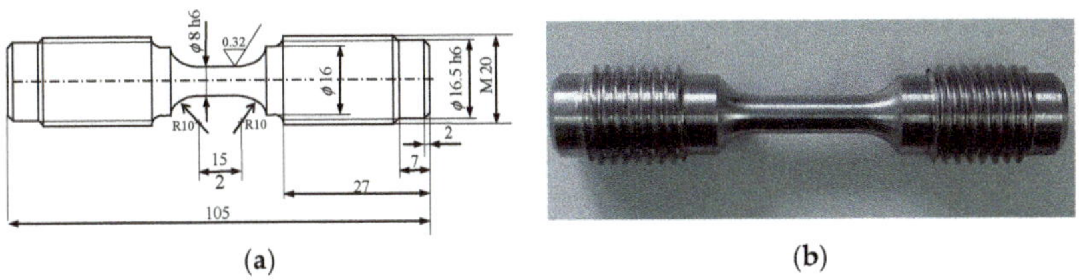

Figure 1. (**a**) Test sample dimensions (in [mm]); (**b**) test sample view.

The load program of experimental tests included: static tensile tests, creep tests, low-cycle fatigue tests, and tests in which the samples were subjected to fatigue with a dwell time in each cycle. The load programs and their symbols are shown in Figure 2.

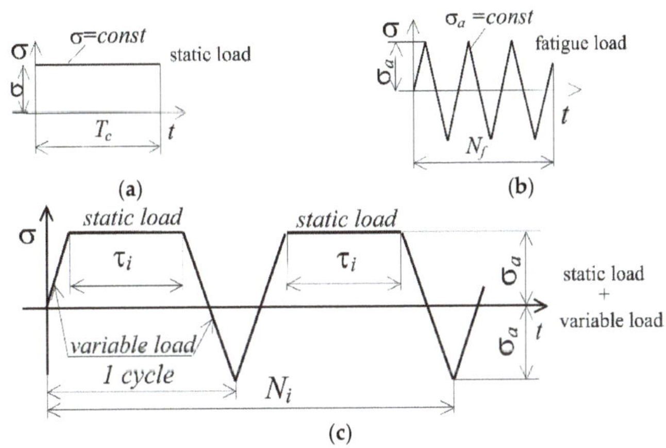

$\sigma_1 = \sigma_{a1} = 219$ MPa, $\sigma_2 = \sigma_{a2} = 225$ MPa, $\sigma_3 = \sigma_{a3} = 249$ MPa, $\sigma_4 = \sigma_{a4} = 258$ MPa

$\tau_1 = 0$ s, $\tau_2 = 5$ s, $\tau_3 = 30$ s

(d)

Figure 2. Load programs: (**a**) static load, (**b**) fatigue load, (**c**) static load + fatigue load, (**d**) load parameters.

Experimental tests (static and fatigue tests) were carried out at a temperature of $T = 600$ °C on an Instron 8502 testing machine, equipped with a heating chamber with a maximum temperature range of 1000 °C (Figure 3). The temperature of the sample was monitored with a thermocouple attached to the sample. The sample deformation was measured with an extensometer with a measurement base of 12.5 mm. Fatigue tests in the conditions σ_a = const were carried out on four stress levels (see Figure 2d), determined based on the analysis of static tension diagrams (Figure 4b). The load frequency was 0.2 Hz. During the tests, the instantaneous values of the force loading the sample and its deformation were recorded. Two fatigue tests were performed at each load level.

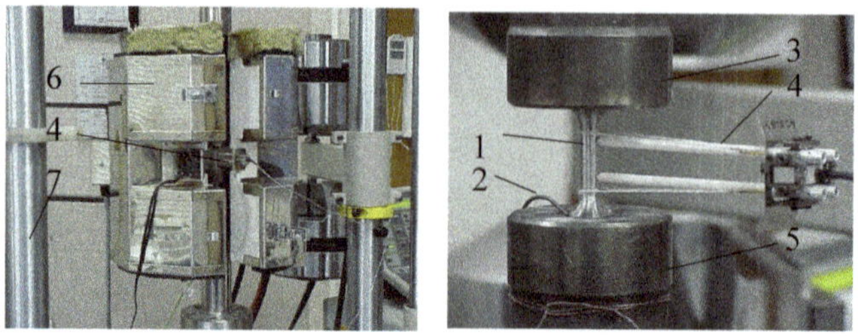

(**a**) (**b**)

Figure 3. Test stand: (**a**) heating chamber; (**b**) sample in grips. 1—sample, 2—thermocouple, 3—upper grip, 4—extensometer, 5—lower grip, 6—heating chamber, 7—machine frame.

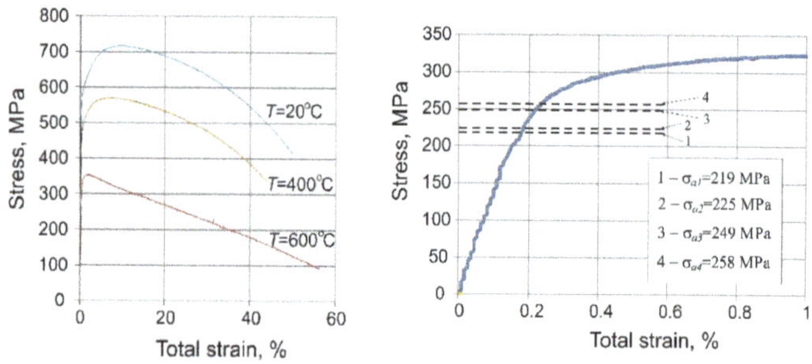

(**a**) (**b**)

Figure 4. Monotonic stress–strain curves: (**a**) for different test temperatures, (**b**) stress amplitude levels chosen for creep-fatigue tests (T = 600 °C).

3. Results

3.1. Static Tension Test

Figure 4a present a stress–strain curve obtained at temperature T = 600 °C. To illustrate the influence of temperature on strength properties, the figure additionally includes the tensile curves of P91 steel obtained at two other temperatures, 20 °C and 400 °C. The stress in the tested sample was calculated as the ratio of the instantaneous force loading the sample and the initial cross-sectional area of the sample (nominal stress). Figure 4b include a fragment of the tensile diagram obtained at the temperature T = 600 °C, with the stress amplitude levels for the creep-fatigue tests indicated in it, see Figure 2d.

The tensile diagrams (Figure 4) were subjected to a detailed analysis aimed at determining the basic strength parameters (see Table 2).

Table 2. Strength parameters of P91 steel at different temperatures.

T, °C	$R_{p0.2}$, MPa	R_m, MPa	Z, %	$A_{12.5}$, %	E, MPa
20	564	716	62	35	209,850
400	463	571	69	44	183,220
600	317	353	72	56	132,890

The comparative analysis of the data in Table 2 confirm the common literature reports on the influence of temperature on the basic strength parameters of P91 steel [25–28]. Based on the analysis of the tensile diagrams (Figure 4) and the values of the strength parameters listed in Table 2, it can be concluded that the increase in temperature leads to a reduction in the yield point ($R_{p0.2}$). A similar relationship is observed in the case of the tensile strength (R_m) and Young's modulus (E), for which a decrease in this parameter is also visible. The decrease in strength properties with temperature is accompanied by an increase in elongation ($A_{12.5}$) and constriction (Z).

3.2. Fatigue Tests

The analysis of the fatigue test results was carried out using the hysteresis loop parameters (σ_a, ε_{ap}, ε_{ac}), which are necessary for the analytical description of the cyclic properties of steel following the standard [24]. Changes in the hysteresis loop parameters in function of the number of a load cycle were observed during all the fatigue tests. As expected, cyclic creep of the sample material was observed under the conditions of controlled stress (σ_a = const). The phenomenon consisted of the shift of the hysteresis loop along the strain axis, increasing the maximum strain on a cycle ε_{max}. Figure 5 show exemplary hysteresis loops at two stress levels (σ_a = 225 MPa, σ_a = 249 MPa). The presented loops refer to a load program in which creep time $\tau = 0$ (see Figure 2).

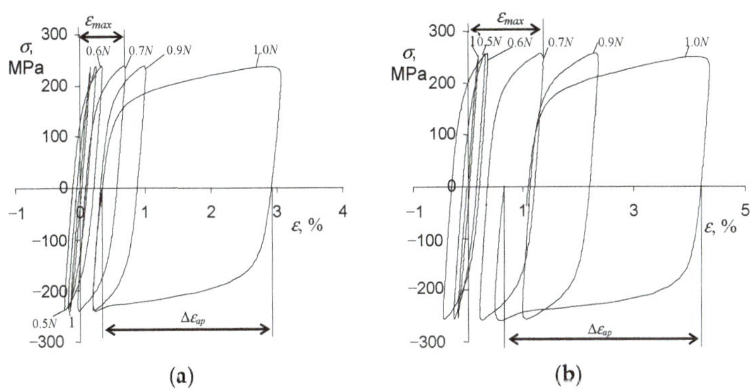

Figure 5. Hysteresis loops recorded in fatigue tests (τ = 0, T = 600 °C): (**a**) σ_a = 225 MPa, (**b**) σ_a = 249 MPa.

Based on the hysteresis loops analysis, it can be concluded that the amount of maximum strain on a cycle (ε_{max}) during the fatigue test is influenced by the stress level σ_a. As expected, this influence is the smallest at the lowest stress levels and increases with increasing stress. It can also be seen that the strain increments are characterized by different rates, low at the beginning of the test and increasing with the number of a load cycle (Figure 6a).

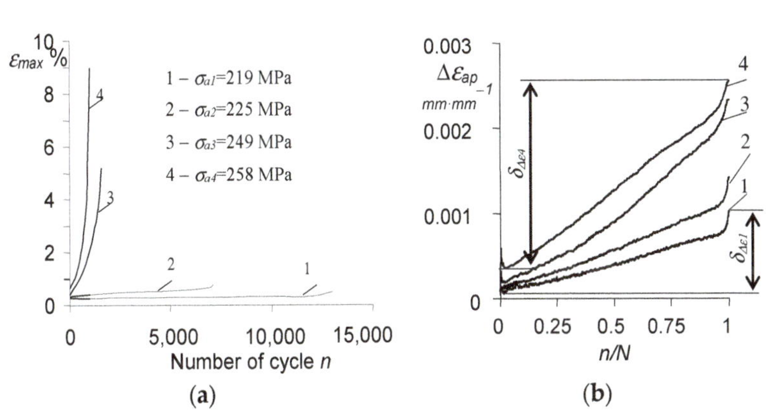

Figure 6. Evolution of: (**a**) maximum strain on cycle; (**b**) plastic strain on cycle $\Delta\varepsilon_{ap}$.

Based on the analysis of the loops shown in Figure 5, it can be concluded that with the increase in the number of the load cycle, the range of plastic deformation $\Delta\varepsilon_{ap}$ also increases at a growing rate. The amount of plastic deformation in a cycle, $\Delta\varepsilon_{ap}$, is also influenced by the stress level σ_a. To illustrate this influence, Figure 6b summarize the changes of $\Delta\varepsilon_{ap}$ in function of a load cycle number n related to the fatigue life N.

The observed increase in the range of plastic deformation proves a clear cyclic softening of the P91 steel at the temperature of 600 °C. To quantify the softening, a coefficient $\delta_{\Delta\varepsilon}$ was introduced (see Figure 6b), defined by the following relationship:

$$\delta_{\Delta\varepsilon} = \Delta\varepsilon_{ap(N)} - \Delta\varepsilon_{ap(1)} \tag{1}$$

where $\Delta\varepsilon_{ap(1)}$ is the range of plastic deformation in the first cycle, while $\Delta\varepsilon_{ap(N)}$ is the range in the last cycle. It can be seen in Figure 6b that with the increase in stress amplitude, the value of the coefficient $\delta_{\Delta\varepsilon}$ also increases ($\delta_{\Delta\varepsilon 4} > \delta_{\Delta\varepsilon 3} > \delta_{\Delta\varepsilon 2} > \delta_{\Delta\varepsilon 1}$).

3.3. Creep Tests

During the creep tests, similarly to the fatigue tests, the instantaneous values of the sample elongation as a function of the creep test time were recorded. As expected, the results of the creep tests are influenced by the constant stress level σ. Figure 7 present the sample elongation vs. time recorded during the creep tests at four stress levels (cf. also [11]).

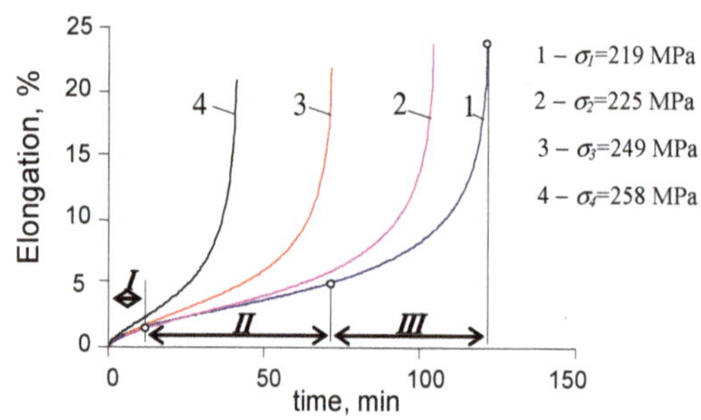

Figure 7. Sample elongation in creep test.

All creep-to-fracture curves exhibit three stages with different elongation rates of the specimen over time. These stages were indicated on the creep curve obtained for the lowest stress (σ_1 = 219 MPa), stage I with a variable elongation rate, stage II with a constant rate, and stage III also with a variable creep rate. The stress level σ affects both the length of these stages as well as the creep rate in the individual stages.

3.4. Fatigue-Creep Tests

As expected, during the programs containing both variable and static loads ($\tau > 0$, see Figure 2c), cyclic creep of the material was observed. It manifested in the horizontal shift of the hysteresis loops and the increase in the maximum strain on a cycle ε_{max}. Figures 8 and 9 present chosen hysteresis loop positions at two stress levels, σ_a = 219 MPa and σ_a = 258 MPa.

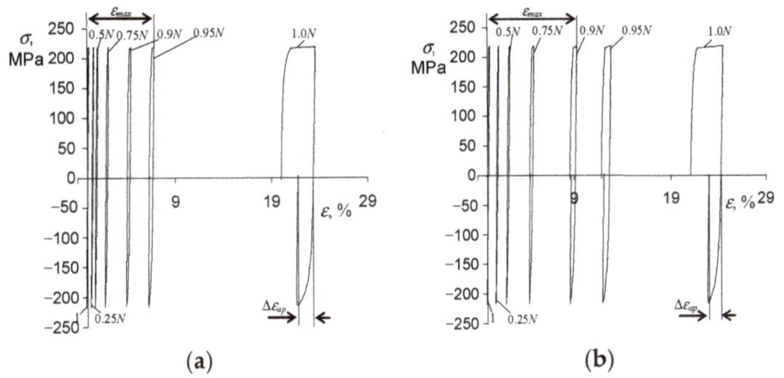

Figure 8. Hysteresis loops in creep-fatigue conditions for σ_a = 219 MPa, and (**a**) τ = 5 s, (**b**) τ = 30 s.

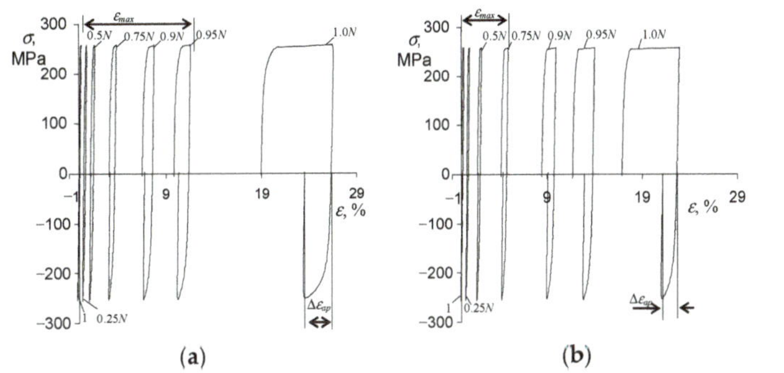

Figure 9. Hysteresis loops in creep-fatigue conditions for σ_a = 258 MPa, and (**a**) τ = 5 s, (**b**) τ = 30 s.

It can be seen from Figures 8 and 9 that the creep time affects both the maximum strain ε_{max} and the range of plastic deformation in a cycle. Figure 10 shows the maximum strains at two stress levels and three values of the dwell time τ.

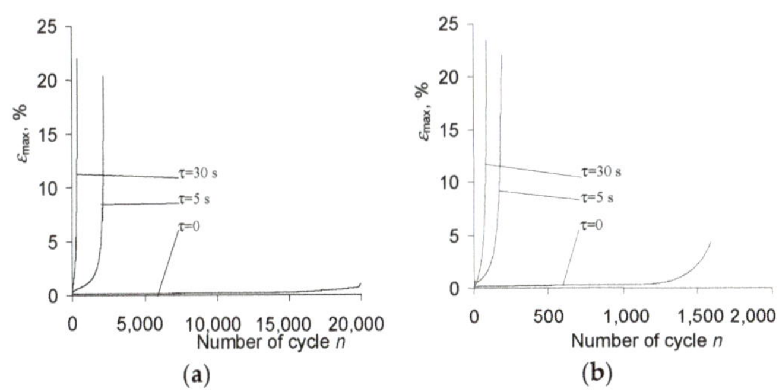

Figure 10. Maximum strain on cycle ε_{max} vs. number of cycles for different dwell times τ and different stress amplitudes: (**a**) σ_a = 219 MPa, (**b**) σ_a = 258 MPa.

On the basis of the above results (Figures 8 and 9), it can be concluded that creep occurring during the variable load ($\tau > 0$) causes a significant increase in the maximum strain ε_{max}. With the increase in a dwell time, the elongation of the sample in the fatigue test increases significantly. The analysis of the position of successive hysteresis loops (Figures 8 and 9) in the function of the number of a cycle allows concluding that under the conditions of combined variable and constant load, P91 steel does not exhibit a stabilization of its cyclic properties.

Regardless of the level of alternating stress amplitude σ_a, and the duration τ of the permanent load, the steel always exhibits significant cyclic softening. This is illustrated in Figure 11 by the increase in plastic strain $\Delta\varepsilon_{ap}$ as a function of the number of a load cycle for various test conditions. In order to compare changes in plastic strains for different dwell times τ, Figure 11b,d present results in the function of relative durability n/N.

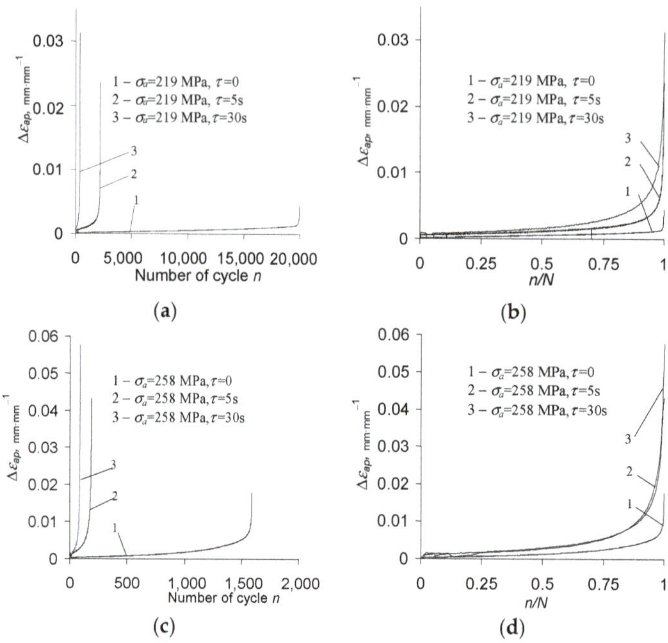

Figure 11. Changes in plastic strain on cycle $\Delta\varepsilon_{ap}$ for different stress amplitude levels: (**a**,**b**) σ_a = 219 MPa; (**c**,**d**) σ_a = 258 MPa.

Based on the analysis of changes in the plastic strain range $\Delta\varepsilon_{ap}$, it can be concluded that the magnitude of changes in this parameter is influenced, among others, by the dwell time τ. At the same amplitude stress levels, under the conditions of combined fatigue and creep loads, plastic strains $\Delta\varepsilon_{ap}$ are larger than under the conditions of pure fatigue σ_a = const ($\delta_{\varepsilon 2} > \delta_{\varepsilon 1}$, see also Figure 6b).

During the experimental tests, a very clear effect on the fatigue life of the dwell time τ was observed. The fatigue life experimental results obtained for various load program sequences are summarized in Figure 12 in the form of fatigue diagrams. The durability results at four stress levels were approximated in the semi-logarithmic coordinate system by a regression equation of the form:

$$\sigma_a = a \log N + b \tag{2}$$

It can be observed that the creep time τ significantly affects the fatigue life. With the increase of time τ, the fatigue life decreases. The results confirm the research described in [21–23].

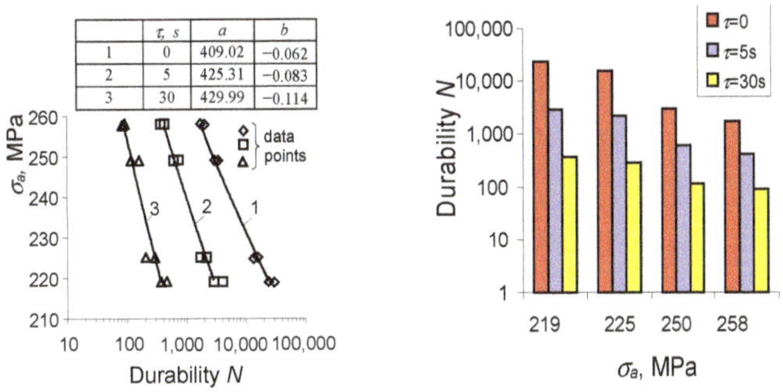

(a) (b)

Figure 12. Durability results (experiment): (**a**) fatigue curves, (**b**) summary of durability results.

3.5. Fractographic Observations

After testing, the samples were subjected to fractographic analysis. The fractographic samples were prepared from the measurement parts of the samples, parallel to the load direction. Figure 13 show the sample surfaces after selected variants of the load program. For comparison, the surface of the P91 steel sample before the test is also included in Figure 13a.

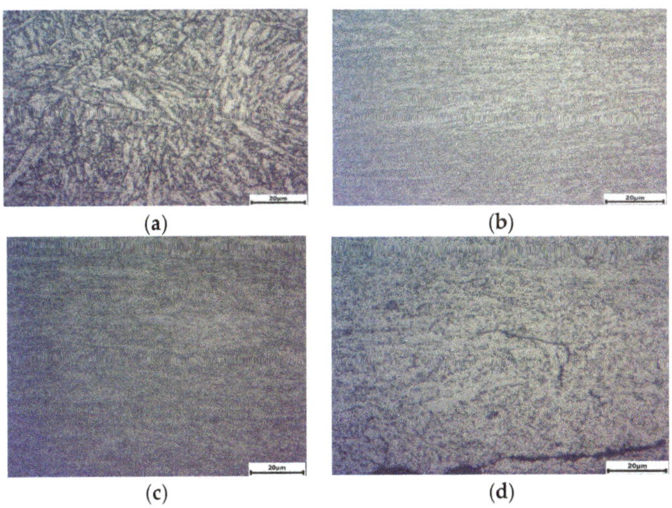

Figure 13. P91 steel microstructure: (**a**) the initial material, (**b**) after uniaxial tension test (T = 600 °C), (**c**) after creep test (T = 600°C, σ = 258 MPa), (**d**) after fatigue test (T = 600 °C, σ_a = 258 MPa).

P91 steel (Figure 13a) is characterized by a typical microstructure for the class of steels containing 8 ÷ 12% Cr; the microstructure of highly tempered martensite with numerous precipitates of carbides and nitrides is observed. A lamellar microstructure of tempered martensite with numerous precipitates visible at the boundaries of the former austenite grains and martensite lamellae are visible. After the tensile test at T = 600 °C (Figure 13b), as expected, a highly deformed microstructure with visible banding is detected—the direction of a material flow corresponds to the direction of the principal stress. Numerous precipitations are visible on the deformation bands.

A similar structure as after the tensile test was observed after the creep test (Figure 13c). The visible structure is also lamellar, with numerous precipitates.

The structure of the P91 steel after the fatigue test is also characterized by lamellar microstructure (Figure 13d); however, the grain deformation is less pronounced than in the creep test or monotonic test. The deformed microstructure shows the process of coagulation of the precipitates. The main crack propagates transcrystalline, while the secondary cracks are formed along the grain boundaries of the former austenite.

The results of microscopic analyses confirm the necessity regard creep damage during the fatigue life calculations. However, the determination of the quantitative impact of constant and variable load on changes in the microstructure requires further extensive research program.

3.6. Durability Assessments

Currently, there are several hypotheses for summing the fatigue damage that evolves under the conditions of simultaneously occurring fatigue and creep loads. These were discussed, inter alia, in [14]. In the present research, a linear model was used for durability calculations, which refers directly to the approach proposed in [13]. For the load program shown in Figure 2, the total damage D_t will be equal to the sum of fatigue damage D_f and creep damage D_c:

$$D_f + D_c = D_t. \tag{3}$$

The experimental verification of the linear damage summation model under the conditions of simultaneous creep and fatigue was then carried out. After realizing n cycles of variable load with additional creep ($\tau_i > 0$) in each load cycle, the total damage D_t will be equal to:

$$\sum \frac{n}{N} + \sum \frac{\tau}{T_c} = D_t \tag{4}$$

where N is the number of cycles to failure in pure fatigue (σ_a = const and $\tau = 0$), while T_c denotes the total lifetime in creep (σ = const). It should be emphasized that the approach described by Equation (4) is insensitive to the order of occurrence of particular types of load and ignores changes in the microstructure of the material under various load conditions. This problem was discussed, inter alia, in papers [10,11].

The results of calculations of damages D_f, D_c, and D_t are summarized in Figure 14.

$\sigma=\sigma_a$, MPa	D_f	D_c	D_t
219 (τ=5 s)	0.11	1.49	1.60
219 (τ=30 s)	0.02	1.53	1.55
225 (τ=5 s)	0.13	1.61	1.74
225 (τ=30 s)	0.02	1.38	1.40
250 (τ=5 s)	0.20	0.81	1.01
250 (τ=30 s)	0.03	0.69	0.72
258 (τ=5 s)	0.24	0.85	1.10
258 (τ=30 s)	0.05	1.10	1.15

(a)

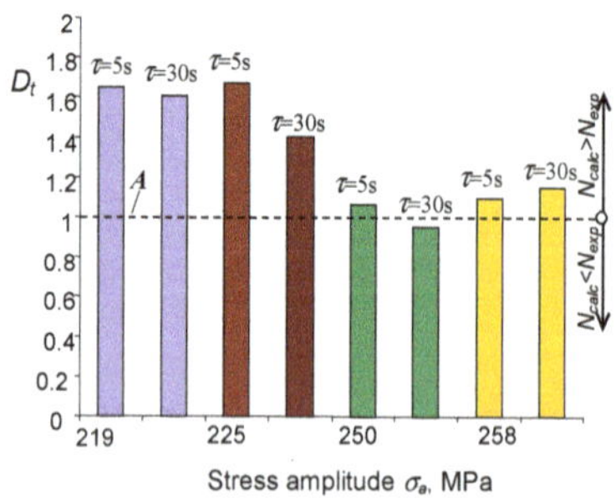

(b)

Figure 14. (**a**) Damage levels resulting from calculations; (**b**) interpretation of results.

Based on the comparative analysis of the values of the total damage components D_t (composed of D_f and D_c), it can be concluded that the dominant component is the creep damage, D_c. The application of the linear model of damage summation under the conditions of simultaneous occurrence of static load (creep) and variable load (fatigue) results in obtaining total damage larger than unity ($D_t > 1$) for most of the analyzed load program sequences. Consequently, the calculated durability is lower than that obtained from the tests.

This conclusion is confirmed in Figure 15, presenting durability diagrams obtained from calculations and experimental tests. The figure includes the following graphs:

(1) a calculation-based diagram taking into account only the fatigue properties (fatigue);
(2) a calculation-based diagram regarding fatigue and creep (fatigue + creep);
(3) a diagram obtained based on experimental results (experiment).

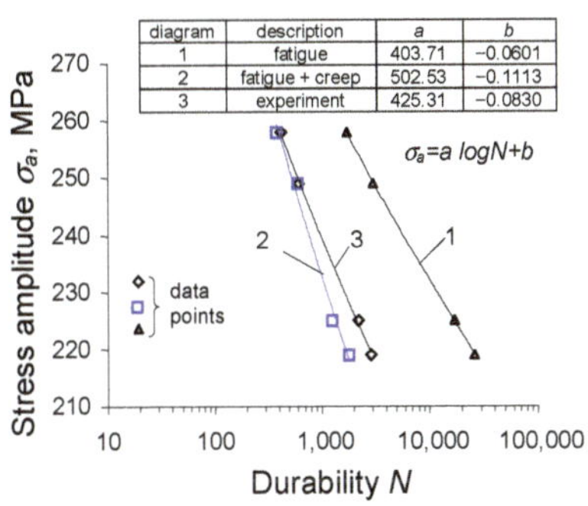

(**a**)

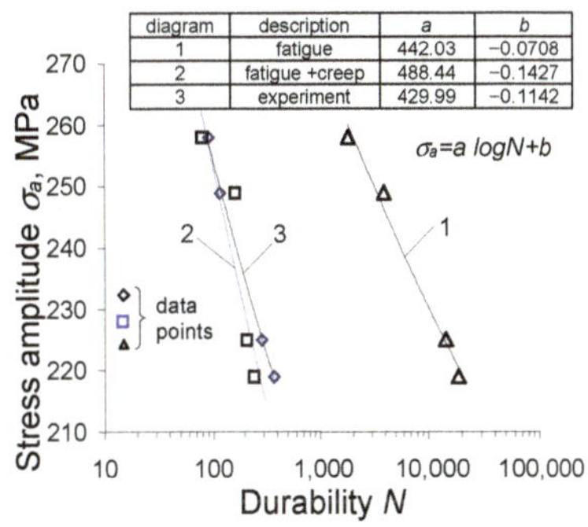

(**b**)

Figure 15. Durability diagrams (experimental and calculated): (**a**) τ = 5 s, (**b**) τ = 30 s.

Analysis of the results presented in Figure 15a,b indicates that regarding creep during the fatigue life calculations significantly improves the efficiency of the durability assessments. If both the creep and fatigue damage is considered within the classical damage summation approach (4), the predicted durability calculations are lower but close to the results of durability obtained from the experimental tests. If only the fatigue damage is considered, the simulated durability is much larger than the real one observed in the experiment.

4. Conclusions

During fatigue tests on P91 steel samples at the temperature of 600 °C under alternating variable and constant loads, cyclic softening of the material was observed, both for the programs containing only the pure fatigue load cycles and for the cases when alternating fatigue and creep took place. The amount of P91 steel softening is influenced by both the load level and the dwell time (creep). With the increase of the dwell time in the cycles, the cyclic softening becomes more pronounced.

The increase in the static load time ($\tau_i > 0$) causes a marked increase in the cyclic creep observed during classic fatigue tests conducted under the conditions of σ_a = const ($\tau_i = 0$), and in the range of inelastic stain on a cycle $\Delta\varepsilon_{ap}$.

The dwell periods occurring during the variable load reduce the fatigue life. The dwell time value influences the durability.

Fatigue life predictions that do not regard the material damage due to creep may lead to a significant overestimation of durability. Taking into account the creep damage in the calculations improves the compliance of the numerical and experimental results.

Author Contributions: S.M.: Supervision, Conceptualization, Methodology, Investigation, Validation, Visualization, Writing—Original draft preparation. Z.L.: Software, Visualization, Investigation. H.E.: Methodology, Writing—Reviewing and Editing, Project administration, Funding acquisition. All authors have read and agreed to the published version of the manuscript.

Funding: This research was funded by the NATIONAL SCIENCE CENTRE of Poland through Grant No. 2017/25/B/ST8/02256.

Institutional Review Board Statement: Not applicable.

Informed Consent Statement: Not applicable.

Data Availability Statement: Not applicable.

Conflicts of Interest: The authors declare no conflict of interest.

References

1. Mroziński, S.; Piotrowski, M.; Egner, H. Effects of fatigue testing on low-cycle properties of P91 steel. *Int. J. Fatigue* **2019**, *120*, 65–72. [CrossRef]
2. Mroziński, S.; Lis, Z. Comparative Analysis of Methods for Determining Cyclic Properties of Metals, Experimental Mechanics of Solids. Materials Research Forum RLC. *Mater. Res. Proc.* **2019**, *12*, 139–145.
3. Wu, D.-L.; Zhao, P.; Wang, Q.-Q.; Xuan, F.-Z. Cyclic behavior of 9–12% Cr steel under different control modes in low cycle regime: A comparative study. *Int. J. Fatigue* **2015**, *70*, 114–122. [CrossRef]
4. Egner, W.; Sulich, P.; Mroziński, S.; Egner, H. Modelling thermo-mechanical cyclic behavior of P91 steel. *Int. J. Plast.* **2020**, *135*, 102820. [CrossRef]
5. Hu, X.; Zhang, Q.; Jiang, Y.; Rao, G.; Miao, G.; He, W.; Nie, X. The effect of cyclic loading on the creep fatigue life and creep strength of a DS superalloy: Damage mechanism and life modeling. *Int. J. Fatigue* **2020**, *134*, 105452. [CrossRef]
6. Xu, L.; Wang, R.-Z.; Wang, J.; He, L.; Itoh, T.; Miura, H.; Zhang, X.-C.; Tu, S.-T. On multiaxial creep–fatigue considering the non-proportional loading effect: Constitutive modeling, deformation mechanism, and life prediction. *Int. J. Plast.* **2022**, *155*, 103337. [CrossRef]
7. Liu, H.; Yang, X.; Jiang, L.; Lv, S.; Huang, T.; Yang, Y. Fatigue-creep damage interaction model of asphalt mixture under the semi-sine cycle loading. *Constr. Build. Mater.* **2020**, *251*, 119070. [CrossRef]
8. Barat, K.; Sivaprasad, S.; Kar, S.K.; Tarafder, S. A novel rate based methodology for creep fatigue life estimation of superalloys. *Int. J. Press. Vessel. Pip.* **2020**, *182*, 104064. [CrossRef]
9. Igumnov, L.A.; Volkov, I.A. Defining relations of mechanics of damaged media effected by fatigue and creep. *Procedia Struct. Integr.* **2020**, *28*, 2086–2098. [CrossRef]
10. Mroziński, S.; Lis, Z. Research on the influence of creep on low-cycle fatigue life. *Energetyka* **2019**, *11*, 760–762(In Polish).
11. Mroziński, S.; Lis, Z.; Egner, H. Energy Dissipated in Fatigue and Creep Conditions. *Materials* **2021**, *14*, 4724. [CrossRef] [PubMed]
12. Tien, J.K.; Nair, S.V.; Nardone, V.C. Creep fatigue Interactions in Structural Alloys. In *Flow and Fracture at Elevated Temperatures*; Raj, R., Ed.; ASM Materials Science Seminar: Philadelphia, PA, USA, 1983; Chapter 6; p. 179.
13. Miner, M. Cumulative Damage in Fatigue. *J. Appl. Mech.* **1945**, *67*, A159–A164. [CrossRef]
14. Zhuang, W.Z.; Swansson, N.S. *Thermo-Mechanical Fatigue Life Prediction: A Critical Review*; DSTO-TR-0609; DSTO Aeronautical and Maritime Research Laboratory: Victoria, Australia, 1998.
15. Ahmadzadeh, G.R.; Varvani-Farahani, A. Concurrent ratcheting–fatigue damage analysis of uniaxially loaded A-516 Gr.70 and 42CrMo Steels. *Fatigue Fract. Eng. Mater. Struct.* **2012**, *35*, 962–970. [CrossRef]
16. Chang, L.; Wen, J.B.; Zhou, C.Y.; Zhou, B.B.; Li, J. Uniaxial ratcheting behavior and fatigue life models of commercial pure titanium. *Fatigue Fract. Eng. Mater. Struct.* **2018**, *41*, 2024–2039. [CrossRef]
17. Samuel, K.G.; Rodriguez, P. An empirical relation between strain energy and time in creep deformation. *Int. J. Press. Vessel. Pip.* **1998**, *75*, 939–943. [CrossRef]
18. Payten, W.M.; Dean, D.W.; Snowden, K.U. A strain energy density method for the prediction of creep–fatigue damage in high temperature components. *Mater. Sci. Eng. A* **2010**, *527*, 1920–1925. [CrossRef]
19. Cheng, H.; Chen, G.; Zhang, Z.; Chen, X. Uniaxial ratcheting behaviors of Zircaloy-4 tubes at 400 C. *J. Nucl. Mater.* **2015**, *458*, 129–137. [CrossRef]
20. Song, G.; Hyun, J.; Ha, J. Creep-fatigue prediction of aged 13CrMo44 steel using the tensile plastic strain energy. *Eur. Struct. Integr. Soc.* **2002**, *29*, 65–73.
21. Zheng, X.T.; Xuan, F.Z.; Zhao, P. Ratcheting-creep interaction of advanced 9–12% chromium ferrite steel with anelastic effect. *Int. J. Fatigue* **2011**, *33*, 1286–1291. [CrossRef]
22. Zhao, Z.; Yu, D.; Chen, X. Creep-ratcheting-fatigue life prediction of bainitic 2.25Cr1MoV steel. *Procedia Struct. Integr.* **2019**, *17*, 555–561. [CrossRef]
23. Zhao, Z.; Yu, D.; Chen, G.; Chen, X. Ratcheting fatigue behaviour of bainite 2.25Cr1MoV steel with tensile and compressed hold loading at 455 °C. *Fatigue Fract. Eng. Mater. Struct.* **2019**, *42*, 1937–1949. [CrossRef]
24. *ASTM E606-92*; Standard Practice for Strain—Controlled Fatigue Testing. ASTM: West Conshohocken, PA, USA, 1998.
25. Skocki, R. Research on the Influence of Elevated Temperature on the Cyclic Properties of P91 Steel. Ph.D. Thesis, Faculty of Mechanics UST, Bydgoszcz, Poland, 2017. (In Polish).
26. Mroziński, S.; Lipski, A. The effects of temperature on the strength properties of aluminium alloy 2024-T3. *Acta Mech. Autom.* **2012**, *6*, 62–66.
27. Oh, Y.-J.; Yang, W.-J.; Jung, J.-G.; Choi, W.-D. Thermomechanical fatigue behavior and lifetime prediction of niobium bearing ferritic stainless steels. *Int. J. Fatigue* **2012**, *40*, 36–42. [CrossRef]
28. Brnic, J.; Turkalk, G.; Canadija, M.; Lanc, D. AISI 316Ti (1.4571) steel—Mechanical, creep and fracture properties vs. temperature. *J. Constr. Steel Res.* **2011**, *67*, 1948–1952. [CrossRef]

materials

Article

Experiment and Theoretical Investigation on Fatigue Life Prediction of Fracturing Pumpheads Based on a Novel Stress-Field Intensity Approach

Yun Zeng [1], Meiqiu Li [1], Han Wu [2], Ning Li [1] and Yang Zhou [3,*]

[1] School of Mechanical Engineering, Yangtze University, Jingzhou 434023, China; mechanicszy@163.com (Y.Z.); limq@yangtzeu.edu.cn (M.L.); lining@yangtzeu.edu.cn (N.L.)
[2] CNOOC Ener Tech-Drilling & Production Co., Shanghai 200444, China; wuhan6@cnooc.com.cn
[3] School of Mechatronic Engineering and Automation, Shanghai University, Shanghai 200444, China
* Correspondence: saber_mio@shu.edu.cn

Abstract: Fracturing pumpheads are typical pressure vessels that experience frequent fatigue failure under the effect of notches in their cross-bore. To enhance the fatigue life of fracturing pumpheads, the study of the notch effect is indispensable and important to establish a reliable mathematical model to predict their fatigue life. In the present paper, two novel fatigue life prediction models are proposed for notched specimens. In these models, two new geometric fatigue failure regions are defined to improve the weight function. Finally, the elaborated novel stress-field intensity approach was applied to three different types of notched specimens. Experiment results indicate that the new SFI approach achieves 47.82%, 39.48%, and 31.85% higher prediction accuracy than the traditional SFI approach, respectively. It was found that the modified SFI approach provided better predictions than the traditional SFI approach and the TCD method. The II-th novel SFI approach had the highest accuracy, and the I-th novel SFI approach was more suitable for sharply notched specimens.

Keywords: fatigue life prediction; notched specimen; S-N curves; fatigue fracture region; stress-field intensity

Citation: Zeng, Y.; Li, M.; Wu, H.; Li, N.; Zhou, Y. Experiment and Theoretical Investigation on Fatigue Life Prediction of Fracturing Pumpheads Based on a Novel Stress-Field Intensity Approach. *Materials* **2022**, *15*, 4413. https://doi.org/10.3390/ma15134413

Academic Editors: Jaroslaw Galkiewicz, Lucjan Śnieżek and Sebastian Lipiec

Received: 13 May 2022
Accepted: 20 June 2022
Published: 22 June 2022

Publisher's Note: MDPI stays neutral with regard to jurisdictional claims in published maps and institutional affiliations.

1. Introduction

In the oil and gas exploration field, with the rapid development of unconventional oil and gas drilling technology, requirements for fracturing and acidizing technology have also gradually increased. Therefore, as important components of the equipment, fracturing pumps require long life and high reliability [1]. However, the average life of pumpheads is much lower than that of other conventional components, and the manufacturing cost of pumphead components is surprisingly high, seriously affecting the economic benefits of shale gas [2,3]. Hence, it is important to accurately estimate the fatigue life of pumpheads to improve the performance of fracturing pumps.

The nominal stress method [4,5], the critical distance method [6–9], and the effective stress volume method [10,11] are most commonly used to estimate the high-cycle fatigue life of notched components. The accuracy of NSM mainly depends on the correction factor (stress concentration factor loading, stress size coefficient, roughness coefficient) and S-N curves of materials. However, owing to the complexity in petroleum engineering, empirical equations are often used to calculate the coefficients of NSM. In addition, it is proven that NSM is not suitable for predicting the fatigue life of complex structural parts [12].

NSM considers the "hot point stress" as the core parameter of fatigue failure; however, some advanced volume methods [13–15] indicate that the "volume stress" in a damaged region should also be considered for fatigue failure. The critical distance method and the effective stress volume method are two main advanced volume methods. The theory of critical distance method was first proposed by Taylor and Tanaka [16–21] based on LEFM.

The average stress of the maximum principal stress field in the critical region near a notch root (critical region can be a point, line, area, or volume) is regarded as a fatigue damage parameter for fatigue life estimation. The formulas for different forms of TCD can be found in [17–20]. The methods are schematically shown in Figure 1, and the formulas of the methods are as follows:

$$\begin{aligned} \text{PointMethod (PM)} : \sigma_{mean} &= \Delta\sigma_1\left(r = \tfrac{L_0}{2}, \theta = 0\right) \\ \text{LineMethod (LM)} : \sigma_{mean} &= \tfrac{1}{2L_0}\int_0^{2L_0} \sigma_1(r, \theta = 0)dr \\ \text{AreaMethod (AM)} : \sigma_{mean} &= \tfrac{2}{1.1\pi L^2{}_0}\int_{-\pi/2}^{\pi/2}\int_0^{L_0} \sigma_1(r, \theta)rdrd\theta \\ \text{VolumeMethod (VM)} : \sigma_{mean} &= \tfrac{3}{2\pi(1.54L_0)^3}\int_0^{2\pi}\int_0^{\pi/2}\int_0^{1.54L_0} \sigma_1(r, \theta, \varphi)r^2 \sin\theta rdrd\theta d\varphi \end{aligned} \quad (1)$$

$$El\ Haddad\ equation\text{:}\ L_0 = \frac{1}{\pi}\left(\frac{\Delta K_{th}}{\Delta\sigma_{-1}}\right)^2 \quad (2)$$

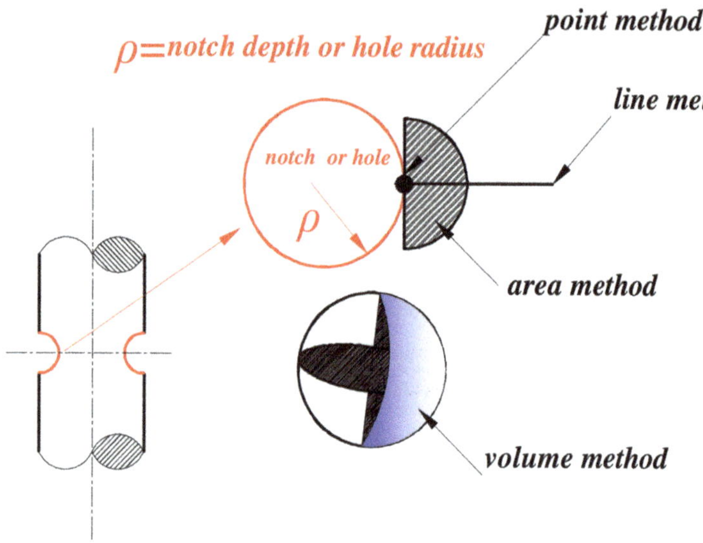

Figure 1. Visualization of the point method, the line method, the area method, and the volume method.

Qylafku [22], Adib [23], and Pluvinage [24] have defined the effective distance as the length between the minimum point of the stress gradient and a notch root. Therefore, there is a certain relationship between fatigue damage and the stress gradient. The effective stress volume method introduces the concept of the effective fatigue failure region, and the calculation is used to obtain the effective stress by integrating the elastic–plastic fatigue crack-opening stress and the weight function of the effective critical damage region (schematically shown in Figure 2). The mathematical definition of the effective stress function (σ_{eff}) is expressed by Equation (3):

$$\begin{aligned} \sigma_{\text{eff}} &= \tfrac{1}{X_{\text{eff}}}\int_0^{X_{\text{eff}}} \sigma_{yy}(x) \times \varphi(x, \chi)\text{d}x \\ \chi &= \left|\tfrac{1}{\sigma_{\max}}\tfrac{\text{d}\sigma_{yy}(x)}{\text{d}r}\right| \end{aligned} \quad (3)$$

where X_{eff} is the effective distance diameter of the fatigue process zone, $\sigma_{yy}(r)$ is the fatigue crack-opening stress, $\varphi(r, \chi)$ is the relative stress gradient, and K_ρ is the notch stress intensity factor.

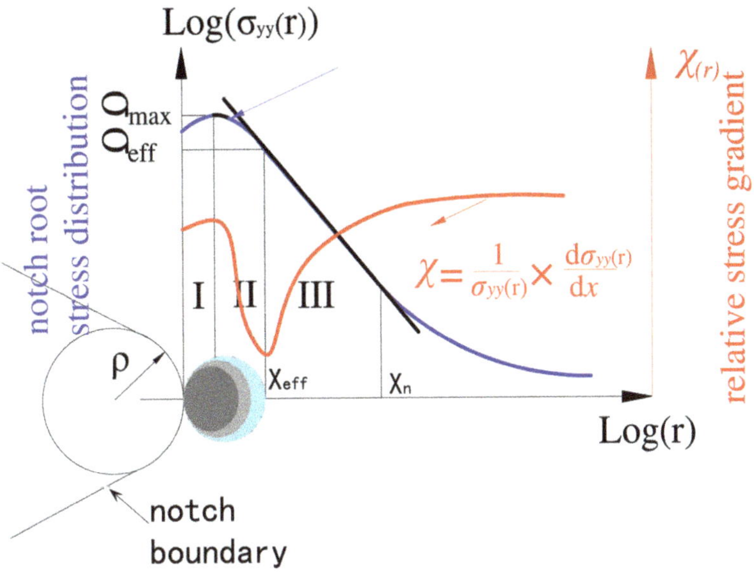

Figure 2. Schematic of elastic–plastic stress distribution along the notch ligament and the relative stress gradient concept.

In addition, these effective stress volume approaches have been verified in simple geometric structures; thus, it is unknown whether they are suitable for fatigue life estimation of thick-walled pressure vessels. The mean value of the fatigue failure parameter, l_0 (critical distance), for aluminum alloys, cast iron, and steel is equal to 0.113 mm [25,26]. The initiation of fatigue cracks depends on grain size, grain dislocations, grain slips, and other features of microstructures. Especially, the initiation of fatigue cracks is a grain movement nucleation process at the mesoscopic scale. To simplify the calculation of the fatigue process zone, the fatigue failure parameter, l_0, can be defined as the field diameter of the fatigue failure region. In addition, the plastic zone is also assumed as the fatigue damage failure region. Numerous empirical formulas are developed to calculate the cyclic plastic zone size [27–33] (Table 1). However, the fatigue plastic yielding zones of specimens and structures are different. In fracture mechanics, the fatigue damage failure region and the plastic yield region are defined as the fracture tip plastic zone and the crack initiation region, respectively [34–38]. To investigate the effects of the cyclic plastic zone on notch tip stress and fatigue life, Zhu et al. [34,35] proposed a novel approach.

Different studies have proposed different fatigue failure parameters to estimate the fatigue life of engineering components; however, these approaches are only verified on notched specimens of regular components. Fracturing pumpheads have a complex circular cross-bore structure; therefore, it is doubtful whether the fatigue failure region size can be defined as a fatigue failure parameter to predict the fatigue life of pumpheads.

In the current study, a new fatigue life prediction model was proposed for notched specimens and a novel cardioid geometric fatigue failure region was defined to improve the weight function. In addition, existing problems of the original SFI approach were corrected. The accuracy of the modified SFI approach was verified in fatigue tests on notched specimens and based on theoretical approaches for improving the fatigue life of pumpheads.

Table 1. Empirical formulas for cyclic plastic zone size calculation.

Reference	Formula
Nicholls and Martin [27]	$r'_y = a\left(\frac{\sigma_{app}^2}{\sigma_y^2-\sigma_{app}^2}\right)$
Bathias and Pelloux [31]	$r'_y = 0.1\left(\frac{\Delta K}{\sigma_y}\right)^2$
Pineau and Pelloux [28]	$r'_y = 0.053\left(\frac{\Delta K}{\sigma_y}\right)^2$
Saxena and Antolovich [29]	$r'_y = \alpha\left(\frac{\Delta K}{\sigma_y}\right)^{2+s}$
Park et al. [30]	$r'_y = \frac{\pi}{144}\left(\frac{\Delta K}{\sigma_y}\right)^2$
Chapetti et al. [21]	$r'_y = \frac{1}{12\pi}\left(\frac{\Delta K}{\sigma_y}\right)^2$
Edmunds and Willis [33]	$r'_y = \frac{1}{24\pi}\left(\frac{\Delta K}{\sigma_y}\right)^2$

σ_{app} represents the applied stress, and α and s denote correlation and variation coefficients, respectively.

2. Theoretical Analysis

2.1. Brief Review of Stress-Field Intensity Approach and Characteristic Parameters

2.1.1. Traditional Stress-Field Intensity Approach

The fatigue failure of materials occurs due to the propagation of fatigue cracks from grains. The formation of cracks can be defined as the cumulative damage of multiple grains in local areas of materials. The SFI model proposed by Yao [39] is displayed in Figure 3, and the mathematical definition of the stress-field intensity function (σ_{FI}) is expressed as:

$$\sigma_{FI} = \frac{1}{V}\int_{\Omega} f(\sigma_{ij})\varphi(\vec{r})\mathrm{d}v = \frac{1}{V}\int_{\Omega} \sigma_{(r-\theta)}\varphi(r,\theta)\mathrm{d}v \tag{4}$$

where σ_{FI} is the stress-field intensity of a notched specimen, Ω is the fatigue failure region, V is the volume of the fatigue failure region, R is the field diameter, $f(\sigma_{ij})$ is the equivalent stress function, and $\varphi(r,\theta)$ is the weight function, which signifies the contribution of stresses at point P to the peak stress at $\left|\vec{r}\right|$.

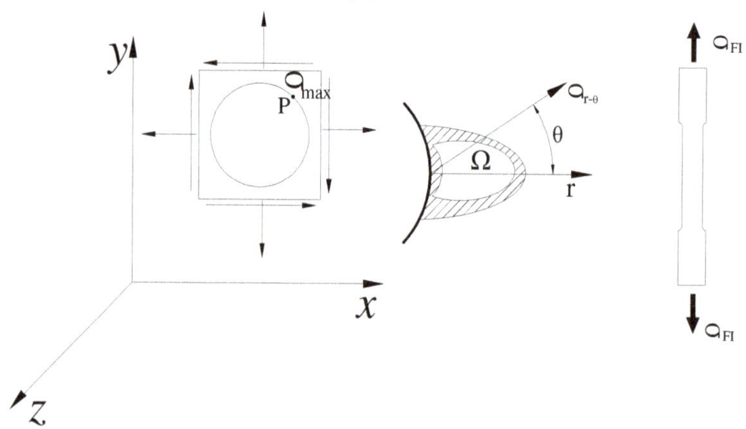

Figure 3. Basic model of the stress-field intensity approach.

2.1.2. Parameters for Critical Fatigue State

All advanced volume methods generally include a "characteristic region parameter". Table 2 summarizes some characteristic region parameters proposed in previous papers [6,9–11,17]. Although these methods were previously successfully verified, they possess some common problems.

Table 2. Fatigue damage regions of different approaches.

Approach	Authors	Fatigue Damage Region	Influence Factors
Neuber's and Peterson's empirical formula	Neuber and Peterson [40,41]	Material constants, a_1, a_2	a_1, a_2 depend on fracture strength σ_b
SED approach	Lazzarin and Berto [42]	Radius of critical volume (area), R_0	R_0 depends on the threshold value of fatigue crack propagation, fatigue limit, Poisson's ratio, and notch shape
SFI approach	Yao [41]	Radius of the fatigue damage field, $\vec{r}$	$\vec{r}$ depends on grain size
Modified SFI approach	Qylafku [22]	Radius of the fatigue damage field, r	r depends on metallic-plastic deformation
TCD and M-TCD	Taylor and Susmel [19]	Intrinsic crack length, l_0	l_0 depends on the threshold value of fatigue crack propagation, fatigue limit, and stress ratio
Advanced volumetric method	Adib-Ramezani [10]	Effective distance, X_{eff}	X_{eff} depends on stress distributions at notch roots

(1) The acquisition of critical distance parameters requires the testing of numerous material constants; however, it takes a long time to experimentally obtain these material constants.

(2) The geometry of a fatigue-damaged failure area is generally defined as a common shape; hence, it is difficult to confirm whether previous advanced volume methods are suitable for calculating the fatigue damage failure region of complex pressure vessels.

(3) Owing to the size effect, the geometry and size of fatigue damage failure regions need to be reconsidered.

2.2. *Determination of Solutions for New Fatigue Failure Regions*

2.2.1. Comparison of Stress Gradient Distributions near Different Notch Roots

Generally, engineering structures have different and complex geometries; thus, it is difficult to select a fatigue failure region. Therefore, the determination of fatigue failure regions for notched components with common geometries is a better choice. According to the research of Susmel and Taylor [19], four different notched specimens were designed (Figure 4). Generally, FEM was applied to calculate the bisector stress distributions of the notched specimens, and LEFM and EPFM were used for calculations.

In the paper, finite element analysis was performed on the specimen in Figure 4, with a fixed constraint on one end and a load on the other end, and the stress distribution in the elastic state of material at the notch root bisector of each specimen was elaborated and represented by a bi-logarithmic coordinate system. Furthermore, the corresponding relative stress gradient was calculated by the volumetric method and the SFI method (Figure 5). Owing to the notch effect, relative stress gradients near the notch root of the four specimens had different forms. However, the variation trends of relative stress gradients for the four specimens were similar. According to the theory of critical distance method (TCD) [14,15,17], the minimum point of a relative stress gradient was defined as the characteristic region parameter. Figure 5 displays the elastic stress distributions and relative stress gradients near the notch root of the four specimens. Some numerical truncation errors appear based on the discrete feature of numerical calculations. Since plastic action is not considered, the relative stress gradient trend of linear elastic analysis can only be used as a reference to judge the trend. In Figure 5a,b, the trend of $\log(\sigma_0/\sigma_{yy})$ is to be linear in zone II. On the contrary, in Figure 5c,d, the trend of $\log(\sigma_0/\sigma_{yy})$ is to be linear in zone III. Therefore, elastoplastic analysis is also required to perform a comparative analysis of the differences between zone II and zone III.

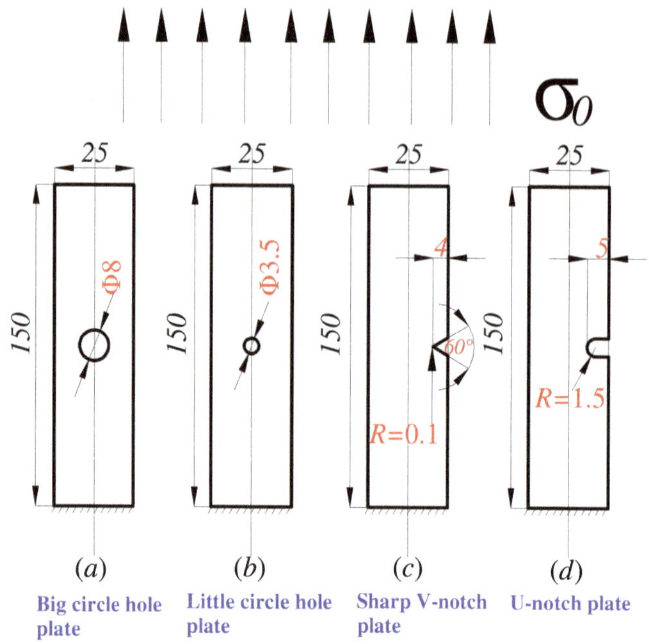

Figure 4. Geometries of different notched specimens (thickness = 6 mm, units: mm).

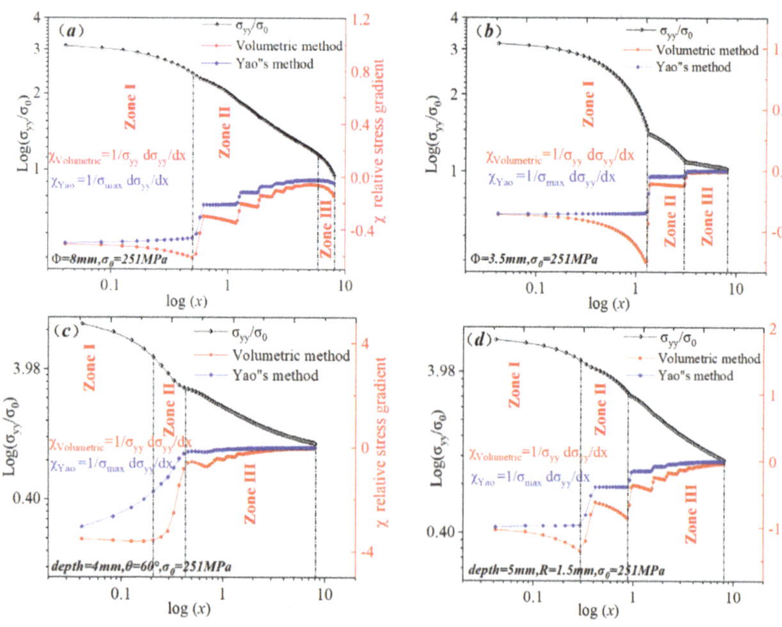

Figure 5. Bi-logarithmic diagrams for stress distributions at the notch bisector of different specimens at the elastic state of material. (**a**) big circle hole plate, (**b**) little circle hole plate, (**c**) sharp V-notch plate, (**d**) U-notch plate.

Moreover, fatigue process zones resulted from the accumulation of the fatigue damage region, where micro- or macro-plastic cyclic strains occur. The fatigue failure region radii of the four specimens were calculated based on stress distributions for the elastic–plastic state at the notch root bisector and relative stress gradients in a bi-logarithmic coordinate system. Figure 6 displays the elastic–plastic stress distributions and relative stress gradients near the notch root of the four specimens. The peak in stress for the elastic state always appeared at the notch root surface (the origin of the X-axis in the Cartesian coordinate system). However, the stress peak in the elastic stress appeared at a certain distance from the notch root. The point of the stress peak and its counterpart distance, X_m, were defined as zone I, which was a completely plastic region in the bi-logarithmic coordinate system. In zone II, the peak stress manifested a decreasing trend. The distribution trends of relative stress gradients in zones II and III were opposite, especially the relative stress gradient, which dropped gradually in zone II. The point at which the relative stress gradient started to manifest an increasing trend was considered as the effective distance. In zone III, the elastic–plastic stress distribution was linear; thus, a gradual plastic to elastic material transition occurred in this region. In zone IV, the elastic–plastic stress distribution was no longer linear, and the relative stress gradient expressed a new decreasing trend. The relative stress gradient was far from the notch root and had a minimum effect on fatigue failure.

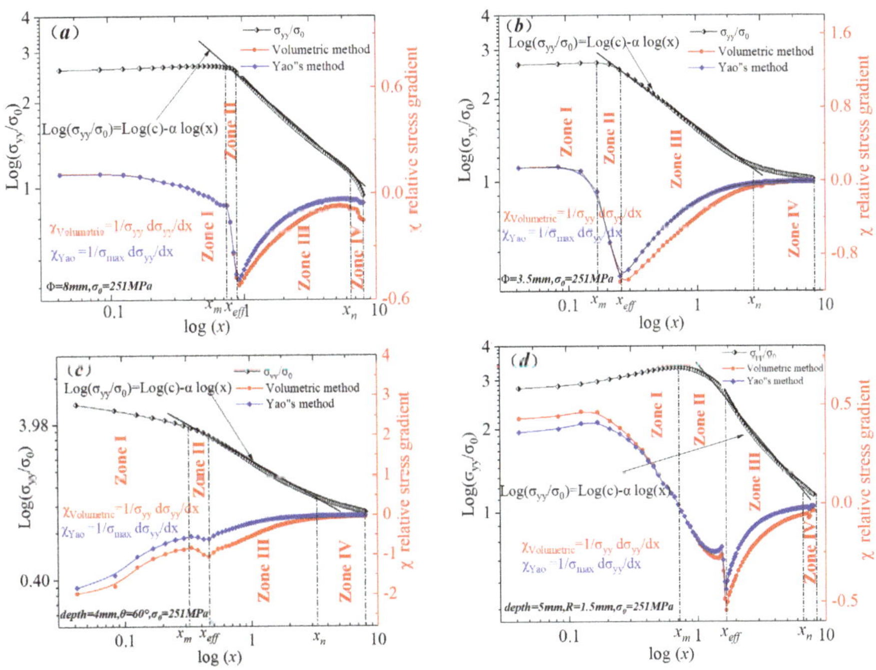

Figure 6. Bi-logarithmic diagrams for elastic–plastic stress distributions at the notch bisector of different specimens. (**a**) big circle hole plate, (**b**) little circle hole plate, (**c**) sharp V-notch plate, (**d**) U-notch plate.

2.2.2. Summary of Fatigue Damage Failure Area Shapes for Different Notched Specimens

It is propounded that in addition to the "stress peak point", the stress field in a specific region should also be considered during fatigue life analysis [6,7,36,43–46]. TCD, the volumetric method, and the SFI approach are defined as macro-mechanical methods; thus, failure criteria of these methods should be combined with the critical damage region. In this work, the SFI approach was used to predict the fatigue life of fracturing pumpheads.

The shape and size of the fatigue damage region are key factors for accurate fatigue life prediction.

Figure 7 displays the stress nephograms for the elastic–plastic stress at the notch root of the four specimens under a pressure of 251 MPa. The high-stress contours of the circular plate specimen had a crescent shape, whereas those of the V-notched and U-notched specimens had a heart shape. Obviously, regions enclosed by these high-stress contour lines were plastic. Elastic–plastic transition zones near high-stress plastic zones also had an important influence on fatigue failure. It was found that elastic–plastic stress contour lines of the four notched specimens manifested roughly similar regions with a heart shape. Therefore, it is feasible to assume that the shape of the fatigue failure region was cardioid.

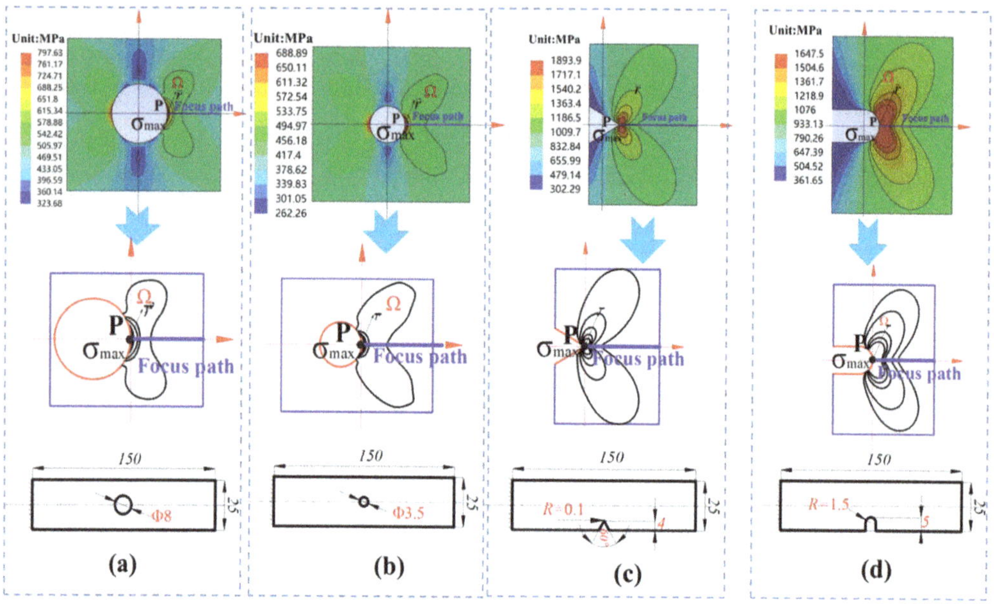

Figure 7. Elastic–plastic stress nephograms at the notch root of the four specimens under pressure of 251 MPa. (**a**) big circle hole plate, (**b**) little circle hole plate, (**c**) sharp V-notch plate, (**d**) U-notch plate.

2.3. *Improvement of the Stress-Field Intensity Approach*

2.3.1. Traditional Stress-Field Intensity Approach

The traditional SFI approach was modified to effectively predict the fatigue life of notched components. However, the traditional SFI approach and other critical damage region approaches have several common problems.

1. Owing to the randomness of grain damage in fatigue damage failure regions and the influence of stress gradients near notch roots, uniform shapes cannot be used to define damage regions in notched components.

2. Micro-plastic cyclic strains are one of the main reasons for fatigue failure. However, macro-mechanical methods do not consider the influence of microcosmic factors. Hence, material parameters at the microstructural scale are essential factors for fatigue failure.

3. The mechanical resistance of microstructure barriers inhibits the propagation of short cracks. However, whether grains near the peak stress point at a notch root play a "contribution" or "hindrance" role in the initiation and propagation of cracks has not been clearly defined.

2.3.2. Optimization of the Traditional Stress-Field Intensity Approach

The accuracy of the traditional SFI approach depends on the failure equivalent stress function, the fatigue failure region, and the weight function. In Section 2.3.1, it is asserted that the effect of inner grains existing far away from a high-stress region also needs to be considered for fatigue life prediction. Therefore, the revised formula can be expressed as:

$$\sigma_{FI} = \frac{1}{V}\int_{\Omega} f(\sigma_{ij})\varphi(\vec{r})\mathrm{d}v = \sigma_{\max} - \xi(\Omega) \tag{5}$$

where $\xi(\Omega)$ is the "auxiliary part", which can be expressed as:

$$\xi(\Omega) = \frac{\int_{\Omega} (\sigma_{\max} - \sigma_{r-\theta})\varphi(r,\theta)\mathrm{d}v}{\int_{\Omega} \varphi(r,\theta)\mathrm{d}v} \tag{6}$$

The novel method proposed in this work is different from the traditional SFI approach and does not require an artificial field diameter. It was considered that the continuous accumulation of damages resulted from the combined effect of the peak stress point and other points in the damage region. The hypothetical concept of an "invisible boundary fatigue failure region" is illustrated in Figure 8. In Figure 8a, r_{pc} is the size of the cyclic plastic zone, and r_{pm} is the size of the monotonic plastic zone.

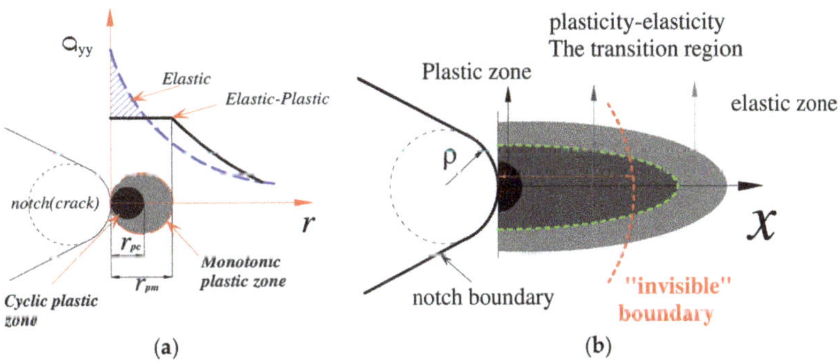

Figure 8. Schematic diagram of the novel elastic–plastic zone at the crack tip (notch): (**a**) monotonic plastic zone and cyclic plastic zone (**b**) invisible boundary region.

The weight function of the traditional SFI approach is expressed as:

$$\varphi(r,\theta) = 1 - \chi r(1+\sin\theta) = 1 - \left|\frac{1}{\sigma_{\max}}\frac{\mathrm{d}\sigma_{r-\theta}}{\mathrm{d}r}\right| r(1+\sin\theta) \tag{7}$$

The weight function of the traditional SFI approach is a generalized monotonically decreasing function about $\left|\vec{r}\right|$. However, it is also reported that as the distance from a notch root increases, the variation trend of the weight function does not comply with the original definition [36]. Therefore, it is important to correct the weight function for accurate fatigue life prediction. It can be inferred from Equation (9) that the weight function is mainly composed of three parts: relative stress gradient, distance function of the notch root, and function of angle θ.

In this work, common problems of the weight function of the traditional SFI approach, similar to those mentioned in Section 2.3.1, are summarized and revised.

(1) The direction angle of the relative stress gradient is not clearly expressed by the traditional weight function. Hence, the variable of distance $\vec{r}$ for a notch root was revised to a distance function $r(\theta)$.

(2) Grain size is an essential factor for fatigue life prediction; therefore, it was introduced as a coefficient in the calculation formula of the weight function of the SFI approach.

(3) The weight function defines the contribution of stresses at point P to the peak stress at $|\vec{r}|$; however, considering the inhibition effect of the "auxiliary part", $(1 + \sin\theta)$ is no longer applicable. Hence, the function $f(\theta)$ of angle θ was proposed to revise the weight function.

(4) Many researchers [43,47,48] have also proven that the grain size is related to value of the stress-intensity field; moreover, the grain size means the scale of the grain dimension. Therefore, the authors consider the grain size of the material to modify the weight function.

Considering the above factors and the function of $\xi(\Omega)$ as a separate part, the revised formula can be expressed as:

$$\varphi(x,\theta) = \left[\left|\frac{1}{\sigma_{\max}}\frac{d\sigma_{r-\theta}}{dr}\right| r(\theta) f(\theta)\right]^{\frac{r(\theta)}{G}} \tag{8}$$

where G is the grain size and $r(\theta)$ is the geometric relationship function of stress contour distribution in a damaged region.

2.4. Hypothesis of the "Cardioid" Fatigue Damage Failure Region

Finite element analysis results revealed that fatigue failure regions of the notched specimens had two geometrical shapes. In Figure 7, the highly stressed damage volume geometry of the plate specimen with holes has a crescent shape, whereas those of the U-notched and V-notched specimens have a heart shape.

2.4.1. Crescent-Shaped Fatigue Failure Region

The relationship between stress contours and the peak point is geometrically expressed in Figure 9. To simplify the calculation, the approximate geometric relationships of some parameters are displayed in Figure 9b, and the corresponding differential equations are expressed as:

$$\begin{cases} OA = OA_1 = OD = r_i \\ OD = OP + PD = r_i = \Delta r + r \\ A_1P = \sqrt{OA_1{}^2 - OP^2} = \sqrt{r_i{}^2 - \Delta r^2} = \sqrt{r^2 + 2r\Delta r} \\ AC = \sqrt{r_i{}^2 - (\Delta r - r_0)^2} \\ AP = \sqrt{r_i{}^2 - (\Delta r - r_0)^2 + r_0{}^2} = \sqrt{r^2 + 2\Delta r(r + r_0)} \end{cases} \tag{9}$$

To solve these differential equations, the results of $r(\theta)$ were considered infinite. Therefore, in Figure 9b, lines A_1P and PD are two special boundary solutions, which can be expressed as:

$$\begin{cases} r(\frac{\pi}{2}) = \sqrt{r_i{}^2 - \Delta r^2}, r(0) = r \\ r'(\theta)|_{\theta=0} = 0 \end{cases} \tag{10}$$

Hence,

$$r(\theta) = (2\sin^2\frac{\theta}{2})(\sqrt{r^2 + 2r\Delta r} - r) + r \tag{11}$$

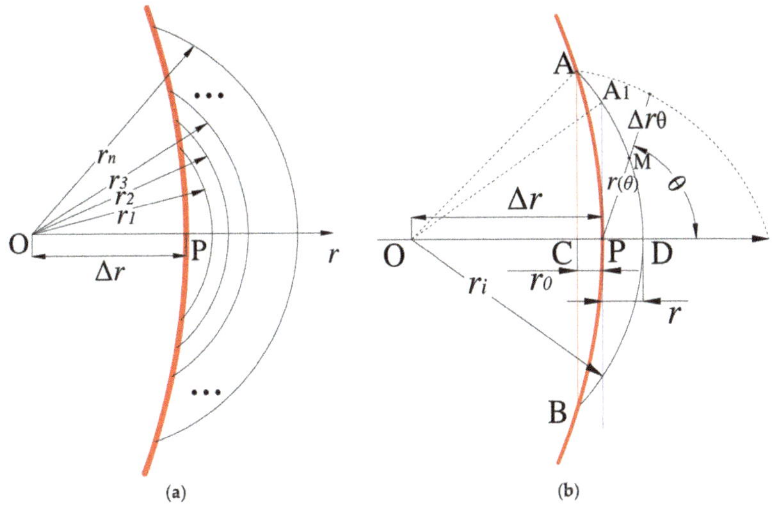

Figure 9. Schematic representation of the notched specimen with holes. (**a**) Stress contour lines at the notch root. (**b**) Relationship between stress contours and the peak point.

2.4.2. Cardioid-Shaped Fatigue Failure Region

According to FEA results, the geometry volume of the fatigue failure region of the U-notched and V-notched specimens in the elastic–plastic stress had a cardioid shape. The relationship between stress contours and the peak point was expressed by periodic distribution (Figure 10).

$$\begin{cases} x^2+y^2+ax = a\sqrt{x^2+y^2} \\ (r(\theta))^2+ar(\theta)\cos\theta = ar(\theta) \\ r(0) = r \end{cases} \quad (12)$$

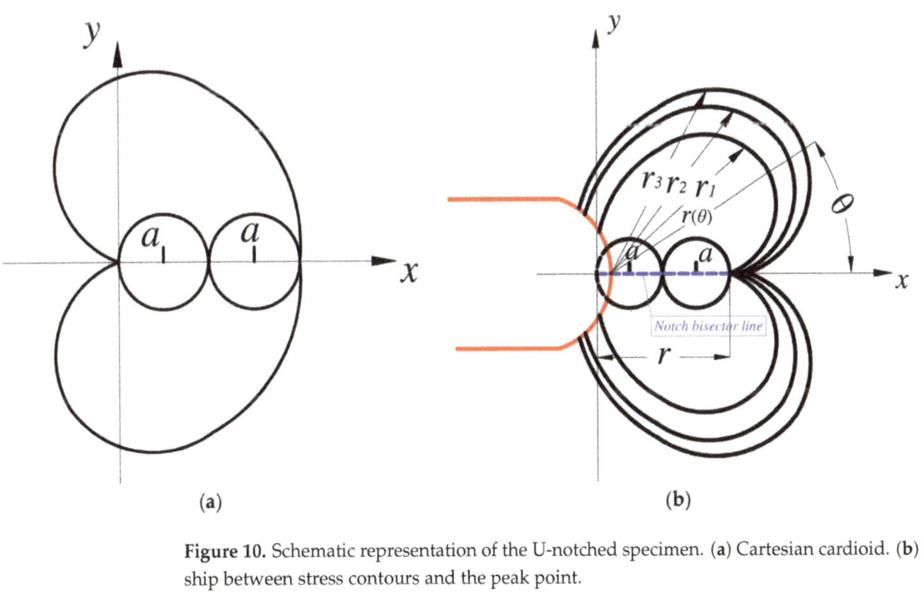

Figure 10. Schematic representation of the U-notched specimen. (**a**) Cartesian cardioid. (**b**) Relationship between stress contours and the peak point.

In addition, according to the concept of the "invisible boundary damage region", the field diameter of an invisible boundary does not require an artificial definition. As the distance increased, the "inhibition" gradient ($d\Omega/dr$) reached zero; thus, the distance was determined as the field diameter. Therefore, the field diameter of the cardioid fatigue failure region can be redefined as:

$$\begin{cases} 2a = r \\ r(\theta) = a(1-\cos\theta) \end{cases} \quad (13)$$

2.4.3. Simplification of the Advanced SFI Approach

In the traditional SFI approach, the function of angle θ is expressed as (1 + sinθ). In addition, the contributions of different stress contours to fatigue failure require different field diameters; however, none of these contributions exceed the corresponding stress values. Therefore, the function of angle θ can be redefined as:

$$f(\theta) = \cos\theta \quad (14)$$

In addition, based on Equations (8), (11)–(14), the field-intensity formula can be expressed as:

$$\sigma_{FI} = \sigma_{\max} - \frac{\int\limits_{\Omega} (\sigma_{\max} - \sigma_{r-\theta}) \left[\left| \frac{1}{\sigma_{\max}} \frac{d\sigma_{r-\theta}}{dr} \right| r(\theta)\cos(\theta) \right]^{\frac{r(\theta)}{G}} dv}{\int\limits_{\Omega} \left[\left| \frac{1}{\sigma_{\max}} \frac{d\sigma_{r-\theta}}{dr} \right| r(\theta)\cos(\theta) \right]^{\frac{r(\theta)}{G}} dv} \quad (15)$$

$$\begin{cases} I - crescent : r(\theta) = (2\sin^2\frac{\theta}{2})(\sqrt{r^2 + 2r\Delta r} - r) + r \\ II - cardioid : r(\theta) = \frac{r}{2}(1-\cos\theta) \end{cases} \quad (16)$$

Considering the symmetry of a notched specimen, the simplified field-intensity calculation formula can be expressed as:

$$\sigma_{FI} = \sigma_{\max} - \frac{\int_0^{\infty}\int_0^{\frac{\pi}{2}} (\sigma_{\max} - \sigma_r) \left[\left| \frac{1}{\sigma_{\max}} \frac{d\sigma_r}{dr} \right| r(\theta)\cos(\theta) \right]^{\frac{r(\theta)}{G}} d\theta dr}{\int_0^{\infty}\int_0^{\frac{\pi}{2}} \left[\left| \frac{1}{\sigma_{\max}} \frac{d\sigma_r}{dr} \right| r(\theta)\cos(\theta) \right]^{\frac{r(\theta)}{G}} d\theta dr} \quad (17)$$

3. Materials and Experiment

3.1. Specimen Geometry

The geometry and dimensions of the smooth solid-bar specimen used for the S-N curve test are presented in Figure 11a. In addition, three notched specimens of different elastic SCFs (kt) were designed to verify the reliability of the new approach, and their geometries and dimensions are presented in Figure 11b–d. All fatigue experiments were conducted on test equipment (Figure 12).

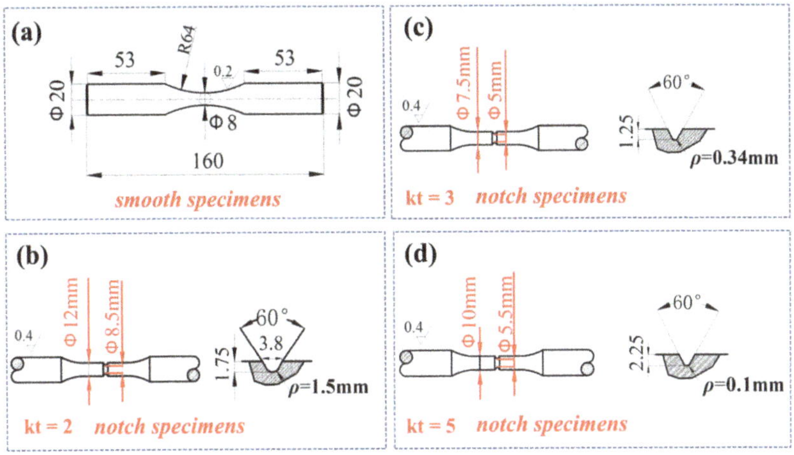

Figure 11. Geometries and dimensions of test specimens (in mm) used for S-N curve and notch specimens tests: (**a**) smooth specimen, and (**b**) circumferentially notched solid-bar specimen with kt = 2, (**c**) kt = 3, and (**d**) kt = 5.

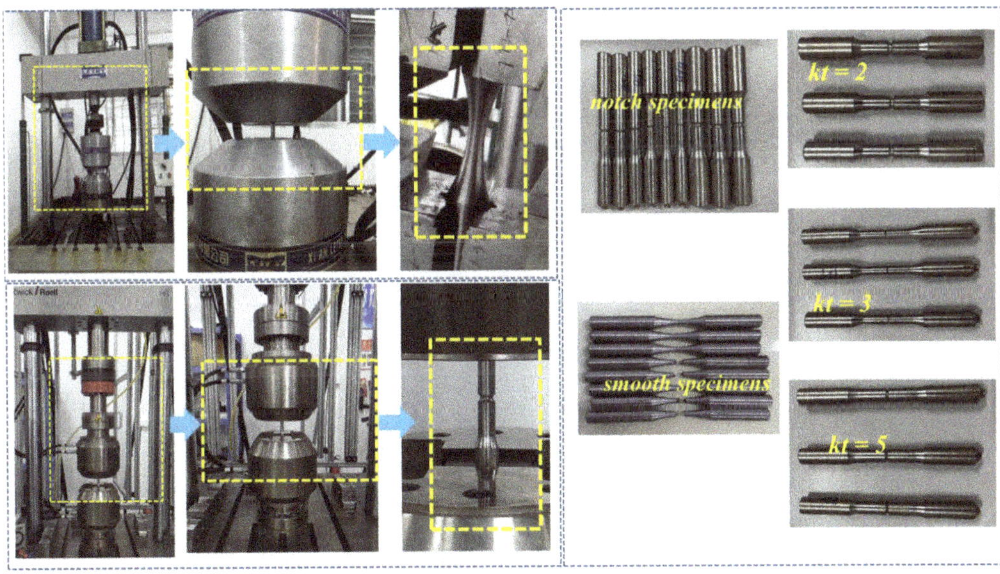

Figure 12. Illustration of the specimen clamping mechanical testing system.

3.2. S-N Curve Test Procedure

3.2.1. Static Mechanical Properties of Test Material

According to the GB/T 3075-2008 standard and ISO 6892:1998, the mechanical properties of the test material (40CrNi$_2$MoVA) were tested prior to the S-N curve experiment (Table 3).

Table 3. Static mechanical properties of nickel-chromium alloy.

Material	Serial Number	Young's Modulus	Poisson's Ratio	Yield Stress (s_Y)	Ultimate Tensile Strength (σ_b)
40Cr Ni_2MoV	1	204 GPa	0.3	980 MPa	1060 MPa
40Cr Ni_2MoV	2	204 GPa	0.3	980 + 3 MPa	1060 + 3 MPa
40Cr Ni_2MoV	3	204 GPa	0.3	980 − 2 MPa	1060 − 3 MPa

3.2.2. Uniaxial Test

Uniaxial fatigue tests were conducted on a PLD-300 electro-hydraulic servo fatigue testing machine (Figure 12), and an axial tensile sinusoidal waveform load was selected to control the process. The stress ratio ($R = -1$) was constant under the varying frequencies of 10–30 Hz. Specimens' buckling was reduced to a minimum because the shapes of measurement distances were represented by hourglass sections and notches that become the weakest regions. It means fatigue damages have occurred in the mentioned zones, while avoiding the other ones (Figure 12). According to the value of ultimate tensile strength (σ_b), 35 specimens were divided into 7 groups with different stress levels. According to the GB/T 24176-2009 standard and ISO 12107:2003, the confidence level of the S-N curve of 40CrNi_2MoV was obtained as 50%. N_f is the fatigue life recorded for each experiment. S-N curve fitting was carried out by using the two-parameter method, and the fitting formula can be expressed as:

$$\sigma^m \times N = C \tag{18}$$

where m and C are fitted values, σ is the test stress, and N is the number of cycles to fracture.

Results of the S-N curve experiment are presented in Table 4, and the fitted S-N curve is plotted in Figure 13.

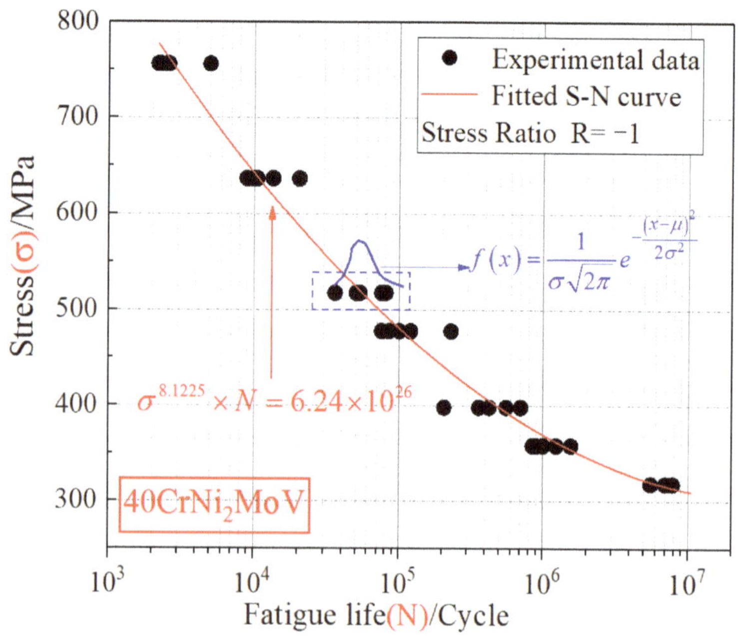

Figure 13. Fatigue data and fitted S-N curve for the pumphead material.

Table 4. Empirical data of different fatigue test groups for the pumphead material.

Material	Stress (S_{max}, MPa)	N_f (Life)					Median ($\bar{x}$)
	Level/count	1	2	3	4	5	——
40CrNi$_2$MoV	756.3	1037	1055	980	957	1123	1030.4
	716.5	5037	4958	5968	7853	6842	6131.6
	636.9	35,600	25,500	20,360	31,186	32,100	28,949.2
	517.5	110,200	175,100	125,900	140,500	152,900	140,920
	437.8	403,100	285,600	605,800	367,500	702,600	472,920
	398.1	321,580	715,600	456,800	555,200	985,600	606,956
	318.4	5,569,800	6,771,100		7,865,200		6.735×10^6

3.3. S-N Curve Test of Notched Specimens

S-N curve tests of three different notched specimens were conducted on a Zwick HB250 fatigue testing machine Figure 12. Based on the actual working condition of the pumphead, axial tensile sinusoidal waveform loading with R = −1 was selected to control the testing machine. The loading conditions and fatigue experimental results of the notched specimens are presented in Table 5.

Table 5. Cyclic loading parameters and results of fatigue test for 40CrNi$_2$MoV notched specimens.

Specimen No.	*kt*	*F (kN)*	*Frequency (Hz)*	*Nf (Cycles)*	*Rσ*	*Median Life*
1-1	2	20 kN	10.0	62,724	−1	
1-2	2	20 kN	10.0	68,563	−1	65,215
1-3	2	20 kN	10.0	64,359	−1	
2-1	2	19 kN	10.0	122,133	−1	
2-2	2	19 kN	10.0	134,572	−1	134,011
2-3	2	19 kN	10.0	145,328	−1	
3-1	2	17 kN	10.0	363,235	−1	
3-2	2	17 kN	10.0	384,527	−1	390,452
3-3	2	17 kN	10.0	423,596	−1	
4-1	3	10 kN	10.0	6680	−1	
4-2	3	10 kN	10.0	6899	1	6949
4-3	3	10 kN	10.0	7268	−1	
5-1	3	7 kN	10.0	85,362	−1	
5-2	3	7 kN	10.0	81,237	−1	82,295
5-3	3	7 kN	10.0	80,198	−1	
6-1	3	5.5 kN	10.0	638,340	−1	
6-2	3	5.5 kN	10.0	534,568	−1	576,630
6-3	3	5.5 kN	10.0	556,982	−1	
7-1	5	8 kN	10.0	36,024	−1	
7-2	5	8 kN	10.0	45,396	−1	37,226
7-3	5	8 kN	10.0	30,258	−1	

Table 5. *Cont.*

Specimen No.	*kt*	*F (kN)*	*Frequency (Hz)*	*Nf (Cycles)*	*Rσ*	*Median Life*
8-1	5	7.5 kN	10.0	104,417	−1	
8-2	5	7.5 kN	10.0	109,853	−1	109,131
8-3	5	7.5 kN	10.0	110,123	−1	
9-1	5	6 kN	10.0	190,823	−1	
9-2	5	6 kN	10.0	198,095	−1	192,826
9-3	5	6 kN	10.0	189,562	−1	

4. Results and Discussion

4.1. Fatigue Life Verification of Notched Specimens

In Section 2, the theoretical SFI approach and the numerical simulation method were introduced. According to experimental loading conditions, three different stress levels were set for each notched specimen prior to the elastic–plastic finite element analysis. The static structural module of ANSYS workbench software was used for numerical calculations. To ensure the accuracy of FEA results, meshes at the notch root of the specimens were refined. In Section 2.4, two different geometrical shapes of the fatigue failure region (crescent and cardioid) were proposed; thus, the improved SFI approach was also divided into two models (Equations (12) and (13)). Extracting stress distribution data along the symmetry line (focus path) of the notched specimens, stress distributions and stress-field intensities were calculated by the novel SFI approach (Figure 14). The stress gradient decreased rapidly in highly stressed volumes, and the "inhibition" gradient presented a similar trend (Figure 14a). However, with the increase of the field diameter, the "inhibition" gradient gradually approached zero and the stress-field intensity gradually changed to a constant value. The same trend was noticed for the other loading cases, and the corresponding results are presented in Figure 14b and Table 6.

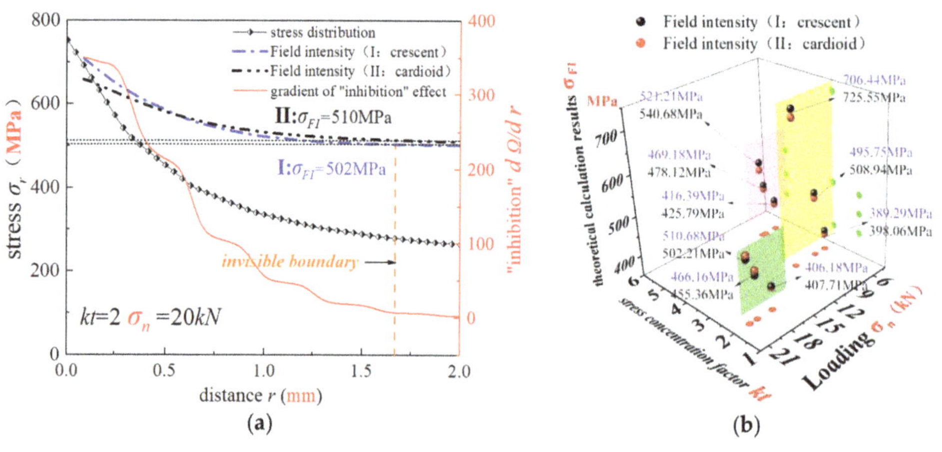

Figure 14. Analysis of stress fields near the notch by the simplified and improved SFI approach: (**a**) loading result for kt = 2 and $\sigma_n = 20kN$, and (**b**) theoretical calculation results for 9 different cases.

Table 6. Theoretical results of fatigue life for three notched specimens, calculated by four approaches.

Stress Concentration Factor	F/kN	Median Life	Novel SFI Approach I		Novel SFI Approach II		Traditional SFI Approach		Traditional TCD (LM)	
			σ_{FI}/ MPa	N_f/ Cycles	σ_{FI}/ MPa	N_f/ Cycles	σ_{FI}/ MPa	N_f/ Cycles	σ_{mean}/ MPa	N_f/ Cycles
kt = 2	20 kN	65,215	502.21	71,072	510.68	63,525	558.3	30,468	572.3	24,911
	19 kN	134,011	455.36	157,695	466.16	132,202	509.5	64,033	529.5	46,846
	17 kN	390,452	407.71	397,452	406.18	404,984	453.2	166,555	482.6	100,499
kt = 3	10 kN	6949	725.55	3649	706.44	4425	762.3	2428	809.2	1497
	7 kN	82,295	508.94	66,667	495.75	80,965	561.3	29,164	592.4	18,966
	5.5 kN	576,630	398.06	476,335	389.29	573,213	452.1	157,535	473.8	117,128
kt = 5	8 kN	37,226	540.68	39,948	521.21	53,424	596.7	17,747	611.5	14,545
	7.5 kN	109,131	478.02	107,568	469.18	125,477	530.4	46,194	562.9	28,506
	6 kN	192,826	425.79	279,319	416.39	332,268	472.9	117,319	498.6	77,084

Based on the aforementioned principle, two other approaches (traditional SFI approach and TCD method) were chosen to calculate the fatigue life of the notched specimens. In comparison to the traditional SFI approach and the TCD method, the novel SFI approach had the narrowest error band (Figure 14), indicating its higher fatigue life prediction accuracy. The error mean absolute percent index parameter (μ) was expressed as:

$$\mu = \frac{1}{n}\sum_{i=1}^{n}\left(\left|\frac{N_P - N_E}{N_E}\right| \times 100\%\right) \tag{19}$$

where N_P is the predicted life and N_E is the experimental life. The index parameters of the notched specimens are presented in Figure 15.

The fatigue life prediction accuracies of different approaches were quantitatively analyzed. The novel SFI approach II had the highest fatigue life prediction accuracy for the kt = 2 notched specimen. For the kt = 2 notched specimen, the deviations between the novel SFI approach II-based calculation results and the experimental fatigue life values under 20, 19, and 17 kN were 2.59%, 1.35%, and 3.72%, respectively. For the kt = 3 notched specimen, the deviations between the novel SFI approach II-based calculation results and the experimental fatigue life values under 10, 7, and 5.5 kN were 36.2%, 1.61%, and 0.59%, respectively. For the kt = 5 notched specimen, the deviations between the novel SFI approach II-based calculation results and the experimental fatigue life values under 8, 7.5, and 6 kN were 43.5%, 14.9%, and 72.3%, respectively. Similarly, for the kt = 2 notched specimen, the deviations between the novel SFI approach II-based calculation results and the experimental fatigue life values under 20, 19, and 17 kN were 8.9%, 17.6%, and 1.79%, respectively. For the kt = 3 notched specimen, the deviations between the novel SFI approach II-based calculation results and the experimental fatigue life values under 10, 7, and 5.5 kN were 47.4%, 18.9%, and 17.3%, respectively. For the kt = 5 notched specimen, the deviations between the novel SFI approach II-based calculation results and the experimental fatigue life values under 8, 7.5, and 6 kN were 7.3%, 1.43%, and 44.8%, respectively.

Therefore, when the stress concentration factors were 2 and 3, the SFI approach II was more accurate in predicting fatigue life; however, when the stress concentration factor was 5, the SFI approach I had better fatigue life prediction efficiency than the SFI approach II. In terms of the absolute percent index parameter (μ), the SFI approach I (25.76%) was more accurate than the SFI approach II; thus, the SFI approach I was more suitable for the fatigue life prediction of sharp notched specimens.

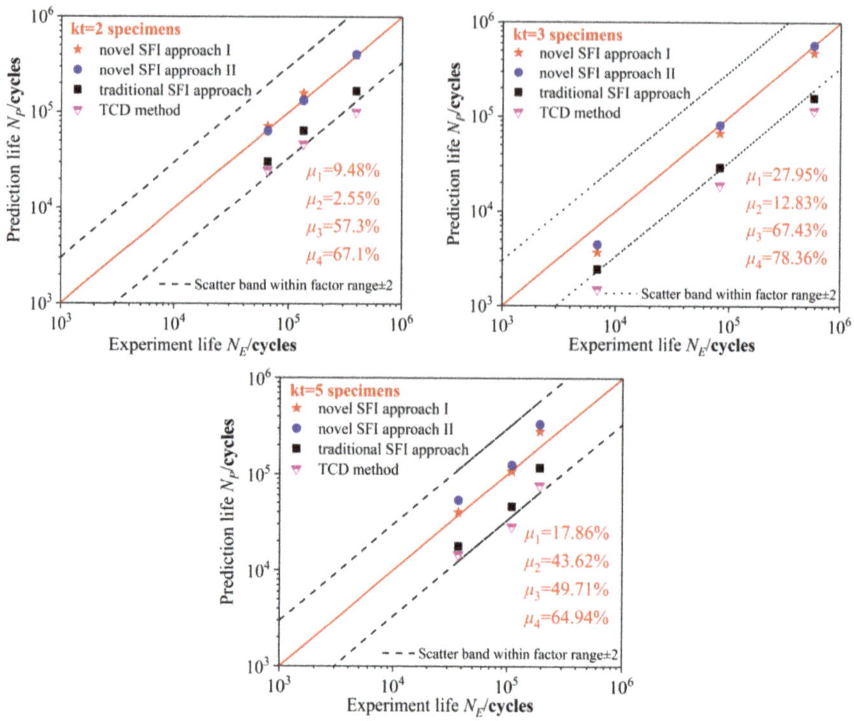

Figure 15. Index parameters of notched specimens calculated by four different approaches.

The deviations between fatigue life values calculated by the traditional SFI approach and the TCD method under different loading conditions were approximately 70%. Therefore, both the novel SFI approaches I and II are highly accurate in predicting high-cycle fatigue life.

4.2. Fatigue Life Verification of Pumphead

To further verify the fatigue life prediction accuracy of the novel SFI approach for the pumphead, numerical simulations for stress distribution and fatigue life were performed in ANSYS workbench software (Figure 16).

For extracting stress distribution data along the symmetry line (focus path) of the pumphead cross-bore, field intensities under four different working conditions were calculated by the novel SFI approach II. The gradient value of "inhibition action" was calculated as shown in Figure 17. As shown in Figure 17, the slope of the "inhibition" effect ($d\Omega/dr$) is close to 0 in almost all the four working conditions when R is close to 1.5 mm. With the increase of the field diameter, the value of field strength does not alter and steadily becomes constant. It can be judged that the damage in this area can be defined as the effective damage area of the ultra-high-pressure pumphead body. The field intensity under 65, 70, 80, and 105 MPa was 318, 340, 389, and 518 MPa, respectively. Further, substituting these four field strengths into the S-N curve of the initial pumphead material, the fatigue life was calculated by the new method.

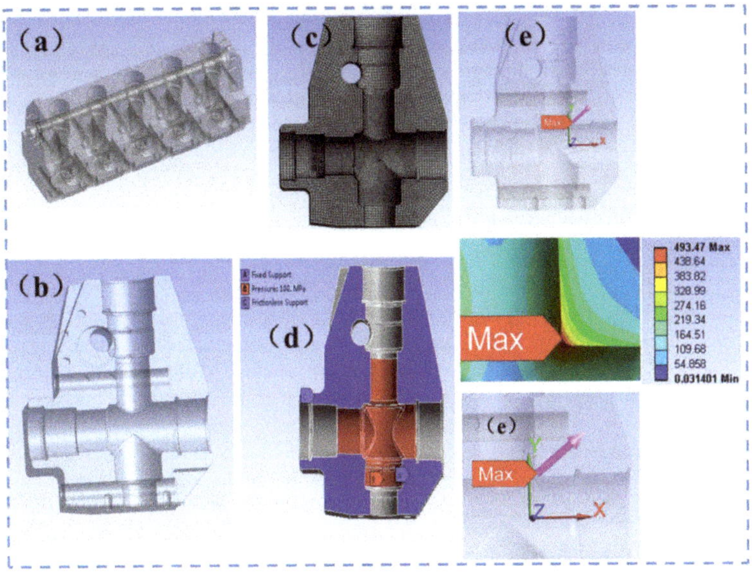

Figure 16. (**a**) Analysis model of the hydraulic part of the fracturing pump, (**b**) 1/2 inner cavity of the pumphead, (**c**) typical mesh structure, (**d**) boundary conditions for numerical simulations, and (**e**) maximum stress contour and focus path.

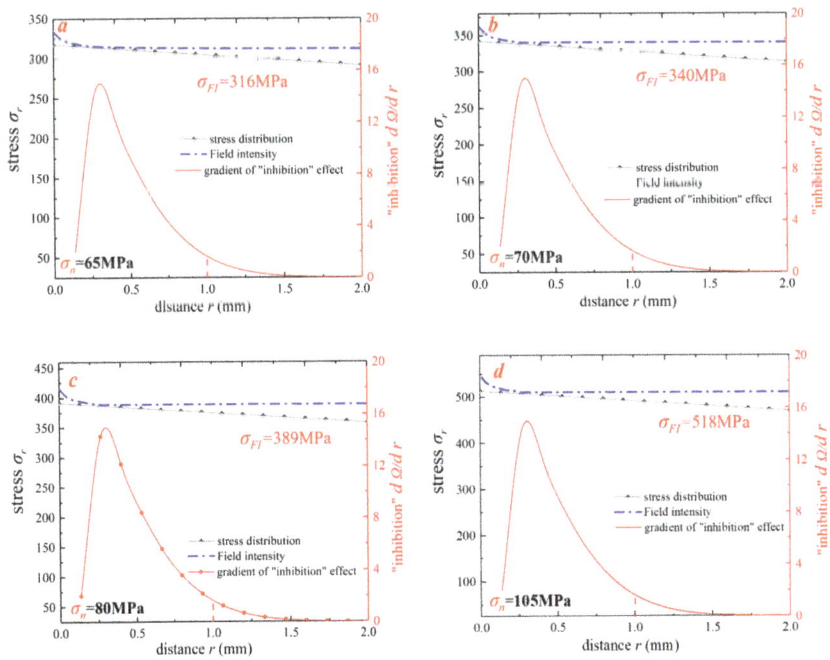

Figure 17. Simplified novel SFI method to analyze the stress field near the concentration point of the intersecting line under the four load conditions: (**a**) 65 MPa, (**b**) 70 MPa, (**c**) 80 MPa, and (**d**) 105 MPa.

According to the actual working conditions of the fracturing pump, the loading frequency of the circular cross-bore was 110 cycles/min. Owing to the component size, material differences, and loading distribution, even under the same operating conditions, the fatigue life of pressure vessels becomes dispersed; thus, fatigue reliability methods are found to be better for data processing [45,46]. To further verify the new method proposed in this paper, the nominal stress method commonly used in engineering was used for comparison. The mathematical definition of the nominal stress method is given in Equation (20):

$$\sigma_n = \frac{\sigma_a}{K_T}\varepsilon\beta C_L \tag{20}$$

where σ_n is the stress of the mechanical structural part in the structural S-N curve, σ_a is the stress of material in the S-N curve, K_T is the stress concentration factor, ε is the size factor, β is the surface processing coefficient, and C_L is the loading coefficient.

In this paper, the value of the size factor, ε, is 0.9, because the specimens are taken from the same nickel-chromium alloy material, so the value of the surface processing coefficient, β, is 1. The loading method of the fracturing pump is the same as in the case of the fatigue test, so the value of the loading coefficient is 1. In addition, based on Equations (18) and (20) and the S-N curve, the fatigue life prediction equation can be expressed as:

$$(4.97\sigma_n)^{8.1225} \cdot N = 6.24 \times 10^{26} \tag{21}$$

Therefore, in 65, 70, 80, and 105 MPa working conditions, combined with the S-N curves, fatigue life is calculated by the SFI approach and NSM, respectively. The fatigue life values of the pumphead under different conditions are displayed in Figure 18. It is evident that the novel SFI approach had the narrowest error band, indicating its higher fatigue life prediction accuracy for the pumphead.

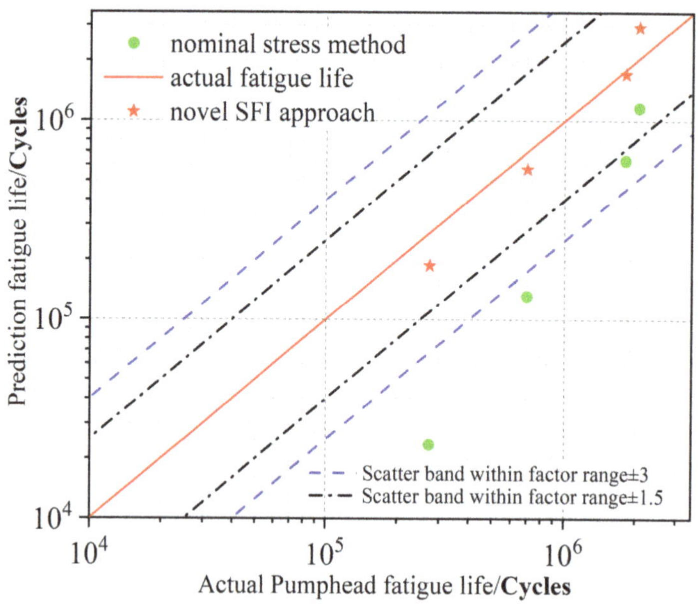

Figure 18. Pumphead fatigue life prediction accuracies of NSM and the novel SFI approach.

5. Conclusions

Based on the theories of the critical distance method and the SFI approach, criteria for defining the fatigue damage failure region of the SFI approach were revisited. The proposed novel SFI approach manifested high fatigue life prediction accuracy for notch specimens and pumpheads. The main findings of this work are summarized below:

(1) Based on numerical simulation results and the concept of the critical volume method, an improved weight function accounting for the fatigue damage failure geometries of cardioid and crescent areas was proposed to improve fatigue life prediction accuracy.

(2) Based on the concept of the "invisible boundary", the effective critical distance for crescent and cardioid areas of the fatigue failure region was revised and simplified. The novel SFI approach was modified considering the effect of grain size.

(3) To verify the fatigue life prediction effectiveness of the novel SFI approaches for notch specimens, fatigue tests were conducted on three different notched bar specimens. The novel SFI approach II had higher fatigue life prediction accuracy.

(4) In comparison with the SFI approach II, the SFI approach I was more suitable for fatigue life prediction of sharply notched specimens.

Nomenclatures

$\sigma_{\max}$	Maximum stress	k_t	Stress concentration factor
σ_{FI}	Stress-field intensity	k_f	Fatigue notch factor
r	Distance to the peak stresspoint	$\overline{\chi}$	Stress gradient
$f(\sigma_{ij})$	Equivalent stress function	χ	Relative stress gradient
$\varphi(\vec{r})$	Weight function	σ_0	Nominal stress
$\varphi(r,\theta)$	Weight function with distance r and deviation angle θ	ε_a	Strain amplitude
σ_a	Stress amplitude	R	Field diameter
σ_n	The load of component weakest region	σ_U	Ultimate stress
σ_b	Ultimate tensile strength	σ_{mean}	Effective stress
ΔK_{th}	Range of threshold value for fatigue crack propagation	$\Delta\sigma_{-1}$	Range of the fully reverse axial fatigue limit
σ_{mean}	Equivalent stress of TCD	σ_1	The maximum principal stress
S_Y	Yield stress	G	Grain size function
$\xi(\Omega)$	Obstructive effect	θ	Angular coordinates in Cartesian coordinates
$\sigma_{yy}(r)$	Fatigue crack-opening stress	L_0	Critical distance
ν	Poisson's ratio		

Abbreviations

SFI	Stress-field intensity	TCD	Theory of critical distance method
LFEM	Linear elastic fracture mechanics	EPFM	Elastic–plastic fracture mechanics
NSM	Nominal stress method	LSSM	Local stress–strain method

Author Contributions: The author H.W. provided help in the preliminary investigation of this article. N.L. provided resources such as experimental equipment. M.L. Completed the numerical simulation. Y.Z. (Yang Zhou) conducted in-depth research on the evaluation the fatigue life of pump heads, then proposed the concept of fatigue failure region. Y.Z. (Yun Zeng) was in charge of the entire research experiment, statistics all the data and sorted it out. Finally wrote and revised the manuscript. All authors have read and agreed to the published version of the manuscript.

Funding: This research received no external funding.

Institutional Review Board Statement: Not applicable.

Informed Consent Statement: Not applicable.

Data Availability Statement: All data are presented in the article.

Acknowledgments: This project was supported financially by the National Natural Science Foundation of China (No. 52174018 and No. 51904181), CNPC Innovation Foundation 2020D-5007-0503, the Project of case Teaching for graduate degree of Yangtze University (No. YAL202106, Application of engineering software analysis), and State Key Science and Technology Projects (No. 2016ZX05038-001-LH002). The authors would like to thank all the members of the project team for their support.

Conflicts of Interest: We declare that we have no financial and personal relationships with other people or organizations that can inappropriately influence our work, there is no professional or other personal interest of any nature or kind in any product, service and/or company that could be construed as influencing the position presented in, or the review of, the manuscript entitled, "Experiment and Theoretical Investigation on Fatigue Life Prediction of Fracturing Pumpheads Based on a Novel Stress-Field Intensity Approach".

References

1. Zhu, W.B. Research on Reciprocating Sealing Performance and Mechanism of Fracturing Pump. *Nat. Gas Industry* **2006**, *60–62*, 161–162.
2. Bokane, A.; Stewart, M.; Sanda, R.; Babar, D.; Katke, N. Personal Safety and Human Factors Risk Assessment during Maintenance of Fracturing Pumps at Well Service Operations. In Proceedings of the International Petroleum Technology Conference, Kuala Lumpur, Malaysia, 10–12 December 2014.
3. Badr, E.A.; Sorem, J.R.; Tipton, S.M.; Yang, S. An analytical procedure for estimating residual stresses in blocks containing crossbores. *Int. J. Press. Vessel. Pip.* **2000**, *77*, 737–749. [CrossRef]
4. Zanetti, M.; Babini, V.; Meneghetti, G. Definition of nominal stress-based FAT classes of complex welded steel structures using the Peak Stress Method. *Procedia Struct. Integr.* **2019**, *19*, 627–636. [CrossRef]
5. Fomichev, P.A.; Zarutskii, A.V. Fatigue Life Prediction for Specimens with an Open Hole with a Pre-Compressed Boundary Via the Nominal Stresses Under Regular Cyclic Loading. *Strength Mater.* **2019**, *51*, 746–752. [CrossRef]
6. Zhu, S.-P.; He, J.-C.; Liao, D.; Wang, Q.; Liu, Y. The effect of notch size on critical distance and fatigue life predictions. *Mater. Des.* **2020**, *196*, 109095. [CrossRef]
7. He, J.-C.; Zhu, S.-P.; Liao, D. Probabilistic fatigue assessment of notched components under size effect using critical distance theory. *Eng. Fract. Mech.* **2020**, *235*, 107150. [CrossRef]
8. Liao, D.; Zhu, S.-P.; Correia, J.A.; De Jesus, A.M.; Berto, F. Recent advances on notch effects in metal fatigue: A review. *Fatigue Fract. Eng. Mater. Struct.* **2020**, *43*, 637–659. [CrossRef]
9. Susmel, L.; Taylor, D. The Theory of Critical Distances to estimate lifetime of notched components subjected to variable amplitude uniaxial fatigue loading. *Int. J. Fatigue* **2011**, *33*, 900–911. [CrossRef]
10. Adib, H.; Gilgert, J.; Pluvinage, G. Fatigue life duration prediction for welded spots by volumetric method. *Int. J. Fatigue* **2004**, *26*, 81–94. [CrossRef]
11. Esmaeili, F.; Zehsaz, M.; Chakherlou, T.N. Investigation the effect of tightening torque on the fatigue strength of double lap simple bolted and hybrid (bolted–bonded) joints using volumetric method. *Mater. Des.* **2014**, *63*, 349–359. [CrossRef]
12. Sobek, J. Proper Nominal Stress Distribution Subjected to Combination of Wedge-Splitting and Bended Geometry Tests. *Procedia Eng.* **2017**, *190*, 406–413. [CrossRef]
13. Ostash, O.P.; Panasyuk, V.V. Fatigue process zone at notches. *Int. J. Fatigue* **2001**, *23*, 627–636. [CrossRef]
14. Susmel, L.; Taylor, D. An elasto-plastic reformulation of the theory of critical distances to estimate lifetime of notched components failing in the low/medium-cycle fatigue regime. *J. Eng. Mater. Technol.* **2010**, *132*, 1–8. [CrossRef]
15. Susmel, L.; Meneghetti, G.; Atzori, B. A simple and efficient reformulation of the classical manson–coffin curve to predict lifetime under multiaxial fatigue loading-Part II: Notches. *J. Eng. Mater. Technol.* **2009**, *131*, 1–9. [CrossRef]
16. Tanaka, K. Engineering formulae for fatigue strength reduction due to crack-like notches. *Int. J. Fract.* **1983**, *22*, 39–46. [CrossRef]
17. Taylor, D. Analysis of fatigue failures in components using the theory of critical distances. *Eng. Fail. Anal.* **2005**, *12*, 906–914. [CrossRef]
18. El Haddad, M.H.; Dowling, N.E.; Topper, T.H.; Smith, K.N. *J* integral applications for short fatigue cracks at notches. *Int. J. Fract.* **1980**, *16*, 15–30. [CrossRef]
19. Susmel, L.; Taylor, D. A novel formulation of the theory of critical distances to estimate lifetime of notched components in the medium-cycle fatigue regime. *Fatigue Fract. Eng. Mater. Struct.* **2007**, *30*, 567–581. [CrossRef]
20. Susmel, L.; Taylor, D. The Modified Wöhler Curve Method applied along with the Theory of Critical Distances to estimate finite life of notched components subjected to complex multiaxial loading paths. *Fatigue Fract. Eng. Mater. Struct.* **2008**, *31*, 1047–1064. [CrossRef]
21. Susmel, L.; Taylor, D. Estimating lifetime of notched components subjected to variable amplitude fatigue loading according to the elastoplastic theory of criticaldistances. *J. Eng. Mater. Technol.* **2015**, *137*, 1–15. [CrossRef]
22. Qylafku, G.; Azari, Z.; Kadi, N.; Gjonaj, M.; Pluvinage, G. Application of a new model proposal for fatigue life prediction on notches and key-seats. *Int. J. Fatigue* **1999**, *21*, 753–760. [CrossRef]
23. Qilafku, G.; Kadi, N.; Dobranski, J.; Azari, Z.; Gjonaj, M.; Pluvinage, G. Fatigue of specimens subjected to combined loading: Role of hydrostatic pressure. *Int. J. Fatigue* **2001**, *23*, 689–701. [CrossRef]
24. Adib-Ramezani, H.; Jeong, J.; Pluvinage, G. Structural integrity evaluation of X52 gas pipes subjected to external corrosion defects using the SINTAP procedure. *Int. J. Press. Vessel. Pip.* **2006**, *83*, 420–432. [CrossRef]
25. Meneghetti, G.; Susmel, L.; Tovo, R. High-cycle fatigue crack paths in specimens having different stress concentration features. *Eng. Fail. Anal.* **2007**, *14*, 656–672. [CrossRef]

26. Ostash, O.P.; Panasyuk, V.V.; Kostyka, E.M. Phenomenological model of fatigue macrocrack initiation near stress concentrators. *Fatigue Fract. Engng. Mater. Struct.* **1999**, *22*, 161–172. [CrossRef]
27. Nicholls, D.J.; Martin, J.W. An examination of the reasons for the discrepancy be-tween long and small fatigue cracks in Al-Li alloys. *Mat. Sci. Eng. A* **1990**, *128*, 141–145. [CrossRef]
28. Pineau, A.G.; Pelloux, R.M. Influence of strain-induced martensitic transformations on fatigue crack growth rates in stainless steels. *Met. Trans.* **1974**, *5*, 1103–1112. [CrossRef]
29. Saxena, A.; Antolovich, S.D. Low cycle fatigue, fatigue crack propagation and substructures in a series of polycrystalline cu-Al alloys. *Met. Trans.* **1975**, *6*, 1809–1828. [CrossRef]
30. Park, H.B.; Kim, K.M.; Lee, B.W. Plastic zone size in fatigue cracking. *Int. J. Press. Vessel. Pip.* **1996**, *68*, 279–285. [CrossRef]
31. Bathias, C.; Pelloux, R.M. Fatigue crack propagation in martensitic and austenitic steels. *Met. Trans.* **1973**, *4*, 1265–1273. [CrossRef]
32. Chapetti, M.D.; Miyatab, H.; Tagawab, T.; Miyatab, T.; Fujiokac, M. Fatigue crack propagation behavior in ultra-fine grained low carbon steel. *Int. J. Fatigue* **2005**, *27*, 235–243. [CrossRef]
33. Edmunds, T.M.; Willis, J.R. Matched asymptotic expansions in nonlinear fracture mechanics—III. In-plane loading of an elastic perfectly-plastic symmetric specimen. *J. Mech. Phys. Solids* **1977**, *25*, 423–455. [CrossRef]
34. Taddesse, A.T.; Zhu, S.-P.; Liao, D.; Huang, H.-Z. Cyclic plastic zone modified critical distance theory for notch fatigue analysis of metals. *Eng. Fail. Anal.* **2021**, *121*, 105163. [CrossRef]
35. Taddesse, A.T.; Zhu, S.-P.; Liao, D.; Keshtegar, B. Cyclic plastic zone-based notch analysis and damage evolution model for fatigue life prediction of metals. *Mater. Des.* **2020**, *191*, 108639. [CrossRef]
36. Wu, Y.-L.; Zhu, S.-P. Assessment of notch fatigue and size effect using stress field intensity approach. *Int. J. Fatigue* **2021**, *149*, 106279. [CrossRef]
37. Ai, Y.; Zhu, S.-P.; Liao, D. Probabilistic modelling of notch fatigue and size effect of components using highly stressed volume approach. *Int. J. Fatigue* **2019**, *127*, 110–119. [CrossRef]
38. Liao, D.; Zhu, S.-P.; Keshtegar, B.; Qian, G.; Wang, Q. Probabilistic framework for fatigue life assessment of notched components under size effects. *Int. J. Mech. Sci.* **2020**, *181*, 105685. [CrossRef]
39. Yao, W. Stress field intensity approach for predicting fatigue life. *Int. J. Fatigue* **1993**, *15*, 243–245.
40. Neuber, H. *Theory of Notch Stress: Principles for Exact Calculation of Strength with Reference to Structural Form and Material*, 2nd ed.; Springer: Berlin/Heidelberg, Germany, 1958.
41. Peterson, R.E.; Sines, G.; Waisman, J.L. (Eds.) *Notch Sensitivity*; Metal Fatigue; McGraw Hill: New York, NY, USA, 1959; pp. 293–306.
42. Berto, F.; Lazzarin, P. The volume-based Strain Energy Density approach applied to static and fatigue strength assessments of notched and welded structures. *Procedia Eng.* **2009**, *1*, 155–158. [CrossRef]
43. Zhao, B.; Song, J.; Xie, L.; Hu, Z.; Chen, J. Surface roughness effect on fatigue strength of aluminum alloy using revised stress field intensity approach. *Sci. Rep.* **2021**, *11*, 19279. [CrossRef]
44. Zeng, Y.; Li, M.; Zhou, Y.; Li, N. Development of a new method for estimating the fatigue life of notched specimens based on stress field intensity. *Theor. Appl. Fract. Mech.* **2019**, *104*, 102339. [CrossRef]
45. Niu, X.-P.; Zhu, S.-P.; He, J.-C.; Shi, K.; Zhang, L. Fatigue reliability design and assessment of reactor pressure vessel structures: Concepts and validation. *Int. J. Fatigue* **2021**, *153*, 106524. [CrossRef]
46. Li, X.-K.; Zhu, S.-P.; Liao, D.; Correia, J.A.F.O.; Berto, F.; Wang, Q. Probabilistic fatigue modelling of metallic materials under notch and size effect using the weakest link theory. *Int. J. Fatigue* **2022**, *159*, 106788. [CrossRef]
47. Zhao, B.; Xie, L.; Song, J.; Ren, J.; Wang, B.; Zhang, S. Fatigue life prediction of aero-engine compressor disk based on a new stress field intensity approach. *Int. J. Mech. Sci.* **2020**, *165*, 105190. [CrossRef]
48. Zhao, H.; Liu, J.; Hua, F.; Ran, Y.; Zi, R.; Li, B. Multiaxial fatigue life prediction model considering stress gradient and size effect. *Int. J. Press. Vessel. Pip.* **2022**, *199*, 104703. [CrossRef]

materials

MDPI

Article

Further Investigations and Parametric Analysis of Microstructural Alterations under Rolling Contact Fatigue

Muhammad Usman Abdullah and Zulfiqar Ahmad Khan *

NanoCorr, Energy & Modelling (NCEM) Research Group, Department of Design & Engineering, Bournemouth University, Poole BH12 5BB, UK

* Correspondence: zkhan@bournemouth.ac.uk

Abstract: Bearing elements under rolling contact fatigue (RCF) exhibit microstructural features, known as white etching bands (WEBs) and dark etching regions (DERs). The formation mechanism of these microstructural features has been questionable and therefore warranted this study to gain further understanding. Current research describes mechanistic investigations of standard AISI 52100 bearing steel balls subjected to RCF testing under tempering conditions. Subsurface analyses of RCF-tested samples at tempering conditions have indicated that the microstructural alterations are progressed with subsurface yielding and primarily dominated by thermal tempering. Furthermore, bearing balls are subjected to static load tests in order to evaluate the effect of lattice deformation. It is suggested from the comparative analyses that a complete rolling sequence with non-proportional stress history is essential for the initiation and progression of WEBs, supported by the combination of carbon flux, assisted by dislocation and thermally activated carbon diffusion. These novel findings will lead to developing a contemporary and new-fangled prognostic model applied to microstructural alterations.

Keywords: bearing steel; diffusion; microstructural alterations; rolling contact fatigue; tempering

Citation: Abdullah, M.U.; Khan, Z.A. Further Investigations and Parametric Analysis of Microstructural Alterations under Rolling Contact Fatigue. *Materials* **2022**, *15*, 8072. https://doi.org/10.3390/ma15228072

Academic Editors: Jaroslaw Galkiewicz, Lucjan Śnieżek and Sebastian Lipiec

Received: 2 October 2022
Accepted: 11 November 2022
Published: 15 November 2022

1. Introduction

Automotive components subjected to severe rolling cyclic loadings not only result in deterioration of interacting surfaces [1,2] but also cause subsurface microstructural damage [3,4]. This subsurface damage is pronounced in bearing steel under rolling contact fatigue (RCF) where the parent microstructure is evolved resulting in formation of microstructural alterations. The subsurface microstructural alterations can be observed under the optical microscope after Nital etching as a ring profile or semi-elliptical shape, depending on the plane of cut. The primary features involved in the altered microstructure are white etching bands (WEBs) preceded by dark etching regions (DERs) [4,5]. It has been reported that the microstructural features are formed due to the decay of parent martensite where carbon migrates within the matrix to form carbon-rich and carbon-depleted areas [6,7]. Extensive numerical and experimental research is available in the literature on bearing steel evolution and microstructural features; however, a comprehensive understanding is still needed to comprehend the general (physical and chemical) formation mechanism of DERs/WEBs.

DERs are considered to be an over-tempered form of martensite constituting a mixture of decayed microstructure, ferrite, and remaining martensite [5]. DERs are governed by the subsurface stress fields and their structure consists of deformed and misoriented martensite plates including reprecipitation of nano carbides, resulting in over-tempering of the microstructure [8]. WEBs are reported as disc-shaped ferrite plates surrounded by carbon-rich areas (also known as lenticular carbides, LCs) [9]. The mechanism by which DERs/WEBs are formed has been debated for the last few decades and redistribution of carbon within the white bands has been reported in the literature [7]. Earlier research [10,11] has indicated that WEBs are caused by stress and are developed as a result of subsurface plastic deformation. The structural changes observed during RCF have been attributed to recrystallisation, in which carbon atoms diffuse from supersaturated ferrite regions to create LCs. Carbon diffuses from ferrite interstitially to LCs and requires instigation energy to break their binding energies. This instigation energy has been reported as dissipated plastic energy during cyclic loadings [12–14]. Another theory [15] proposed that the carbon depletion within ferrite bands is caused by dislocation annihilation. Recently, Fu et al. [16] proposed a carbon migration theory based on dislocation gliding which indicated that carbon diffuses from ferrite bands to thicken neighbouring Lenticular Carbides. However, a bulk of the research has been centred on the migration of carbon within the ferrite matrix and could be restrained by either thermally driven diffusion or gliding dislocations [17]. A clear understanding of band formation is still required to resolve the ambiguity regarding white band formation.

Current research presents a systematic approach to understanding the development and formation mechanism of DERs/WEBs during RCF. Standard martensitic AISI 52100 bearings balls are tempered in a muffle furnace. The tempered and untempered ball samples are then subjected to specific RCF conditions in a rotary tribometer. Additionally, the RCF tests are performed and then subjected to similar tempering conditions. It is observed that the bearing pre-tempered RCF-tested ball samples showed significant microstructural alterations under the contact track as compared with untempered RCF-tested samples. The initially conducted RCF test exhibited the formation of DERs accompanying WEBs. After tempering, little to no difference in the WEB thickening is observed, however, randomly distributed white patch formation in the vicinity of DERs can be seen. Besides tempering, the static loading analysis is conducted to understand the effect of non-proportional loading histories. The bearing ball samples are subjected to the static load RCF test to evaluate the effect of lattice widening at extreme loading conditions. The mechanistic approach employed in current research is highly significant in evaluating RCF-induced subsurface alterations and can be employed for improving efficiency of analytical modelling applied in DER and WEB formation.

2. Experimental Setup

RCF tests were conducted with the help of a four-ball configuration in a high-speed PLINT rotary tribometer, where three lower balls were kept (free to roll) in a cup which worked like an outer race of a bearing. In a four-ball test configuration, the upper ball (test sample) rotates on top of lower three balls and experiences 2.25 stress cycles [18]. A synthetic industrial ISO-Vg-320 oil was used in bath lubrication in order to avoid asperities interaction. Standard martensitic AISI 52100 (also known as 100Cr6) bearing balls of 12.7 mm diameter were used as test samples. Once the test was suspended, the upper ball in the collet was demounted, conditioned in an ultrasonic conditioning ultrasonic bath, and then sectioned into extremely thin slices in both axial and circumferential directions. A cubic boron nitride (CBN) cutter in a precision sectioning machine was employed with extensive cooling water to avoid the machining effects on the sample surface. After machining, the samples were cleaned, preserved via cold mounting, and polished with a 1 μm diamond suspension. Figure 1 demonstrates the sample preparation and resulting micrographs after Nital etching. Further details on the RCF experimentation, four-ball test setup, and respective sample preparation can be found in the authors' previous research [18,19]. The acquired subsurface micrographs reveal the microstructure features as a semi-elliptical or as a ring-shaped geometrical entity via bright field microscopy after etching with Nital solution (3% nitric acid in ethanol).

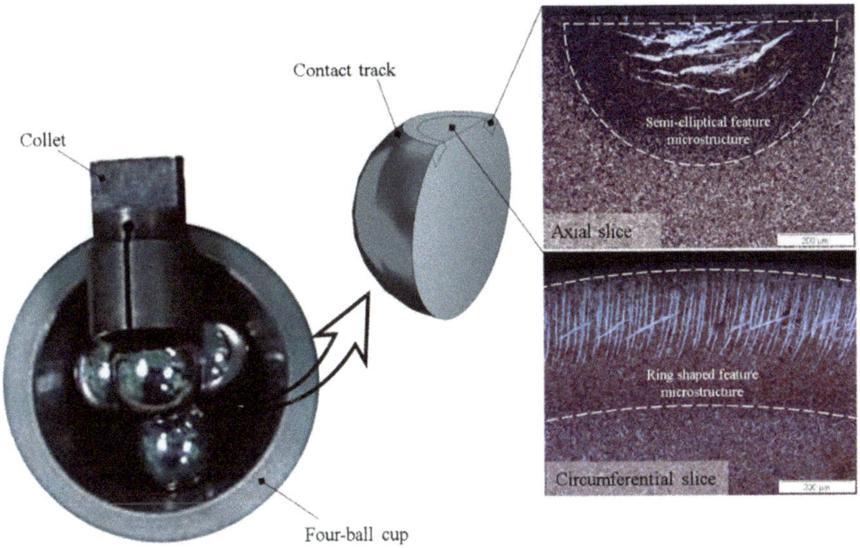

Figure 1. Rolling contact fatigue testing in a rotary tribometer.

3. Tempering Analysis

Standard AISI 52100 bearing ball samples were tested at 6 GPa peak contact pressure with 100 °C operational temperature, and 4051 rpm speed. Previous research reported [18] that the formation of microstructure features, especially the Lower Angle Bands (LABs), starts to form after 37 million cycles at 6 GPa and 100 °C. Therefore, it can be suggested that the initiating point for white bands formation exists near 37 million rolling cycles. The initial plastic shakedown of bearing elements during rolling cyclic fatigue is believed to occur in the very early stage of RCF, usually about 0.1% of total bearing life cycle [20]. Therefore, the bearing ball samples were RCF tested till 37 million and 0.02 million rolling cycles. The full details of conducted tests for tempering analysis are mentioned in Table 1. Figure 2 shows the optical micrographs of RCF-tested samples in axial and circumferential orientations. It can be seen that the bearing sample exhibits formation of LABs, nearly 10–20 µm in length, along with the formation of DER as observed in the circumferential orientation; refer to Figure 2a.

Table 1. Experimentation conducted for tempering analysis.

Sample	Max Contact Pressure	Lubricant Temperature	Spindle Speed	Elapsed Cycles	Tempering Condition
1	6 GPa	100 °C	4051 rpm	37 million	380 °C for 240 min after RCF test
2	6 GPa	100 °C	500 rpm	0.02 million	380 °C for 240 min after RCF test
3	4 GPa	100 °C	4051 rpm	37 million	No heat treatment
4	4 GPa	100 °C	4051 rpm	37 million	380 °C for 240 min before RCF test
5	5 GPa	100 °C	4051 rpm	37 million	No heat treatment
6	5 GPa	100 °C	4051 rpm	37 million	380 °C for 240 min before RCF test

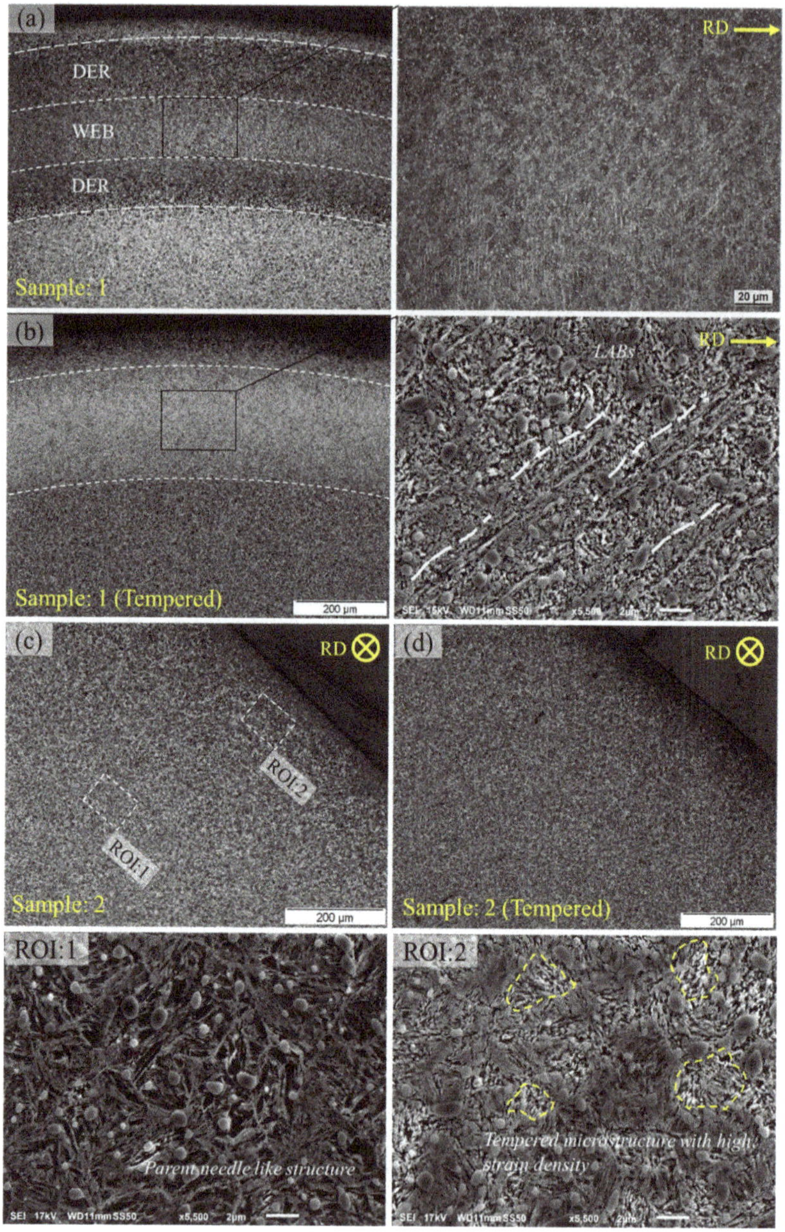

Figure 2. (**a**) Subsurface micrographs obtained at 6 GPa, 100 °C, after 37 million rolling cycles. (**b**) Tempering of RCF-tested sample shown in (**a**). (**c**) Micrographs obtained at 6 GPa, 100 °C, after 0.02×10^6 rolling cycles. (**d**) Tempering of RCF-tested sample shown in (**c**). ROI:1 and ROI:2 represent regions of interest for parent structure and RCF damaged area respectively.

To understand the effects of thermal diffusion, the RCF-tested samples (1 and 2) were tempered in a muffle furnace at a temperature of 380 °C for 240 min. Standard martensitic bearing steel at such elevated temperatures is expected to undergo thermal tempering resulting in more carbon atoms available in the solute leading to enhanced formation of microstructural alterations. After tempering in a muffle furnace, the same sample was again polished and etched to remove any oxidation layer resulting from heat treatment. Figure 2b shows the formation of randomly distributed white patches in the vicinity of DERs. This white etching matter can be characterised as WEB. The inset of Figure 2a,b shows no difference in WEBs formation before and after tempering of bearing samples. The tempering temperature, as high as 380 °C resulted in the reprecipitation of the free carbon in the ferrite matrix which was segregated during the martensite–ferrite phase transformation. Additionally, the tempering of the RCF-tested sample also caused the overall reduction of DER which can only be seen in the upper and lower boundaries of white etching matter. However, no thickening of LABs or formation of higher angle bands (HABs) can be observed after heat treatment.

Figure 2c represents RCF investigated sample at 6 GPa maximum Hertzian contact pressure with 100 °C operational temperature till 0.02 million rolling cycles. In order to achieve such low stress cycles in a high-speed rotary tribometer, the spindle speed was reduced from 4051 rpm to 500 rpm. It should be noted that no microstructure features have been observed under such conditions as seen in the optical micrograph. It is difficult to justify the shakedown limit of certain bearing steel, however, it is speculated that under severe operating conditions, i.e., 6 GPa contact pressure, substantial plastic deformation will occur, which in turn will lead to a high dislocation density after several thousand rolling cycles. A similar sample was tempered to investigate the influence of carbon diffusion within areas of high dislocation density. It can be observed in Figure 2d that the heat treatment of RCF-tested samples does not influence the formation of DER which is a result of an over-tempered form of martensite, as suggested previously [21,22]. The regions of interest (ROI:1 and ROI:2) from Figure 2c are also displayed for parent martensitic structure and RCF-induced damaged area. It is evident from scanning electron micrographs that the original needle-like microstructure has been lost during the early shakedown stage of rolling fatigue cycles. The tempered martensite exists in the form of high dislocation regions due to subsurface plastic deformation.

The bearing ball samples were also tempered prior to RCF testing. For heat treatment, the ball samples were initially heat treated in a muffle furnace at 380 °C temperature for 240 min which resulted in a hardness drop of nearly 250 HV. The baseline hardness of the through-hardened bearing samples decreased from 782–810 HV (10 Kgf) to 525–560 HV (10 Kgf) after the heat treatment process. The tempered and untempered bearing ball samples were then subjected to rolling cyclic loading in the rotary tribometer discretely at 4 GPa and 5 GPa maximum contact pressure with 100 °C working temperature and a constant cyclic frequency of 4051 rpm.

It can be observed in Figure 3a,b that the bearing ball sample, after RCF testing at 4 GPa contact pressure, shows little to no formation of microstructural alterations, whereas the pre-tempered RCF test exhibits the development of microstructural features under the contact track. Correspondingly, the untempered bearing ball tested at 5 GPa contact pressure shows formation of DER in a very localised area under the centreline contact track, refer to Figure 3c, whereas the pre-tempered RCF-tested sample exhibits enhanced formation of white bands which can be observed in axial orientation as well as in circumferential orientation in Figure 3d,e. Scanning electron micrographs show these elongated band structures making steeper angles with the rolling direction (RD) and are described as premature HABs. The formation of premature HABs before LABs has been reported in the authors' previous work [19] where stress as high as 6 GPa contact pressure has been reported to be a cause of WEBs sequence reversal. The development of HABs before LABs can be explained by considering white bands formation to be yield controlled rather than stress controlled as suggested by previous researchers [23]. It has been suggested that a threshold yield limit exists, beyond which the formation of white bands takes place in a preferable orientation, which is steeper to the RD.

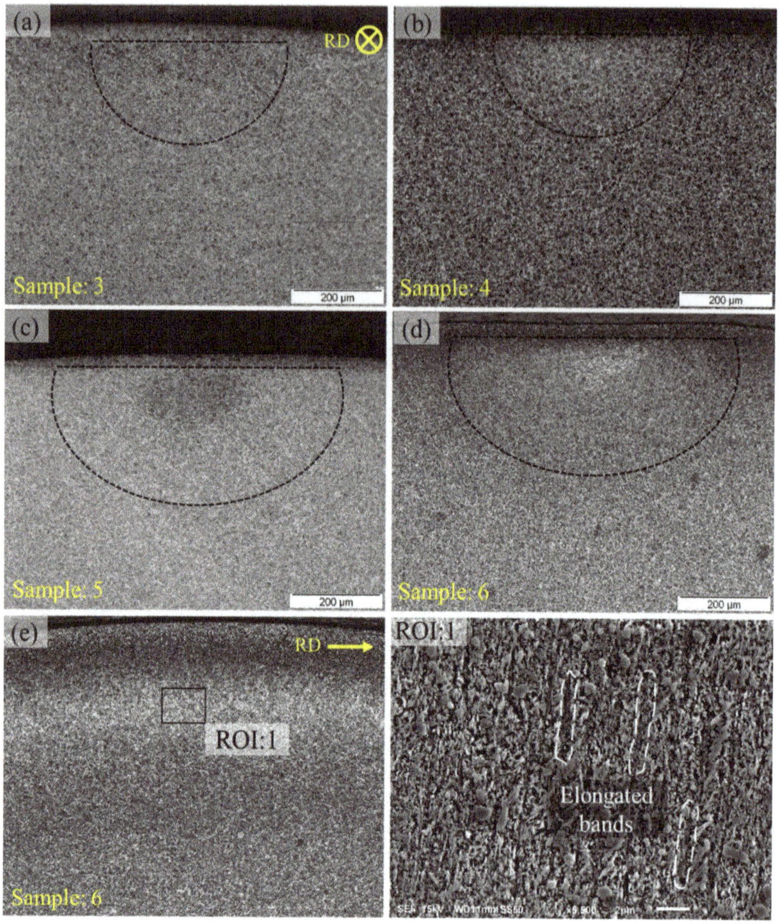

Figure 3. (**a**) Untempered and (**b**) tempered RCF-tested sample at 4 GPa, 100 °C, 4051 rpm. (**c**) Untempered and (**d**) tempered RCF-tested samples at 5 GPa, 100 °C, and 4051 rpm. The inset of circumferentially sliced sample (**e**) is shown in ROI:1.

The accelerated growth of white bands for pre-tempered RCF-tested samples suggests a significant effect of lowering of yield stress on microstructural alterations, resulting in subsequent subsurface plastic deformation. Moreover, the white bands are considered to be a softer structure as compared to the parent matrix [9], which can help promote accumulation of plastic strains. The change in yield stress and evolution of localised flow curves during the material phase transformation is presented in Section 6.

4. Effect of Thermal Diffusion

Figure 4a represents a schematic for carbon diffusion in a 5 μm thick ferrite plate. The initial carbon concentration of the matrix is assumed to be 6.52×10^{-6} mol/um^3 as shown and the boundaries of the plate are assumed to be carbon-free, i.e., 0% carbon concentration. Equation (1) represents the generalised form of carbon flux J expressed as [24]

$$J = -sD = \left[\frac{\partial\varphi}{\partial x} + k_s\frac{\partial}{\partial x}(\ln(\theta - \theta^z)) + k_p\frac{\partial p}{\partial x}\right] \tag{1}$$

where D and s represent diffusivity and solubility, θ and p represent temperature and stress, where k_s and k_p denote the temperature-driven and pressure-driven carbon diffusion factors, respectively. Assuming isotropic behaviour of carbon diffusion in alpha ferrite and neglecting the temperature and stress gradients in Equation (1), the change in carbon concentration of ferrite bands can be obtained relative to time. The diffusivity of carbon in α ferrite is defined in Equation (2) [25], where T represents temperature in kelvin.

$$\log \mathrm{D} = -0.9064 - 0.5199\left(\frac{10^4}{\mathrm{T}}\right) + 1.61 \times 10^{-3}\left(\frac{10^4}{\mathrm{T}}\right)^2 \quad (2)$$

Figure 4b,c represent the estimated diffusion of carbon at 100° and 160 °C respectively. Complete depletion of a 5µm thick ferrite band is obtained at 160 °C temperature just after 69 h. It can be noted that the initial formation of WEBs (LABs/HABs) is observed after ~69 h of RCF testing under given conditions. The enhanced formation of HABs at 160 °C operating temperature is an indication of high carbon flux due to thermally activated carbon along with a large number of mobile dislocations. On the other hand, at 100 °C operating temperature, it takes more than 720 h (1 month) to diffuse out carbon in a 5 µm thick ferrite plate. It can be understood that at such slow rates of carbon diffusion, the transportation of carbon by dislocation glide becomes decisive. However, a typical run time for bearing components during service can reach tens of thousands of hours, where thermal diffusion of carbon should be accounted for in the formation of WEBs.

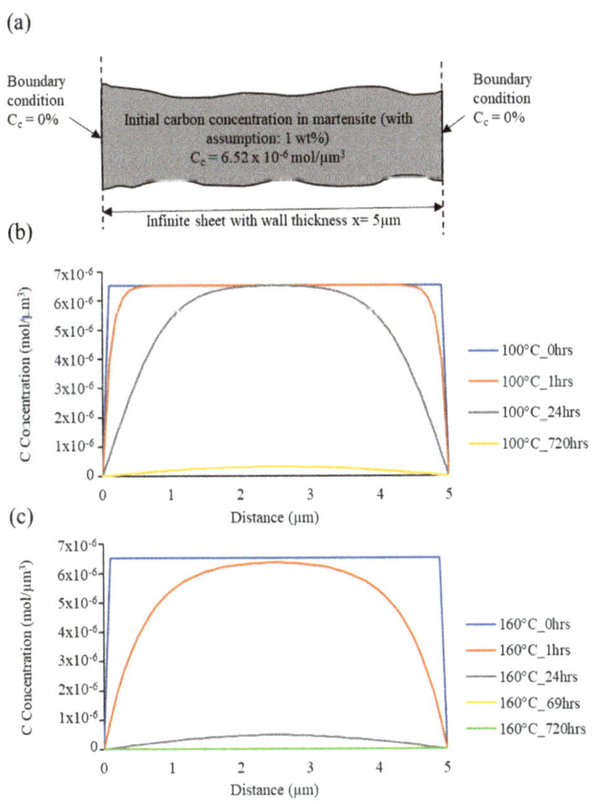

Figure 4. (**a**) Schematic for carbon diffusion in a 5 µm thick ferrite plate. Change in carbon concentration due to diffusion at (**b**) 100 °C and (**c**) 160 °C.

5. Static Load Analysis

To refute the migration of carbon as a result of dislocation gliding and understanding the carbon diffusion process during cyclic loading, static load analyses were carried out. The static load analysis follows a sequence where bearing balls are initially subjected to rolling cyclic loadings in a rotary tribometer followed by a static loading condition. During static loading, the upper ball in contact with the lower three balls was held static with the driver motor turned off. The contact track formed on the upper ball during rolling cyclic loading was marked precisely at the points of interactions. The time required to load the sample under static loading conditions was estimated from the numerical analysis discussed in Section 4. After the RCF test, the bearing ball sample was serially sectioned to investigate the statically loaded and offloaded subsurface regions. The testing parameters employed in the static load analysis are presented in Table 2.

Table 2. Test conducted for static load analysis.

Test ID	Loading Condition	Peak Contact Pressure	Operating Temperature	Spindle Speed	Time for Static Load	Stress Cycles
ST0.02M	Rolling cyclic	6 GPa	160 °C	500 rpm	-	20,034
	Static			-	69 h	-
ST22M-1	Rolling cyclic			4051 rpm	-	22,001,035
	Static			-	69 h	-
ST22M-2	Rolling cyclic			4051 rpm	-	22,067,262
	Static			-	100 h	-

Figure 5 represents the optical micrographs obtained at 6 GPa maximum contact pressure with 160 °C operational temperature. It was suggested previously [18] that DERs start to form at extreme testing conditions well before one million cycles. However, no feature microstructure can be observed till 20,000 rolling cycles. Having said that, at such extreme testing conditions, the bearing material undergoes subsurface plastic deformation with high dislocation density, as demonstrated previously in Figure 2. The RCF-tested bearing ball sample was then subjected to static loading conditions for 69 h, as illustrated in Table 2. As suggested by numerical simulation of the carbon flux across a thin ferrite plate, diffusion of carbon is highly likely to occur at 160 °C temperature after 69 h. Despite the numerical prediction and high dislocation density, no feature microstructure can be observed in the loaded and offloaded regions of ST0.02M sample. It should be noted that the optical micrograph for RCF testing and subsequent static loading analysis is acquired from different bearing ball samples.

Correspondingly, the optical micrographs obtained for 6 GPa, 160 °C, and 22 million rolling cycles are represented in Figure 6 which shows the formation of early HABs in the rolling direction. These initially developed HABs span nearly 100 μm in length and their microstructure contains deformed shear bands as shown in the inset image. These shear bands are formed due to the shear localisation and plastic deformation resulting in the martensite–ferrite phase transformation. As the maximum carbon solubility in the ferrite phase is nearly 0.02% [24], it is considered that the free carbon from such microstructural phase transformation can segregate at the boundaries and form carbon-rich plates termed as lenticular carbides (LCs). The transport of excess carbon atoms from the deformed ferrite matrix towards LCs can be accelerated with the help of carbon diffusion where thermal effects and lattice widening during static loading could lead to the thickening of pre-existing LCs. ST22M-1 and ST22M-2 represent the static loading analysis of RCF-tested samples till 69 h* and 100 h of testing time (* 69 h represents the testing time where initial WEBs can be observed under given loading conditions). It can be seen that despite the numerical prediction from Section 4, no thickening of LCs can be observed for either of the samples despite the prolonged tempering of the RCF-tested sample under static loading.

Moreover, the static loaded regions do not show any notable difference in the growth of LCs as compared to the offload regions when observed in the high-contrast images from cross-polarised light microscopy.

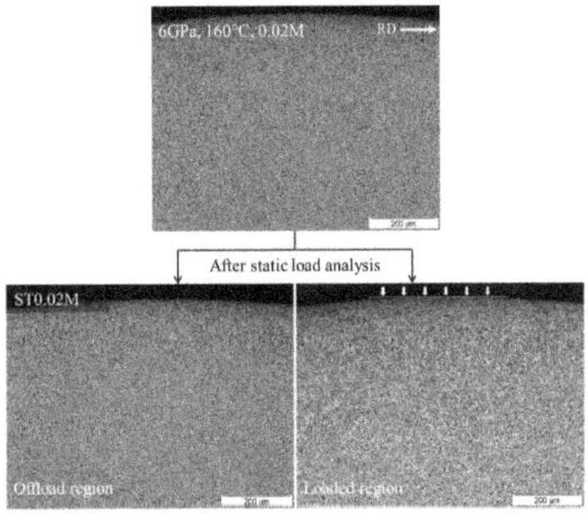

Figure 5. Circumferentially sliced micrograph acquired at 6 GPa, 160 °C after 0.02M rolling cycles. The static loaded and offload regions are also shown for ST0.02M.

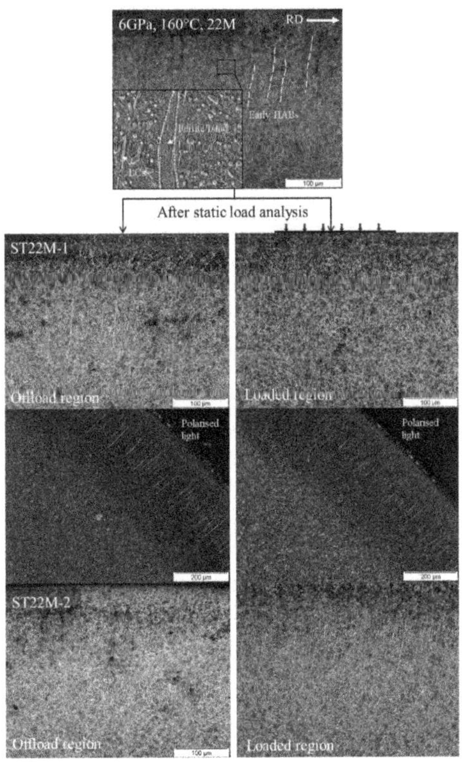

Figure 6. Circumferentially sliced micrographs obtained after 6 GPa, 160 °C, 22 M rolling cycles test, where the inset of figure shows microstructure of an early HAB. The loaded and offload regions of ST22M-1 and ST22M-2 are also shown after static load analysis. The high-contrast images are acquired from cross-polarised light.

The additional RCF tests were carried out at 6 GPa, 100 °C till 0.75 million rolling cycles with variation in spindle speed, i.e., 4051 rpm and 289 rpm. Figure 7a,b representing the subsurface micrograph indicate that a higher cyclic speed results in decreased DERs formation. DERs formation can be quantified by a global DER%, characterising the fractional area of overall dark patches in the region of RCF-induced subsurface damage [19]. The global DER% calculated for slow cyclic speed gives an overall DER percentage of 95% which reduces to 40% at a higher cyclic speed. Likewise, at 5 GPa and 100 °C, a DER% of 50% can be observed at 289 rpm, whereas no DER can be seen at 4051 rpm speed. At low-stress frequency, more time is available for carbon to diffuse from ferrite to surrounding boundaries. In addition, the low-stress frequency promotes more carbon to be captured by dislocation during the gliding motion as suggested previously [8].

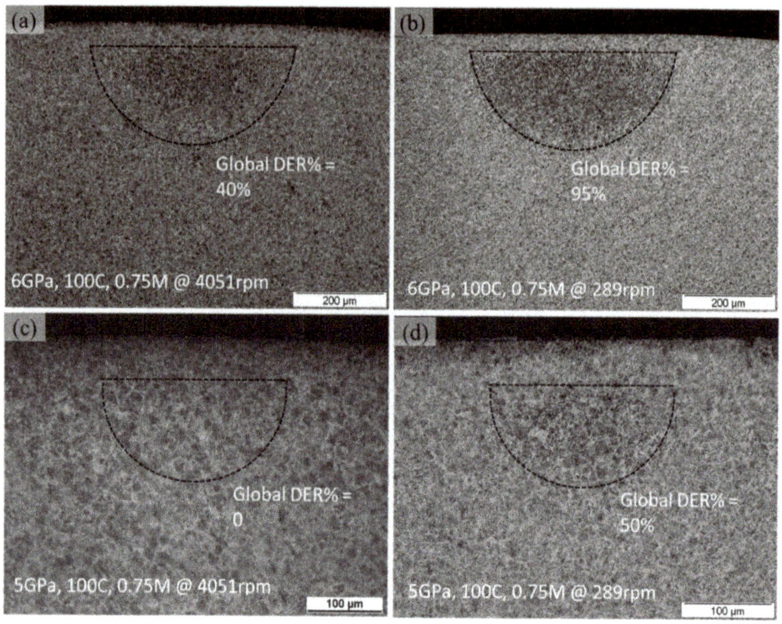

Figure 7. Axially sliced micrographs acquired at 6 and 5 GPa, 100 °C till 0.75 million cycles with (**a**,**c**) 4051 rpm (**b**,**d**) 289 rpm speed, respectively.

6. Discussions

It is believed that the development of microstructure features involves degeneration of the parent structure and the microstructural phase transformation during plastic deformation. The subsurface plastic deformation takes place as a result of stress localisation and forms shear bands in the preferred planes due to non-proportional loading histories. Figure 8a represents the subsurface micrograph of feature microstructures containing HABs and LABs alongside the martensitic matrix. In order to evaluate the micromechanical properties of feature microstructure, a spherical conical indenter of 9 μm in diameter was employed. The loading was applied in a cyclic loading-partial unloading manner with a total penetration depth of 400 nm (assuming homogenous depth distribution). Each loading cycle yielded an elastic/plastic response of the target surface depending upon penetration depth. The semi-empirical approach to computing flow stress and representative strain from spherical indentation is described in Appendix A and further details are available in [26]. The localised flow curves are acquired from the indentation targets, shown in Figure 8a, and are plotted in Figure 8b). It is apparent from flow curves that bearing steel exhibits considerable strain hardening under cyclic loadings. The spherical indents Ind1, Ind3, and Ind4 demonstrate the martensitic structure's stress–strain curve. Ind2 and Ind5 show LCs stress–strain curves, whereas Ind 6 and Ind7 depict stress–strain curves of LABs and HABs respectively. The martensitic matrix exhibits a baseline flow curve of bearing steel while significant hardening and softening can be observed in the regions of LCs and WEBs, respectively. A decrease in hardness in the carbon-depleted areas (ferritic grains) and an increase in hardness have been reported for carbon-rich areas (LCs) [9]. Moreover, a localised hardness map of WEBs has revealed hardness variation within a white band [19]. The variation in localised hardness arises due to grain size deviation, crystallography, change in chemical composition, and indenter penetration depth. The constitutive stress–strain curves obtained from spherical indentation suggest that the localised hardness can vary within a WEB, however, the global response of a WEB exhibit overall softening of the microstructure. In addition, the lowering of yield stress (softening), as suggested by the constitutive flow curves in the localised regions, indicates the ferritic nature of WEBs, which is the manifestation of RCF-induced accumulated plastic strain. The accumulation of plasticity within WEBs results in the formation of ferrite bands comprising dislocation cells.

(a)

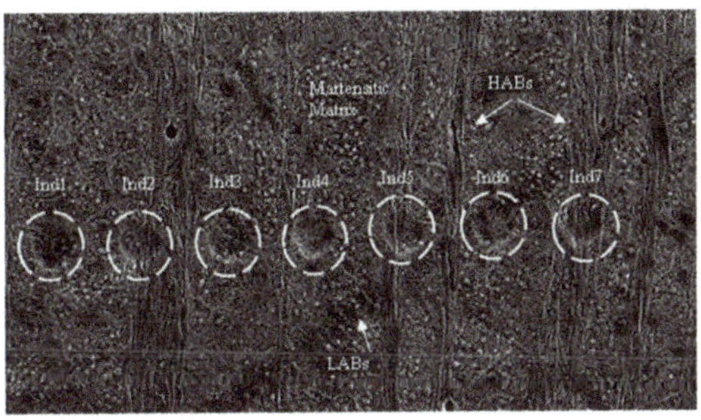

(b)

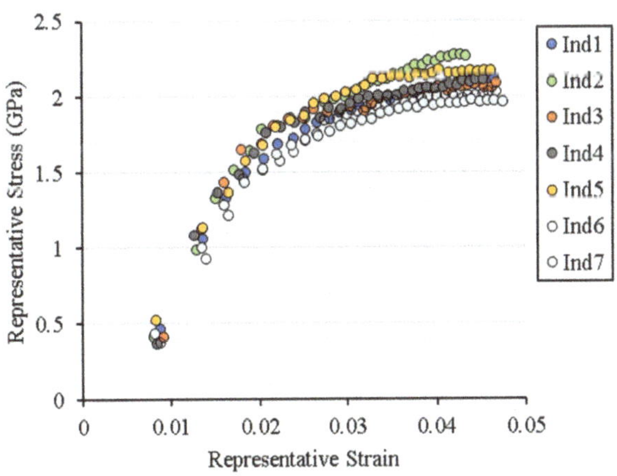

Figure 8. (**a**) Circumferentially sliced micrograph representing white etching bands (WEBs) and spherical nanoindentation performed. (**b**) Flow curves obtained from nanoindentation analysis of WEBs.

The martensite–ferrite transformation and the formation of dislocation cells are an indication of high dislocation density in the areas of WEBs. With progressive loading cycles, the subsurface plasticity accumulates, and forms elongated ferrite bands, refer to Figure 6. These ferrite bands are oriented in the preferred slip system due to complex stress histories during rolling cyclic loading. The preferential planes of elongated ferrite bands have been described as the maximisation of normal relative stress across the contact track and have been explained in previous research work [19]. The carbon transport from ferrite to the grain edge becomes favourable. The free carbon in the ferrite region diffuses towards boundaries and precipitates as LCs. With continuous cyclic loadings, more and more carbon atoms become available in the ferrite matrix due to the dissolution of primary carbides and its transport to thicken the pre-existing carbide plates. To understand the transport of carbon, tempering and static load analysis have been conducted.

The heat treatment of bearing ball samples results in the reprecipitation of carbides. The post-tempering results also showed that the thermally activated carbon during thermal tempering cannot lead to formation of WEBs without external stress histories. It is noticeable that the pre-tempered RCF-tested sample showed the formation of white bands after 37 million rolling cycles along with DERs, whereas no white band is visible for the untempered RCF-tested sample. The poorly bonded carbides to the martensitic matrix during thermal tempering favour the formation of WEBs instead of DERs. The RCF test conducted with variation in stress frequency illustrated that a lower cyclic speed increases the time available for carbon to diffuse from ferrite to surrounding boundaries. Nevertheless, the low stress time also promotes more carbon to be captured by dislocation during the gliding motion. From Section 4, the numerical results of the carbon concentration gradient within supersaturated regions of ferrite predicted the diffusion of carbon from ferrite bands towards the surrounding matrix. However, the tempering results and static load analysis of bearing ball samples did not reveal the thickening of pre-existing LCs despite the prolonged tempering of RCF-tested samples under static loading. The static load analysis confirms that the WEBs development mechanism is determined by the non-proportional stress histories during cyclic loading which results in accumulated plastic strains to forming elongated ferrite bands.

It is evident from the comparative analyses that microstructural alterations are formed as a result of carbon flux, which is dislocation-assisted, in consort with thermally activated carbon diffusion. However, it is quite difficult to differentiate the individual effects of carbon diffusion and plastic deformation on the development of a feature microstructure. It is recommended to employ further material characterisation techniques to quantify the grain misorientations and lattice deformations within microstructural alterations. The mechanistic approach employed in the current study is highly significant in terms of evaluating the microstructural response of RCF-affected regions and can be used to improve the efficiency of analytical models for DERs/WEBs formation.

7. Conclusions

Bearing balls are tested systematically in a rotary tribometer for further evaluation of microstructural alterations. After RCF testing and subsequent parametric analyses, the following points are highlighted,

- The formation mechanism of WEBs accelerates at a temperature above the tempering condition of bearing steel. The excessive undissolved carbon arising due to thermal tempering is suggested to enhance the growth of WEBs rather than form DERs.
- The static load RCF test has validated that a complete rolling mechanism with a non-proportional stress history is essential for the development and unique orientation of WEBs.
- The localised hardness within WEBs can vary, however, the global response of white bands indicates softening of bearing material, as illustrated by localised flow curves.
- The development of a feature microstructure is a holistic effect of carbon diffusion and plasticity; and cannot be described solely with dislocation gliding or the carbon diffusion phenomenon, as presented previously in various research studies.

Author Contributions: M.U.A. contributed to planning, scheduling of experimental work, data acquisition, analysis, and manuscript writing. Z.A.K. led conception of idea in terms of ordinality and significance, organised funding, resources, and project management. All authors have read and agreed to the published version of the manuscript.

Funding: The work was funded by Schaeffler Technologies AG & Co., KG, Germany (grant ID: 10187) to conduct this research at Bournemouth University, United Kingdom.

Institutional Review Board Statement: Not applicable.

Informed Consent Statement: Not applicable.

Data Availability Statement: Not applicable.

Acknowledgments: Authors would like to acknowledge Schaeffler Technologies AG & Co., KG, Germany, and in particular Wolfram Kruhoeffer, for in-kind support to conduct this research.

Conflicts of Interest: The authors declare no conflict of interest.

Appendix A

Nanoindentation analysis is performed at a single point with continuously increasing multiple cyclic loading and unloading. For this, a 9 μm spherical indenter is used with NanoTest VANTAGE Micromaterials. The contact area is determined by using the data from spherical indentation analysis and the given radius of diamond indenter, as demonstrated in Figure A1. Subsequently, hardness H, Young's modulus E, stress Y, and strain are calculated. This cyclic loading–unloading procedure is expressed as the Field & Swain analysis [27]. The spherical indentation has been employed previously [28,29] for mechanical property evaluation of small volumes of materials.

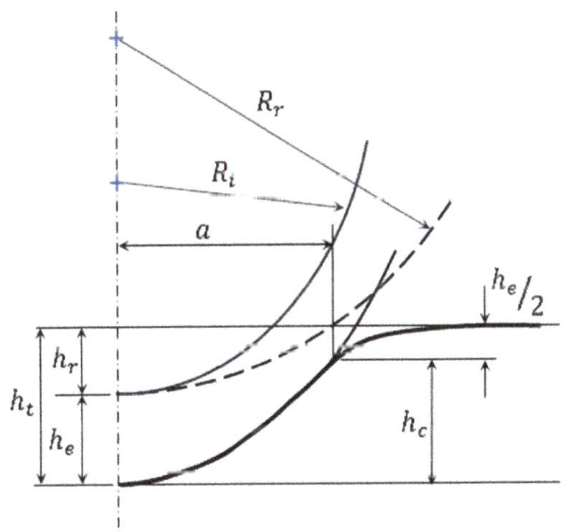

Figure A1. Schematic for spherical indentation.

With multiple loading cycles during spherical indentation, the total penetration depth h_t (elastic + plastic component) depends on total applied load P_t. During partial unloading, the recovered depth h_s depends upon the reduced load P_s. Once the entire load is removed, there will be an associated residual depth h_r which can be expressed as [29],

$$h_r = \frac{r \cdot h_s - h_t}{r - 1} \tag{A1}$$

Whereas r is a factor described as,

$$r = \left(\frac{P_t}{P_s}\right)^{\frac{2}{3}} \tag{A2}$$

The elastic depth (h_e) and plastic depth (h_p) during an individual loading cycle can be computed as,

$$h_e = h_t - h_r \tag{A3}$$

$$h_p = \frac{h_t + h_r}{2} \tag{A4}$$

Plastic depth and radius of spherical indenter R are used to calculate contact circle radius 'a' as given below, which is in the original surface plane at every indentation cycle.

$$A = \sqrt{2\,R\,h_p - h_p^2} \tag{A5}$$

Hardness H, the yield/flow stress Y_r, and subsequent representative strain ε_r can be computed for load step i = 1ton as follow,

$$H_i = \frac{P_{t_i}}{A_i} = \frac{P_{t_i}}{\pi a_i^2} \tag{A6}$$

$$Y_{r_i} = \frac{H_i}{3} \tag{A7}$$

$$\varepsilon_{r_i} = 0.2\frac{a_i}{R} \tag{A8}$$

Equations (A1)–(A8) [27,29] can be used to extract micromechanical properties from the indented surface with the help of cyclic loading–unloadings using a conical spherical indenter.

References

1. Arshad, W.; Hanif, M.A.; Bhutta, M.U.; Mufti, R.A.; Shah, S.R.; Abdullah, M.U.; Najeeb, M.H. Technique developed to study camshaft and tappet wear on real production engine. *Ind. Lubr. Tribol.* **2017**, *69*, 174–181. [CrossRef]
2. Ţălu, Ş. *Coperta Thread Rolling Technology 2019*; Napoca Star Publishing House: Cluj-Napoca, Romania, 2019.
3. Abdullah, M.U.; Shah, S.R.; Bhutta, M.U.; Mufti, R.A.; Khurram, M.; Najeeb, M.H.; Arshad, W.; Ogawa, K. Benefits of wonder process craft on engine valve train performance. *Proc. Inst. Mech. Eng. Part D J. Automob. Eng.* **2018**, *233*, 1125–1135. [CrossRef]
4. Abdullah, M.U.; Khan, Z.A. A Multiscale Overview of Modelling Rolling Cyclic Fatigue in Bearing Elements. *Materials* **2022**, *15*, 5885. [CrossRef] [PubMed]
5. El Laithy, M.; Wang, L.; Harvey, T.J.; Vierneusel, B.; Correns, M.; Blass, T. Further understanding of rolling contact fatigue in rolling element bearings—A review. *Tribol. Int.* **2019**, *140*, 105849. [CrossRef]
6. Yin, H.; Wu, Y.; Liu, D.; Zhang, P.; Zhang, G.; Fu, H. Rolling Contact Fatigue-Related Microstructural Alterations in Bearing Steels: A Brief Review. *Metals* **2022**, *12*, 910. [CrossRef]
7. Voskamp, A.P.; Österlund, R.; Becker, P.C.; Vingsbo, O. Gradual changes in residual stress and microstructure during contact fatigue in ball bearings. *Met. Technol.* **1980**, *7*, 14–21. [CrossRef]
8. Fu, H.; Song, W.; Galindo-Nava, E.I.; Rivera-Díaz-Del-Castillo, P.E. Strain-induced martensite decay in bearing steels under rolling contact fatigue: Modelling and atomic-scale characterisation. *Acta Mater.* **2017**, *139*, 163–173. [CrossRef]
9. Šmeļova, V.; Schwedt, A.; Wang, L.; Holweger, W.; Mayer, J. Electron microscopy investigations of microstructural alterations due to classical Rolling Contact Fatigue (RCF) in martensitic AISI 52100 bearing steel. *Int. J. Fatigue* **2017**, *98*, 142–154. [CrossRef]
10. Martin, J.; Borgese, S.F.; Eberhardt, A.D. Eberhardt, Microstructural alterations of rolling—Bearing steel undergoing cyclic stressing. *ASME J. Basic Eng.* **1966**, *88*, 555–565. [CrossRef]
11. Bhargava, V.; Hahn, G.T.; Rubin, C.A. Rolling contact deformation, etching effects, and failure of high-strength bearing steel. *Met. Mater. Trans. A* **1990**, *21*, 1921–1931. [CrossRef]
12. Schlicht, H. Material Properties Adapted to the Actual Stressing in a Rolling Bearing. In *Ball and Roller Bearing Engineering*; SKF Industries Inc.: Philadelphia, PA, USA, 1981; pp. 24–29.
13. Mitamura, N.; Hidaka, H.; Takaki, S. Microstructural Development in Bearing Steel During Rolling Contact Fatigue. In *Materials Science Forum*; Trans Tech Publication: Stafa-Zurich, Switzerland, 2007.
14. Warhadpande, A.; Sadeghi, F.; Evans, R.D. Microstructural Alterations in Bearing Steels under Rolling Contact Fatigue: Part 2—Diffusion-Based Modeling Approach. *Tribol. Trans.* **2014**, *57*, 66–76. [CrossRef]
15. Polonsky, I.; Keer, L.M. On white etching band formation in rolling bearings. *J. Mech. Phys. Solids* **1995**, *43*, 637–669. [CrossRef]
16. Fu, H.; Galindo-Nava, E.; Rivera-Díaz-Del-Castillo, P. Modelling and characterisation of stress-induced carbide precipitation in bearing steels under rolling contact fatigue. *Acta Mater.* **2017**, *128*, 176–187. [CrossRef]
17. Abdullah, M.U. Finite Element Modelling of Deep Zone Residual Stresses in Rolling Contact Bearing Elements. Ph.D. Thesis, Bournemouth University, Bournemouth, UK, 2022.

18. Abdullah, M.U.; Khan, Z.A.; Kruhoeffer, W. Evaluation of Dark Etching Regions for Standard Bearing Steel under Accelerated Rolling Contact Fatigue. *Tribol. Int.* **2020**, *152*, 106579. [CrossRef]
19. Abdullah, M.U.; Khan, Z.A.; Kruhoeffer, W.; Blass, T.; Vierneusel, B. Development of white etching bands under accelerated rolling contact fatigue. *Tribol. Int.* **2021**, *164*, 107240. [CrossRef]
20. Abdullah, M.U.; Khan, Z.A.; Kruhoeffer, W.; Blass, T. A 3D Finite Element Model of Rolling Contact Fatigue for Evolved Material Response and Residual Stress Estimation. *Tribol. Lett.* **2020**, *68*, 122. [CrossRef]
21. Kang, J.H.; Hosseinkhani, B.; Rivera-Díaz-del-Castillo, P.E. Rolling contact fatigue in bearings: Multiscale overview. *Mater. Sci. Technol.* **2012**, *28*, 44–49. [CrossRef]
22. Bush, J.J.; Grube, W.L.; Robinson, G.H. Microstructural and residual stress changes in hardened steel due to rolling contact. *Trans. ASM* **1961**, *54*, e412.
23. Warhadpande, A.; Sadeghi, F.; Evans, R.D. Microstructural alterations in bearing steels under rolling contact fatigue part 1—Historical overview. *Tribol. Trans.* **2013**, *56*, 349–358. [CrossRef]
24. Bhadeshia, H.; Honeycombe, R. *Steels: Microstructure and Properties*; Butterworth-Heinemann: Oxford, UK, 2017.
25. Shabalin, I.L. *Ultra-High Temperature Materials. I, Carbon (Graphene/Graphite) and Refractory Metals*; Springer: Dordrecht, The Netherlands, 2014.
26. Abdullah, M.U.; Khan, Z.A.; Kruhoeffer, W. Nanoindentation Analysis of Evolved Bearing Steel under Rolling Contact Fatigue (RCF). In Proceedings of the 2020 STLE Tribology Frontier Virtual Conference Co-sponsored by ASME Tribology Division, Cleveland, OH, USA, 9–13 November 2020.
27. Field, J.; Swain, M. Determining the mechanical properties of small volumes of material from submicrometer spherical indentations. *J. Mater. Res.* **1995**, *10*, 101–112. [CrossRef]
28. He, L.H.; Fujisawa, N.; Swain, M.V. Elastic modulus and stress–strain response of human enamel by nano-indentation. *Biomaterials* **2006**, *27*, 4388–4398. [CrossRef] [PubMed]
29. Kalidindi, S.R.; Pathak, S. Determination of the effective zero-point and the extraction of spherical nanoindentation stress–strain curves. *Acta Mater.* **2008**, *56*, 3523–3532. [CrossRef]

materials

MDPI

Article

Thermo-Mechanical Fatigue Behavior and Resultant Microstructure Evolution in Al-Si 319 and 356 Cast Alloys

Kun Liu [1], Shuai Wang [1], Lei Pan [2] and X.-Grant Chen [1,*]

1 Department of Applied Science, University of Quebec at Chicoutimi, Saguenay, QC G7H 2B1, Canada
2 Arvida Research and Development Centre, Rio Tinto Aluminum, Saguenay, QC G7S 4K8, Canada
* Correspondence: xgrant_chen@uqac.ca

Abstract: The out-of-phase thermo-mechanical fatigue (TMF) behavior of the two Al-Si cast alloys most widely used for engine applications (319 and 356) were investigated under temperature cycling (60–300 °C) and various strain amplitudes (0.1–0.6%). The relationship between the microstructural evolution and TMF behavior was closely studied. Both alloys exhibited asymmetric hysteresis loops with a higher portion in the tensile mode during TMF cycling. The two alloys showed cyclic softening behavior with regard to the maximum stress, but an earlier inflection of cyclic stress was found in the 356 alloy. The TMF lifetime of the 319 alloy was generally longer than that of the 356 alloy, especially at higher strain amplitudes. All the precipitates (β′-MgSi in 356 and θ′-Al_2Cu in 319) coarsened during the TMF tests; however, the coarsening rate per cycle in the 356 alloy was significantly higher than that in the 319 alloy. An energy-based model was applied to predict the fatigue lifetime, which corresponded well with the experimental data. However, the parameters in the model varied with the alloys, and the 356 alloy exhibited a lower fatigue damage capacity and a higher fatigue damage exponent.

Keywords: Al-Si casting alloys; thermo-mechanical fatigue; precipitates; fatigue life prediction; fracture

Citation: Liu, K.; Wang, S.; Pan, L.; Chen, X.-G. Thermo-Mechanical Fatigue Behavior and Resultant Microstructure Evolution in Al-Si 319 and 356 Cast Alloys. *Materials* **2023**, *16*, 829. https://doi.org/10.3390/ma16020829

Academic Editors: Lucjan Śnieżek, Jaroslaw Galkiewicz and Sebastian Lipiec

Received: 22 December 2022
Revised: 9 January 2023
Accepted: 12 January 2023
Published: 15 January 2023

1. Introduction

Al-Si cast alloys are widely used in modern automotive industries to replace cast iron-fabricated engine components (engine blocks and cylinder heads) due to their high strength-to-weight ratio, excellent castability, and thermal conductivity [1–4]. In combustion engine applications, components undergo complex loading changes and dramatic temperature gradients during start-up and shutdown cycles. For example, engines are often heated from ambient temperature—or even lower in cold winters (−25 °C)—to 250–300 °C and, vice versa, rapidly cooled down in start–stop cycles [5]. Due to the large thermal gradient, significant thermal and mechanical stresses occur in the engine components because of the changing thermal expansion/contraction of various engine components [6]. As a result, cyclic stress and temperature changes can result in thermo-mechanical fatigue (TMF) [4,6], which can lead to serious failures, significantly limiting components' service life. Therefore, understanding TMF behavior is increasingly important for the design, reliability assessment, and lifecycle management of critical engine components during their industrial application. It has been reported [4–6] that the most significant damage mechanism in engine components is out-of-phase (OP) TMF cycling, where the maximum mechanical strain occurs at the minimum temperature.

Several studies have evaluated the TMF behavior in Al-Si cast alloys [7–13]. Huter et al. [11] performed TMF tests on Al-Si-Cu and Al-Si-Mg cast alloys with varying Si, Cu, Fe, and Sr contents and reported that the nucleation and propagation of cracks were predominantly influenced by the eutectic Si and Fe-rich intermetallics, which lowered the fatigue resistance. Sehitoglu et al. [14] reported similar results for the Al-Si 319 alloy, finding that the alloy presented worse TMF resistance at higher Fe contents due to a lower stress level and higher softening speed. It was also found that the rapidly solidified material with a small secondary dendrite arm space (SDAS) presented the highest cyclic stress–strain amplitude, whereas the stress levels decreased gradually as the SDAS levels increased [14]. It has also been reported that a finer SDAS, lower porosity level, and lower brittle intermetallic content are beneficial for better thermal resistance and, hence, the fatigue life of defect-free A356 alloys can be improved by at least one order of magnitude [15,16].

A recent study employed ultrasonic melt treatment for Al-Si cast alloys to refine the grain size, the eutectic Si, and the Fe-rich intermetallics, which significantly enhanced the fatigue life [17]. However, contradictory results regarding the influence of precipitates on the TMF resistance of Al-Si cast alloys can be found in the literature. For example, Azadi et al. [8,9] studied the effect of heat treatment and strain rate on the TMF cyclic behavior of the A356 alloy and found that the TMF life was not significantly affected by the heat treatment in the as-cast and T6 conditions. In contrast, Moizumi et al. [18] reported that the thermal fatigue life of an A356 + 1%Cu alloy in the over-aged condition was doubled compared to the peak-aged condition. Hunter et al. [11] also reported that the TMF resistance of an Al-Si-Cu cast alloy was improved compared to that of a Cu-free Al-Si alloy due to the Cu-containing precipitates. Beck et al. [19] studied the influence of high-cycle fatigue (HCF) loading on the TMF life of Al-Si-Mg/Cu cast alloys in the T6 condition and concluded that TMF was not significantly affected by HCF loading at low strain amplitudes and that precipitation in the T6 condition was effected by HCF. Toda et al. [13] reported that precipitates preferentially oriented perpendicular to the loading direction can effectively prolong the in-phase TMF life, but the study did not comment on the OP-TMF life. Toyoda et al. [20] investigated the microstructural change in an Al-Si-Cu-Mg cast alloy under thermo-mechanical loading conditions and reported that the precipitates were observed to align perpendicularly to the compressive stress during heating, but they did not further discuss how these precipitates with preferential orientations affected the TMF behavior. Therefore, the systematic evolution of precipitates during TMF under various strain amplitudes, which is one of the significant factors influencing the TMF resistance, is still not well-documented in Al-Si cast alloys.

Contemporary engine components are designed to increase the maximum operating temperature and pressure in order to further reduce greenhouse gas emissions and improve their fuel efficiency [5,19,21,22]. However, the TMF behaviors reported in the literature are limited to relatively low maximum temperatures of 200 or 250 °C [9,11,16,18,19], and the TMF behavior at higher maximum temperatures (approximately 300–350 °C) is seldom reported. Moreover, the microstructural evolution during TMF over a wider temperature range, especially precipitation in age-hardening aluminum alloys, is still not fully understood. This study selected two of the most widely used Al-Si cast alloys—namely, Al-Si 356 and Al-Si-Cu 319—to perform OP-TMF under temperature cycling (60–300 °C) and various strain amplitudes (0.1–0.6%). The TMF behaviors were evaluated using the strain and stress cyclic response, softening rate, and fatigue life assessment. The evolution of the precipitates during TMF cycling was characterized using transmission electron microscopy (TEM) to establish the relationship between the microstructure and TMF behavior. Additionally, the TMF damage mechanism is discussed and an energy-based model for the prediction of the TMF lifetimes of the two alloys is introduced.

2. Materials and Methods

Two Al-Si cast alloys (319 and 356) were prepared using commercially pure aluminum (99.7%), pure Mg (99.9%), and Al-50%Si, Al-25%Mn, Al-50%Cu, Al-10%Sr, and Al-5%Ti-1%B master alloys. The raw materials were batched in clay-graphite crucibles and melted using an electric resistance furnace. After reaching 780 °C, the melt was held for 30 min and then degassed with high purity argon gas for 15 min. The melt was then cast into a permanent wedge mold pre-heated at 250 °C. The chemical compositions of the two alloys were analyzed using optical emission spectroscopy, and the results are listed in Table 1. All the as-cast samples were subjected to a two-step solution treatment followed by water quenching. Artificial aging was then performed at 200 °C for 5 h to achieve the T7 temper condition. Table 2 summarizes the parameters of the heat treatment applied to the alloy samples.

Table 1. Chemical compositions of the alloys studied (wt.%).

Alloy	Si	Cu	Mg	Mn	Fe	Ti	Sr	Al
319	5.93	3.34	0.12	0.284	0.307	0.11	0.0106	Bal.
356	7.27	0.60	0.34	0.206	0.109	0.21	0.0113	Bal.

Table 2. Heat treatment parameters for the alloys.

Alloy	Solution Treatment	T7 Aging Treatment
319	495 °C/4 h + 515 °C/2 h	200 °C/5 h
356	500 °C/4 h + 540 °C/2 h	

After the T7 heat treatment, the TMF specimens were machined to a parallel gauge length of 75 mm and diameter of 10 mm and a 5 mm diameter hole was machined through the specimens for air circulation, as schematically shown in Figure 1. In the present work, OP-TMF, where the highest mechanical strain is reached at the lowest temperature, was applied to the experimental alloys (as shown in Figure 1a). OP-TMF tests were performed using a Gleeble 3800 thermal–mechanical simulator system with a strain-controlled loading mode. The samples were heated using a Joule heating system and cooling was achieved by using compressed air delivered through a hollow tube specimen. The TMF temperature was varied between 60 and 300 °C for a cycle using heating and cooling rates of 5 °C/s. The temperature was measured and controlled using thermocouples attached to the center of the gauge length. The applied mechanical strain amplitudes were 0.1, 0.2, 0.4, and 0.6%. The TMF tests were automatically terminated when either a total of 2000 fatigue cycles was reached or a decrease of 30% in the initial maximum tensile/compression stress was detected. When the TMF test was terminated without full fracture of the sample, the sample was pulled to full fracture in the Gleeble machine at room temperature for a later fracture analysis. More details on the TMF test procedure, such as the process to reach zero stress at the beginning of the test and the isolation of the mechanical strain from the total strain during the TMF test, can be found in our previous study [23].

The microstructures of the TMF samples were characterized using optical and electron microscopies. Optical microscopy was used to observe the as-cast and T7 grain structures and intermetallics, whereas a scanning electron microscope (SEM) equipped with an energy-dispersive X-ray spectroscope was used to identify the intermetallic phases and to observe the fracture faces after the TMF tests. TEM was used to observe the evolution of the precipitates during TMF testing. TEM samples were cut from the cast ingot after the T7 treatment and from the gauge zone of the samples (close to fracture) after the TMF tests at various strain amplitudes. The TEM samples were initially punched into 3 mm diameter disks and then ground and polished to approximately 30 μm. The samples were then twin-jet electrochemically polished using a solution of 75 mL HNO3 in 250 mL methanol at −30 °C to create TEM observation regions. Image analysis was used to quantify the characteristics of the intermetallics and precipitates, such as their area fraction and number density, under various conditions.

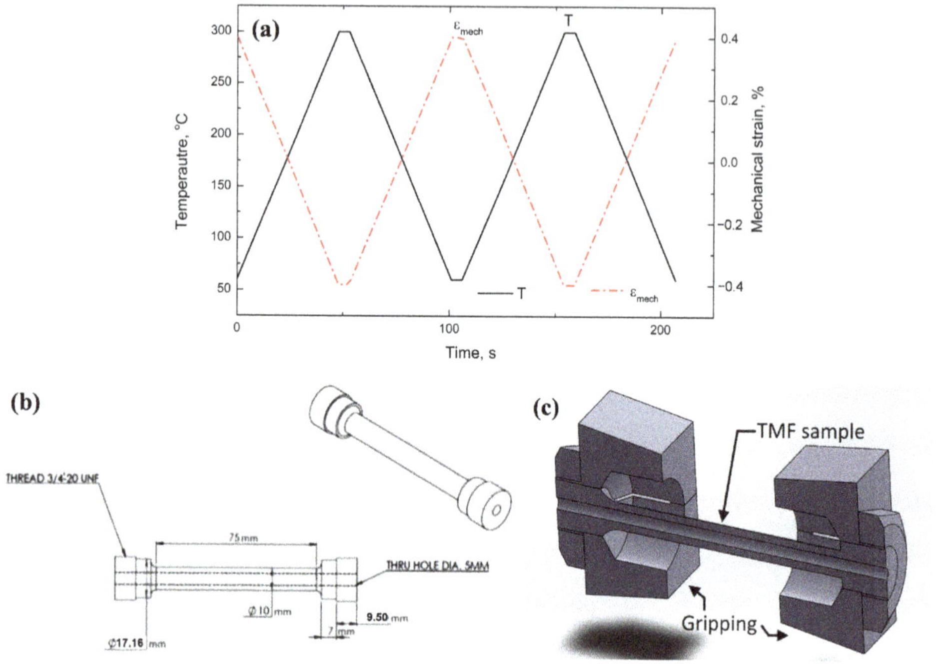

Figure 1. Schematic illustration of OP-TMF at 0.4% strain amplitude (**a**), TMF sample dimensions (**b**), and the setup with gripping blocks (**c**).

3. Results

3.1. TMF Behaviors

Figure 2 shows the stress–strain hysteresis loops of the initial, mid-life, and end-life cycles of the two alloys at the mechanical strain amplitudes of 0.1, 0.2, 0.4, and 0.6%. Both alloys exhibited some common phenomena. First, both alloys exhibited asymmetrical hysteresis loops at all strain amplitudes, and the tensile portions were higher than the compression portions, indicating a higher tensile stress than compression stress at the same tensile/compression strain, which has also been reported in the literature [8,11,14,18,23]. This can be attributed to the characteristics of OP-TMF, where the maximum tensile strain is reached at the lowest temperature (60 °C) and the compression strain increases to a maximum at the highest temperature (300 °C). The mechanical properties of Al-Si cast alloys worsen with increasing temperature [5,24,25]. Therefore, the stress needed to reach the desired strain was higher in the tensile mode due to its low temperature compared to the compression mode at high temperature, leading to asymmetrical hysteresis loops with higher portions in the tensile mode (stress > 0 MPa in Figure 2). Second, the loop areas were significantly larger and the peak tensile/compression stress increased with increasing strain amplitude. As shown in Figure 2a,c,e,g, the loop area of the 319 alloy gradually increased with an increase in the strain amplitude from 0.1 to 0.6%, whereas the mid-life peak tensile stress increased from 100 MPa at 0.1% to 124, 198, and 211 MPa at 0.2, 0.4, and 0.6%, respectively. A comparison of the mid- and end-life cycles with the first cycle showed that the shape of the hysteresis loop became flatter with increasing fatigue life at a fixed strain amplitude, indicating an increasing plastic strain, which can be calculated using the strain range at zero stress. Additionally, the maximum tensile and compression stresses decreased at all strain amplitudes, indicating cyclic softening behavior [10,11,23]. Taking the TMF cycles of alloy 319 at 0.4% strain amplitude as an example (Figure 2e), the plastic strain increased from 0.36% after the second cycle to 0.45 and 0.51% at the mid-life and final cycles, respectively, while the maximum tensile stress decreased from 236 MPa after the second cycle to 198 and 156 MPa at the mid-life and final cycles, respectively.

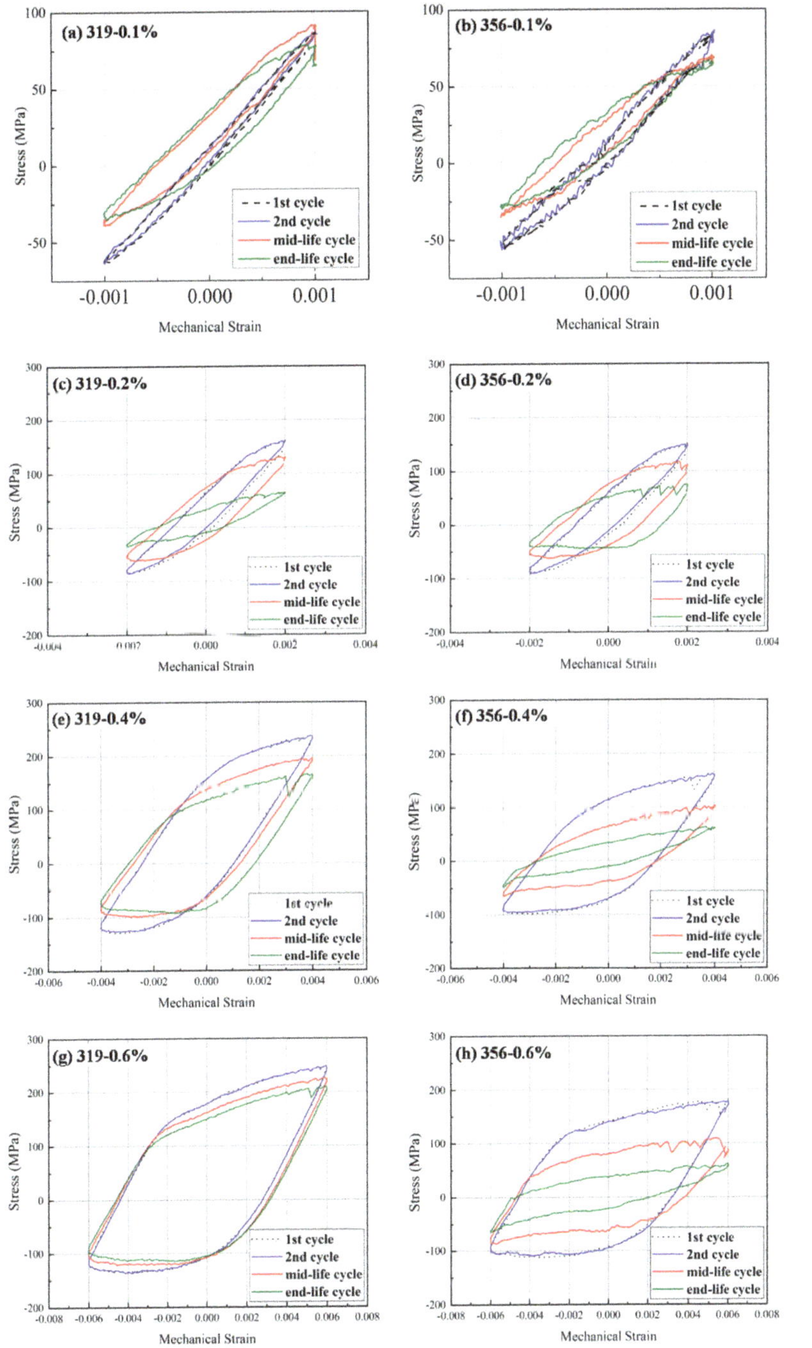

Figure 2. OP-TMF stress–strain hysteresis loops for experimental alloys at different strain amplitudes of 0.1, 0.2%, 0.4%, and 0.6%: (**a**,**c**,**e**,**g**) 319 alloy; (**b**,**d**,**f**,**h**) 356 alloy.

On the other hand, differences between the two experimental alloys were also observed. Figure 2 shows that the 319 alloy always exhibited higher maximum/minimum stress than the 356 alloy, whereas the extent of the strain range at zero stress (plastic strain) was always greater for the 356 alloy than for the 319 alloy. For example, the maximum and minimum stresses of the 319 alloy during the mid-life cycle under a 0.6% strain amplitude (Figure 2g,h) were 206 and 121 MPa, respectively, which were significantly higher than the respective values of 96 and 86 MPa obtained for the 356 alloy. In contrast, the plastic strain in the 356 alloy was 0.8%, higher than the value of 0.6% in the 319 alloy. These differences can be attributed to the different mechanical properties of the two alloys, with the 319 alloy with a high Cu content exhibiting higher strength but lower elongation than the 356 alloy [5,24–26].

Figure 3 shows the evolution of the maximum tensile and compression stresses of the 319 and 356 alloys as a function of the fatigue cycles at different strain amplitudes. The maximum tensile stress for both alloys was always higher than the compression stress, which was due to the higher strength at lower temperatures in OP-TMF [11,23]. Additionally, the maximum tensile/compression stress generally decreased with increasing numbers of cycles, indicating cyclic softening for both alloys, which is also reflected in Figure 2.

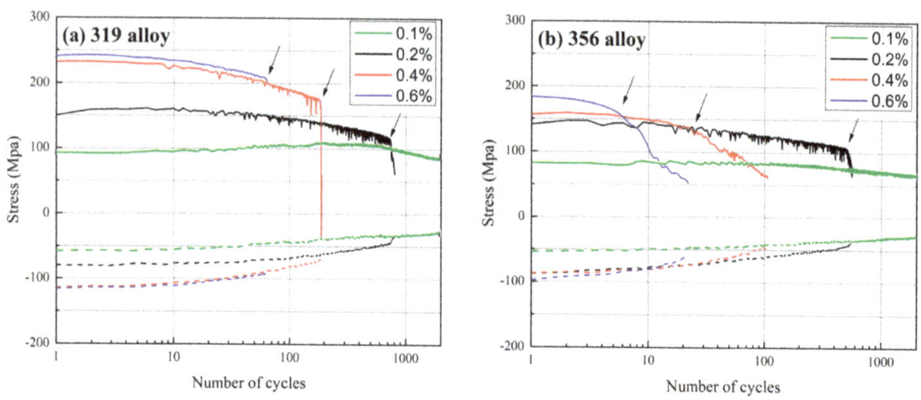

Figure 3. Evolution of the maximum tensile and compression stresses as a function of the cycle number at different strain amplitudes: (**a**) 319 alloy; (**b**) 356 alloy.

In addition to the common phenomena in the two alloys, Figure 3 also shows some differences between the two. First, similarly to Figure 2, the maximum tensile/compression stresses of the 319 alloy were always higher than those of the 356 alloy. For example, the initial maximum tensile and compression stresses of the 319 alloy at a strain amplitude of 0.6% were 245 and 110 MPa, respectively, which were higher than the respective values of 190 and 97 MPa obtained for the 356 alloy. Second, the decrease in stress as a function of the number of cycles differed for the two alloys. As indicated by the arrows in Figure 3, the point where the softening rate exhibited a sharp change was defined as the inflection point at which the TMF could generally be divided into two different stages. Stage I spanned from the initial fatigue cycle to the inflection point. In this stage, the stress level initially increased during the first few cycles and then gradually decreased with a low softening rate. Stage II spanned from the inflection point to the point where the specimen failed, at which point the softening rate significantly increased. In this stage, the stress sharply decreased with increasing numbers of TMF cycles, which can be attributed to the occurrence of large cracks that exceeded the critical crack length [11]. The 356 alloy always exhibited an earlier inflection point than the 319 alloy, especially under higher strain amplitudes (0.4 and 0.6%), which resulted in early fatigue damage in the 356 alloy and, hence, reduced total fatigue life. For an improved comparison, the proportions of stage I in the total lifetimes of the two alloys were calculated, and the results are listed in Table 3. The table shows that stage I accounted for more than 90% of the total lifetime of the 319 alloy at all the tested strain amplitudes, even reaching 100% at the strain amplitude of 0.6%, which may indicate that there was little macro-damage before the end of the TMF test. However, the proportion of stage I in the 356 alloy was considerably lower, especially at higher strain amplitudes. As shown in Table 3, stage I accounted for only 22 and 27% of the total lifetime of the 356 alloy at the 0.4 and 0.6% strain amplitudes, respectively, indicating that the 356 alloy may have been subjected to earlier severe macro-damage during TMF cycling.

Table 3. Proportion of stage I in total TMF lifetimes of experimental alloys.

Strain Amplitude	319 Alloy	356 Alloy
0.2%	90%	85%
0.4%	96%	22%
0.6%	100%	27%

Figure 4 summarizes the OP-TMF lifetimes of the 319 and 356 alloys under different strain amplitudes. Generally, the fatigue life decreased with increasing strain amplitude, and both alloys exhibited approximately linear relations when plotted on a log–log scale. Both alloys reached the limit of 2000 cycles at a strain amplitude of 0.1%; however, the 319 alloy exhibited increased fatigue life starting at the 0.2% strain amplitude, especially at higher strain amplitudes (0.4 and 0.6%). As shown in Figure 4, the average fatigue life of the 319 alloy at a 0.2% strain amplitude was 700 cycles, which was slightly higher than the 580 cycles obtained for the 356 alloy. The 319 alloy exhibited a significantly longer average fatigue life than the 356 alloy at higher strain amplitudes, demonstrating two- and threefold longer average TMF lifetimes at the 0.4 (183 versus 90) and 0.6% strain amplitudes (62 versus 20), respectively. Therefore, the TMF resistance of the 319 alloy was generally higher than that of the 356 alloy, especially when exposed to higher strain amplitudes.

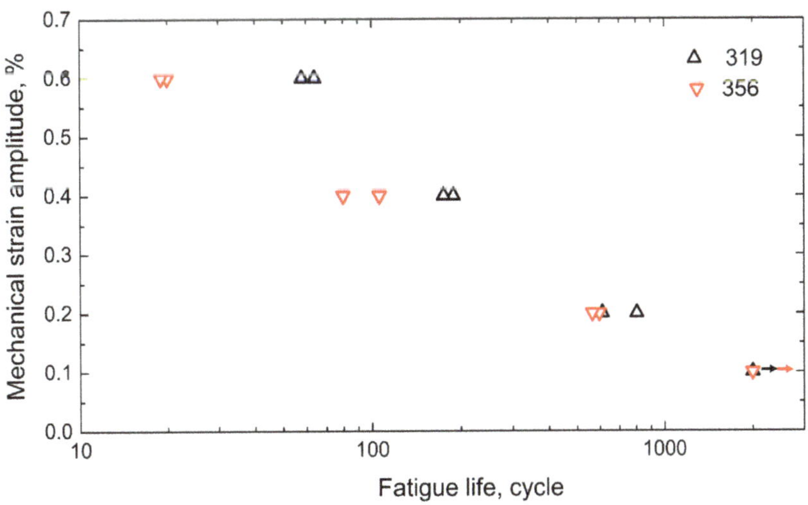

Figure 4. OP-TMF lifetimes of two experimental alloys under various strain amplitudes.

3.2. Fracture Analysis

Figure 5 shows overall views of the fracture surfaces of the 319 and 356 alloys after TMF cycling at 0.2% strain amplitude. Porosity was observed on both fracture surfaces, as marked by the red circles in Figure 5. The porosity in the 319 alloy was relatively low and evenly spread across the sample. Crack ridges were found on the fracture surface of the 319 alloy, indicating that the cracks may have nucleated at multiple initiation sites and merged together during fatigue failure. Large and continuous pores were observed on the fracture surface of the 356 alloy, and these could have served as crack nucleation sites and provided easier propagation paths for the crack [11,15–17], which was one of the likely reasons for the earlier inflection point in the 356 alloy (see Figure 3). It is worth noting that the left part in Figure 5b (356 alloy, marked by two red lines) was not the original TMF fracture surface but the tensile fracture surface from the pulling process after the TMF test, as the sample was not fully fractured when the TMF was terminated. As explained in the Materials and Methods section, the sample that remained unfractured upon termination of the test was pulled to full fracture.

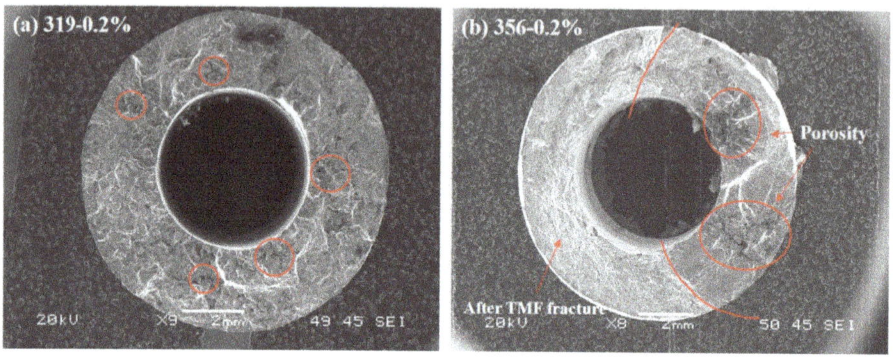

Figure 5. Fracture surface after 0.2% TMF test: (**a**) 319 alloy; (**b**) 356 alloy.

Figure 6 shows the fracture faces at the crack propagation areas in the 319 and 356 alloys at 0.2% strain amplitude. As shown in Figure 6a,d, both fracture faces appeared as ductile fractures with dimples in the matrix. However, the dimples in the 319 alloy were smaller than those in the 356 alloy, indicating its lower plastic strain. The SEM backscatter images in Figure 6b,e show that a number of bright phases were found on the fracture surface, as marked by the arrows, which were identified as Fe-rich intermetallics and eutectic Si particles [24,25]. The 319 alloy (Figure 6b) exhibited more Fe-rich intermetallics than the 356 alloy (Figure 6e), which was attributed to its higher Fe content (0.3 and 0.1% in the 319 and 356 alloys, respectively (Table 1)). As reported in [5,11,17], these hard intermetallic phases could serve as nucleation sites for cracks as a result of their detachment from the aluminum matrix or by rupturing themselves. This was confirmed by the observation of the cross-sectional fracture surface in Figure 7, in which several broken eutectic Si particles and Fe-rich intermetallic phases were observed in the 319 and 356 alloys after the TMF tests. Table 4 summarizes the characteristics of the Fe-rich intermetallics and eutectic Si particles in both alloys in the T7 condition (before TMF). The eutectic Si in the 356 alloy was more globular and had a higher area fraction, whereas the area fraction of the Fe-rich intermetallics in the 319 alloy was double that in the 356 alloy. However, the total area fractions of the eutectic Si and Fe-rich intermetallics in both alloys were similar, indicating their similar contributions to the nucleation and propagation of cracks during TMF deformation. Additionally, striations were observed in both the 319 (Figure 6c) and 356 (Figure 6e) alloys. The interspacing of the striations in the 356 alloy was larger than that in the 319 alloy, indicating faster fatigue propagation during TMF cycling [27].

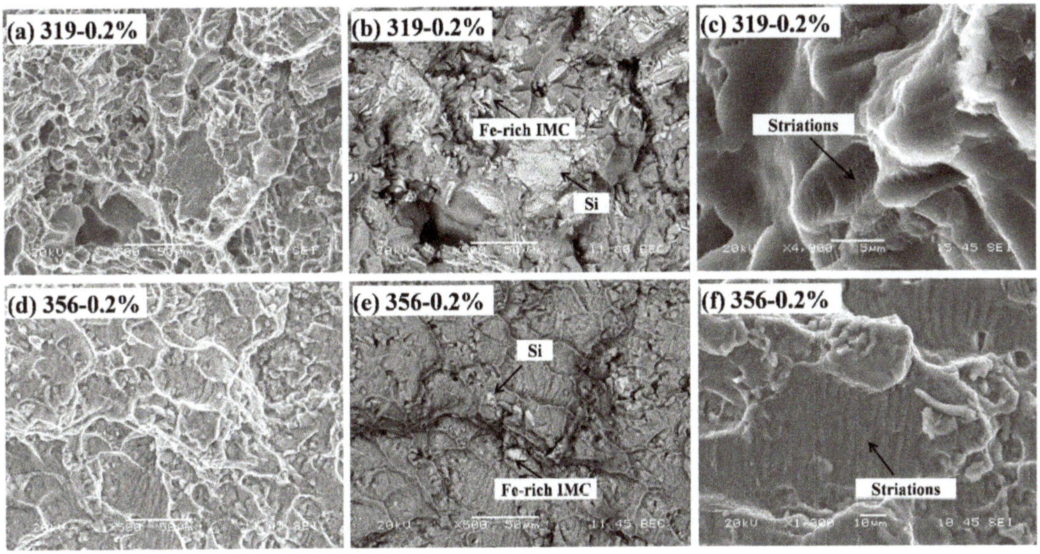

Figure 6. Fractography of fractured specimens after 0.2% TMF test: (**a**–**c**) 319 alloy; (**d**,**e**) 356 alloy; (**a**,**c**,**d**,**f**) SEM secondary electron images; (**b**,**e**) SEM backscatter images.

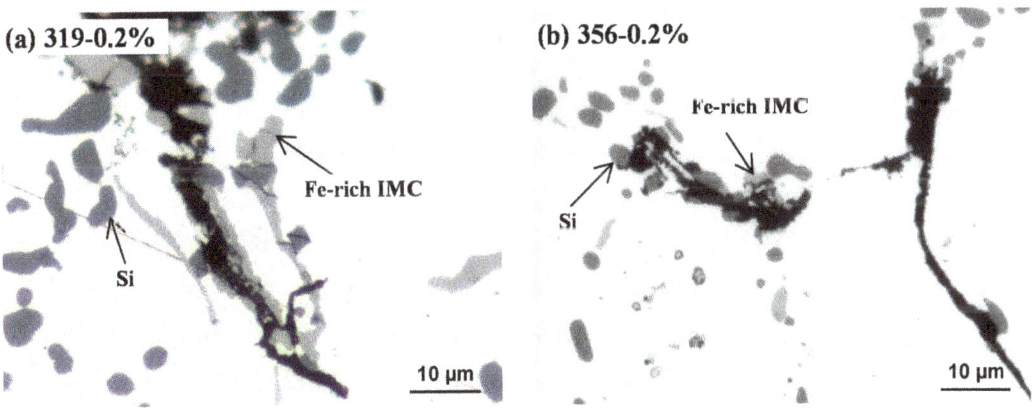

Figure 7. Optical cross-section images of the fracture surfaces showing the cracks in the (**a**) 319 alloy and (**b**) 356 alloy after the 0.2% TMF test.

Table 4. Microstructure characteristics of experimental alloys under T7 condition (before TMF).

	319 Alloy	356 Alloy
Morphology of eutectic Si	Lamellar	Globular
Diameter of eutectic Si	4.0 μm	3.5 μm
Area fraction of eutectic Si	6.6%	8.8%
Area fraction of intermetallics (total)	1.6%	0.8%

4. Discussion

4.1. Microstructural Evolution during TMF

As shown in Figure 3, the maximum tensile and compression stresses for both alloys gradually decreased with the increasing numbers of cycles at all strain amplitudes. This cyclic softening behavior during TMF was likely related to the evolution of the precipitates during the TMF cycling [12,16,18,20,23]. As typical precipitation-strengthening alloys, the precipitates in both alloys play an important role in their mechanical properties. Figure 8 shows the initial precipitate microstructures of the two alloys in the T7 condition (before TMF). A large number of nano-sized precipitates were visible in the aluminum matrixes of both alloys, which were identified as mainly the plate-like θ'-Al_2Cu phase in the 319 alloy and the needle-like β'-MgSi phase in the 356 alloy, consistent with reports from the literature [5,16,25].

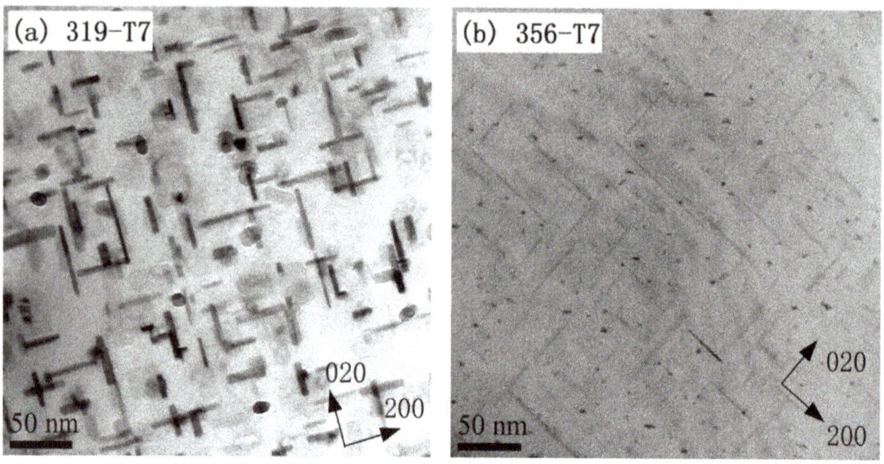

Figure 8. Bright-field TEM images showing precipitate microstructure after T7 (before TMF) for (**a**) θ'-Al_2Cu in the 319 alloy and (**b**) β'-MgSi in the 356 alloy.

Figure 9 presents the precipitate microstructures of the two alloys after TMF cycling at various strain amplitudes. In general, all the precipitates in the two alloys coarsened during TMF cycling. Figure 9 shows that, after TMF cycling, the plate-like θ'-Al_2Cu in the 319 alloy became rod-like θ'-Al_2Cu, whereas the needle-like β'-MgSi coarsened to rod-like β'-MgSi. However, their number densities decreased significantly as their size increased. As shown in Table 5, the number density of the precipitates in the 319 alloy after TMF cycling at 0.2% strain amplitude decreased from 4451 to 501 μm^{-1} and the length increased from 48 to 115 nm compared to the T7 condition (before TMF). A similar phenomenon was also observed in the 356 alloy. Though both the θ'-Al_2Cu and β'-MgSi coarsened during the TMF, different coarsening behaviors were observed for the two alloys. Compared to the evolution of θ'-Al_2Cu in the 319 alloy shown in Figure 9, the size and number-density changes for the β'-MgSi during TMF were significantly larger, especially at the higher strain amplitude (0.6%) (compare Figure 9e,f). As shown in Table 5, after TMF cycling at the 0.6% strain amplitude, the length of the θ'-Al_2Cu in the 319 alloy increased from 48 to 80 nm and the number density decreased from 4451 to 691 μm^{-1}. On the other hand, the size of the β'-MgSi in the 356 alloy increased from 70 to 205 nm and the number density significantly decreased from 19,144 to 394 μm^{-1}. Additionally, a small part of the β'-MgSi in the 356 alloy transformed into equilibrium β-Mg_2Si after TMF cycling, as indicated by the red arrows in Figure 9b,d,f. The rod-like θ'-Al_2Cu remained the predominant precipitate in the 319 alloy after TMF cycling, which indicated that the precipitates in the 356 alloy exhibited a higher coarsening rate than those in the 319 alloy.

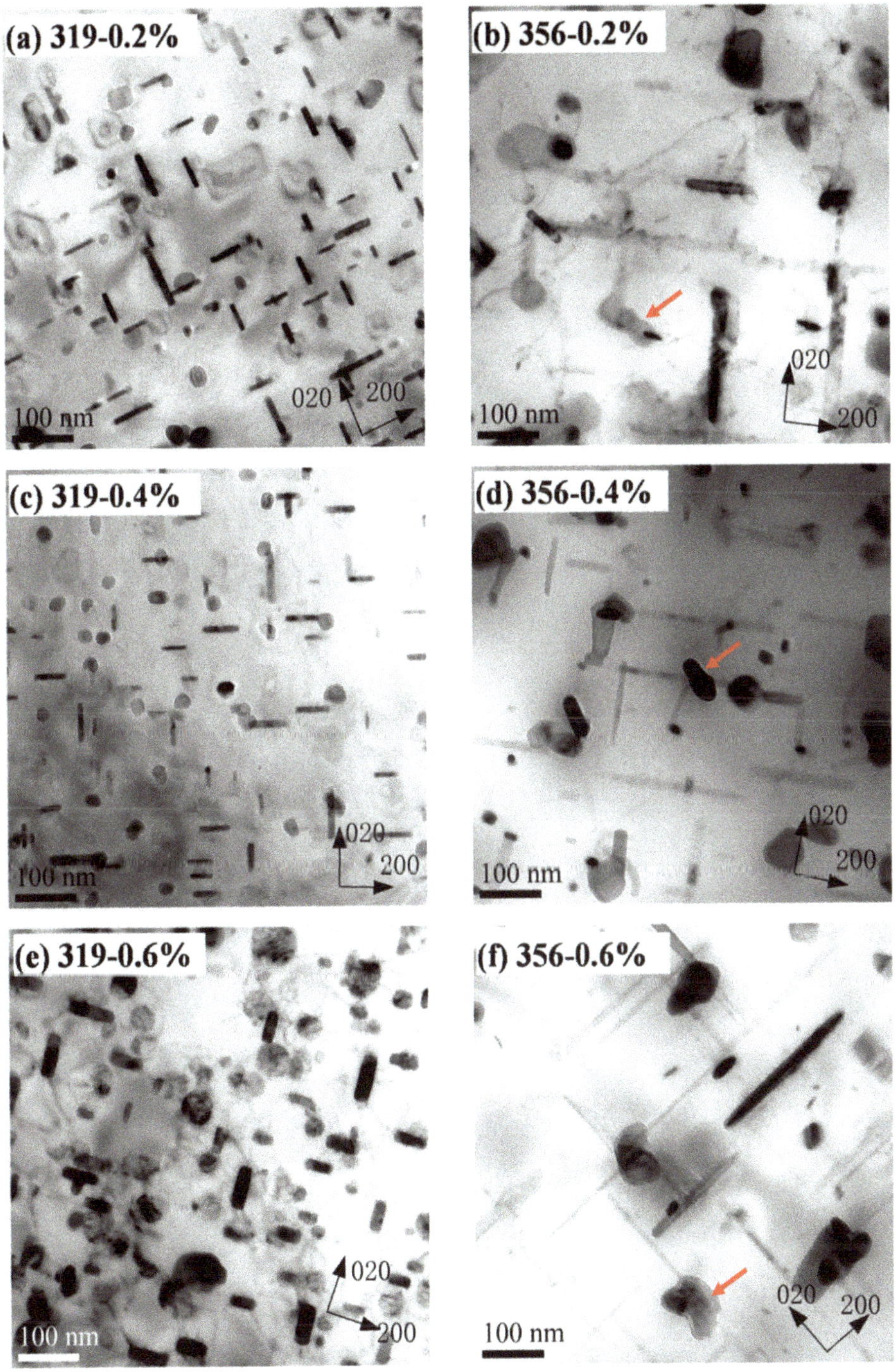

Figure 9. Bright-field TEM images showing precipitate coarsening after TMF at various strain amplitudes: (**a**,**c**,**e**) 319 alloy and (**b**,**d**,**f**) 356 alloy.

Table 5. Characteristics of precipitates of the alloys before and after TMF.

Alloys	Conditions	Length (nm)	Width (nm)	Number Density (μm^{-1})
319 (θ'-Al_2Cu)	T7 before TMF	48.1	4.3	4451.2
	After 0.2%	115.9	12.1	501.4
	After 0.4%	83.9	10.0	789.8
	After 0.6%	80.1	18.1	690.6
356 (β'-MgSi)	T7 before TMF	70.1	2.8	19144.8
	After 0.2%	331.1	14.9	191.1
	After 0.4%	204.7	14.6	253.7
	After 0.6%	204.8	12.6	394.0

The coarsening of precipitates during TMF cycling was mainly related to their TMF behavior. In this study, the coarsening rates of both the precipitates were evaluated using the classical Lifshitz–Slyozov–Wagner model, which is expressed by Equation (1) [28]:

$$L^n - L_0^n = k \cdot t \quad (1)$$

where L is the average half-length of the precipitates after TMF cycling, L_0 is the average half-length of the precipitates in the T7 condition (before TMF), k is the coarsening rate constant, t is the time, and n is the temporal exponent. It has been reported [29,30] that the growth of θ'-Al_2Cu and β'-MgSi precipitates is prone to obeying the $t^{1/2}$ law (n = 2) under the dominant coarsening of the interface reaction. Thus, n = 2 was adopted in this study. Additionally, the value of t is normally taken as the time elapsed at constant temperature. However, in this study, the temperature in each fatigue cycle changed, meaning that the coarsening rate per second could not be applied here due to the temperature variation between 60 and 300 °C in each cycle. For simplification, the value of t was modified to the number of TMF cycles in this study, and the value of k in Equation (1) was calculated as the coarsening rate for the precipitates per cycle. The k values of the precipitates in the two alloys were calculated using the data in Table 5, and the results are summarized in Table 6.

Table 6. Coarsening rate during TMF under various strain amplitudes.

Strain Amplitude	k Value	
	319 Alloy	356 Alloy
0.2%	3.45	43.85
0.4%	6.22	87.32
0.6%	16.05	463.01

As shown in Table 6, k generally increased with an increasing strain amplitude, indicating that a higher strain amplitude resulted in a higher precipitate coarsening rate. This was attributed to the heavy dislocations in the matrix, which were generated under the high-strain-amplitude conditions, accelerating the coarsening of the precipitates [31,32]. However, the 319 alloy exhibited a significantly lower coarsening rate than the 356 alloy at each strain amplitude, and the differences in k between the two alloys also increased with the strain amplitude. For example, k in the 319 alloy at a 0.2% strain amplitude was 3.45, which was only one tenth of the value of 43.85 obtained for the 356 alloy. It was even found that k in the 319 alloy at the 0.4 and 0.6% strain amplitudes decreased to levels 1/5th (6.2 versus 87.3) and 1/30th (16 versus 463), respectively, of those in the 356 alloy.

The lower coarsening rate for θ'-Al_2Cu in the 319 alloy resulted in a slower decrease in the maximum stress during TMF (Figure 3), which has also been reported in the literature [11,18]. The lower coarsening of the precipitates and slower decrease in the maximum stress could stabilize both the matrix and stress level and, hence, enhance the TMF endurance of the 319 alloy (Figure 4). The coarsening rate of the precipitates per cycle in the 319 alloy, especially at the high strain amplitude (0.6%), was significantly lower than that in the 356 alloy, leading to a longer TMF life for the 319 alloy. In contrast, the plastic strain in the 356 alloy during TMF cycling was higher than that in the 319 alloy, and the rate of decrease in the maximum stress was also higher (Figure 3), which can result in a greater portion of the sample deforming during TMF loading, as well as the buckling effect in the gauge area, which further produces early failures [11]. This phenomenon occurred more frequently at the higher strain amplitudes. As shown in Figure 3, the stress inflection point of the 356 alloy at the strain amplitude of 0.6% occurred after only a few cycles, resulting in a significantly shorter TMF life relative to that of the 319 alloy.

4.2. TMF Lifetime Prediction

Thermo-mechanical fatigue lifetimes are difficult to predict due to the complex deformation under both strain and temperature cycling during TMF. Various models have been developed based on different mechanisms, including the Neu–Sehitoglu model based on creep, fatigue, and oxidation damage [33]; the Miller model based on the accumulation of the damage rate [34]; the J-integral model based on fracture mechanics [35]; and the energy-based model based on the dissipated energy per cycle [8,36–38]. Among these models, the energy-based model is widely accepted as a more suitable approach to predict TMF lifetimes [8,36–40]. The hysteresis energy in the energy-based model involves both the strain and stress amplitudes and can be approximated as the product of the plastic strain range ($\Delta\varepsilon_p$) and the stress range ($\Delta\sigma$) using Equation (2) [8,40]:

$$W_i = \int \sigma d\varepsilon \approx \Delta\varepsilon_p \cdot \Delta\sigma \quad (2)$$

where W_i represents the hysteresis energy variable of the ith cycle. The evolution of the hysteresis energy (W, calculated using the hysteresis loop area) with increasing numbers of cycles under various strain amplitudes in the two alloys, as determined using Equation (2), is shown in Figure 10. The hysteresis energy for both alloys involved both the strain and stress amplitudes and exhibited cyclic stability after a short period of adaptation for most of the strain amplitudes applied, confirming the greater rationality of the energy-based model. However, an early energy loss occurred in the 356 alloy after the first few cycles at the 0.6% strain amplitude, which may have been related to early crack formation during TMF loading.

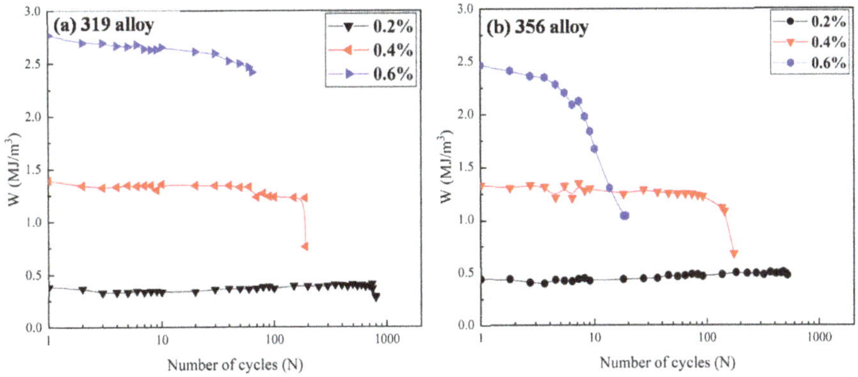

Figure 10. Evolution of hysteresis loop area (W) as a function of cycles for (**a**) 319 alloy and (**b**) 356 alloy.

On the other hand, the hysteresis energy (equal to the plastic strain energy) is also related to the fatigue life (N_f), as shown in Equation (3) [8,36]:

$$W_s = W_0 \cdot N_f^{-1/\beta} \tag{3}$$

where W_s is the saturation hysteresis plastic strain energy. As shown in Figure 10, the plastic strain energy reached the saturation stage after only a few cycles, which thus allowed for W_s to be determined from the fatigue data after that point, and it was generally obtained at the mid-life cycle of the completed TMF tests. W_0 and β are material parameters representing the fatigue damage capacity and fatigue damage exponent, respectively. They can be calculated from the $\log W_s - \log N_f$ relationship [36,39]. In this way, the W_0 and β calculated from the known fatigue life and plastic strain energy under a selection of strain amplitudes can be used with Equation (3) to estimate the TMF life under other strain amplitudes after only implementing a few cycles, such as 10; thus, the saturation plastic strain energy can be obtained, which is more time-efficient than running a full TMF test at each strain amplitude.

Due to the limited data in this study, the TMF behaviors at the 0.2 and 0.6% strain amplitudes were used to calculate W_0 and β. The TMF life at the 0.4% strain amplitude was then estimated using Equation (3) and compared with the experimental value to validate the model. The data for both alloys after 2000 cycles at the 0.1% strain amplitude were obtained by triggering the settled limit of the test and could not be used to determine the W_0 and β. Table 7 lists the calculated W_0 and β values, which are similar to the reported values for Al-Si cast alloys [36]. The 319 alloy exhibited a higher fatigue damage capacity value (W_0) than the 356 alloy. The exponent value (β) represents the relation between the TMF lifetime and hysteresis energy, and the higher β value for the 356 alloy indicates that the TMF lifetime decreased significantly with increasing hysteresis energy.

Table 7. Material parameters calculated with energy-based model.

	W_0	β
319 alloy	107.7	1.09
356 alloy	5.27	2.28

The predicted fatigue lifetime at the 0.4% strain amplitude was calculated using Equations (2) and (3), as well as the data in Table 7, and compared with the experimental fatigue life, as listed in Table 8. The results illustrate that the predicted lifetime corresponded well with the experimentally measured lifetime. Figure 11 presents the experimental and predicted TMF lifetimes at 0.2–0.6% strain amplitudes. All the predicted TMF lifetimes correlated with the low life prediction factor (LPF; approximately 1.3) obtained from the experimental life [36], confirming that the energy-based model was suitable for the prediction of the TMF lifetimes of the two alloys and had acceptable accuracy. Using the parameters from Table 7 and the TMF behavior shown in Figure 3, the TMF lifetimes of the 319 and 356 alloys at the strain amplitude of 0.1% were estimated to be 4152 and 4730 cycles, respectively, both of which were significantly higher than 2000 cycles.

Table 8. Experimental and predicted TMF lifetimes under 0.4% strain amplitude for the two alloys.

	319		**356**	
	Test 1	**Test 2**	**Test 1**	**Test 2**
Experimental	189	175	80	106
Predicted	163	155	61	92

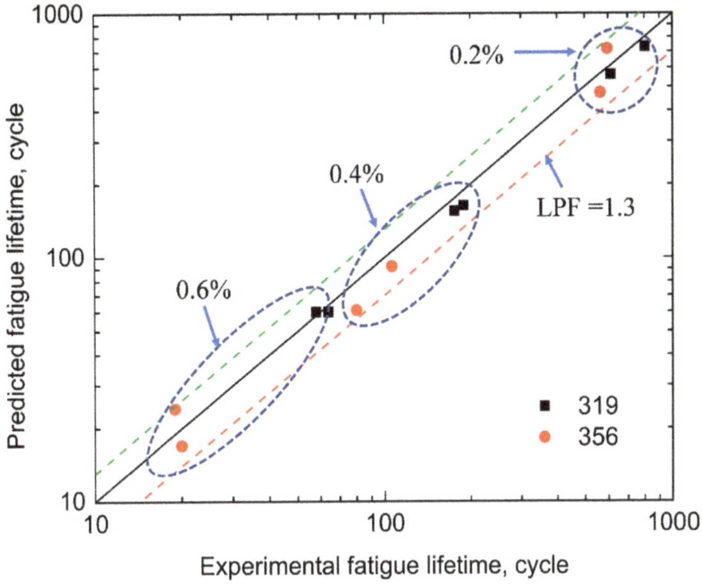

Figure 11. Comparison of predicted and experimental fatigue lifetimes for 356 and 319 alloys.

5. Conclusions

This study investigated and compared the OP-TMF behaviors of two typical Al-Si cast alloys (319 and 356) and the following conclusions were drawn:

1. During TMF loading, asymmetrical hysteresis loops were observed in both alloys and the tensile portion was higher than the compression portion due to out-of-phase cycling;
2. Cyclic softening behavior was observed for the maximum stress in both alloys, but the rate of decrease in the cyclic stress in the 356 alloy was greater than that in the 319 alloy, especially at higher strain amplitudes. Moreover, the 356 alloy presented an earlier inflection point in the cyclic evolution of the cyclic stress;
3. The TMF resistance decreased with increasing strain amplitudes. The 319 alloy exhibited a longer TMF lifetime than the 356 alloy, especially at higher strain amplitudes;
4. The precipitates in both alloys coarsened during TMF cycling. The coarsening rate of the β'-MgSi in the 356 alloy per cycle was higher than that of the θ'-Al_2Cu in the 319 alloy and increased at higher strain amplitudes;
5. The fatigue lifetime predicted using the energy-based model corresponded well with the experimental results, exhibiting a low life prediction factor of 1.3. However, the material parameters varied with the alloys, and the 356 alloy exhibited a lower fatigue damage capacity (W_0) and a higher fatigue damage exponent (β).

Author Contributions: K.L.: conceptualization, methodology, validation, writing—original draft preparation; S.W.: investigation, formal analysis; L.P.: validation, review and editing; X.-G.C.: conceptualization, methodology, validation, review and editing, supervision, funding acquisition. All authors have read and agreed to the published version of the manuscript.

Funding: This research was funded by the Natural Sciences and Engineering Research Council of Canada (NSERC) under grant no. CRDPJ 514651-17.

Institutional Review Board Statement: Not Appliable.

Informed Consent Statement: Not Appliable.

Data Availability Statement: Supporting data are available upon reasonable request.

Acknowledgments: The authors would like to acknowledge the financial support of the Natural Sciences and Engineering Research Council of Canada (NSERC) under grant no. CRDPJ 514651-17 and Rio Tinto Aluminum through the Research Chair in the Metallurgy of Aluminum Transformation at the University of Quebec in Chicoutimi. Special thanks are given to Jian Qin for performing part of the TMF tests.

Conflicts of Interest: The authors declare no conflict of interest.

References

1. Ye, H. An overview of the development of Al-Si-Alloy based material for engine applications. *JMEP* **2003**, *12*, 288–297. [CrossRef]
2. Cole, G.; Sherman, A. Light weight materials for automotive applications. *Mater. Charact.* **1995**, *35*, 3–9. [CrossRef]
3. Garat, M.; Laslaz, G. Improved Aluminium Alloys for Common Rail Diesel Cylinder Heads. 2007. Available online: https://www.researchgate.net/publication/284890980_Improved_aluminum_alloys_for_common_rail_diesel_cylinder_headsJ (accessed on 30 June 2022).
4. Javidani, M.; Larouche, D. Application of cast Al–Si alloys in internal combustion engine components. *Int. Mater. Rev.* **2014**, *59*, 132–158. [CrossRef]
5. Grieb, M.; Christ, H.-J.; Plege, B. Thermomechanical fatigue of cast aluminium alloys for cylinder head applications–experimental characterization and life prediction. *Procedia Eng.* **2010**, *2*, 1767–1776. [CrossRef]
6. Zhang, S.; Wang, Z.; Han, Y.; Zheng, Y.; Zhang, D. Experimental and theoretical studies on thermo-mechanical fatigue test for aluminium cast alloy. *Fatigue Fract. Eng. Mater. Struct.* **2020**, *43*, 110–118. [CrossRef]
7. Mattos, J.; Uehara, A.; Sato, M.; Ferreira, I. Fatigue properties and micromechanism of fracture of an alsimg0.6 cast alloy used in diesel engine cylinder head. *Procedia Eng.* **2010**, *2*, 759–765. [CrossRef]
8. Azadi, M. Effects of strain rate and mean strain on cyclic behavior of aluminum alloys under isothermal and thermo-mechanical fatigue loadings. *Int. J. Fatigue* **2013**, *47*, 148–153. [CrossRef]
9. Azadi, M.; Shirazabad, M. Heat treatment effect on thermo-mechanical fatigue and low cycle fatigue behaviors of A356.0 aluminum alloy. *Mater. Des.* **2013**, *45*, 279–285. [CrossRef]
10. Engler-Pinto, C.; Sehitoglu, H.; Maier, H.; Foglesong, T. Thermo-Mechanical Fatigue Behavior of Cast 319 Aluminum Alloys. In *European Structural Integrity Society*; Rémy, L., Petit, J., Eds.; Elsevier: Amsterdam, The Netherlands, 2002; pp. 3–13.
11. Huter, P.; Oberfrank, S.; Grün, F.; Stauder, B. Thermo-mechanical fatigue influence of copper and silicon on hypo-eutectic Al–Si–Cu and Al–Si–Mg cast alloys used in cylinder heads. *Int. J. Fatigue* **2016**, *88*, 142–155. [CrossRef]
12. May, A.; Belouchrani, M.; Taharboucht, S.; Boudras, A. Influence of heat treatment on the fatigue behaviour of two aluminium alloys 2024 and 2024 plated. *Procedia Eng.* **2010**, *2*, 1795–1804. [CrossRef]
13. Toda, H.; Fukunaga, T.; Kobayashi, M. Improvement of thermomechanical fatigue life in an age-hardened aluminum alloy. *Scr. Mater.* **2009**, *60*, 385–387. [CrossRef]
14. Sehitoglu, H.; Emgler-Pinto, C.; Maier, H.; Foglesong, T. Thermomechanical deformation of AL319 Alloys with different iron contents. In Proceedings of the CAMP 2002—High Temperature Fatigue, Bad Lippspringe, Germany, 3–4 April 2002; pp. 76–83.
15. Wang, Q.; Apelian, D.; Lados, D. Fatigue behavior of A356-T6 aluminum cast alloys. Part I. Effect of casting defects. *J. Light Met.* **2001**, *1*, 73–84. [CrossRef]
16. Firouzdor, V.; Rajabi, M.; Nejati, E.; Khomamizadeh, F. Effect of microstructural constituents on the thermal fatigue life of A319 aluminum alloy. *Mater. Sci. Eng. A* **2007**, *454–455*, 528–535. [CrossRef]
17. Wang, M.; Pang, J.; Liu, X.; Wang, J.; Liu, Y.; Li, S.; Zhang, Z. Optimization of Thermo-Mechanical Fatigue Life for Eutectic Al–Si Alloy by the Ultrasonic Melt Treatment. *Materials* **2022**, *15*, 7113.
18. Moizumi, K.; Mine, K.; Tezuka, H.; Sato, T. Influence of Precipitate Microstructures on Thermal Fatigue Properties of Al-Si-Mg Cast Alloys. *Mater. Sci. Forum* **2002**, *396–402*, 1371–1376. [CrossRef]
19. Beck, T.; Löhe, D.; Luft, J.; Henne, I. Damage mechanisms of cast Al–Si–Mg alloys under superimposed thermal–mechanical fatigue and high-cycle fatigue loading. *Mater. Sci. Eng. A* **2007**, *468–470*, 184–192. [CrossRef]
20. Toyoda, M.; Toda, H.; Ikuno, H.; Kobayashi, T.; Kobayashi, M.; Matsuda, K. Preferential orientation of precipitates during thermomechanical cyclic loading in an aluminum alloy. *Scr. Mater.* **2007**, *56*, 377–380. [CrossRef]
21. Tabibian, S.; Charkaluk, E.; Constantinescu, A.; Szmytka, F.; Oudin, A. TMF–LCF life assessment of a Lost Foam Casting A319 aluminum alloy. *Int. J. Fatigue* **2013**, *53*, 75–81. [CrossRef]
22. Tsuyoshi, T.; Sasaki, K. Low cycle thermal fatigue of aluminum alloy cylinder head in consideration of changing metrology microstructure. *Procedia Eng.* **2010**, *2*, 767–776. [CrossRef]
23. Qin, J.; Racine, D.; Liu, K.; Chen, X.-G. Strain-controlled thermo-mechanical fatigue testing of aluminum alloys using the gleeble 3800 system. In Proceedings of the 16th International Aluminum Alloys Conference (ICAA 16), Canadian Institute of Mining, Metallurgy & Petroleum, Montreal, QC, Canada, 17–21 June 2018; pp. 1–9, Paper no. 401895.
24. Chen, S.; Liu, K.; Chen, X. Precipitation behavior of dispersoids and elevated-temperature properties in Al–Si–Mg foundry alloy with Mo addition. *J. Mater. Res.* **2019**, *34*, 3071–3081. [CrossRef]
25. Jin, L.; Liu, K.; Chen, X. Evolution of dispersoids and their effects on elevated-temperature strength and creep resistance in Al-Si-Cu 319 cast alloys with Mn and Mo additions. *Mater. Sci. Eng. A* **2020**, *770*, 138554. [CrossRef]

26. Farkoosh, A.; Chen, X.G.; Pekguleryuz, M. Dispersoid strengthening of a high temperature Al–Si–Cu–Mg alloy via Mo addition. *Mater. Sci. Eng. A* **2015**, *620*, 181–189. [CrossRef]
27. Liu, K.; Mirza, F.; Chen, X. Effect of Overaging on the Cyclic Deformation Behavior of an AA6061 Aluminum Alloy. *Metals* **2018**, *8*, 528. [CrossRef]
28. Baldan, A. Review Progress in Ostwald ripening theories and their applications to nickel-base superalloys Part I: Ostwald ripening theories. *J. Mater. Sci.* **2002**, *37*, 2171–2202. [CrossRef]
29. Chen, Y.; Doherty, R. On the growth kinetics of plate-shaped precipitates in aluminium-copper and aluminium-gold alloys. *Scr. Metall.* **1977**, *11*, 725–729. [CrossRef]
30. Liu, G.; Zhang, G.; Ding, X.; Sun, J.; Chen, K. Modeling the strengthening response to aging process of heat-treatable aluminum alloys containing plate/disc- or rod/needle-shaped precipitates. *Mater. Sci. Eng. A* **2003**, *344*, 113–124. [CrossRef]
31. Nakajima, T.; Takeda, M.; Endo, T. Accelerated coarsening of precipitates in crept Al–Cu alloys. *Mater. Sci. Eng. A* **2004**, *387–389*, 670–673. [CrossRef]
32. Yassar, R.; Field, D.; Weiland, H. The effect of cold deformation on the kinetics of the β″ precipitates in an Al-Mg-Si alloy. *Metall. Mat. Trans. A* **2005**, *36*, 2059–2065. [CrossRef]
33. Neu, R.; Sehitoglu, H. Thermomechanical fatigue, oxidation, and Creep: Part II. Life prediction. *MTA* **1989**, *20*, 1769–1783. [CrossRef]
34. Miller, M.; McDowell, D.; Oehmke, R. A Creep-Fatigue-Oxidation Microcrack Propagation Model for Thermomechanical Fatigue. *J. Eng. Mater. Technol.* **1992**, *114*, 282–288. [CrossRef]
35. Eichlseder, W.; Winter, G.; Farrahi, G.; Azadi, M. The Effect of Various Parameters on Out-of-phase Thermo-mechanical Fatigue Lifetime of A356.0 Cast Aluminum Alloy. *Int. J. Eng.* **2013**, *26*, 1461–1470.
36. Wang, M.; Pang, J.; Zhang, M.; Liu, H.; Li, S.; Zhang, Z. Thermo-mechanical fatigue behavior and life prediction of the Al-Si piston alloy. *Mater. Sci. Eng. A* **2018**, *715*, 62–72. [CrossRef]
37. Gocmez, T.; Awarke, A.; Pischinger, S. A new low cycle fatigue criterion for isothermal and out-of-phase thermomechanical loading. *Int. J. Fatigue* **2010**, *32*, 769–779. [CrossRef]
38. Riedler, M.; Leitner, H.; Prillhofer, B.; Winter, G.; Eichlseder, W. Lifetime simulation of thermo-mechanically loaded components. *Meccanica* **2007**, *42*, 47–59. [CrossRef]
39. Wang, M.; Pang, J.; Liu, H.; Li, S.; Zhang, M.; Zhang, Z. Effect of constraint factor on the thermo-mechanical fatigue behavior of an Al-Si eutectic alloy. *Mater. Sci. Eng. A* **2020**, *783*, 139279. [CrossRef]
40. Farrahi, G.; Azadi, M.; Winter, G.; Eichlseder, W. A new energy-based isothermal and thermo-mechanical fatigue lifetime prediction model for aluminium–silicon–magnesium alloy. *Fatigue Fract. Eng. Mater. Struct.* **2013**, *36*, 1323–1335. [CrossRef]

materials

MDPI

Article

Mechanical Behavior of Austenitic Steel under Multi-Axial Cyclic Loading

Abhishek Biswas [1,†], Dzhem Kurtulan [2], Timothy Ngeru [2], Abril Azócar Guzmán [1,‡], Stefanie Hanke [2] and Alexander Hartmaier [1,*]

1 Interdisciplinary Centre for Advanced Materials Simulation (ICAMS), Ruhr-Universität Bochum, Universitätstraße 150, 44801 Bochum, Germany
2 Chair of Materials Science and Engineering, Universität Duisburg-Essen, 47057 Duisburg, Germany
* Correspondence: alexander.hartmaier@rub.de
† Current address: VTT Technical Research Centre of Finland Ltd., Vuorimiehentie 2, FI-02150 Espoo, Finland.
‡ Current address: Institute for Advanced Simulations, Forschungszentrum Jülich GmbH, 42525 Jülich, Germany.

Abstract: Low-nickel austenitic steel is subjected to high-pressure torsion fatigue (HPTF) loading, where a constant axial compression is overlaid with a cyclic torsion. The focus of this work lies on investigating whether isotropic J2 plasticity or crystal plasticity can describe the mechanical behavior during HPTF loading, particularly focusing on the axial creep deformation seen in the experiment. The results indicate that a J2 plasticity model with an associated flow rule fails to describe the axial creep behavior. In contrast, a micromechanical model based on an empirical crystal plasticity law with kinematic hardening described by the Ohno–Wang rule can match the HPTF experiments quite accurately. Hence, our results confirm the versatility of crystal plasticity in combination with microstructural models to describe the mechanical behavior of materials under reversing multiaxial loading situations.

Keywords: multi-axial fatigue; crystal plasticity; micromechanical modeling; austenitic steel

Citation: Biswas, A.; Kurtulan, D.; Ngeru, T.; Azócar Guzmán, A.; Hanke, S.; Hartmaier, A. Mechanical Behavior of Austenitic Steel under Multi-Axial Cyclic Loading. *Materials* **2023**, *16*, 1367. https://doi.org/10.3390/ma16041367

Academic Editor: Carlos Garcia-Mateo

Received: 10 December 2022
Revised: 31 January 2023
Accepted: 2 February 2023
Published: 6 February 2023

1. Introduction

Austenitic stainless steels, like 316L and Rex 734, exhibit remarkable corrosion resistance and formability, which makes them suitable for medical implants. However, the leaching of alloying metals like chromium, cobalt, and nickel ion in the human body environment can cause a strong allergic response and is even found to be carcinogenic in rats [1–3]. This led to the development of a low nickel and high nitrogen austenitic stainless steel 1.3808 with the commercial name P558, which exhibits remarkable mechanical properties and good bio-compatibility [4–7] with targeted applications in medical implants. In view of the excellent properties of P558, many advanced manufacturing techniques like liquid-phase sintering [8] and metal powder injection molding [9] have been explored to manufacture complex components. However, the existing research of austenitic stainless steel mechanical properties is mainly focused on stainless steel grades, like 316L or 304L, and few publications report even very basic mechanical properties like yield stress, surface roughness, or hardness of P558 [9,10].

The understanding of material behavior under multiaxial mechanical loading is of great importance for the application of structural materials because components are oftentimes exposed to multiaxial stress states. In addition, the loading is typically time-dependent and reversing. Experimental investigations under such complex loading conditions are very laborious, such that our understanding of the mechanisms leading to plastic deformation and failure of materials under multiaxial reversing loading is quite insufficient. This is particularly important for a newly developed steel, like P558, and the present work aims at closing this gap in our understanding of the mechanical behavior of this new austenitic steel under complex mechanical loading.

To investigate the kinematic hardening behavior of P558 austenitic stainless steel under multiaxial reversing loading, the steel specimens are subjected to combined compression and cyclic torsional load. In preliminary investigations, it has been observed experimentally that, under combined compression and cyclic torsional loading, the specimens undergo plastic axial strain although the compressive load is well below the yield stress of the material. This time-dependent deformation under small stresses can be considered as room-temperature creep behavior, triggered by the multiaxial loading. This compressive creep occurs as soon as a critical angle for the cyclic torsion is exceeded, such that it remains unclear whether an equivalent stress of the applied stress tensor can be defined that describes this deformation mode with sufficient accuracy. Ngeru et al. [11] reported similar behavior in a high nitrogen stainless steel subjected to similar loading conditions. It is noted here, that this kind of creep deformation is purely mediated by dislocation motion, whereas diffusive or other thermally activated mechanisms do not play a role. The term creep is merely used here to describe time-dependent plastic deformation occurring at a constant stress below the yield stress of the material.

Numerical simulations of the mechanical behavior of metals under cyclic loading have been reported frequently in the literature, e.g., Cruzado et al. [12] performed crystal plasticity finite element model (CPFEM) simulations of a polycrystalline microstructure and showed a good agreement between simulation and experiments. In other work, the effects of monotonic-cyclic loading on the behavior of high-alloyed steel and pure copper were investigated and an axial stress-drop was reported for combined cyclic torsion and monotonic axial tensile loading [13].

To incorporate kinematic hardening in constitutive models, mathematical descriptions are required that can predict both the stress–strain hysteresis as well as the relaxation of the mean stress that occurs under cyclic loading [14]. During strain-controlled cyclic loading, the mean stress is defined as the average between the maximum and the minimum stress resulting from the cyclic material response. If the upper and the lower value of the applied strain are not equal with respect to their absolute value, i.e. for asymmetric loading with amplitude ratios different from $R = -1$, the mean stress typically takes a rather high positive or negative values at the beginning of the cyclic experiment, but then it usually relaxes towards zero. The importance of the correct description of kinematic hardening in constitutive models for multiaxial cyclic loading has been demonstrated for investigations on thin-walled tubular specimens under axial tension, and cyclic torsion [15]. To describe kinematic hardening in general, the empirical kinematic hardening models of Armstrong–Fredrick [16], Chaboche [17], and Ohno–Wang [18] are most widely used. Several works implying J2 plasticity [18,19] or crystal plasticity [14] have noted that the Armstrong–Fredrick kinematic hardening model failed to predict the mean stress relaxation that occurs under non-symmetric loading conditions, where the amplitude ratio $R \neq -1$. Schäfer et al. [20] showed a comparison between different kinematic hardening models in CPFEM with loading cases having both symmetric loading $R = -1$ and non-symmetric loading $R = 0$. The results indicate that the Ohno–Wang model is most adept at predicting cyclic mechanical properties. Sajjad et al. [21] used both, J2 as well as CPFEM, with the Chaboche kinematic hardening model for predicting the hysteresis loop and the mean stress relaxation under uniaxial cyclic loading. Based on the aforementioned publications, in this work, the Chaboche kinematic hardening model is used with J2 plasticity, and the Ohno–Wang kinematic hardening model is selected for the CPFEM simulations due to their ability to predict complex cyclic properties for different loading cases and R-values.

The outline of the present work is as follows: Section 2 describes the experimental setup for the EBSD and fatigue experiments. Section 3 explains the modeling strategy for the estimation of elastic stiffness constants, the micromechanical modeling based on experimental input, and the material models used to mimic the material behavior in the numerical model. Section 4 shows the comparison of the results from the numerical model and experiments and provides a discussion of the results and observations. Finally, in Section 5, the conclusions of this work are laid out. To support the reading of the mathematical equations, a list of variables/symbols is provided in the Nomenclature section.

2. Material Characterization

2.1. Specimen Preparation and Setup of EBSD

P558 is the commercial name for the austenitic stainless steel 1.3808 sold by voestalpine BÖHLER Edelstahl GmbH & Co KG (Kapfenberg, Austria). The raw material was delivered in the form of cylindrical bars from which the specimens were produced as specified below. The chemical analysis of the tested material is given in Table 1. Due to the absence of nickel in the alloy, a combination of excellent mechanical properties and high corrosion resistance is achieved, as well as good bio-compatibility. To bring the material into a well-defined state and to remove all cold-deformation from the sample manufacturing process, the material was solution annealed at 1150 °C for 45 min. After that, tensile tests of the material resulted in an ultimate tensile strength of 937 MPa and a yield stress of 595 MPa, where the results of three standard tensile tests have been averaged.

Table 1. Chemical composition of P558.

C	Si	Mn	P	S	Cr	Ni	Mo	Cu	V	Nb	N
0.19	0.41	12.20	0.01	0.00	17.15	0.02	2.89	0.02	0.01	0.03	0.54

For a microscopic analysis, specimens from bar material were cut both in longitudinal and cross-sectional directions according to Figure 1. Consequently, the material was embedded with conductive resin Technotherm 3000 (Kulzer GmbH, Hanau, Germany), ground up to 800 grit, and polished with silica on RotoPol-31 (Struers GmbH, Willich, Germany). The final preparation step for EBSD was vibration polishing of specimens at a frequency of 90 Hz for 3 h on Saphir Vibro (ATM Qness GmbH, Mammelzen, Germany).

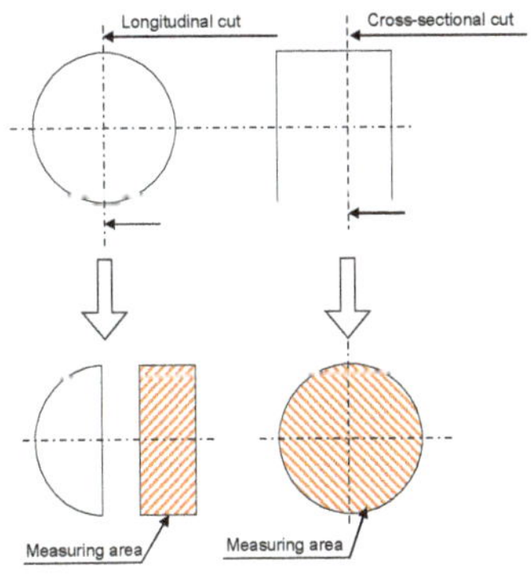

Figure 1. Schematic view of the cutting directions of the specimens for EBSD Analysis.

A Field Emission Gun Scanning Electron Microscope (LEO Gemini 1530 from Carl Zeiss AG) was used to capture micrographs with the electron backscatter detector (EBSD) at a voltage of 20 kV and an aperture size of 120 μm. The working distance was 15 mm, and the spot size varied between 0.5 μm and 2 μm. The EBSD data sets were initially analyzed with the help of the analysis software OIM 6.2 from EDAX (Mahwah, NJ, USA), due to the large number of EBSD data sets, MATLAB scripting (with MTEX toolbox) was used for the post-processing of data sets.

2.2. Microstructure Analysis Using EBSD

In this section, we describe the microstructural features observed in the EBSD data sets, which provide the key inputs for the generation of the virtual microstructure used for

CPFEM simulations. To properly characterize the microstructure across the bulk material, 23 EBSD measurements were performed at different locations of the longitudinal and cross-sectional cuts and for different magnifications, resulting in the characterization of ≈19,000 grains in the underlying microstructure.

Figure 2 shows the (Inverse pole figure) IPF-Z plot of one representative EBSD map and the contour plot of the pole figure of the orientation distribution function (ODF) estimated by combining 23 EBSD data sets. The ODF is estimated using a kernel density estimate [22].

$$f(q) = \sum_{i=1}^{N_g} w_i \psi_\Omega(q_i) \quad (1)$$

where the function f is the ODF, q is the crystallographic orientation, N_g is the number of grains, and $w_i = a_i / \sum a_i$ is the weighting factor for each grain estimated considering the area a_i of the grain. The grains are estimated using a disorientation threshold of 5°, and to avoid experimental/estimation errors the grains smaller than 3 μm are discarded. The grain size is represented using the equivalent diameter, which is the diameter of a circle with the same area ($= 2\sqrt{(a_i/\pi)}$).

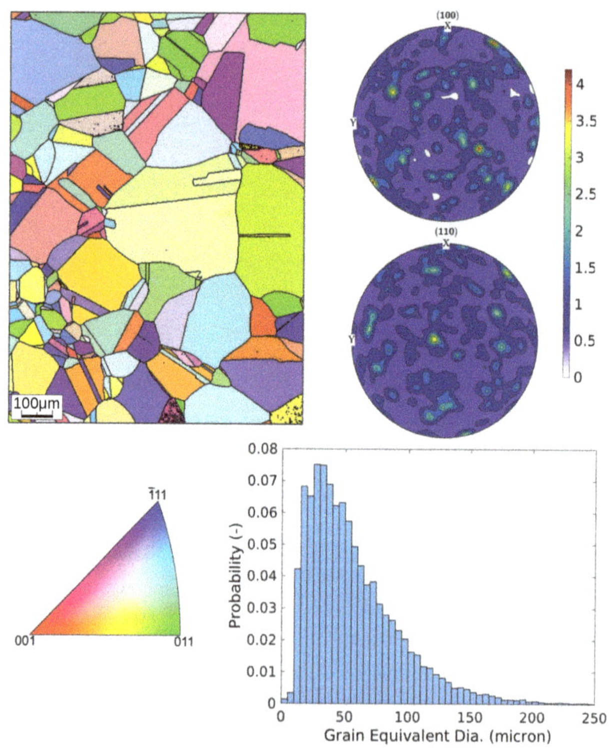

Figure 2. (**top left**) IPF-Z plot of one of the EBSD data sets, (**bottom left**) corresponding IPF key, (**top right**) pole figure contour plot of the ODF estimated by combining all EBSD data sets (i.e., ≈19,000 grains), (**bottom right**) the grain equivalent diameter histogram.

The microstructure exhibits mostly equiaxed grains containing annealing twins, which are formed due to the re-crystallization during the annealing process and which show a density that is proportional to the strain level achieved before the annealing process [23]. Usually, the annealing twins are of Σ3 type with both a symmetrical and asymmetrical grain boundary tilt [24].

2.3. Fatigue Test

Both uniaxial and high-pressure torsion fatigue (HPTF) experiments were conducted on cylindrical double-cone specimens, presented in Figure 3a,b. While uniaxial fatigue specimens have a shorter gauge length, the HPTF specimens possess a thicker gauge section to prevent the buckling of the specimens due to the multiaxial loading case. Before experiments, the gauge section of all specimens was ground with increasingly finer grades of sandpaper and polished with 1 μm diamond paste to remove turning marks. After polishing, the arithmetic mean roughness (R_a) of the specimen surface was 46 ± 6 nm.

A Bionix 858 (MTS Corporation, Eden Prairie, MN, USA) servo-hydraulic universal testing machine was used for all fatigue tests. With the stated testing setup, the specimen can be loaded with 50 kN of maximum axial force and 300 Nm of maximum torsion moment. Uniaxial fatigue tests were performed at 20 ± 2 °C with 40–60% relative humidity under total strain-controlled tension-compression with sinusoidal loading at $R = -1$. The frequency of the experiments was 2 Hz. All experiments were conducted until failure or run out at 2×10^6 cycles.

The frequency of experiments was also controlled, limiting the temperature increase of specimens to 50 K for uniaxial tests (2.5 Hz for $\epsilon < 1.5\%$ and 0.5 Hz for $\epsilon = 1.5\%$). During the HPTF tests, however, the temperature increase may reach values over 200 K and saturated after 400–500 cycles.

Multiaxial loading of the specimens was rotation angle controlled torsion (0° to 5–20°) with sinusoidal loading at $R = 0$ or -1 and a frequency of 2.5 Hz, combined with a monotonous constant compressive stress 250 or 350 MPa. Figure 3 visualizes the special multiaxial loading of the specimen during experiments. The axial force necessary to apply the stated axial stress is related to the smallest cross-section of specimens, measured with a laser thickness gauge with a precision of 10 μm. The experiments are terminated with the observation of clearly visible cracks on the surface of the specimen. During the HPTF experiments, the specimens exhibit compressive axial strains under constant load below the yield stress, which is referred to as axial creep deformation in this work.

To have a proper database for the modeling of the material behavior under HPTF loading for different R values and strains, three different loading conditions were selected: ($R = -1$, $\theta = 7.5°$), ($R = 0$, $\theta = 10°$), and ($R = 0$, $\theta = 15°$) where θ is the amplitude of the cyclic torsion. Figure 4 shows the experimental results for the stress–strain hysteresis loops after the saturation of the mean stress and the axial creep for the selected HPTF loading cases. The equivalent true strain shown in Figure 4 is estimated using

$$\epsilon^{\text{true}} = \log\left(1 + \epsilon^{\text{eng}}\right), \tag{2}$$

where the engineering strain ϵ^{eng} for torsion is defined as

$$\epsilon^{\text{eng}} = \theta \frac{r_g}{l_g}, \tag{3}$$

with the radius r_g and the length l_g of the gauge section. This equivalent true strain, thus, corresponds to the strain reached at the surface of the gauge section. Similarly, the engineering shear stress τ_{eng} at the surface of the gauge section is defined in a linear approximation as

$$\tau^{eng} = T\frac{r_g}{J}, \tag{4}$$

where $J = 0.5\pi r_g^4$ is the polar moment of inertia for the circular cross-section and T is the experimentally measured torque. The general expression for true stress is defined as

$$\sigma^{true} = \sigma^{eng}\left(1 + \epsilon^{eng}\right). \tag{5}$$

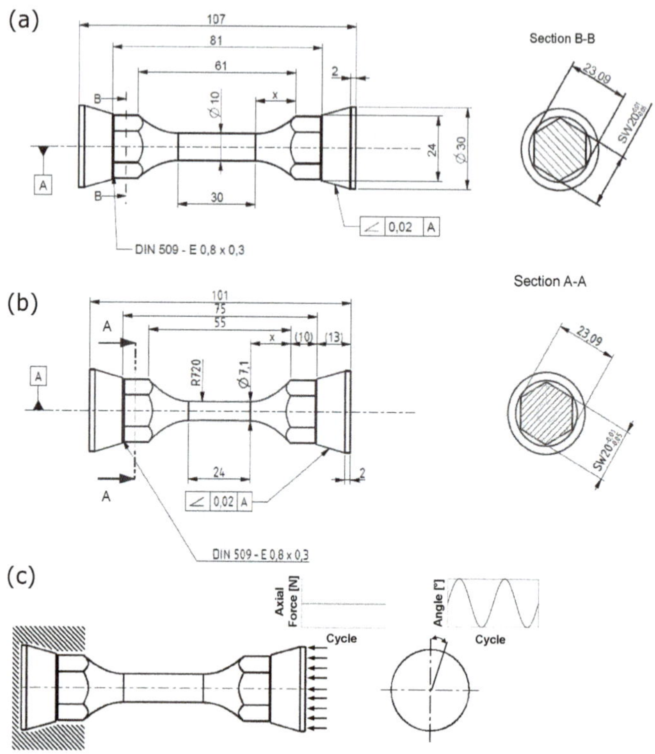

Figure 3. Figure showing the dimensions and surface finish for (**a**) HPTF specimen, (**b**) uniaxial fatigue specimen, and (**c**) HPTF loading of the specimen, indicating the constant axial force and the cyclic torsional angles.

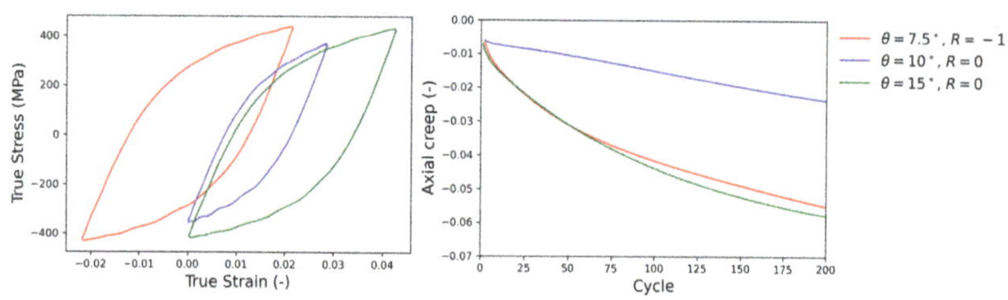

Figure 4. (**left**) Experimental stress–strain hysteresis loops recorded for uniaxial fatigue after the saturation of the mean stress and (**right**) axial creep for the HPTF loading cases given in the legend, where θ is the amplitude of the cyclic torsion loading, and R is the ratio of the minimum vs. maximum torsion angle.

Table 2 summarizes the details of the fatigue experiments performed in this work, where θ denotes the amplitude of torsion cycles for HPTF tests and E_{33} is the loading amplitude in the axial direction (z-axis) for uniaxial fatigue tests.

Table 2. Details of uniaxial fatigue and high pressure torsion fatigue (HPTF) experiments. Axial stress and strain components are denoted by indices "33".

Type	R	Amplitude	Frequency	Axial Pressure
Uniaxial	-1	$E_{33} = 0.55\%$	2.5 Hz	-
HPTF	-1	$\theta = 7.5°$	2.5 Hz	$\sigma_{33} = 250$ MPa
HPTF	0	$\theta = 10°$	2.5 Hz	$\sigma_{33} = 250$ MPa
HPTF	0	$\theta = 15°$	2.5 Hz	$\sigma_{33} = 250$ MPa

3. Micromechanical Model

In this section, the workflow of the micromechanical modeling is described. Due to the recent development of the P558 austenitic stainless steel, the stiffness constants essential for numerical modeling are unavailable in the published literature. Therefore, density functional theory (DFT) calculations are employed to estimate the elastic constants, as detailed in Appendix B. To incorporate the influence of various microstructural features such as inclusions, defects, etc., the ideal stiffness tensor estimated by the aforementioned DFT method using a pristine molecular structure is scaled by a scalar constant λ, such that $\mathbb{C}_{\text{eff}} = \lambda\mathbb{C}_{\text{DFT}}$). The parameter λ is calibrated to match the elastic part of the experimental hysteresis loop with corresponding FE simulations. $\mathbb{C}_{\text{eff}}$ is used to describe anisotropic elastic behavior in CPFEM and, in the case of the J2 model, isotropic elastic constants are determined by homogenizing the elastic stiffness tensor according to the crystallographic texture as quantified by $f(q)$ [25]. Table 3 shows the ideal elastic stiffness parameters obtained with DFT calculations, homogenized Young's modulus (Y) (without scaling), and the scaling factor λ. The values obtained from DFT simulations are comparable to the elastic modulus of 155–220 GPa reported for similar nickel-free stainless steels [26,27].

Table 3. Ideal elastic stiffness parameters C_{11}, C_{12}, and C_{44} obtained by DFT calculations, resulting homogenized Young's modulus Y, and scaling factor λ.

C_{11} (GPa)	C_{12} (GPa)	C_{44} (GPa)	Y (GPa)	λ (-)
263.159	122.644	76.506	185.184	0.63

The microstructure characterization of the present material with EBSD reveals the presence of annealing twins. Castelluccio and McDowell [28] indicate that the annealing twins can influence the fatigue crack initiation. Furthermore, Pande et al. [29] demonstrate that size effects can occur due to annealing twins in the microstructure. However, this work focuses on cyclic property prediction rather than aiming at predicting fatigue failure. Therefore, for the sake of simplicity, annealing twins in the micromechanical model and application of gradient-based plasticity models to describe size effects are disregarded in the present work.

3.1. Representative Volume Element

Due to the high computational effort required for CPFEM simulations with kinematic hardening, a simple representative volume element (RVE) is selected in which each grain is represented by a cubic shape. To optimize the number of elements in the model, a parametric study of the RVE focusing on the number of elements used for the discretization of each grain is performed. This study indicates that an RVE with 512 grains and 8 elements per grain, see Figure 5, provides the optimum balance between mechanical property prediction and computational effort. The edge length of each cubic grain is taken as 0.05 mm, which is estimated as the mean equivalent grain diameter from the EBSD analysis, see Figure 2.

The required number of 512 discrete orientations for the grains constituting the RVE are systematically sampled from the EBSD data for a total of 19,000 grains, using the method in Biswas et al. [30]. The results of this sampling method are shown in Figure 5, and the quality of the sampling is judged using the error function $\|f(q) - \hat{f}(q)\|_1$, where $\|\cdot\|_1$ is the L_1 norm of the difference of experimental ODF $f(q)$ and the reduced ODF $\hat{f}(q)$ used for the RVE simulations. It is seen that sampling of 512 discrete orientations is achieved within an error margin of 10%. Since this work focuses on the macroscopic properties, the grain boundary disorientation distribution is neglected [31].

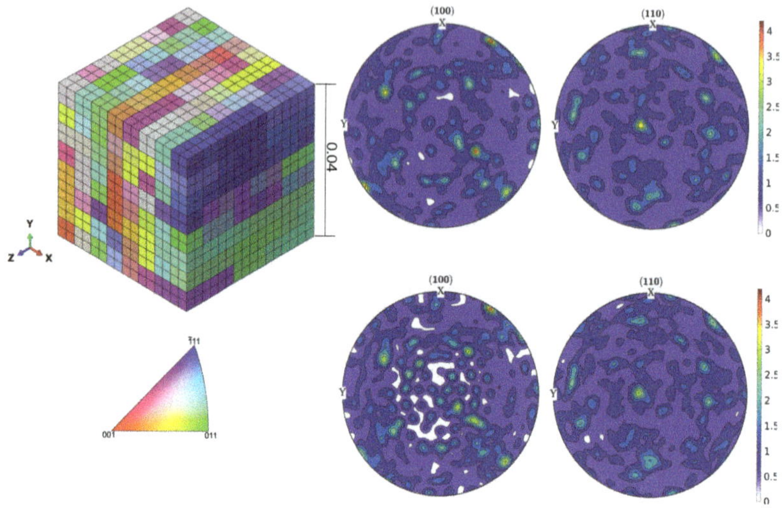

Figure 5. (**top left**) Graphical representation of the used RVE with 512 cubic grains and 8 finite elements per grain, (**bottom left**) grains are colored according to their orientation and the corresponding IPF key. Contour plots of the ODF pole figures from the orientation set obtained by combining all the EBSD data sets of about 19,000 grains (**top right**) and 512 discrete orientations in the RVE (**bottom right**).

3.2. *Material Model*

3.2.1. J2/Von Mises Plasticity

In this section, the J2 model with the isotropic hardening/softening and the Chaboche kinematic hardening model [17] are described briefly. For a detailed description please refer to ABAQUS documentation [32]. The yield function

$$f(\sigma) = \sqrt{\frac{3}{2}(\sigma^{\text{dev}} - \kappa^{\text{dev}}) : (\sigma^{\text{dev}} - \kappa^{\text{dev}})} - \sigma^m, \tag{6}$$

depends on the stress tensor σ and incorporates kinematic hardening via the back-stress κ. The scalar yield stress of the material is denoted as σ^m, and the superscript "dev" indicates that the corresponding deviatoric tensor is addressed. Plastic yielding sets in when the yield function vanishes, i.e., $f(\sigma) = 0$ is used as the yield criterion. If the value of the yield function is larger than zero, a plastic strain increment is calculated that reduces the elastic strain and, thus, relaxes the stress to a value where the yield function is again zero. In this so-called return mapping algorithm, the plastic strain increments are calculated as

$$\mathrm{d}\epsilon_{\text{pl}} = \mathrm{d}\lambda\,\frac{\partial f}{\partial \sigma} = \frac{3}{2}\mathrm{d}\lambda\,\frac{\sigma^{\text{dev}} - \kappa^{\text{dev}}}{\sqrt{\frac{3}{2}(\sigma^{\text{dev}} - \kappa^{\text{dev}}) : (\sigma^{\text{dev}} - \kappa^{\text{dev}})}}, \tag{7}$$

where the direction of the plastic strain is given by the gradient of the yield function with respect to the stress tensor and the magnitude of the plastic strain is quantified by the plastic strain multiplier $\mathrm{d}\lambda$. It is noted here, that for this associated flow rule the direction of the plastic strain increment corresponds to the deviatoric stress tensor reduced by the back stress tensor. For details, the reader is referred to the basic literature on continuum plasticity, e.g. the textbook of de Borst et al. [33].

The isotropic hardening/softening is modeled using an exponential law as

$$\sigma^m = \sigma^m + Q(1 - e^{-b\epsilon_{eq}}), \tag{8}$$

where Q and b are material parameters that can be estimated by matching experimental results, and $\epsilon_{\text{eq}} = \sqrt{(2/3)\epsilon_{\text{pl}} : \epsilon_{\text{pl}}}$is the equivalent plastic strain. The evolution of $\boldsymbol{\kappa}$ is given by

$$\dot{\boldsymbol{\kappa}}_i = C_i\,\dot{\epsilon}_{\text{eq}}\,\frac{1}{\sigma^m}(\boldsymbol{\sigma} - \boldsymbol{\kappa}_i) - g_i\dot{\epsilon}_{\text{eq}}\,\boldsymbol{\kappa}_i\,, \tag{9}$$

where C_i, g_i are the material parameters, and $\dot{\epsilon}_{\text{eq}}$ is the equivalent plastic strain rate. The index $i \in \{1,2\}$ addresses two different back-stresses terms considered here. The total back-stress is computed as $\boldsymbol{\kappa} = \sum_i \boldsymbol{\kappa}_i$.

3.2.2. Crystal Plasticity

This section presents a brief description of the crystal plasticity model used in FE simulations of the mechanical behavior of the RVE. For the sake of brevity, we mainly focus on the modeling of kinematic hardening, please refer to [34] and references therein for a detailed description of the CPFEM method. The total deformation gradient is described using multiplicative decomposition $\mathbf{F} = \mathbf{F}^{\text{e}}\mathbf{F}^{\text{p}}$, where $\mathbf{F}^{\text{e}}$ is the elastic part representing a reversible lattice deformation and rotation, and $\mathbf{F}^{\text{p}}$ is the plastic part representing an irreversible lattice deformation.

The plastic deformation is assumed to result from the motion of dislocations on crystallographic slip systems described by the slip direction $\mathbf{d}^\alpha$ and slip plane normal $\mathbf{n}^\alpha$ where $\alpha = 1, ..., N_s$ is the slip system index, and N_s is the number of slip systems. The driving force for the motion of dislocations is given by the resolved shear stress (τ^α), which is calculated by mapping the second Piola-Kirchhoff stress tensor $\mathbf{S} = \frac{\mathbb{C}_{\text{eff}}}{2}(\mathbf{F}^{\text{e}\,\text{T}}\mathbf{F}^{\text{e}} - \mathbf{I})$ on the slip system α. $\mathbf{I}$ is second rank unit tensor.

In this work, a phenomenological constitutive model is used, in which the kinematic hardening is modeled by including an additional back-stress τ_b^α in the flow rule

$$\dot{\gamma}^\alpha = \dot{\gamma}_0 \left| \frac{\tau^\alpha - \tau_b^\alpha}{\tau_c^\alpha} \right|^{p_1} \text{sign}(\tau^\alpha - \tau_b^\alpha) \tag{10}$$

where $\dot{\gamma}^\alpha$ is the shearing rate, τ_c^α is the slip resistance for isotropic hardening. In this work, the Ohno–Wang model is used, which is adapted from the original formulation to the crystal plasticity model by defining the evolution of the back-stress as

$$\dot{\tau}_b^\alpha = \eta\dot{\gamma}^\alpha - \mu\frac{|\tau_b^\alpha|}{(\eta/\mu)^m}\tau_b^\alpha|\dot{\gamma}^\alpha| \tag{11}$$

following the work of McDowell [35]. Hennessey et al. [14] used two components for τ_b^α using sets of $\{\eta_i, \mu_i, m_i\}$ with $(i = 1, 2)$. For the sake of simplicity, in this work τ_b^α is limited to a single component with only one set of $\{\eta, \mu, m\}$ as parameters. The isotropic hardening is given by the evolution of the slip resistance, described as

$$\dot{\tau}_c^\alpha = \sum_{\beta=1}^{N_s} h_0\xi_{\alpha\beta}\left(1 - \frac{\tau_c^\beta}{\tau_c^f}\right)^{p_2}|\dot{\gamma}^\beta| \tag{12}$$

where τ_c^α is the saturation slip resistance and $\xi_{\alpha\beta}$ is the cross-hardening matrix in which the diagonal elements representing the coplanar slip systems are set to 1.0, and off-diagonal elements representing the non-coplanar slip systems are set to 1.4.

4. Results and Discussion

In this section, the results obtained from numerical models are compared with experimental findings. Due to the multiaxial and heterogeneous stress and strain fields in the gauge section of the specimens, the material parameters cannot be assessed directly from the results of the HPTF test, but the constitutive parameters for J2 and CPFE model must be obtained using an inverse procedure instead, as shown in Appendix C. In the first step, a FE model of the gauge section of the test specimens is applied to characterize the material response under HPTF loading using a J2/von Mises plasticity model with Chaboche kinematic hardening and isotropic softening. In the second step, a representative volume element (RVE) is introduced that considers the mean grain size and crystallographic texture resulting from the EBSD analysis of the tested specimens, as described above. In FE simulations of this RVE, boundary conditions are applied that mimic the HPTF loading on a volume element close to the surface of the gauge section of the experimental specimens. The plastic deformation of each grain in the RVE is described by the CP model introduced above. It is noted here that in the present study, we do not attempt to model the grain shapes or even microstructural details as annealing twins, but we focus on the determination of the material parameters, which requires the use of a numerically efficient model. Furthermore, a parametric study of the CP kinematic hardening parameters is shown in Appendix A, emphasizing the importance of axial creep experiments for the calibration of parameters.

4.1. J2 Plasticity

The parameters for the J2 model are calibrated using data from uniaxial fatigue experiments. The parameters controlling the stress–strain hysteresis loop, i.e., C_1, C_2, g_1, and g_2, are obtained in an iterative process in which the difference between experimental and numerical hysteresis loops are minimized by systematically varying those material parameters. For this purpose, a representative hysteresis loop is chosen from the regime of load cycles in which the mean stress has already saturated. In contrast, the isotropic softening parameters, i.e., σ^m, Q, and b, are gained from the comparison of the maximum stress obtained during each load cycle between simulation and experiment. To obtain the hysteresis loop in the regime where the amplitude is constant, the experimental hysteresis loop from the 5000th cycle is used for calibration of the kinematic hardening parameters. The isotropic softening parameters are calibrated by simulating the initial 300 cycles.

Figure 6 shows the comparison of the saturated hysteresis, i.e., the hysteresis loop in the regime where the stress amplitude is constant, and the maximum stress obtained from simulation and experiments; the corresponding calibrated J2 material parameters are given in Table 4.

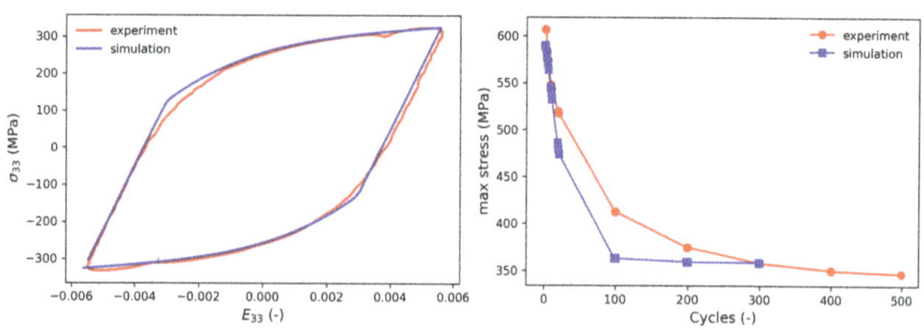

Figure 6. (**left**) Stress–Strain hysteresis loop and (**right**) reduction of the maximum stress in uniaxial fatigue tests with a strain amplitude of $E_{33} = 0.55\%$ and an amplitude ratio of $R = -1$.

Table 4. Calibrated J2 model parameters.

C_1 (MPa)	C_2 (MPa)	g_1 (-)	g_2 (-)	σ^m (MPa)	Q (MPa)	b (-)
43,106.44	4192.81	513.95	0.0	545.82	−300	3.17

Keeping the J2 material parameters constant, HPTF simulations are performed with the same FE model, see Figure 7, in which the resulting axial creep strain E_{33} obtained from the experiment and the HPTF simulations also are shown. It is seen that the simulations result in a similar axial creep behavior under different HPTF loading conditions, which stands in stark contrast to the experiment where the axial creep behavior is very sensitive to the torsional loading. Furthermore, the axial creep strains obtained from the FE simulations with material parameters fitted to uniaxial fatigue experiments grossly overestimate the experimentally found creep strains under HPTF loading, which points out the limitations of the J2 plasticity model. The reason for these deficits in the J2 model in predicting the axial creep behavior correctly can be attributed to the associated flow rule used here, see Equation (7), where the plastic strain increment is proportional to the deviatoric stress reduced by the back stress. Hence, as long as constant stress is applied along the axial direction, the plastic strain in this direction will not be sensitive to the amplitude of the cyclic torsional load, which is clearly seen in the simulation results for axial creep in Figure 7. Since this is a fundamental restriction, no attempt is made to test other modeling strategies for J2 plasticity, although in the literature, it has been demonstrated that an associated flow rule in combination with a three-surface hardening-recovery model for kinematic hardening could be successfully parameterized to describe the material behavior of thin walled tubular specimens under monotonic-cyclic loading [15].

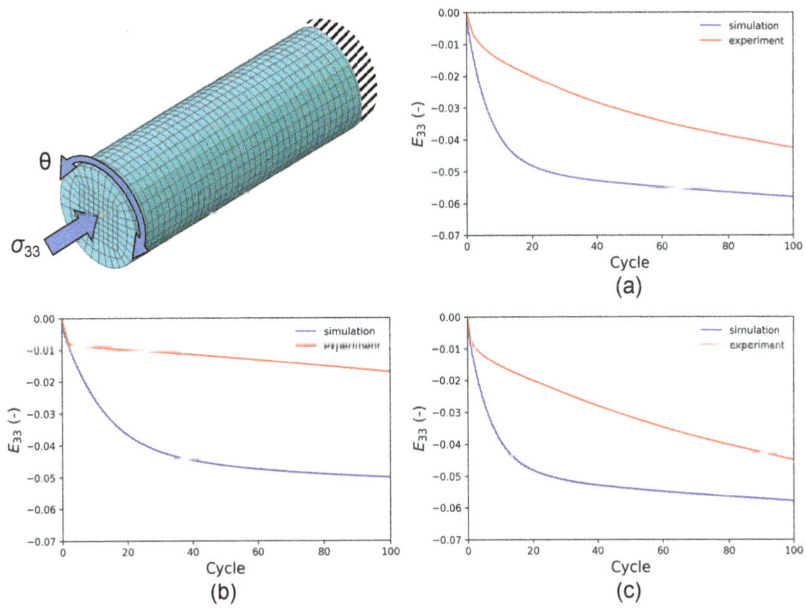

Figure 7. (**top left**) FE model used to simulate HTPF loading with a J2 plasticity model and resulting axial creep strain (E_{33}) of simulation and experiment vs. cycle number. The axial pressure amounted to $\sigma_{33} = 250$ MPa and three different combinations of amplitude ratios R and cyclic torsion amplitudes θ are shown: (**a**) $R = -1$, $\theta = 7.5°$, (**b**) $R = 0$, $\theta = 10°$, and (**c**) $R = 0$, $\theta = 15°$.

To understand the distributions of axial stress and axial plastic strain in the J2 model, these quantities are shown along a cut through the gauge section in Figure 8. The contour plots indicate that both, axial stress and plastic strain, are rather constant in each cross-section of the model. The axial stress, however, exhibits elevated compressive stresses close to the symmetry axis, where high hydrostatic stress components are observed.

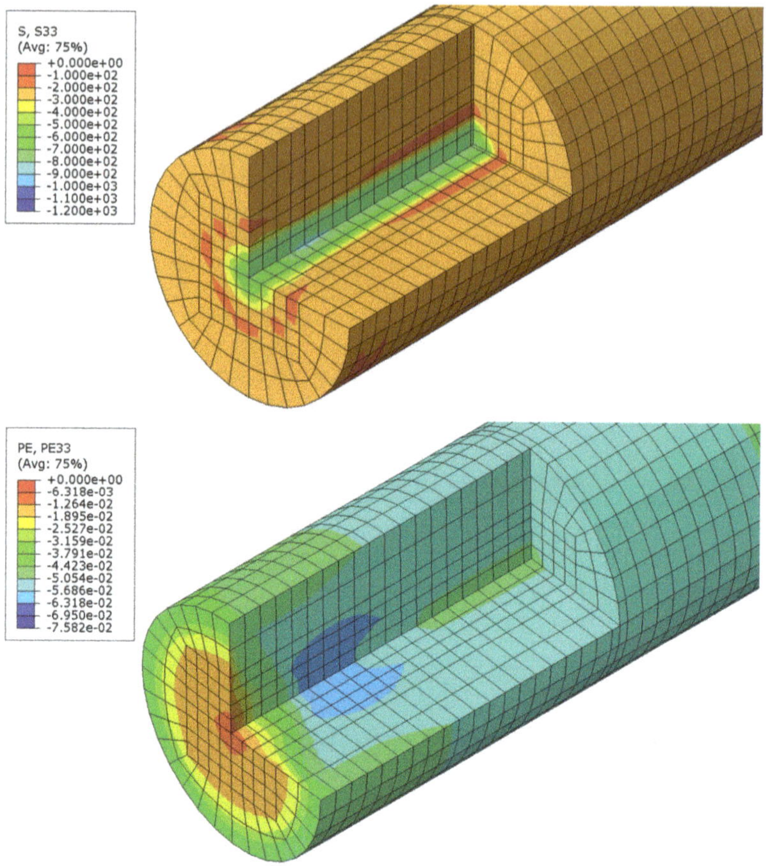

Figure 8. Contour plot of the axial stress in MPa (**top**) and the axial plastic strain (**bottom**) along a cut through the J2 model.

4.2. *Crystal Plasticity*

Since the RVE with 512 grains is much smaller than the dimension of the experimental specimens, we cannot directly apply torsional loading on this RVE. Instead, we consider a small volume element close to the surface of the cylindrical FE model representing the gauge section of the experimental specimens, which has been used for the J2 plasticity calculations, see Figure 9. The distortions experienced by this volume element during the different HPTF loading cases are applied as boundary conditions to the CPFEM model to achieve a comparable loading situation on the length scale of the RVE. The major distortion occurs in the X–Z and Y–Z shear directions (tensor components E_{13} and E_{23}, respectively). The RVE is subject to displacement boundary conditions resulting in X–Z shear, and a distributed force boundary condition in the Z direction, the magnitude of the force is adjusted such that the homogenized stress in the RVE matches the required axial pressure for the given HPTF load, and similarly, the shear displacement is adjusted to match the strain obtained from the saturated hysteresis loop of the HPTF experiment. Furthermore, as shown in the J2 model, the axial plastic strain and the axial stress are rather homogeneous throughout the gauge section, which justifies the application of the distortions of a volume element close to the surface as boundary conditions to the RVE, as the plastic strains observed in the RVE will be representative for the entire gauge section.

It is noted here, that the RVE is exposed to periodic boundary conditions to avoid free surfaces, which would severely influence the stress field within the RVE and produce finite size effects. This simplification is justified because we do not attempt to study crack initiation, for which free surface effects might play a crucial role.

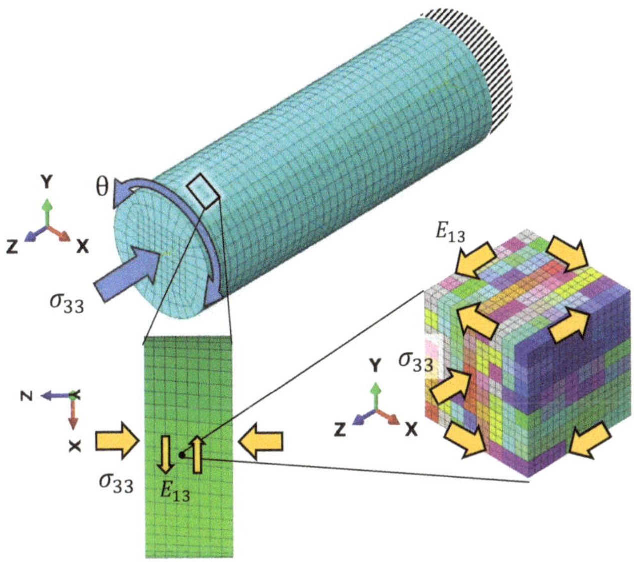

Figure 9. To mimic HPTF loading conditions on the length scale of the RVE used for CPFEM simulations, the distortions of a small volume element close to the surface of the cylindrical FE model, representing the gauge section of the HPTF specimens, are applied as boundary conditions to the RVE.

In the first step to calibrate the CP parameters, the HPTF experiment with $R = -1$ and $\theta = 7.5°$ is used. Only the kinematic hardening parameters η and μ can be calibrated in this way, whereas the isotropic hardening parameters τ_0^f, h_0, and p_2 are temporarily set to zero and will be calibrated in a second step as described below. To validate the correctness and generality of the CP parameters obtained by this procedure, the calibrated model is verified by comparing the simulation results with experiments for $R = 0, \theta = 10°$ and $R = 0, \theta = 15°$. Figure 10 shows that the stress–strain hysteresis loops obtained from HPTF simulations using the RVE model closely match the experiments for all loading cases, which confirms the predictive capability of the parameterized model.

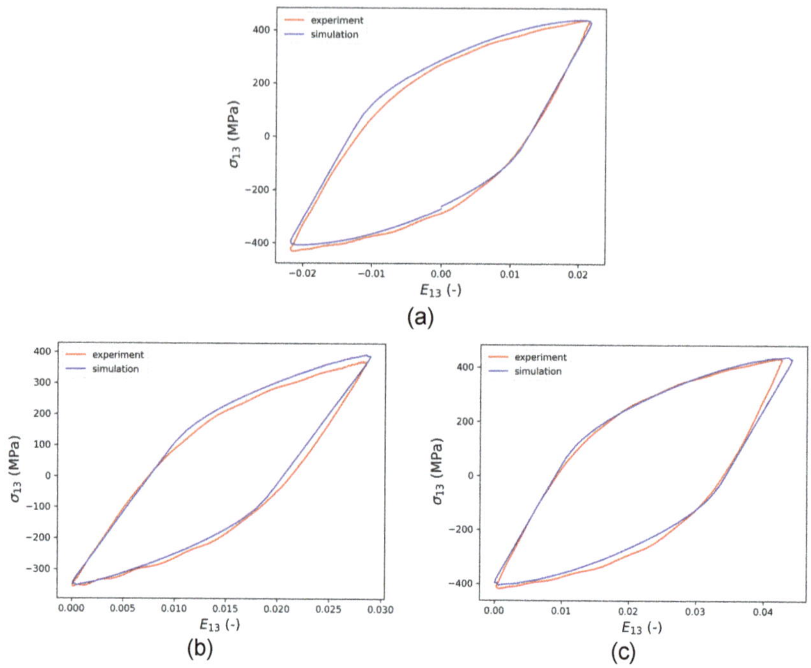

Figure 10. Comparison of saturated HPTF stress–strain hysteresis loops between simulation and experiment; the constant axial pressure amounts to $\sigma_{33} = 250$ MPa in all cases and cyclic shear loading E_{13} is applied to mimic the cyclic torsion in the experiment. (**a**) The result of calibration for HPTF loading with $R = -1$, $\theta = 7.5°$, $E_{13} = 0.02$. The two subfigures in the bottom represent predictions of the model for (**b**) $R = 0$, $\theta = 10°$, $E_{13} = 0.028$, and (**c**) $R = 0$, $\theta = 15°$, $E_{13} = 0.04$.

The calibration of the isotropic hardening parameters (τ_c^f, h_0, and p_2) and the remaining kinematic hardening parameter m, requires matching the evolution of maximum and minimum stress recorded during all cycles between simulation and experiment. Again, this is done for loading with ($R = -1$, $\theta = 7.5°$), and the result is validated by comparison with the other cases. As seen in Figure 11, a satisfying agreement between simulation and experiment could be achieved in this way. All resulting CP parameters are given in Table 5 and referred to as set 1.

Table 5. Calibrated CPFE model parameters (set 1).

λ (-)	$\dot{\gamma}_0$ (1/s)	p_1 (-)	τ_0 (MPa)	τ_c^f (MPa)	h_0 (MPa)	p_2 (-)	η (MPa)	μ (-)	m (-)
0.63	0.001	25	235	184.2	−25.0	2	16,888.0	125.0	2.0

As indicated in the parametric study shown in Appendix A, the axial creep behavior is also employed for the calibration of the CP parameters. Figure 12 shows the axial creep in the RVE during HPTF simulations. The results indicate that the CPFE model does not only predict the cyclic properties but also the axial creep behavior for the different HPTF loading cases very well.

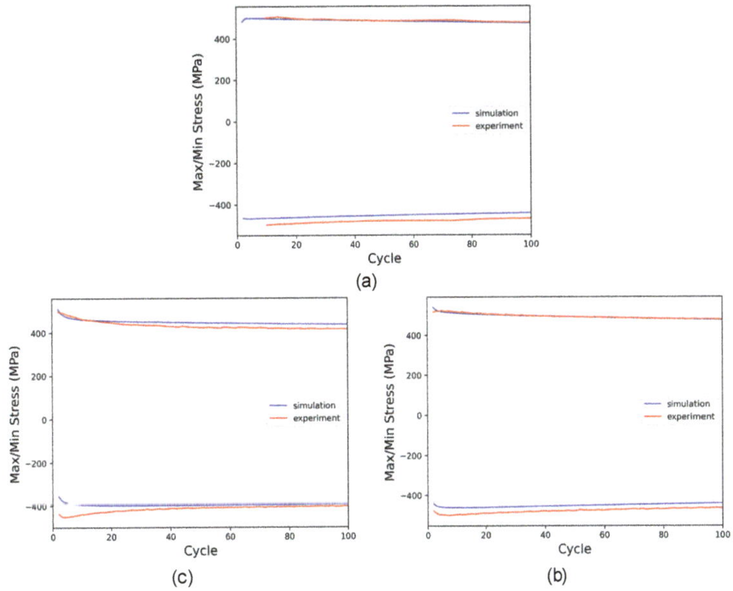

Figure 11. Comparison of maximum and minimum stress obtained in experiment and simulation during torsional cycles with (**a**) $R = -1,\ \theta = 7.5°$, comparison of cyclic softening behavior for the loading cases (**b**) $R = 0,\ \theta = 10°$, and (**c**) $R = 0,\ \theta = 15°$.

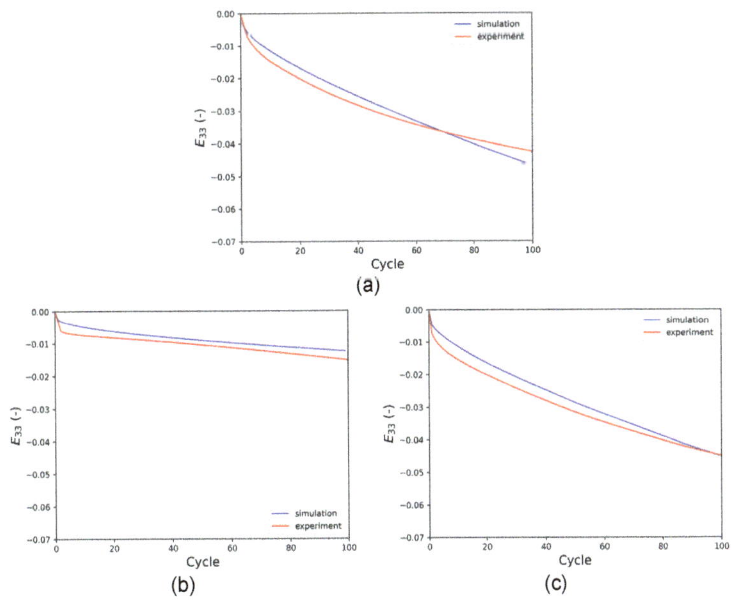

Figure 12. Axial creep strain (E_{33}) vs. the number of cycles obtained from HPTF simulations with the CPFE model and from experiment for a constant axial pressure of $\sigma_{33} = 250$ MPa and cyclic torsional loading: (**a**) $R = -1,\ \theta = 7.5°$) used for calibration, (**b**) $R = 0,\ \theta = 10°$, and (**c**) $R = 0,\ \theta = 15°$. Cases (**b**,**c**) represent predictions of the model.

In the next step, the CPFE model calibrated with HPTF experiments (set 1) is used to predict material behavior under uniaxial fatigue loading. It is pointed out here that it is expected that the material parameters fitted to more complex loading situations should be able to describe the material behavior under uniaxial loading (see, for example, the study of de Castro e Susa [36], where the authors tested various protocols to determine material parameters for cyclic plasticity using inverse procedures. However, the comparison of the simulation and experiments shown in Figure 13a exhibits the limitations of this approach, as the predicted and measured stress–strain hysteresis loops for uniaxial loading differ significantly from the parameters of set 1 that yield good comparability between experiment and simulation for the HPFT case, as seen in Figure 13b.

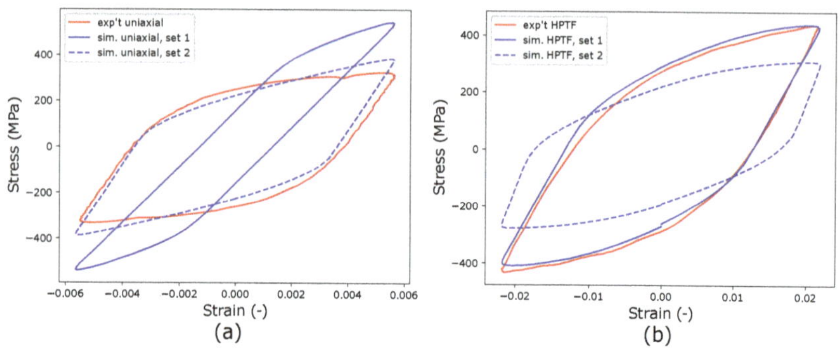

Figure 13. Comparison of experimental and simulated stress–strain hysteresis loops (**a**) for uniaxial fatigue with the strain amplitude $E_{33} = 0.55\%$ and (**b**) under HPTF loading with ($R = -1$, $\theta = 7.5°$, $E_{13} = 0.02$). The material parameters for set 1 (Table 5) were obtained using an inverse analysis of the HPTF experiments, whereas the parameters of set 2 (Table 6) were fitted to uniaxial experiments.

To understand the reason for this discrepancy in the prediction of the CPFE model, the experimental stress–strain hysteresis loops for HPTF and uniaxial fatigue tests are analyzed more closely. Figure 14 shows the comparison of the experimental results for saturated hysteresis loops under HPTF loading ($\theta = 7.5°$, $R = -1$) and uniaxial fatigue loading ($E_{33} = 0.55\%$, $R = -1$). Since the elastic stiffness in the former case is dominated by the materials' shear modulus and in the latter case by Young's modulus, the stress is normalized by the respective elastic quantities, which are estimated from the stiffness parameters given in Table 3 as Young's modulus $Y = 185.2$ GPa and shear modulus $G = 76.5$ GPa.

It is seen that the effective slopes of the elastic branches of the HPTF and uniaxial hysteresis loops are significantly different, even after the proper scaling is applied, and that the elastic response of the material under uniaxial fatigue is approximately 1.46 times stiffer than under HPTF loading.

Due to this difference in the elastic slope, the effective elastic parameters required for the uniaxial fatigue simulations are different from those used for the HPTF simulation, which can be controlled by the scaling factor λ in this work. A similar change in the stress–strain relationship due to pre-straining has been reported in the experimental work of Vrh et al. [37]. A degradation of Young's modulus by more than 20% during fatigue tests was reported for dual-phase steels in [38]. It can be interpreted from the experimental work of Wilshire and Willis [39], Stout et al. [40], and Schneider et al. [41] that the magnified effect of defects (especially dislocation structures) due to pre-straining in HPTF experiments can lead to the change in the stress–strain behavior. It is also pointed out here that the temperature increase during the HPTF experiments is significantly higher than that during the uniaxial tests (200 K vs. 50 K), which might also lead to a reduction in the elastic stiffness and the yield stress of the austenitic steel.

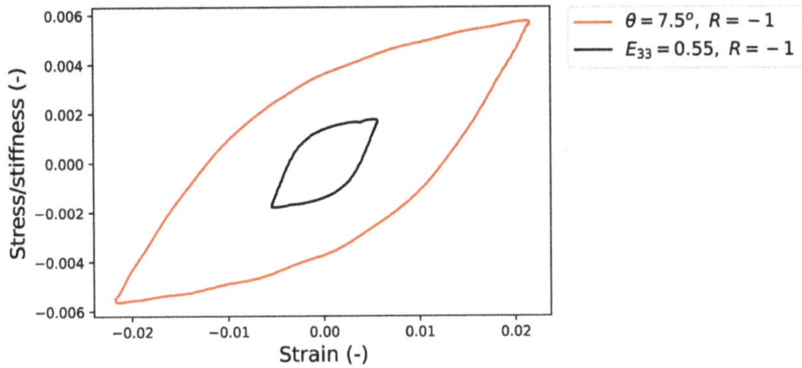

Figure 14. Comparison of the hysteresis loops in the regime where the stress amplitude is constant for the uniaxial fatigue experiment ($E_{33} = 0.55\%$, $R = -1$) and the HPTF loading case ($\theta = 7.5°$, $R = -1$), where θ and E_{33} are the amplitudes of the cyclic deformation. The values of all stresses in the plot are normalized with the corresponding stiffness parameters, i.e., Young's Modulus $Y = 185.2$ GPa for uniaxial fatigue and shear modulus $G = 76.506$ GPa for the torsional loading during HPTF.

To demonstrate the possibility of calibrating the material parameters used for CPFE simulations to describe uniaxial fatigue behavior, the scaling parameter for elastic properties λ, the critical resolved shear stress τ_0, and the isotropic hardening parameters ($\{\tau_c^f, h_0\}$) are re-calibrated to match uniaxial experiments, while keeping the kinematic hardening parameters unchanged. The re-calibrated uniaxial parameters, referred to as set 2, are given in Table 6. The CPFE results obtained for both loading cases are plotted in Figure 13 as set 2. It is seen that this data set describes the stress–strain hysteresis under uniaxial loading very well, whereas the results for HPTF are not represented in an acceptable manner for set 2. A closer analysis of the material parameters of set 1 and set 2 reveals that the ratio of the fitted elastic constants in the CP models of both loading cases amounts to 1.51, which is close to the ratio of the effective stiffness values read from the experimental curves, which takes a value of 1.46. Furthermore, the scaling parameter λ used to adapt the elastic constants from DFT calculations to fit the uniaxial experiments is 0.95 and thus close to unity, which demonstrates the predictive capabilities of DFT methods concerning the calculation of elastic parameters. Generally, this finding that the transfer of crystal parameters obtained by inverse analysis of data referring to one experiment cannot be simply transferred to predict another experiment indicates a limitation in our current constitutive models and their parameters.

However, this analysis also reveals, that the kinematic hardening parameters obtained from uniaxial fatigue or HPTF loading are consistent, whereas the effective elastic parameters and, with them, some basic plastic properties, such as critical resolved shear stress and isotropic hardening parameters, take different values for the two different loading cases. The difference in these parameters can possibly be attributed to the higher temperatures that occur during HPTF testing or to different dislocation structures that result from multiaxial vs. uniaxial deformations. However, more research is required to draw a firm conclusion and to derive a model that describes these differences quantitatively. It is also pointed out here, that care needs to be taken when transferring material parameters obtained for particular loading cases to different scenarios.

Table 6. Re-calibrated CPFE model parameters for uniaxial fatigue loading (set 2). The remaining material parameters are unchanged with respect to set 1 and given Table 5.

λ (-)	τ_0 (MPa)	τ_c^f (MPa)	h_0 (MPa)
0.95	188	92	−130.0

5. Conclusions

In this work, we investigated the mechanical behavior of austenitic stainless steel 1.3808 (commercial name P558), exposed to high-pressure torsion fatigue (HPTF), i.e., overlaid constant axial and cyclic torsional loading. Experimental tests revealed that under such multiaxial reversed loading, the material undergoes plastic deformation in the axial direction if the amplitude of the cyclic torsion exceeds a critical value, even if the axial stress is well below the yield stress of the tested material. Since this time-dependent axial deformation occurs under a constant axial stress that is smaller than the yield stress, it can be considered a low temperature creep phenomenon, although its mechanisms are the same as those for plastic deformation and do not involve diffusive processes. Such material behavior has not yet been widely described in the literature. However, as multiaxial loading occurs frequently in technical systems or medical implants, it is essential to understand the mechanisms leading to this kind of plastic deformation. In order to accomplish this, two different numerical models, J2 and crystal plasticity (CP), were employed in finite element simulations of the HPTF experiments. The results indicate that the J2 model with an associated flow rule is not suitable to describe material behavior under such multi-axial cyclic loading cases. In contrast, the CP model with an Ohno–Wang kinematic hardening model reveals a good agreement with the experiment and even was able to predict the material response for different HPTF loading cases after fitting the material parameters to experimental data from one individual test.

A comparison between uniaxial fatigue experiments and HPTF loading demonstrated that there was a significant difference in the elastic branches of the respective stress–strain hysteresis loops. The origin of this discrepancy might lie either in the significantly higher temperature that builds up in the material during HPTF testing, in different dislocation structures that develop during the different test conditions, or in a combination of both. Consequently, the CP model calibrated by an inverse procedure based on HPTF data is incapable of matching the results of uniaxial fatigue experiments. However, it was demonstrated that a re-calibration of the material stiffness tensor and isotropic hardening parameters by a scalar factor, while leaving the kinematic hardening parameters unchanged, leads to a good description of the data obtained from uniaxial fatigue experiments. To understand the correlation between the constitutive parameters and the resulting material behavior in more detail, a parametric study of the CP kinematic hardening model under HPTF loading was performed, and the resulting stress–strain hysteresis loops and axial creep deformation were analyzed. The most important result of this case study is that a reliable parameter identification by an inverse modeling approach based on experimental data is possible, and that including axial deformation into the parameter fitting procedure improves the calibration of the parameters for kinematic hardening. Generally, our results indicate that material parameters that are identified by an inverse analysis of experimental data from more complex situations can be used to describe simpler loading conditions, whereas material parameters identified from simpler tests typically cannot be used to describe more complex loading situations. However, it must also be pointed out here that there are cases where material parameters obtained by inverse analysis of some test conditions simply cannot be generalized to other situations, which is a clear limitation of current constitutive models and the parameters they rely on.

The finite element model used for the simulation of the material response based on crystal plasticity employed a simple representative volume element (RVE) that mimics the crystallographic texture obtained from the EBSD analysis of the P558 steel specimens used in the experiment. This RVE contained 512 cubic grains, thus ignoring microstructural features like grain shape or annealing twins—a simplification that was necessary to minimize the numerical effort in this work. In forthcoming studies, it is planned to investigate the influence of grain shapes, annealing twins, and crystallographic texture on the material behavior under HPTF loading in more detail by considering more realistic grain geometries in the RVE simulations.

Author Contributions: Conceptualization, all authors; methodology, A.B., A.H.; software, A.B., A.A.G. and A.H.; validation, A.B., D.K. and T.N.; formal analysis, A.B., D.K., T.N. and A.A.G.; investigation, D.K., T.N. and S.H.; resources, S.H., A.H.; writing—original draft, all authors; writing—review & editing, all authors; visualization, A.B., D.K., T.N. and A.H.; supervision, S.H., A.H.; project administration, S.H., A.H.; funding acquisition, S.H., A.H. All authors have read and agreed to the published version of the manuscript.

Funding: This research was funded by the Deutsche Forschungsgemeinschaft (DFG, German Research Foundation) under grant number 441180620.

Institutional Review Board Statement: Not applicable.

Informed Consent Statement: Not applicable.

Data Availability Statement: Data will be made available upon request.

Acknowledgments: The authors thank voestalpine BÖHLER Edelstahl GmbH & Co KG, Kapfenberg, Austria, and in particular Rainer Fluch, for supporting this project and providing the steel investigated here.

Conflicts of Interest: The authors declare no conflict of interest.

Nomenclature

Variable	Description
σ, σ_{ij}	Cauchy stress
$\mathbf{S}$	Second Piola Kirchhoff stress
$\mathbf{E}, E_{ij}$	Green-Lagrange strain
$\epsilon_{\mathrm{pl}}, \epsilon_{\mathrm{eq}}$	Plastic strain tensor, equivalent plastic strain
ϵ_{eng}	Engineering strain
$\dot{\gamma}^{\alpha}$	Shear rate of dislocation on slip system α
γ_0	Initial shear rate
τ^{α}	Resolved shear stress for slip system α
τ_0	Initial guess for the slip resistance
$\{\sigma_m, Q, b\}$	Isotropic hardening material parameter for J2 plasticity model
$\{C_i, g_i\}$	Chaboche kinematic hardening material parameter for J2 plasticity model
$\{\tau_c^f, h_0, p_2\}$	Isotropic hardening parameters for CP model
$\{\eta, \mu, m\}$	Onho-Wang Kinematic hardening parameters for CP model
Y, G	Young's modulus and shear modulus
ϵ_{eq}	Equivalent strain
$\mathbb{C}$	Stiffness tensor
$\mathbf{F}$	Deformation gradient
$\mathbf{F}^{\mathrm{e}}$	Elastic deformation gradient
$\mathbf{F}^{\mathrm{p}}$	Plastic deformation gradient
$\mathbf{I}$	Unit tensor
q	Crystallographic orientation
w_i	Weight of ith crystallographic orientation
f	Orientation distribution function from experimental data
$\hat{f}$	Orientation distribution function from reduced discrete orientations
ψ_{Ω}	Kernel density function with half-width Ω
R	Ratio of maximum and minimum strains in cyclic hysteresis loop
θ	Amplitude of torsional cyclic deformation

Appendix A. Parametric Study of CPFE Parameters

In this section, a parametric study of the CP parameters for kinematic hardening is performed, we focus particularly on the kinematic hardening parameters $\{\eta, \mu, m\}$, while keeping other parameters $\{ \dot{\gamma}_0, p_1, p_2, h_0, \tau_0, \tau_c^f \}$ constant. As reported by Schäfer et al. [20] the CP kinematic hardening parameters η, μ mainly influence the stress–strain hysteresis loop, and m controls the mean stress relaxation. Therefore the parametric study is divided into two parts: (i) characterize the influence of η and μ on saturated stress–strain hysteresis loop, and (ii) quantify the influence m on the mean stress relaxation in combination with the axial creep behavior during HPTF loading. This study indicates that several sets of CP parameters for kinematic hardening can produce a similar stress–strain hysteresis loop with minor differences, however, parameters η and μ have a strong effect on the axial creep behavior.

For the parametric study, the hysteresis stress–strain loop (without isotropic hardening) and the axial creep values are compared for the loading case $\theta = 7.5°$, $R = -1$. To ensure the saturation of the mean stress value, the hysteresis loop is taken after 25 cycles, and the isotropic hardening is neglected. In the case of the axial creep simulation, the isotropic hardening parameters are kept constant (as reported in Table 5). Figure A1 shows the effect of the η and μ on the stress–strain hysteresis loop and the axial creep behavior. The results indicate that both the stress–strain hysteresis loop and the axial creep behavior are important for the calibration of the CP kinematic hardening parameters η and μ.

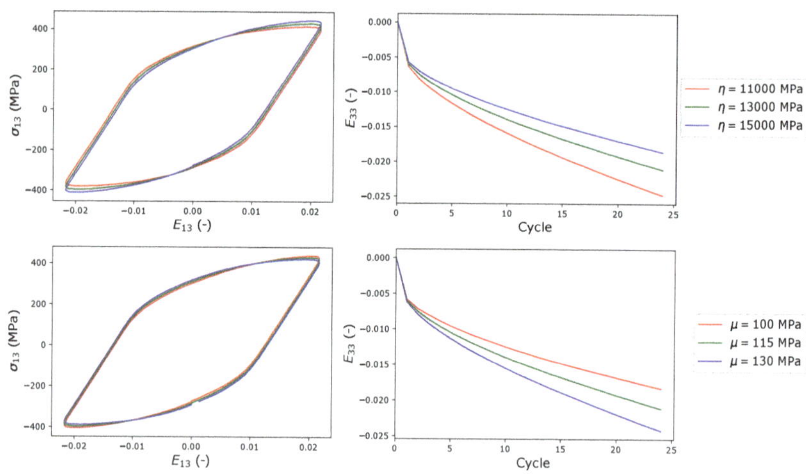

Figure A1. Effect of CP kinematic hardening parameters η and μ on the stress–strain hysteresis loop and the axial creep behavior.

To study the effect of the kinematic hardening parameter m on the mean stress relaxation and the axial creep deformation during HPTF the loading case $\theta = 10°$, $R = 0$ is chosen. Figure A2 shows the results from the simulation and illustrates that a reduction of m leads to lower values of the mean stress and increases the increments of axial creep per torsion cycle. Although the general trend in the change of the mean stress and the axial deformation with respect to the change of the parameter m is similar, the mean stress changes more uniformly (rather linearly) with the respect to m in comparison to axial deformation, which has a progressive effect for smaller values of m.

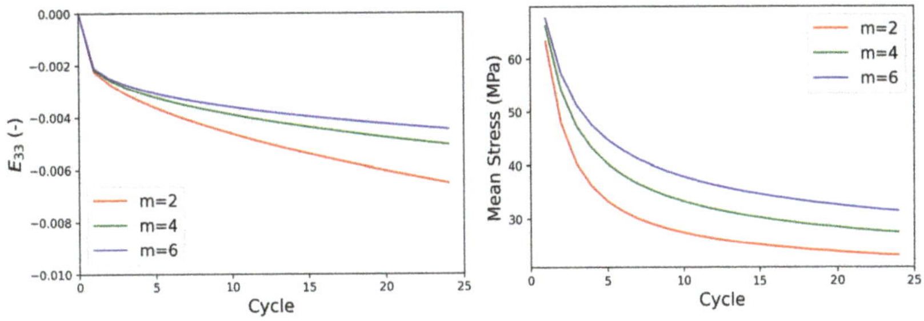

Figure A2. Effect of the kinematic hardening parameter *m* on the axial creep behavior and the mean stress relaxation during HPTF loading.

Appendix B. Elastic Constants Estimation

Density functional theory (DFT) calculations were employed to estimate the elastic constants using the VASP [42–44] software. The calculations were performed with the projector augmented-wave (PAW) method [45], as implemented in VASP, and the generalized gradient approximation, which used a Perdew-Burke-Ernzerhof (PBE) parametrization [46]. The structure of face-centered cubic Fe was constructed with repetitions of $4 \times 2 \times 2$ of the four-atom unit cell. Correspondingly, a $4 \times 8 \times 8$ k-point mesh of the Monkhorst-Pack [47] type was created. The substitutional alloying elements were added randomly, assuming a disordered structure. The final composition of the supercell was $Fe_{44}Cr_{11}Mn_7Mo_2C_1$, where carbon occupies octahedral lattice sites. Spin-polarized calculations were carried out to account for a ferromagnetic Fe state. The plane–wave basis cut-off energy was 500 eV.

In a cubic system, the independent elastic constants are C_{11}, C_{12}, and C_{44}. Applying uniform isotropic straining to the lattice and fitting the obtained data to the Murnaghan equation of state allows the calculation of the bulk modulus, B.

$$B = \frac{1}{3}C_{11} + 2C_{12} \tag{A1}$$

In an isotropic material $C' = C_{11}$ - C_{12}. The values of C' and C_{44} can be found considering volume-conserving strains applying distortions, δ, in a $\pm 2\%$ range. To obtain C', an orthorhombic strain as follows was used,

$$\epsilon_{\text{ortho}} = \begin{pmatrix} \delta & 0 & 0 \\ 0 & -\delta & 0 \\ 0 & 0 & \delta^2/(1-\delta^2) \end{pmatrix} \tag{A2}$$

In the case of C_{44}, a monoclinic strain was considered:

$$\epsilon_{\text{mono}} = \begin{pmatrix} 0 & \delta/2 & 0 \\ \delta/2 & 0 & 0 \\ 0 & 0 & \delta^2/(4-\delta^2) \end{pmatrix} \tag{A3}$$

Respectively, the elastic constants' values can be obtained from:

$$E_{\text{ortho}} = E_0 + C'V\delta^2 + O[\delta^4] \tag{A4}$$

$$E_{\text{mono}} = E_0 + C_{44}V\delta^2 + O[\delta^4], \tag{A5}$$

where E_0 is the energy of the system in its equilibrium state.

The numerical values of the elastic constants determined by this procedure are shown in Table 3.

Appendix C. Material Parameter Optimization Algorithm

The optimization algorithm is implemented using python script (refer Figure A3), the minimize function (with Nelder-Mead [48] method) is used from the python Scipy optimize library. The function used for minimization takes in material parameters as input, performs FEM simulations with the parameters, homogenizes the solution obtained from FEM simulation, and provides the error between the simulation and experiments. The mean square difference is used as the error function $\frac{1}{n}\sum_{i=1}^{n}(\sigma_{13}^{sim}|_i - \sigma_{13}^{exp}|_i)^2$, where n is the number of time step in simulation, $\sigma_{13}^{sim}|_i$ is the 13 (or xz) component of the result from the simulation, and $\sigma_{13}^{exp}|_i$ is the 13 (or xz) component of the result from the experiment.

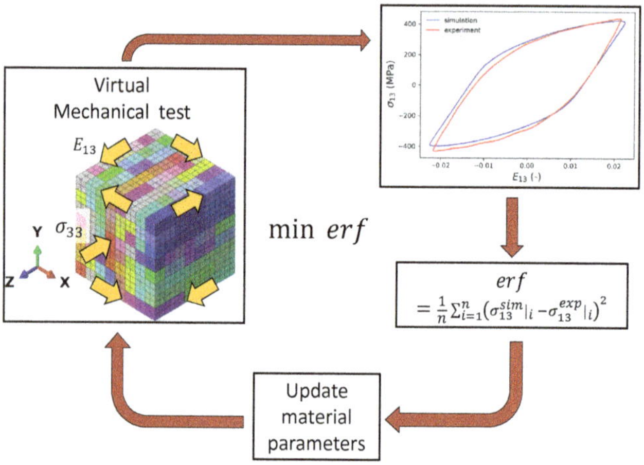

Figure A3. Flow diagram of the optimization algorithm used for optimizing the material parameters.

References

1. Thomann, U.I.; Uggowitzer, P.J. Wear-corrosion behavior of biocompatible austenitic stainless steels. *Wear* **2000**, *239*, 48–58. . [CrossRef]
2. Uggowitzer, P.J.; Magdowski, R.; Speidel, M.O. Nickel free high nitrogen austenitic steels. *ISIJ Int.* **1996**, *36*, 901–908. [CrossRef]
3. Williams, D. Tissue-biomaterial interactions. *J. Mater. Sci.* **1987**, *22*, 3421–3445. [CrossRef]
4. Fini, M.; Giavaresi, G.; Giardino, R.; Lenger, H.; Bernauer, J.; Rimondini, L.; Torricelli, P.; Borsari, V.; Chiusoli, L.; Chiesa, R.; et al. A new austenitic stainless steel with a negligible amount of nickel: An in vitro study in view of its clinical application in osteoporotic bone. *J. Biomed. Mater. Res. Part B Appl. Biomater.* **2004**, *71*, 30–37. [CrossRef] [PubMed]
5. Montanaro, L.; Cervellati, M.; Campoccia, D.; Prati, C.; Breschi, L.; Arciola, C.R. No genotoxicity of a new nickel-free stainless steel. *Int. J. Artif. Organs* **2005**, *28*, 58–65. [CrossRef] [PubMed]
6. Torricelli, P.; Fini, M.; Borsari, V.; Lenger, H.; Bernauer, J.; Tschon, M.; Bonazzi, V.; Giardino, R. Biomaterials in orthopedic surgery: Effects of a nickel-reduced stainless steel on in vitro proliferation and activation of human osteoblasts. *Int. J. Artif. Organs* **2003**, *26*, 952–957. [CrossRef]
7. Niinomi, M.; Nakai, M.; Hieda, J. Development of new metallic alloys for biomedical applications. *Acta Biomater.* **2012**, *8*, 3888–3903. [CrossRef]
8. Salahinejad, E.; Hadianfard, M.J.; Ghaffari, M.; Mashhadi, S.B.; Okyay, A.K. Fabrication of nanostructured medical-grade stainless steel by mechanical alloying and subsequent liquid-phase sintering. *Metall. Mater. Trans. A* **2012**, *43*, 2994–2998. [CrossRef]
9. Carreno-Morelli, E.; Zinn, M.; Rodriguez-Arbaizar, M.; Bassas, M. Nickel-free P558 stainless steel processed from metal powder–PHA biopolymer feedstocks. *Eur. Cells Mater.* **2016**, *32*, 32.
10. Fini, M.; Aldini, N.N.; Torricelli, P.; Giavaresi, G.; Borsari, V.; Lenger, H.; Bernauer, J.; Giardino, R.; Chiesa, R.; Cigada, A. A new austenitic stainless steel with negligible nickel content: an in vitro and in vivo comparative investigation. *Biomaterials* **2003**, *24*, 4929–4939. [CrossRef]
11. Ngeru, T.; Kurtulan, D.; Karkar, A.; Hanke, S. Mechanical Behaviour and Failure Mode of High Interstitially Alloyed Austenite under Combined Compression and Cyclic Torsion. *Metals* **2022**, *12*, 157. [CrossRef]
12. Cruzado, A.; LLorca, J.; Segurado, J. Modeling cyclic deformation of inconel 718 superalloy by means of crystal plasticity and computational homogenization. *Int. J. Solids Struct.* **2017**, *122*, 148–161. [CrossRef]

13. Kowalewski, Z.; Szymczak, T.; Maciejewski, J. Material effects during monotonic-cyclic loading. *Int. J. Solids Struct.* **2014**, *51*, 740–753. [CrossRef]
14. Hennessey, C.; Castelluccio, G.M.; McDowell, D.L. Sensitivity of polycrystal plasticity to slip system kinematic hardening laws for Al 7075-T6. *Mater. Sci. Eng. A* **2017**, *687*, 241–248. [CrossRef]
15. Mróz, Z.; Maciejewski, J. Constitutive modeling of cyclic deformation of metals under strain controlled axial extension and cyclic torsion. *Acta Mech.* **2018**, *229*, 475–496. [CrossRef]
16. Armstrong, P.J.; Frederick, C.O. *A Mathematical Representation of the Multiaxial Bauschinger Effect*; Berkeley Nuclear Laboratories: Berkeley, CA, USA, 1966; Volume 731.
17. Lemaitre, J.; Chaboche, J.L. *Mechanics of Solid Materials*; Cambridge University Press: Cambridge, UK , 1994.
18. Ohno, N.; Wang, J.D. Kinematic hardening rules with critical state of dynamic recovery, part I: Formulation and basic features for ratchetting behavior. *Int. J. Plast.* **1993**, *9*, 375–390. [CrossRef]
19. Bari, S.; Hassan, T. Anatomy of coupled constitutive models for ratcheting simulation. *Int. J. Plast.* **2000**, *16*, 381–409. [CrossRef]
20. Schäfer, B.J.; Song, X.; Sonnweber-Ribic, P.; Hartmaier, A. Micromechanical modelling of the cyclic deformation behavior of martensitic SAE 4150—A comparison of different kinematic hardening models. *Metals* **2019**, *9*, 368. [CrossRef]
21. Sajjad, H.M.; Hanke, S.; Güler, S.; Fischer, A.; Hartmaier, A. Modelling cyclic behaviour of martensitic steel with J2 plasticity and crystal plasticity. *Materials* **2019**, *12*, 1767. [CrossRef]
22. Hielscher, R. Kernel density estimation on the rotation group and its application to crystallographic texture analysis. *J. Multivar. Anal.* **2013**, *119*, 119–143. [CrossRef]
23. Jin, Y.; Bernacki, M.; Rohrer, G.S.; Rollett, A.D.; Lin, B.; Bozzolo, N. Formation of annealing twins during recrystallization and grain growth in 304L austenitic stainless steel. In *Materials Science Forum*; Trans Tech Publications Ltd.: Baech, Switzerland , 2013; Volume 753, pp. 113–116.
24. Ren, S.; Sun, Z.; Xu, Z.; Xin, R.; Yao, J.; Lv, D.; Chang, J. Effects of twins and precipitates at twin boundaries on Hall–Petch relation in high nitrogen stainless steel. *J. Mater. Res.* **2018**, *33*, 1764–1772. [CrossRef]
25. Mainprice, D.; Hielscher, R.; Schaeben, H. Calculating anisotropic physical properties from texture data using the MTEX open-source package. *Geol. Soc.* **2011**, *360*, 175–192. [CrossRef]
26. Buhagiar, J.; Qian, L.; Dong, H. Surface property enhancement of Ni-free medical grade austenitic stainless steel by low-temperature plasma carburising. *Surf. Coatings Technol.* **2010**, *205*, 388–395. [CrossRef]
27. Heidari, L.; Tangestani, A.; Hadianfard, M.; Vashaee, D.; Tayebi, L. Effect of fabrication method on the structure and properties of a nanostructured nickel-free stainless steel. *Adv. Powder Technol.* **2020**, *31*, 3408–3419. [CrossRef]
28. Castelluccio, G.M.; McDowell, D.L. Effect of annealing twins on crack initiation under high cycle fatigue conditions. *J. Mater. Sci.* **2013**, *48*, 2376–2387. [CrossRef]
29. Pande, C.S.; Rath, B.; Imam, M. Effect of annealing twins on Hall–Petch relation in polycrystalline materials. *Mater. Sci. Eng. A* **2004**, *367*, 171–175. [CrossRef]
30. Biswas, A.; Vajragupta, N.; Hielscher, R.; Hartmaier, A. Optimized reconstruction of the crystallographic orientation density function based on a reduced set of orientations. *J. Appl. Crystallogr.* **2020**, *53*, 178–187. [CrossRef] [PubMed]
31. Biswas, A.; Prasad, M.R.; Vajragupta, N.; ul Hassan, H.; Brenne, F.; Niendorf, T.; Hartmaier, A. Influence of microstructural features on the strain hardening behavior of additively manufactured metallic components. *Adv. Eng. Mater.* **2019**, *21*, 1900275. [CrossRef]
32. Smith, M. *ABAQUS/Standard User's Manual*, Version 6.14; Dassault Systèmes Simulia Corp: Johnston, RI, USA, 2014.
33. de Borst, R.; Crisfield, M.A.; Remmers, J.J.C.; Verhoosel, C.V. *Non-Linear Finite Element Analysis of Solids and Structures*; Wiley: Chichester, UK, 2012.
34. Roters, F.; Eisenlohr, P.; Hantcherli, L.; Tjahjanto, D.D.; Bieler, T.R.; Raabe, D. Overview of constitutive laws, kinematics, homogenization and multiscale methods in crystal plasticity finite-element modeling: Theory, experiments, applications. *Acta Mater.* **2010**, *58*, 1152–1211. [CrossRef]
35. McDowell, D. Stress state dependence of cyclic ratchetting behavior of two rail steels. *Int. J. Plast.* **1995**, *11*, 397–421. [CrossRef]
36. de Castro e Sousa, A.; Suzuki, Y.; Lignos, D. Consistency in solving the inverse problem of the Voce-Chaboche constitutive model for plastic straining. *J. Eng. Mech.* **2020**, *146*, 04020097. [CrossRef]
37. Vrh, M.; Halilovič, M.; Štok, B. The evolution of effective elastic properties of a cold formed stainless steel sheet. *Exp. Mech.* **2011**, *51*, 677–695. [CrossRef]
38. ul Hassan, H.; Maqbool, F.; Güner, A.; Hartmaier, A.; Ben Khalifa, N.; Tekkaya, A.E. Springback prediction and reduction in deep drawing under influence of unloading modulus degradation. *Int. J. Mater. Form.* **2016**, *9*, 619–633. [CrossRef]
39. Wilshire, B.; Willis, M. Mechanisms of strain accumulation and damage development during creep of prestrained 316 stainless steels. *Metall. Mater. Trans. A* **2004**, *35*, 563–571. [CrossRef]
40. Stout, M.; Martin, P.; Helling, D.; Canova, G. Multiaxial yield behavior of 1100 aluminum following various magnitudes of prestrain. *Int. J. Plast.* **1985**, *1*, 163–174. [CrossRef]
41. Schneider, A.; Kiener, D.; Yakacki, C.; Maier, H.; Gruber, P.; Tamura, N.; Kunz, M.; Minor, A.; Frick, C. Influence of bulk pre-straining on the size effect in nickel compression pillars. *Mater. Sci. Eng. A* **2013**, *559*, 147–158. [CrossRef]
42. Kresse, G.; Hafner, J. Ab initio molecular dynamics for open-shell transition metals. *Phys. Rev. B* **1993**, *48*, 13115–13118. [CrossRef] [PubMed]

43. Kresse, G.; Furthmüller, J. Efficient iterative schemes for ab initio total-energy calculations using a plane-wave basis set. *Phys. Rev. B* **1996**, *54*, 11169–11186. [CrossRef]
44. Kresse, G.; Joubert, D. From ultrasoft pseudopotentials to the projector augmented-wave method. *Phys. Rev. B* **1999**, *59*, 1758–1775. [CrossRef]
45. Blöchl, P.E. Projector augmented-wave method. *Phys. Rev. B* **1994**, *50*, 17953–17979. [CrossRef]
46. Perdew, J.P.; Burke, K.; Ernzerhof, M. Generalized gradient approximation made simple. *Phys. Rev. Lett.* **1996**, *77*, 3865–3868. [CrossRef] [PubMed]
47. Pack, J.D.; Monkhorst, H.J. Special points for Brillouin-zone integrations. *Phys. Rev. B* **1977**, *16*, 1748–1749. . PhysRevB.16.1748. [CrossRef]
48. Gao, F.; Han, L. Implementing the Nelder-Mead simplex algorithm with adaptive parameters. *Comput. Optim. Appl.* **2012**, *51*, 259–277. [CrossRef]

materials MDPI

Article

Effects of Micro-Shot Peening on the Fatigue Strength of Anodized 7075-T6 Alloy

Chih-Hang Su [1], Tai-Cheng Chen [2], Yi-Shiun Ding [3], Guan-Xun Lu [1] and Leu-Wen Tsay [1,*]

1 Department of Optoelectronics and Materials Technology, National Taiwan Ocean University, Keelung 20224, Taiwan
2 Nuclear Fuels and Materials Division, Institute of Nuclear Energy Research, Taoyuan 32546, Taiwan
3 Material Research Group, Asia Development Center, SRAM LLC, Taichung 40765, Taiwan
* Correspondence: b0186@mail.ntou.edu.tw; Tel.: +886-2-24622192 (ext. 6405)

Abstract: Micro-shot peening under two Almen intensities was performed to increase the fatigue endurance limit of anodized AA 7075 alloy in T6 condition. Compressive residual stress (CRS) and a nano-grained structure were present in the outermost as-peened layer. Microcracks in the anodized layer obviously abbreviated the fatigue strength/life of the substrate. The endurance limit of the anodized AA 7075 was lowered to less than 200 MPa. By contrast, micro-shot peening increased the endurance limit of the anodized AA 7075 to above that of the substrate (about 300 MPa). Without anodization, the fatigue strength of the high peened (HP) specimen fluctuated; this was the result of high surface roughness of the specimen, as compared to that of the low peened (LP) one. Pickling before anodizing was found to erode the outermost peened layer, which caused a decrease in the positive effect of peening. After anodization, the HP sample had a greater fatigue strength/endurance limit than that of the LP one. The fracture appearance of an anodized fatigued sample showed an observable ring of brittle fracture. Fatigue cracks present in the brittle coating propagated directly into the substrate, significantly damaging the fatigue performance of the anodized sample. The CRS and the nano-grained structure beneath the anodized layer accounted for a noticeable increase in resistance to fatigue failure of the anodized micro-shot peened specimen.

Keywords: AA 7075-T6 alloy; anodizing; micro-shot peening; nanograin; rotating bending fatigue

Citation: Su, C.-H.; Chen, T.-C.; Ding, Y.-S.; Lu, G.-X.; Tsay, L.-W. Effects of Micro-Shot Peening on the Fatigue Strength of Anodized 7075-T6 Alloy. *Materials* **2023**, *16*, 1160. https://doi.org/10.3390/ma16031160

Academic Editors: Lucjan Śnieżek, Jaroslaw Galkiewicz and Sebastian Lipiec

Received: 31 December 2022
Revised: 18 January 2023
Accepted: 27 January 2023
Published: 29 January 2023

1. Introduction

Surface technologies are extensively applied to components in the aerospace, automobile, and power industries to protect against corrosion and wear. Thermal spray coating, electroplating, and chemical and physical vapor deposition are employed to modify the surface characteristics of the components. The most typical approach to increase the wear and corrosion resistance of Al alloys is to coat them with anodic oxide film [1,2]. An oxide film formed on the surface of anodized Al alloys consists of a thin compact inner layer and a porous outer layer [2]. A major concern about the use of AA 7075 Al alloy is the possible fatigue failure of structural components [3–5]. The anodization process is reported to obviously degrade the fatigue performance of high-strength Al alloys [6–11].

The effect of anodization on the fatigue strength/life of Al alloys is counted on several process variables. Moreover, the fatigue strength of a soft anodized coating is superior to that of a hard anodized one [12]. Chromic acid anodizing is less harmful to fatigue strength than sulfuric acid anodizing [13,14]. Increasing the thickness of the anodizing coating lowers the fatigue strength/life of AA 7075 alloy [10]. Moreover, a coating on an ultrafine-grained substrate has shown enhanced resistance to anodizing-induced and fatigue-induced cracking of AA 6682 alloy [12]. It is reported that the microcracks in the anodized film and irregularities beneath the film are the reasons for the decrease in fatigue strength [10]. The loss in fatigue strength of anodized AA 2219 and AA 2024 alloys is related to the induced microcracks at the etch pits, which are formed from the preferential dissolution of intermetallics in pickling performed before anodizing [7]. In one study, Al_2O_3 coatings were applied on the AA 6061-T6 alloy by hard anodizing, micro-arc oxidation, detonation spray, and air plasma spray, and hard anodizing was found to have the most damaging effect on fatigue endurance [15].

Shot peening is employed to upgrade the fatigue performance of the Al alloys [16–18]. The resulting compressive residual stress (CRS), strain hardening, and fine-grained microstructure in the shot-peened layer are the main causes of improved fatigue properties. Successful fatigue enhancement depends on a compromise between the CRS and the detrimental effect on surface quality [19]. Advanced shot peening technologies, such as laser [20,21], ultrasonic [22,23], and waterjet peening [24,25], are employed to reduce the irregularity and upgrade the fatigue life of Al alloys. It is noted that shot peening can mitigate the harmful effect of an anodized coating on the fatigue strength of high-strength Al alloys [13,21,26]. With shot peening, the fatigue strength of anodized AA 6082-T651 alloy is even higher than that of the unanodized substrate [26]. In 3.5% NaCl solution, the fatigue lives of AA 7075-T73 alloy can be markedly increased by shot peening treatments as compared to an electropolished sample [27]. To reduce surface irregularity, the use of fine particles for peening [28,29] is more economical and practical than other advanced peening processes.

AA 7075-T6 alloy has been widely used in the aerospace industry, which can be heat-treated to achieve various microstructures and properties [30]. A decrease in the surface roughness of machined AA 7075 alloy can be achieved by increasing the cutting speed, increasing the cutting depth, or decreasing the feeding rate [30]. The aim of this work was to explore the application of micro-shot peening as a pretreatment before anodizing, which provided a means to restore the fatigue performances of anodized AA 7075 alloy. The fatigue limit of the specimens was evaluated by a rotating bending machine at room temperature. The effect of hard anodizing on the fatigue limit/life of the unpeened and peened AA 7075 alloy was investigated. Plastic deformation of the micro-shot peened specimen was measured by using the Almen intensity, and the outer roughness was detected with a 3D contour profiler. The surface morphologies of ruptured specimens after fatigue tests were examined with a scanning electron microscope (SEM). The refined structure in the peened samples was identified by the inverse pole figure (IPF) map.

2. Material and Experimental Procedures

2.1. Sample Preparation

The fatigue tests were performed on AA 7075 bar with 10 mm diameter. The chemical composition (wt.%) of the alloy bar was as follows: 1.462 Cu, 2.504 Mg, 5.854 Zn, 0.029 Mn, 0.224 Cr, 0.07 Si, 0.117 Fe, 0.023 Ti, and the residual Al. Moreover, AA 7075 plate with a thickness of 8 mm was used for surface metrology examination and residual stress determination of the samples. The composition in wt.% of the plate was as follows: 1.294 Cu, 2.420 Mg, 5.616 Zn, 0.015 Mn, 0.292 Cr, 0.040 Si, 0.192 Fe, 0.020 Ti, and the balance Al. All the samples were solution-annealed at 743 K for 1 h, followed by water-quenching, and then aged at 393 K for 24 h. The solution-annealed and aged AA 7075 alloy was named the base metal (BM) sample. The BM had a yield strength, tensile strength, and tensile elongation of 618 MPa, 668 MPa, and 9%, respectively [31]. The tested specimens were ground with SiC paper before micro-shot peening. Micro-shot peening was performed on the ground samples using amorphous powders with sizes of 50–80 μm under 200% surface coverage. The two micro-shot peen intensities employed in this work were determined from the height of the N-type Almen specimen to be 0.110 mm (low peen, LP) and 0.204 mm (high peen, HP). The micro-shot peened samples were distinguished as LP and HP according to the Almen intensity.

2.2. Sulfuric Hard Anodizing

Some of the unpeened and peened samples were subjected to anodizing treatment. Prior to the hard anodization process, the sample surface was thoroughly cleaned by degreasing, acid activation, and chemical polishing to totally remove the surface contaminants.

Pickling acts chemically to remove the oxides, inclusions, and compounds from the sample surface. Hard anodizing to a thickness of 30 µm was performed in an electrolyte (20 wt.% sulfuric electrolyte + 5 wt.% aluminum sulfate) at a bath temperature of −3 to 3 °C under a varying voltage (25 to 45 V). Sealing was performed in nickel acetate solution at 85–92 °C. For anodized samples, the symbol A affixed to the specified sample (e.g., HPA denoted the anodized high peened sample).

2.3. Hardness Measurement and Fatigue Testing

An MVK-G1500 Vickers hardness tester (Mitutoyo, Kawasaki, Japan) was applied to determine the hardness of the samples. The anodized coating was loaded at 10 gf for 15 s. Moreover, a Hysitron TI 980 nanoindenter (Bruker, Billerica, MA, USA) loaded at 2000 µN was used to determine the change in hardness around the interface between the anodized and peened layers. Surface metrology of the samples with and without micro-shot peening was performed with a Contour GT-K 3D optical profiler (Bruker, Billerica, MA, USA), which provided noncontact surface measurements. Figure 1 displays the dimensions of the dog-bone samples for fatigue tests. Rotary bending fatigue tests were conducted at room temperature and a frequency of 33.3 Hz at R = −1 (fully reversed). Fatigue stress (S) vs. number of cycles (N) to failure of the tested samples was measured, and the results presented herein are the averages of three samples, although individual values are also reported.

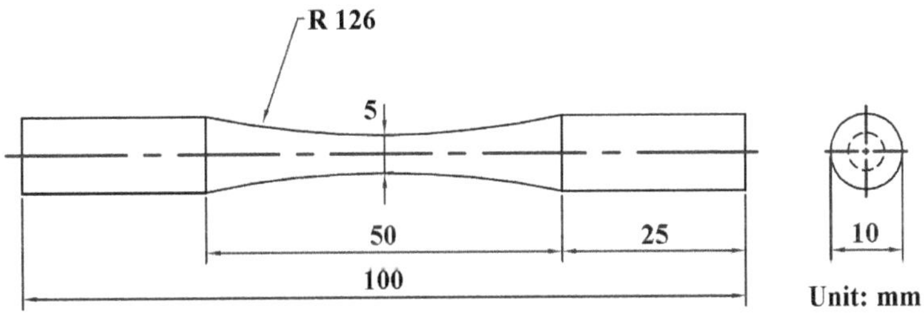

Figure 1. Dimensions of the fatigue test specimen used in this study.

2.4. Microstructural and Fracture Surface Observations

The microstructures of the samples after metallurgical preparation were examined using an S-4800 SEM (Hitachi, Tokyo, Japan). The samples were also detected using an SEM equipped with a NordlysMax^2 electron backscatter diffraction (EBSD, Oxford Instruments, Abingdon, UK) detector to identify the refined structure of the inspected specimens. Moreover, the strain fields around the interface between the coating and substrate of the peened samples after anodizing were analyzed using the HKL Channel 5 software (Oxford Instruments, Abingdon, UK) to process the original data obtained by the EBSD. The macro-fracture appearance and the detail of the fracture features of the fatigue-fractured specimens were examined using an S-3400N SEM (Hitachi, Tokyo, Japan).

2.5. Residual Stress Measurement

The µ-X360s (Pulstec, Hamamatsu, Japan), a residual stress analyzer, was applied to determine the distribution of residual stress in a micro-shot peened sample. The standard settings of the X-ray source were using Cr target Kα radiation (wavelength 2.291 Å) at an X-ray tube voltage of 30 kV with 1.5 mA current. The device for measuring residual stress was based on the cos α method. The full width at half maximum (FWHM) of the (311) peak was related to the distortion of the lattice. The distribution of residual stress in the thickness direction was obtained by removing the surface layer of the sample using an EP-3 electrochemical polisher (Pulstec, Hamamatsu, Japan).

3. Results

3.1. Micro-Shot Peening and Morphology

The morphology of the amorphous shot and the surface appearance of the micro-shot peened sample are displayed in Figure 2. The individual pellet of the amorphous shot ranged from 50 to 80 μm in size (Figure 2a). As reported in previous studies [28], the amorphous pellet had a Vickers microhardness of about HV 1150. The surface morphology of the micro-shot peened samples (Figure 2b) had indentations of 10–30 μm in size. AA 7075 BM was ground with SiC paper before micro-shot peening. The surface roughness of the BM and two micro-shot peened samples (LP and HP), determined by a 3D contour profiler, are listed in Table 1. The ground sample had Sa, Sp, and Sv values of 0.217, 1.494, and −1.211 μm, respectively. Continuous bombardments with amorphous shot increased the surface roughness of the peened sample, especially that of the HP sample. The Sa, Sp, and Sv roughness values of the LP one were 0.385, 1.396, and −1.506 μm, accordingly. This revealed that the LP specimen had a slightly higher surface roughness than the unpeened sample. The HP sample had an obviously higher surface roughness than the other two samples; the Sa, Sp, and Sv were 1.303, 6.400, and −7.519 μm, respectively. In general, such a high surface roughness of the HP sample was expected to degrade the resistance to fatigue cracking.

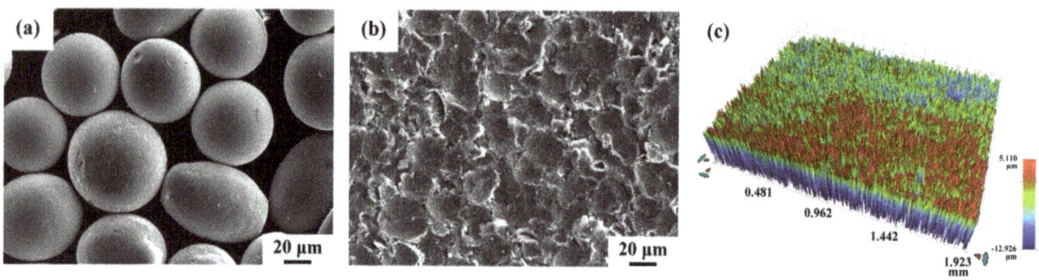

Figure 2. Typical appearance of the (**a**) amorphous pellet, (**b**) surface morphology, and (**c**) surface roughness of the micro-shot peened specimen.

Table 1. The surface roughness values of the distinct specimens (unit: μm).

Specimen	Sa [1]	Sp [2]	Sv [3]
Ground sample	0.217	1.494	−1.211
LP sample	0.385	1.396	−1.506
HP sample	1.303	6.400	−7.519

[1] Sa—arithmetical mean height of the surface. [2] Sp—maximum peak height of the surface. [3] Sv—maximum pit depth of the surface.

Figure 3 displays the top and cross-sectional views of anodized samples with or without micro-shot peening. The top surface features of the unpeened anodized sample showed regular rectangular cracks (Figure 3a). In the HPA sample, shallow dents with vague cracks were visible on the top surface (Figure 3b). It was obvious that the surface topography and roughness of the anodized samples were affected by the surface texture of the substrate. The cross-sectional views of the anodized layers of different samples are shown in Figure 3c,d, revealing that the thickness of the anodized layer was about 30 μm in both samples. Without peening, the top surface and the interface between the anodized layer and the substrate were quite flat (Figure 3c). Some deep microcracks and fine defects of uneven sizes were present in the anodized layer (Figure 3c). Overall, the anodized peened samples (HPA and LPA) and the BMA had similar microstructures within the anodized zone. The difference between them was that the straight interfaces of the unpeened sample (Figure 3c) were replaced by a slightly tortuous profile of the peened one (Figure 3d). It was found that micro-shot peening did not reduce the number of defects in the anodized zone. Within the anodized zone, the open crack width of the peened sample seemed to be narrower than that of the unpeened sample, but the difference was hard to distinguish.

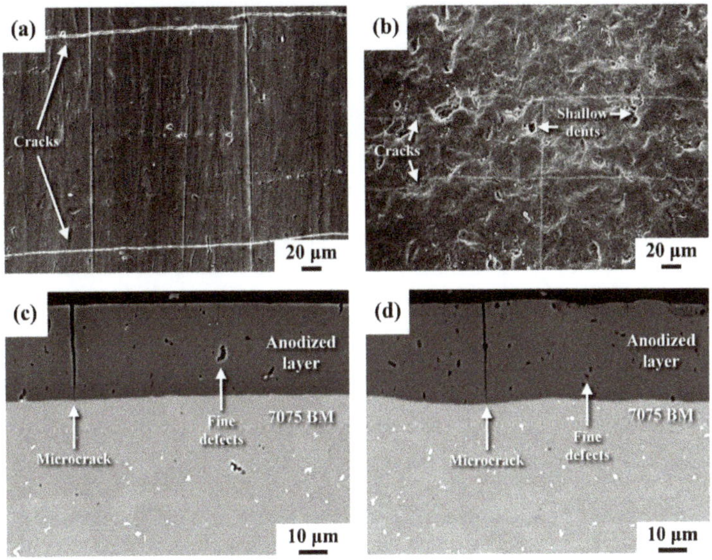

Figure 3. (**a**,**b**) Top view and (**c**,**d**) cross-sectional view of the anodized sample: (**a**,**c**) the unpeened sample; (**b**,**d**) the micro-shot peened sample.

3.2. Hardness Measurements

Figure 4 shows the hardness value of the anodized film and the hardness distribution around the interface between the micro-shot peened zone and anodized layer, as determined by a Vickers hardness tester and nanoindenter, respectively. As shown in Figure 4a, the anodized layer had an apparently higher hardness (around HV 400) than that of the substrate (around HV 180). Thus, the anodized layer improved the resistance to corrosion and wear of the AA 7075 alloy. The hardness indenters were also applied to the specific sites around the interface of the anodized sample. The hardness values determined using a nanoindenter on the right side of the interface (Figure 4b), i.e., the substrate side, were 1.74, 1.65, and 1.70 GPa, respectively. On the other side, hardness values of 3.83 and 2.92 GPa were obtained; these two high values likely belonged to the anodized layer. In prior work [30], the hardness of the micro-shot peened layer was found to exceed 2.94 GPa (HV 300). Thus, the strain hardening and nanocrystalline structure introduced by micro-shot peening could be partly removed by pickling before anodizing.

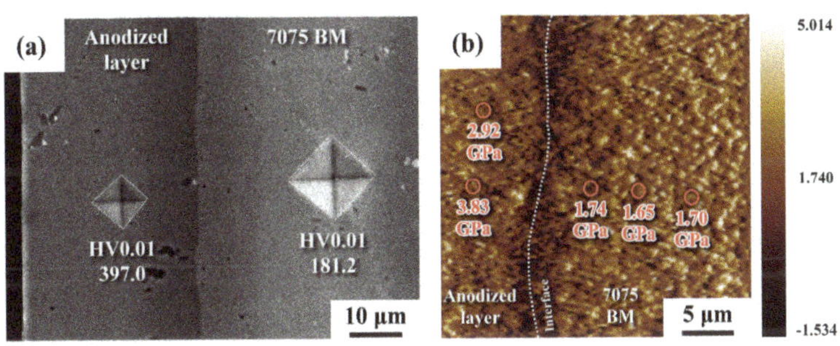

Figure 4. The hardness indentations: (**a**) the micro-Vickers indentations of the anodized sample; (**b**) the nano-hardness distribution around the anodized interface of the micro-shot peened sample.

3.3. Microstructural Observations

A highly deformed layer should consist of nanograins in the outermost surface and micron-size elongated grains in the subsurface of the micro-shot peened sample [31]. Figure 5 presents the IPF and strain maps of the HPA and LPA samples, showing the changes in the granular microstructures and strain fields around the interfaces between the substrates and anodized layers. It should be noticed that no Kikuchi pattern was detected within the anodized layer due to its amorphous structure. Because of the limited resolution of the EBSD analysis, the ultrafine grains could not be distinguished and displayed in different colors to show the individual grain boundaries or orientations. The IPF map shows a nanogranular structure present within a depth of 10 μm below the interface of the HPA sample (Figure 5a); this structure was associated with the original micro-shot peened zone. As shown in Figure 5b, no fine granular structure was seen in the LPA sample. The results indicated that only some fine grains occasionally nucleated along the boundaries of coarse grains in the LPA sample. Those refined grains in the micro-shot peened layer were caused by the dynamic recrystallization of the deformed surface layer during the bombardment with fine shot. The original deformed and recrystallized layer disappeared in the LPA sample (Figure 5b), which was attributed to the results of pickling before anodizing. Figure 5c,d show the strain distribution maps, which are in cross-sectional view around the anodizing interfaces of the HPA and LPA specimens. These two figures refer to the IPF maps shown in Figure 5a,b. The high-strain zone is shown in red; the low-strain zone appears in blue. The HPA sample exhibited high strain (in red) within a depth of 20 μm beneath the interface (Figure 5c), which was not seen in the LPA sample (Figure 5d). Those red zones in the HPA sample (Figure 5c) could be related to the severely deformed zones during peening. Moreover, the LPA sample had low strain beneath the interface (Figure 5d). It was confirmed that, in the HPA sample, part of the micro-shot peened zone remained after pickling.

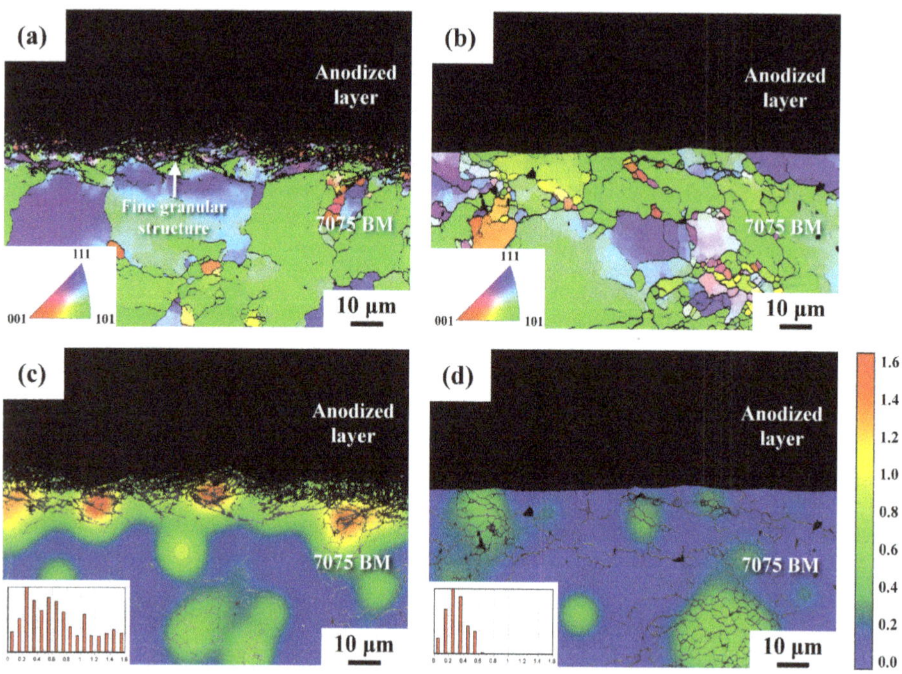

Figure 5. (**a**,**b**) The IPF maps and (**c**,**d**) the strain maps of the inspected samples in cross-sectional view: (**a**,**c**) the HPA sample; (**b**,**d**) the LPA sample.

3.4. Fatigue Evaluation

Figure 6 demonstrates the results of fatigue tests conducted in laboratory air for up to 10^7 cycles. The fatigue life of the tested samples was sensitive to the loading condition. Without peening and anodizing, the endurance limit of the AA 7075 BM was about 275 MPa [31] (Figure 6a). As revealed in prior work [31], the endurance limit of the LP sample was about 500 MPa, which was much greater than that of the AA 7075 BM (Figure 6a). It was noticed that the fatigue properties of the HP sample were not better than those of the LP specimen (Figure 6a). Furthermore, the HP sample exhibited high fluctuation in fatigue strength/life during testing. The fatigue curves of anodized specimens are shown in Figure 6b. The endurance limit of the anodized specimen (BMA) decreased to about 180 MPa (Figure 6b). It was obvious that hard anodizing had a strongly adverse effect on the fatigue strength of AA 7075 alloy in the high-cycle region. Peening at a low Almen intensity improved the fatigue performance of the LPA sample, as compared with that of the BMA sample (Figure 6b). The endurance limit of the LPA specimen was about 300 MPa, if the BM sample was pretreated with low peening intensity before anodizing. Peening at a high Almen intensity raised the endurance limit of the HPA specimen to about 400 MPa, which was about 100 MPa greater than that of the BM. The results indicated that the peening and pickling strongly influenced the fatigue performance of anodized AA 7075 alloy. Moreover, micro-shot peening could improve the fatigue performance of anodized AA 7075-T6 alloy.

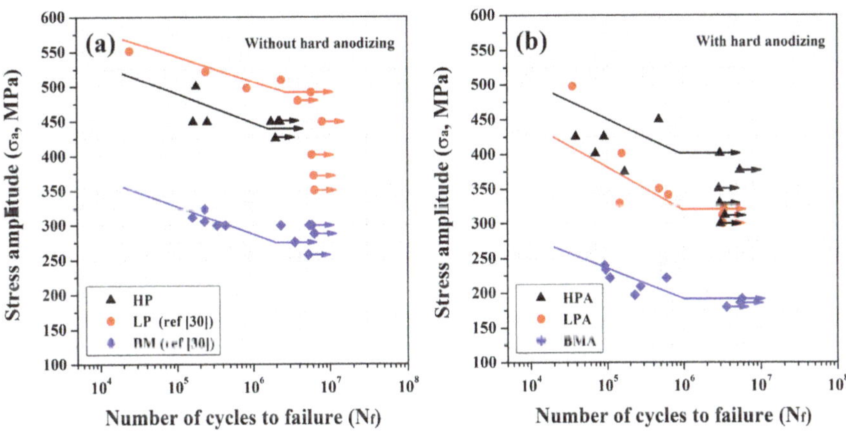

Figure 6. Fatigue stress (S) vs. cycle (N) curves of the samples (**a**) without and (**b**) with a hard anodizing coating.

3.5. Fractured Surface Examinations

The fatigue-fractured morphologies of the anodized BM sample (BMA) are shown in Figure 7. The backscattering electron (BSE) image is more likely to reveal the microcracks and delamination at the interface, whereas the secondary electron (SE) image can show the detailed surface feature of the fatigue-cracked sample. Macroscopically, the main crack grew from the outer surface and into the interior in a radial crack path (Figure 7a). In the BSE image, a thin brittle case was found to decorate the outer profile of the BMA sample (Figure 7b). The anodized layer contained some deep fine cracks (Figure 3) and had a flat fracture appearance. This shows that the secondary cracks grew in the direction normal to the external surface of the sample. Therefore, the cracking of the anodized layer resulted in degradation of the fatigue resistance of the anodized AA 7075-T6 alloy. Moreover, the presence of surface cracks within the anodized layer indicated that the crack-initiation stage should be unnecessary. Examining the fracture features around the anodizing layer at higher magnification, the fatigue crack propagated across the anodized coating and into the substrate, and strong bonding between the anodized coating and substrate was observed (Figure 7c). Moreover, a transgranular crack extended through the straight interface and showed the trace of the crack path (parallel stripes) just beneath the anodized layer (Figure 7d). This event confirmed the continuous progress of the crack without facing the barrier. It is obvious that strong bonding between the anodized layer and the substrate caused the fatigue crack growth rapidly through the interface (Figure 7c,d). In addition, a typical transgranular cleavage-like fracture was observed as the fatigue crack extended into the AA 7075 substrate (Figure 7d). Furthermore, a dimple fracture mixed with small facets was seen in the final fractured zone (not shown here).

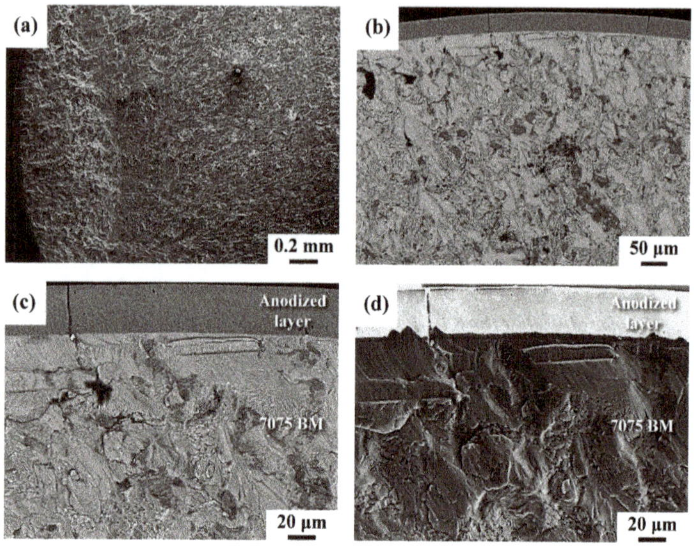

Figure 7. Fatigue fracture appearance of the BMA sample: (**a**,**d**) SE images; (**b**,**c**) BSE images. (**a**) Macro-fractured appearance of the crack initiation site; (**b**) enlarged view of the anodized layer; (**c**,**d**) fracture features around the anodized interface.

SEM photographs of the fractured appearance of the HPA specimen after fatigue tests are presented in Figure 8. The results indicated that the macro-appearance of the fractured HPA sample was similar to that of the BMA sample, consisting of a thin brittle case decorating the outer profile of the specimen (Figure 8a,b). It was noticed that the flat fracture of the anodized layer of the BMA sample was replaced by a chopped and smashed layer in the HPA sample (Figure 8c). Moreover, relatively large numbers of fine cracks and fragmental debris present in the anodized coating could be attributed to the brittle nature of the coating under high fatigue loading of the test. It was noticed that micro-shot peening before anodizing assisted the formation of a tortuous interface between the anodized layer and the AA 7075 substrate. Numerous microcracks assisted in dividing the anodized layer into many fine fragments after the fatigue test (Figure 8c). Moreover, the interface tended to delaminate as the crack growth passed it (indicated by the arrows in Figure 8c). In addition, a rubbed fracture feature was observed beneath the anodized layer (Figure 8d). It was noticed that the fatigue-fractured feature revealed a microscopic change in crack growth direction that occurred as the fatigue crack growth passed through the interface (Figure 8d). It was deduced that the interface separation seemed to deflect the crack growth direction (Figure 8d), which was beneficial in impeding the fatigue crack growth. The cleavage-like brittle fracture in the BMA sample was replaced by the squeezed rubbed feature in the HPA sample. Therefore, the CRS in the micro-shot peened specimen caused crack closure and had a great effect on retarding the fatigue crack growth of the anodized specimen.

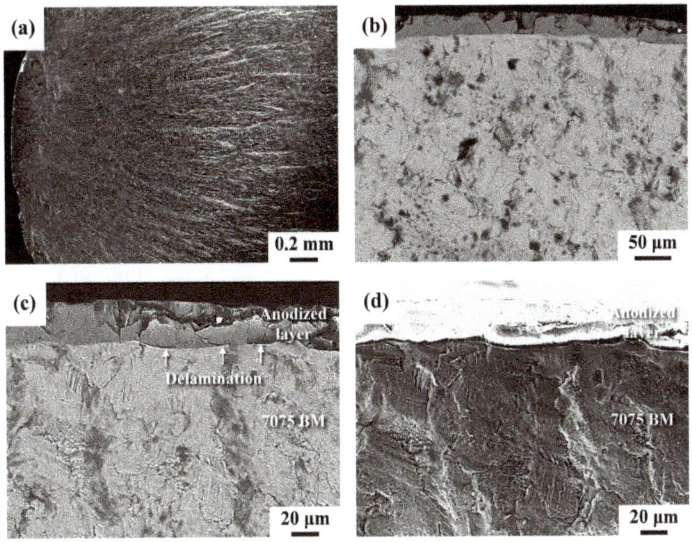

Figure 8. Fatigue fracture appearance of the HPA sample: (**a**,**d**) SE images; (**b**,**c**) BSE images. (**a**) Macro-fractured appearance of the crack initiation site; (**b**) enlarged view of the anodized layer; (**c**,**d**) fracture features around the anodized interface.

Figure 9 shows the changes in residual stress and FWHM intensity in the thickness direction of the LP [31] and HP samples from the surface to the interior. As mentioned previously, the increase in peening intensity caused an increase in surface roughness but was expected to increase the strength and depth of the residual stress field. Peak CRS was found in the subsurface zones of both peened samples. Moreover, the increase in peening intensity caused an obvious increase in depth of the CRS field. A steep decrease in residual stress to −48 MPa resulted in the LP specimen at a depth 50 μm below the peened surface. However, under the same CRS, the depth was increased to about 100 μm in the HP one. With the increase in peening intensity, the peak CRS increased from about −350 MPa to over −400 MPa, and the stress field also increased. The increased CRS field was definitely beneficial to the fatigue resistance to cracking. Moreover, the narrow CRS field meant that only a very limited deformed depth was introduced by the micro-shot peening.

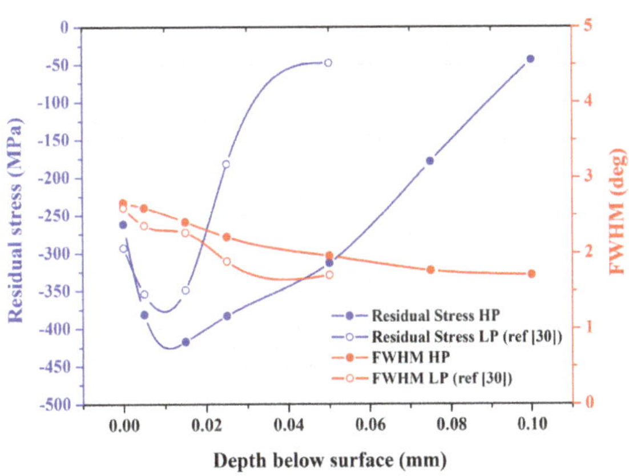

Figure 9. The changes in the residual stress and FWHM intensity in the X-ray diffraction pattern in the thickness direction of the LP and HP samples from the surface to the interior.

4. Discussion

Although the anodized layer consisted of a few deep microcracks and fine defects of uneven sizes (Figure 3), the anodized layer was much harder (around HV 400) than the AA 7075-T6 substrate (around HV 180), as revealed in Figure 4. Thus, the anodized layer improved the resistance to corrosion and wear of AA 7075 alloy. The straight boundary between the anodized layer and the substrate of the unpeened specimen was replaced by a slightly tortuous trace of the peened one (Figure 3). It was noticed that the shallow dents with vague cracks were seen in the top surface of the micro-shot peened specimen. By contrast, scratches with rectangular microcracks were observed in the top surface of anodized unpeened specimen (Figure 3). Thus, the surface morphology and roughness of the anodized sample were affected by the surface texture of the original condition. In prior work, the micro-shot peened layer of AA 7075-T6 alloy comprised ultrafine grains in the outermost layer and micron-order elongated grains in the subsurface of the micro-shot peened specimen [31]. With increased peening intensity, the depth of the residual stress field and the surface roughness are expected to be increased due to deep deformation. The IPFs showed a nanogranular structure within a depth 10 μm below the anodized interface of the HPA sample (Figure 5a), but no such structure was detected in the LPA specimen (Figure 5b). It was clear that pickling before anodizing completely removed the fine-grained structure of the LPA sample. The retained fine structure, which was caused by the recrystallization of the heavily deformed zone during peening, was linked with the detectable strain within a depth of 20 μm beneath the anodized interface of the HPA sample (Figure 5c).

The results of fatigue tests in air revealed that the fatigue strengths/lives of the micro-shot peened (LP and HP) samples were much greater than that of the BM, particularly that of the LP one (Figure 6a). The high fluctuation in the fatigue strength/life of the HP sample in air, relative to that of the LP specimen, could be attributed to its inherent high surface roughness. Similar results have been reported: increasing the machined roughness of AA 7010-T7451 alloy from 0.6 to 3.2 μm caused a 32% drop in endurance limit in a low-stress state [11]. It was noted that hard anodizing had a very detrimental effect on the high-cycle fatigue of AA 7075 alloy shown in Figure 6b. The fatigue strength/life will be obviously shortened by the defects, especially those of surface defects [32]. A probabilistic fatigue life prediction model using the calibrated weakest-link theory has been proposed considering the superiority of the notch and surface defect [32,33]. With the presence of surface microcracks in the anodized coating, the fatigue limit of the BMA sample decreased to about 180 MPa. It was interesting that the fatigue strengths/lives of the peened anodized samples (LPA and HPA), especially that of the HPA, were much higher than that of the unpeened anodized (BMA) samples, as shown in Figure 6b. With micro-shot peening, the harmful effects of microcracks present in the anodized coating on the fatigue performance of AA 7075 alloy were mitigated. The difference in the outermost microstructure between the LPA and HPA samples was due to the retained nanocrystal structure of the latter (Figure 5a), which was hard to find in the former (Figure 5b). It was obvious that pickling almost completely removed the nano-grained structure of the LPA sample. It was deduced that precise control of the pickling process would highly alter the fatigue performance of anodized Al alloys, especially that of the micro-shot peened samples.

Micro-cracks initiate at corrosion pits around the surface and etch pits of AA 7050 alloy under cyclic loading [3]. Furthermore, fatigue failure of anodized Al alloy is often induced by the nucleation and growth of those micro-cracks formed in the anodic oxide film and through the interface into the substrate [9,10]. The stress concentration at the microcracks in the coating is responsible for the decreased fatigue properties of hard anodized 7475-T6 alloy [9]. It has been reported that pickling deteriorates the fatigue properties due to localized dissolution around the intermetallic compounds and/or inclusions [7,34,35], especially in the very-long-life regime [34]. Therefore, pickling was one of the controlling factors for the fatigue performance of the anodized Al alloys. As shown in Figure 7, fatigue cracks present in the brittle coating of an anodized sample grew through the coating/substrate interface and propagated directly into the substrate without the need of micro-crack initiation. This event illustrated the poor fatigue resistance of the BMA sample, as compared with that of the BM sample. Since micro-shot peening only influenced a thin external layer of the impacted material, the difference in the damaging mechanisms between the anodized samples with and without peening should only exist in the stage of retardation of crack growth below the anodized layer. The nano-grained structure and the CRS beneath the anodized layer were helpful in delaying or retarding fatigue crack propagation, which accounted for the higher fatigue strengths/lives of the peened anodized samples than that of the unpeened anodized one.

The macro-fractured appearances of the anodized unpeened and peened specimens were similar, as shown in Figures 7 and 8. However, differences in the fracture features could still be distinguished between the samples in detail. Under high loading, the anodized layer of the peened sample was cracked into small patches (Figure 8c), in contrast to the overall flat fracture of the anodized layer of the unpeened sample (Figure 7c). Moreover, a change in the fatigue fracture appearance around the interface between the unpeened and micro-shot peened samples was observed. The crack propagated directly into the substrate of the unpeened sample, showing a cleavage-like fracture therein (Figure 7d). As the crack growth passed the interface of the micro-shot peened sample, delamination of the interface occurred (Figure 8c,d); this delamination could cause a change in the crack growth direction. It was noticed that a rubbed fracture feature instead of cleavage-like fracture was observed just below the anodized interface of the micro-shot peened sample (Figure 8d). The CRS around the anodized interface enhanced the crack closure, retarding the fatigue crack growth and leading to the rubbed fracture feature.

5. Conclusions

1. Micro-shot peening under two Almen intensities was performed to increase the fatigue strength/life of the anodized AA 7075-T6 alloy. Micro-cracks in the anodized layer significantly deteriorated the fatigue performance of AA 7075 alloy. Under high cycle fatigue, the endurance limit of the BMA sample was lowered to less than 200 MPa. Micro-shot peening could improve the fatigue strength/life of the anodized sample to the level of the unanodized substrate (about 300 MPa). Without anodizing, the fatigue performance of the HP sample was worse than that of the LP one. Moreover, fluctuation in the fatigue strength of the HP sample, relative to the LP sample, was attributed to the inferior effect of high surface roughness.
2. With an anodized layer, the fatigue strength of the HP sample was higher than that of the LP one. Pickling before anodizing eroded the outermost nanograins in the peened layer, which degraded the positive effect of micro-shot peening. Therefore, the HPA sample had higher fatigue resistance than the LPA one did. The fracture appearance of the anodized fatigued samples consisted of an observable ring of brittle fracture. Without any need for crack initiation, fatigue cracks present in the anodized layer propagated directly into the unpeened substrate, significantly reducing its fatigue strength/life. By contrast, the presence of CRS beneath the anodized layer of the micro-shot peened sample could retard the fatigue crack growth and was responsible for a noticeable increase in its fatigue performance. Moreover, interfacial separation between the anodized layer and peened surface could possibly deflect the crack path, which also contributed to the increased resistance to fatigue crack growth of the micro-shot peened sample.

Author Contributions: Conceptualization, L.-W.T.; methodology, L.-W.T.; formal analysis, C.-H.S., T.-C.C. and G.-X.L.; investigation, C.-H.S., T.-C.C. and G.-X.L.; resources, L.-W.T.; writing—original draft preparation, L.-W.T.; writing—review and editing, T.-C.C. and Y.-S.D.; visualization, T.-C.C.; supervision, L.-W.T.; project administration, L.-W.T.; funding acquisition, L.-W.T. All authors have read and agreed to the published version of the manuscript.

Funding: This research was funded by the Ministry of Science and Technology, R.O.C., grant number MOST 110-2622-E-019-008.

Institutional Review Board Statement: Not applicable.

Informed Consent Statement: Not applicable.

Data Availability Statement: Data available on request due to restrictions eg privacy or ethical.

Acknowledgments: The authors are grateful to the Ministry of Science and Technology (National Taiwan University) for the assistance in TEM (EM003800) examinations and EPMA analysis (EPMA000300). The authors are grateful to the Core Facility Center of National Cheng Kung University (OTHER002200) for granting access to the 3D optical profiler. The authors also thank the Ministry of Science and Technology (National Sun Yat-sen University) for the assistance in sample preparations using a focused ion beam (EM025100). The authors would also like to thank Likuan Technology Corp. for their great help in determining the residual stress of the micro-shot peened sample and Vincent Vacuum Tech. for performing the micro-shot peening.

Conflicts of Interest: The authors declare no conflict of interest.

References

1. Lerner, L.M. Hard Anodising of Aerospace Aluminum Alloys. *Trans. Inst. Met. Finish.* **2010**, *88*, 21–24. [CrossRef]
2. Matinez-Viademonte, M.P.; Abrahami, S.T.; Hack, T.; Burchardt, M.; Terryn, H. A Review on Anodizing of Aerospace Aluminum Alloys for Corrosion Protection. *Coatings* **2020**, *10*, 1106. [CrossRef]
3. Barter, S.A.; Sharp, P.K.; Holden, G.; Clark, G. Initiation and Early Growth of Fatigue Cracks in an Aerospace Aluminum Alloy. *Fatigue Fract. Eng. Mater. Struct.* **2002**, *25*, 111–125. [CrossRef]
4. Ma, J.; Wang, Q.; Yang, Y.; Yang, F.; Dong, B.; Che, X.; Cao, H.; Zhang, T.; Zhang, Z. Anisotropic Low Cycle Behavior of the Extruded 7075 Al Alloy. *Materials* **2021**, *14*, 4506. [CrossRef] [PubMed]
5. Turkmen, H.S.; Loge, R.E.; Dawson, P.R.; Miller, M.P. On the Mechanical Behaviour of AA 7075-T6 during Cyclic Loading. *Int. J. Fatigue* **2003**, *25*, 267–281. [CrossRef]
6. Shiozaw, K.; Kobayash, H.; Terada, M.; Matsui, A. Effect of Anodized Coatings on Fatigue Strength in Aluminum Alloy. *Trans. Eng. Sci.* **2001**, *33*, 398–406. [CrossRef]
7. Rateick, R.G., Jr.; Griffith, R.J.; Hall, D.A.; Thompson, K.A. Relationship of Microstructure to Fatigue Strength Loss in Anodised Aluminum-Copper Alloys. *Mater. Sci. Technol.* **2005**, *25*, 1227–1235. [CrossRef]
8. Sadeler, R. Effect of a commercial hard anodizing on the fatigue property of a 2014-T6 aluminium alloy. *J. Mater. Sci.* **2006**, *41*, 5803–5809. [CrossRef]
9. Lonyuk, B.; Apachitei, I.; Duszczyk, J. The Effect of Oxide Coatings on Fatigue Properties of 7475-T6 Aluminium Alloy. *Surf. Coat. Technol.* **2007**, *201*, 8688–8694. [CrossRef]
10. Cirik, E.; Genel, K. Effect of anodic oxidation on fatigue performance of 7075-T6 alloy. *Surf. Coat. Technol.* **2008**, *202*, 5190–5201. [CrossRef]
11. Shahzad, M.; Chaussumier, R.; Chieragatti, R.; Mabru, C.; Rezai-Aria, F. Influence of Anodizing Process on Fatigue Life of Machined Aluminium Alloy. *Procedia Eng.* **2010**, *2*, 1015–1024. [CrossRef]
12. Hockauf, K.; Winter, L.; Händel, M.; Halle, T. The Effect of Anodic Oxide Coating on the Fatigue Behaviour of AA6082 with an Ultrafine-Grained Microstructure. *Mater. Werkst.* **2011**, *42*, 624–631. [CrossRef]
13. Camargo, A.; Voorwald, H. Influence of Anodization on the Fatigue Strength of 7050-T7451 Aluminium Alloy. *Fatigue Fract. Eng. Mater. Struct.* **2007**, *30*, 993–1007. [CrossRef]
14. Inoue, A.K.; Tagawa, T.; Ishikawa, T. The Effect of Anodizing on the Fatigue Life of Aluminum Alloy 7050. *J. Jpn. Inst. Met.* **2012**, *76*, 365–374. [CrossRef]
15. Sundararajan, G.; Wasekar, N.P.; Ravi, N. The Influence of the Coating Technique on the High Cycle Fatigue Life of Alumina Coated Al 6061 Alloy. *Trans. Indian Inst. Met.* **2010**, *63*, 203–208. [CrossRef]
16. Oguri, K. Fatigue Life Enhancement of Aluminum Alloy for Aircraft by Fine Particle Shot Peening (FPSP). *J. Mater. Process. Technol.* **2011**, *21*, 1395–1399. [CrossRef]
17. Abdulstaar, M.; Mhaede, M.; Wollmann, M.; Wagner, L. Investigating the Effects of Bulk and Surface Severe Plastic Deformation on the Fatigue, Corrosion Behaviour and Corrosion Fatigue of AA5083. *Surf. Coat. Technol.* **2014**, *254*, 244–251. [CrossRef]
18. Benedetti, M.; Fontanari, V.; Bandini, M.; Savio, E. High- and Very High-Cycle Plain Fatigue Resistance of Shot Peened High-Strength Aluminum Alloys: The Role of Surface Morphology. *Int. J. Fatigue* **2015**, *70*, 451–462. [CrossRef]
19. Trsko, L.; Guagliano, M.; Bokuvka, O.; Novy, F.; Jambor, M.; Florkova, Z. Influence of Severe Shot Peening on the Surface State and Ultra-High-Cycle Fatigue Behavior of an AW 7075 Aluminum Alloy. *J. Mater. Eng. Perform.* **2017**, *26*, 2784–2797. [CrossRef]
20. Gao, Y.K. Improvement of Fatigue Property in 7050-T7451 Aluminum Alloy by Laser Peening and Shot Peening. *Mater. Sci. Eng. A* **2011**, *528*, 3823–3828. [CrossRef]

21. Luong, H.; Hill, M.R. The Effects of Laser Peening and Shot Peening on High Cycle Fatigue in 7050-T7451 Aluminum Alloy. *Mater. Sci. Eng. A* **2010**, *527*, 699–707. [CrossRef]
22. Pandey, V.; Chattopadhyay, K.; Srinivas, N.C.S.; Singh, V. Role of Ultrasonic Shot Peening on Low Cycle Fatigue Behavior of 7075 Aluminium Alloy. *Int. J. Fatigue* **2017**, *103*, 426–435. [CrossRef]
23. Singh, V.; Pandey, V.; Kumar, S.; Srinivas, N.C.S. Effect of Ultrasonic Shot Peening on Surface Microstructure and Fatigue Behavior of Structural Alloys. *Trans. Indian Inst. Met.* **2015**, *69*, 295–301. [CrossRef]
24. Muruganadhan, R.; Mugilvalavan, M.; Thirumavalavan, K.; Yuvaraj, N. Investigation of Water Jet Peening Process Parameters on AL6061-T6. *Surf. Eng.* **2018**, *34*, 330–340. [CrossRef]
25. Mahmoudi, A.H.; Jamali, A.M.; Salahi, F.; Khajeian, A. Effects of Water Jet Peening on Residual Stresses, Roughness, and Fatigue. *Surf. Eng.* **2021**, *37*, 972–981. [CrossRef]
26. Hadzima, B.; Novy, F.; Trško, L.; Pastorek, S. Shot Peening as a Pre-Treatment to Anodic Oxidation Coating Process of AW 6082 Aluminum for Fatigue Life Improvement. *Int. J. Adv. Manuf. Technol.* **2017**, *93*, 3315–3323. [CrossRef]
27. Mhaede, M. Influence of Surface Treatments on Surface Layer Properties, Fatigue and Corrosion Fatigue Performance of AA7075 T73. *Mater. Des.* **2012**, *41*, 61–66. [CrossRef]
28. Chung, Y.-H.; Chen, T.-C.; Lee, H.-B.; Tsay, L.-W. Effect of Micro-Shot Peening on the Fatigue Performance of AISI 304 Stainless Steel. *Metals* **2021**, *11*, 1408. [CrossRef]
29. Sawada, T.; Yanagitani, A. Properties of Cold Work Tool Steel Shot Peened by 1200 HV-Class Fe-Cr-B Gas Atomized Powder as Shot Peening Media. *Mater. Trans.* **2010**, *51*, 735–739. [CrossRef]
30. Zhang, P.; Liu, J.P.; Gao, Y.R.; Liu, Z.H.; Mai, Q.Q. Effect of heat treatment process on the micro machinability of 7075 aluminum alloy. *Vacuum* **2023**, *207*, 111574. [CrossRef]
31. Su, C.-H.; Chen, T.-C.; Tsay, L.-W. Improved Fatigue Strength of Cr-electroplated 7075 Al by Micro-Shot Peening. *Int. J. Fatigue* **2023**, *167*, 107354. [CrossRef]
32. Li, X.-K.; Zhu, S.-P.; Liao, D.; Correia, J.A.F.O.; Berto, F.; Wang, Q.Y. Probabilistic fatigue modelling of metallic materials under notch and size effect using the weakest link theory. *Int. J. Fatigue* **2022**, *159*, 106788. [CrossRef]
33. He, J.C.; Zhu, S.-P.; Luo, C.Q.; Niu, X.P.; Wang, Q.Y. Size effect in fatigue modelling of defective materials: Application of the calibrated weakest-link theory. *Int. J. Fatigue* **2022**, *165*, 107213. [CrossRef]
34. Wang, Y.Q.; Kawagoishi, N.; Chen, Q. Effect of Pitting Corrosion on Very High Cycle Fatigue Behavior. *Scr. Mater.* **2003**, *49*, 711–716. [CrossRef]
35. Shahzad, M.; Chaussumier, M.; Chieragatti, R.; Mabru, C.; Rezai-Aria, F. Surface Characterization and Influence of Anodizing Process on Fatigue Life of Al 7050 Alloy. *Mater. Des.* **2011**, *32*, 3328–3335. [CrossRef]

materials

MDPI

Article

Statistical Characterization of Strain-Controlled Low-Cycle Fatigue Behavior of Structural Steels and Aluminium Material

Žilvinas Bazaras and Vaidas Lukoševičius *

Faculty of Mechanical Engineering and Design, Kaunas University of Technology, Studentų Str. 56, 51424 Kaunas, Lithuania

* Correspondence: vaidas.lukosevicius@ktu.lt

Abstract: Probabilistic evaluation of the resistance to low-cycle deformation and failure of the critical components in the equipment used in the energy, engineering, metallurgy, chemical, shipbuilding, and other industries is of primary importance with the view towards their secure operation, in particular, given the high level of cyclic loading acting on the equipment during its operation. Until recently, systematic probabilistic evaluation has been generally applied to the results of statistical and fatigue investigations. Very few investigations applying this approach to the low-cycle domain. The present study aims to substantiate the use of probabilistic calculation in the low-cycle domain by systematic probabilistic evaluation of the diagrams of cyclic elastoplastic deformation and durability of the materials representing the major types of cyclic properties (hardening, softening, stabilization) and investigation of the correlation relationships between mechanical properties and cyclic deformation and failure parameters. The experimental methodology that includes the calculated design of the probabilistic fatigue curves is also developed and the curves are compared to the results of the experiment. Probabilistic values of mechanical characteristics were determined and calculated low-cycle fatigue curves corresponding to different failure probabilities, to assess them from the probabilistic perspective. A comparison of low-cycle fatigue curves has shown that the durability curves generated for some materials using analytical expressions are not accurate. According to the analysis of the relative values of experimental probabilities of low-cycle fatigue curves, the use of analytical expressions to build the curves can lead to a significant error. The results obtained allow for the revision of the load bearing capacity and life of the structural elements subjected to cyclic elastoplastic loading in view of the potential scattering of mechanical properties and resistance parameters to low-cycle deformation and failure. In addition, the results enable determination of the scatter tolerances, depending on the criticality of the part or structure.

Keywords: correlation; durability; low-cycle fatigue; regression; probability; strain-controlled loading

Citation: Bazaras, Ž.; Lukoševičius, V. Statistical Characterization of Strain-Controlled Low-Cycle Fatigue Behavior of Structural Steels and Aluminium Material. *Materials* **2022**, *15*, 8808. https://doi.org/10.3390/ma15248808

Academic Editors: Thomas Niendorf, Jaroslaw Galkiewicz, Lucjan Śnieżek and Sebastian Lipiec

Received: 14 November 2022
Accepted: 7 December 2022
Published: 9 December 2022

1. Introduction

Individual structural elements of machinery and units have been subjected to elastoplastic, static, or cyclic deformation caused by operation under high stress conditions related to the attempt to achieve maximum performance indicators (capacity, output, speed) while maintaining minimum metal consumption [1,2]. For greater operational reliability and safety of different structures and products, probabilistic methods are used increasingly extensively to calculate their strength and durability. These methods are based primarily on the use of statistical information on the mechanical properties and durability of the material under cyclic loading [3–5].

The probabilistic approach to the calculation of structures and the determination of the design characteristics of materials has been developed for more than 80 years. Weibull made a considerable contribution to the development of probabilistic methods, probabilistic substantiation of the permissible stresses, and strength safety margins for the calculation of static and cyclic strength [6,7]. The 'weakest link hypothesis' developed by Weibull allowed the scholar to build the theory of probabilistic weak fracture of bodies under the action of stress. This helped to solve the issues of fatigue failure theory. The strain-controlled failure strength was investigated by Coffin [8], Manson [9], and Langer [10]. The study by Iida and Inoue [11] provided findings on the probability of low-cycle fatigue failure. The results of the life distribution under strain-controlled loading in low-cycle fatigue tests were investigated using the normal, log-normal, and Weibull distributions. Low-cycle fatigue durability distributions were found to be fairly consistent with the log-normal distribution and Weibull distribution rather than with the normal distribution. Scattering of the crack initiation life was found to generally exceed scattering of the durability to failure. The concept of fatigue design curves was addressed in the investigation of the strength and life reduction factors. A statistical evaluation of the fatigue characteristics of the sample tested at the same level primarily requires addressing the issue of the distribution law. Different distribution functions were proposed by Freudenthal and Gumbel [12,13] as well as other researchers in relation to this issue. Serensen, Kogayev, Shneiderovich [14,15], Stepnov [16], revised and upgraded probabilistic fatigue calculations under low-cycle fatigue strain-controlled loading, performed the analysis of random deviations of the acting stress-strain forces, and analyzed the distribution of the probability of fatigue failure according to the durability characteristics.

Several researchers contributed considerably to the calculation of probabilistic methods for the mechanical and low-cycle properties. Makhutov et al. [17] presented the results of experimental investigations and durability and life calculations of low alloy and austenitic steels with different mechanical properties. Daunys et al. [18,19] investigated the dependence of low-cycle fatigue durability on the mechanical properties of steel welds in nuclear power plants. Timofeev et al. [20] and Raslavičius et al. [21] investigated the probabilistic low-cycle fatigue life and the dependence of low-cycle durability on the mechanical properties WWER-type reactor of the nuclear power plant made of steel 15Cr2MoVA and structural steel C45. Another study [22] involved the analysis of the mechanical reliability of magnesium alloys and systematic evaluation using the statistical Weibull analysis. The results obtained are very important in terms of the safety and reliability evaluation of the magnesium alloys for lightweight structures. Zhu et al. [23] have developed a probabilistic methodology to predict low cycle fatigue life by using an energy-based damage parameter built on the Bayes theorem. Strzelecki [24] presented the characteristics of the S-N curve that employ the 2-parameter and 3-parameter Weibull distributions for fatigue limit and limited life, respectively. The parameters of the proposed model were evaluated under the maximum likelihood method. In addition, the article presented the solution to the problem of estimating the initial values of the likelihood function. Fekete [25] proposed a new model for low-cycle fatigue prediction based on strain energy to account for just the part of the microstructure of the strain energy stored in the material that causes the fatigue damage.

The above review of the work has demonstrated that, until now, statistical evaluation under low-cycle loading has only been applied to durability [26–28]. There are no studies dedicated to the statistical evaluation of the strain diagrams or the parameters thereof. The evaluation of the durability to the final rupture or crack initiation has generally been performed under low-cycle fatigue conditions by applying the laws of normal, log-normal, and Weibull distribution. The following could be concluded in relation to the findings of the investigation of static characteristics and low-cycle fatigue:

- The mechanical characteristics and the characteristics of resistance to low-cycle deformation and failure have not been investigated to the extent that would be enable comprehensive application of its results to the probabilistic calculations.
- In the case of low-cycle loading, there is lack of systematic data on the investigation of the laws of distribution of durability for materials with different cyclic properties (hardening, softening, stabilization), while static investigations of resistance to low-cycle deformation under strain-controlled loading are non-existent.
- It is known that the curves of low-cycle fatigue under strain-controlled loading can be built using the mechanical characteristics, for example, σ_{ys}, σ_u, ψ etc. However,

there are no systematic investigations that would be dedicated to determination of correlation relations between mechanical characteristics and durability.

Based on the above review of the scientific studies and the conclusions thereof, the following objectives were pursued in the study presented here: (1) Definition of the laws of distribution of the mechanical characteristics and low-cycle (deformation and durability) characteristics for the materials that contrast to each other by their cyclic properties (hardening, softening, stable) with the aim of subsequent reliable static assessment (by using the confidence intervals) of the mechanical and resistance properties of low-cycle fatigue; (2) Definition of the correlation relations between the key mechanical characteristics and parameters of low-cycle fatigue that enable building of the low-cycle fatigue curves by using the mechanical characteristics; (3) Development of the methodology for determination of the probabilistic curves of low-cycle failure on the basis of the investigation conducted.

2. Materials, Specimens and Methods

2.1. Materials

Due to the statistic nature of the fatigue failure as well as static failure, there is scattering of the material properties, which may also depend on the cyclic properties of materials and the character of failure under low-cycle loading (fatigue, quasistatic, transient failure). Therefore, the static investigations conducted in the study used the specimens produced of the materials that contrasted with each other by cyclic properties: cyclically softening low alloy steel 15Cr2MoVA, cyclically stable medium carbon steel C45, and cyclically hardening aluminium alloy D16T1. The same cast was used for production of the specimens.

The 15Cr2MoVA specimens were cut from 120 mm thick rolled stock making sure that the direction of the rolled stock corresponded to the specimen axis. Steel specimens C45 and aluminium alloy D16T1 were cut out of 50 mm diameter bar stock. The 15Cr2MoVA forging and aluminium alloy were subject to thermal processing in the following modes: hardening in oil by heating up to 1000 °C followed by two subsequent quenching sequences at 700 °C for 14 h and in 670 °C for the 70 h for steel 15Cr2MoVA and standard hardening and quenching for the aluminium alloy D16T1. The chemical composition [29,30] and mechanical characteristics of the investigated materials are presented in Tables 1 and 2.

Table 1. Chemical composition of the materials.

Material	C	Si	Mn	Cr	Ni	Mo	V	S	P	Mg	Cu	Al
						%						
15Cr2MoVA (GOST 5632-2014)	0.18	0.27	0.43	2.7	0.17	0.67	0.30	0.019	0.013	-	-	-
C45 (GOST 1050-2013)	0.46	0.28	0.63	0.18	0.22	-	-	0.038	0.035	-	-	-
D16T1 (GOST 4784-97)	-	-	0.70	-	-	-	-	-	-	1.6	4.5	9.32

Table 2. Mechanical properties of the materials.

Material	e_{pr}	σ_{pr}	σ_{ys}	σ_u	σ_f	ψ
	%		MPa			%
15Cr2MoVA (GOST 5632-2014)	0.200	280	400	580	1560	80
C45 (GOST 1050-2013)	0.260	340	340	800	1150	39
D16T1 (GOST 4784-97)	0.600	290	350	680	780	14

Table 2 provides the expected value of the mechanical characteristics of the investigated materials, while Figure 1 provides the single deformation representing these characteristics, that is, the curves corresponding to the 50% probability.

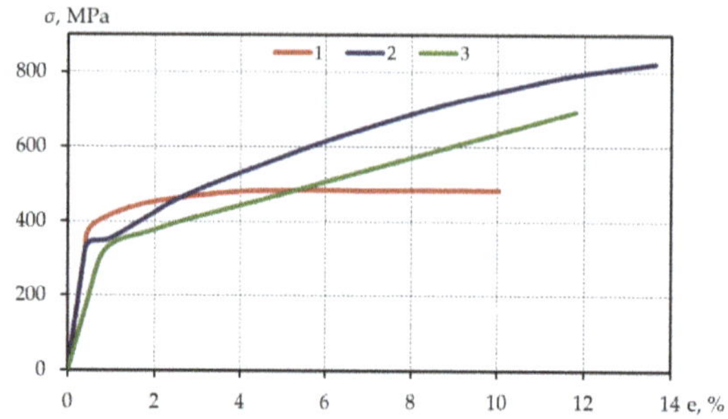

Figure 1. Single uniaxial tensile deformation curves: 1—aluminium alloy D16T, 2—structural steel C45, 3—alloyed steel 15Cr2MoVA.

Based on Table 2, steels 15Cr2MoVA and C45, especially steel 15Cr2MoVA ($\psi = 80\%$) are plastic materials. The diagram of single uniaxial tensile deformation of steel C45 is characterized by the yield plateau (up to $e = 0.6 - 0.65$), while the diagrams of single uniaxial tensile deformation of steel 15Cr2MoVA and aluminum alloy D16T1 do not contain a yield plateau. The chosen materials represent key types of cyclic properties: cyclic hardening, stabilization, and failure. Therefore, the experimental and theoretical findings obtained can be applied to the assessment of a wide range of materials used in structures subject to low-cycle loading.

2.2. Specimens

For the purpose of cyclic experiments, specimens with a cylindrical deformable part: (length—23 mm, diameter—10 mm) were selected. The cutting modes of processing were chosen to avoid traces of crushing and vibrations on the working surface. The specimens were processed with a special 60mm diameter form turning tool for all the materials investigated. A detailed drawing of the specimen is presented in Figure 2.

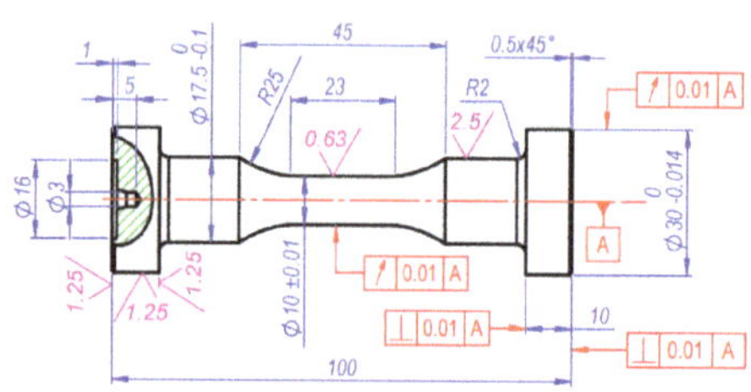

Figure 2. Specimens for low-cycle fatigue experiments (units in mm).

The specimens made of steel 15Cr2MoVA or C45 were machined with a 0.1 mm for grinding on the main part and with allowance 0.15 mm on the surface of the deformable part. The grinding of the deformable part and the base surfaces of the specimens was performed on the external grinding machine with a rounding radius of 25 mm. The minimum radial and face runout of the specimen heads in relation to the cylindrical deformable part was achieved by grinding of the surfaces on the basis of the central holes. The samples made of aluminium alloy D16T1 were processed by turning only. The final passages were performed with cutting models that provided roughness of the working and base surfaces in the detailed drawing (Figure 2).

After fatigue tests the fractured specimens were used as workpiece materials to produce monotonous tensile specimens with the aim of reaching the material properties nearly identical to the properties of the material subjected to cyclic loading. The detailed drawing of the specimen is presented in Figure 3. The specimens for uniaxial tension experiments have a diameter of 5 mm and 25 mm length. They were taken from the parts of cyclic test specimens (part of 17.5 mm diameter, Figure 2), that had not been subjected to plastic deformation.

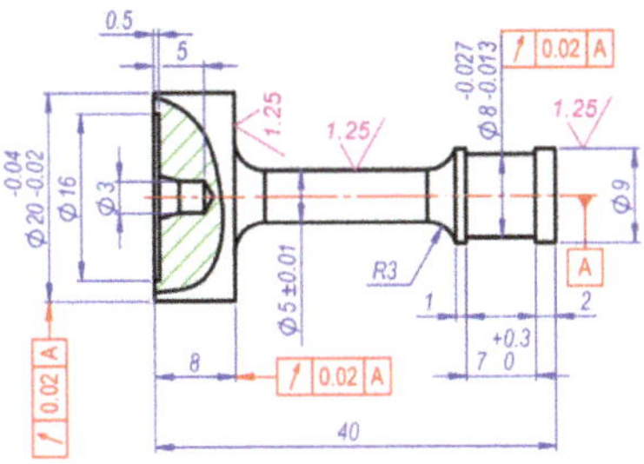

Figure 3. Specimens for single uniaxial tension experiments (units in mm).

To eliminate the elastoplastic bending of the specimen during production as a result of the cutting forces, the cutting depth and feed were decreasing as the diameter to be machined decreased. For the same purpose, the processing was performed using special high-speed steel cutters the geometry whereof was chosen with the view towards minimum cutting forces.

2.3. Methods

The experiments were carried out in the Laboratory of the Faculty of Mechanical Engineering and Design of Kaunas Technology University. The experimental equipment consisted of a 50 kN UMM-5T low-cycle tension-compression test machine (Kaunas University of Technology, Kaunas, Lithuania) and an electronic device designed to record stress-strain diagrams, cycles, and load reversal. Mechanical characteristics were measured with an error that did not exceed ±1% of the deformation scale or ±0.01% of the maximum load. GOST 25502-79 standard (Strength analysis and testing in machine building. Methods of metals mechanical testing. Methods of fatigue testing) [31] was used to perform low-cycle fatigue tests. Statistical characteristics were calculated according to the GOST 22015-76 standard (Quality of product, Regulation, and statistical quality evaluation of metal materials and products on speed-torque characteristics) [32].

As mentioned above, to define the patterns of statistical distribution of mechanical characteristics and parameters of low-cycle loading, three materials with contrasting cyclic properties were investigated in the study: softening steel 15Cr2MoVA, mildly softening (virtually stable) steel C45, and hardening aluminium alloy D16T1. The experiments were conducted under low-cycle strain-controlled symmetric tension and compression. Low-cycle fatigue curves under strain-controlled loading in relative coordinate system 'total strain amplitude $\bar{e}_0$—durability to complete fracture N_f' are presented in Figure 4.

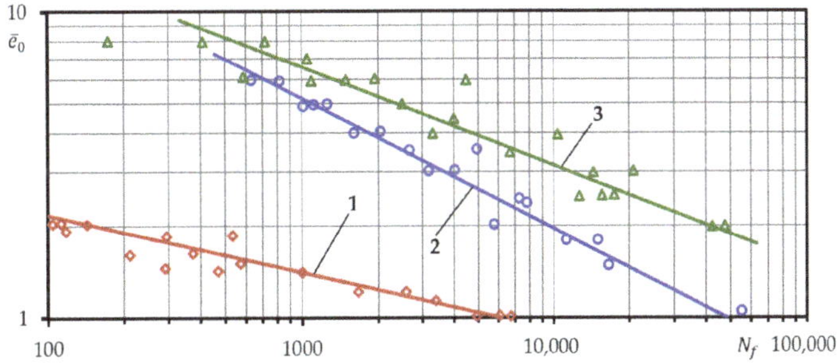

Figure 4. Low-cycle fatigue durability curves under strain-controlled loading: 1—aluminum alloy D16T, 2—alloyed steel 15Cr2MoVA, 3—structural steel C45.

The curves were built according to the guidelines established in the Rules and Norms in Nuclear Power Engineering (PNAE) [33]. According to the guidelines, low-cycle fatigue curves must be built after at least 15 tests on the specimens under different levels of uniformly distributed load.

The levels of loading and the number of specimens under the static investigation of each of them are presented in Table 3.

Table 3. Specification of low-cycle strain-controlled loading ($\bar{e}_0$ = constant).

Material	Loading Level, $\bar{e}_0$	Number of Specimens, Pcs.
15Cr2MoVA	1.8	40
	3.0	80
	5.0	40
C45	2.5	50
	4.0	120
	6.0	50
D16T1	1.0	20
	1.5	80
	2.0	20

According to Table 3, static investigations under strain-controlled loading were conducted at three levels of loading. The three levels of deformation under strain-controlled loading were chosen to be uniform throughout the investigation range. Although only fatigue failure was observed under strain-controlled loading, all three levels were considered equal in relation to the type of failure. However, the number of specimens of the materials investigated at the medium level was increased in order to identify the effect of the number of specimens on the statistical characteristics of the deformation and durability diagrams.

3. Results

3.1. Statistical Investigation of the Relationship between the Mechanical Characteristics and Parameters of Resistance to Cyclic Deformation and Failure

Under strain-controlled loading, only fatigue failure is attainable due to the limitation on total deformation of the specimen due to the conditions of the experiment. Strain-controlled loading tests are generally used for the definition of failure characteristics. The most widely used Coffin—Manson equation has been applied to the strength calculation, as it defines the relationship between the size of the plastic deformation and the number of cycles to failure [8,9]:

$$e_a {N_f}^m = C_\psi \tag{1}$$

where m and C_ψ are the constants of material that, according to the Coffin data, have the following values for the majority of materials: $m = 0.5$, $C_\psi = \frac{1}{2} \ln \frac{1}{1-\psi}$.

Equation (1) expresses the linear relationship between plastic deformation and the number of cycles to failure in coordinate system $\lg e_p$–$\lg N_f$. Plastic deformation changes in the process of strain-controlled loading and is constant only for cyclically stable materials. Hence, in Equation (1), the authors recommend using value e_a, that corresponds to 50% of the loading cycles to failure, i.e., when the process of width stabilization of the elastoplastic hysteresis loop starts. Manson, when testing Equation (1), found that, for 29 materials with contrasting cyclic properties, constant $m = 0.6$. Manson expressed the relationship between total elastoplastic deformation and the number of cycles to failure as a single dependency. The amplitude of total deformation e_a was calculated as a sum of amplitudes of plastic e_p and elastic strain e_y, i.e.,

$$e_a = e_p + e_y = \frac{1}{2}\left(\ln \frac{100}{100-\psi}\right)^{0.6} {N_f}^{-0.6} + 1.75\frac{\sigma_u}{E}{N_f}^{-0.12} \tag{2}$$

Based on the equation by Langer [10], to determine the failure amplitudes of deformation e_a and conditional stresses σ_a^* under strain-controlled symmetric loading, the following dependencies were proposed:

$$e_a = \frac{1}{4e_t}\frac{1}{N_f^m}\ln\frac{100}{100-\psi} + 0.4\frac{\sigma_u}{Ee_t} \quad (3)$$

$$\sigma_a^* = \frac{1}{4}\frac{E}{N_f^m}\ln\frac{100}{100-\psi} + 0.4\sigma_u \quad (4)$$

where m—constant equal to 0.5 under $\sigma_u \leq 687$ (MPa).

The above dependencies are often used by designers for the calculation of heavily loaded parts and structures under low-cycle deformation conditions.

The prepared version of PNAE proposes calculating the elastoplastic deformation by using the dependency:

$$e_a = \frac{0.5\ln\frac{100}{100-\psi}}{(4N_f)^{0.5}} + \frac{\sigma_u}{E(4N_f)^{0.05}} \quad (5)$$

According to the investigations by Daunys [34], the following could be written down for the majority of materials:

$$e_a {N_f}^{\alpha_{1p}} = C_{1p} \quad (6)$$

In contrast to Coffin–Manson Equation (1), in this case $\alpha_{1p} < m$ and $C_{1p} < C_\psi$. Constants α_{1p} and C_{1p} can preliminarily be defined according to the mechanical properties of the material:

$$\alpha_{1p} = 0.17 + 0.55\psi\frac{\sigma_{ys}}{\sigma_u},\ C_{1p} = 0.75\alpha_{1p}\ln\frac{100}{100-\psi} \quad (7)$$

Similarly, the same study [34] attempted to link the parameters of the generalized curve of cyclic deformation A_1, A_2, $\propto$, c, $\overline{S}_T$ to the mechanical properties of material. The following was obtained:

$$A_1 = 0.3 + 0.6\psi\frac{\sigma_{ys}}{\sigma_u} \quad (8)$$

$$A_2 = 0.32\left(\frac{\sigma_u}{\sigma_{ys}}\right)^{-7} + A_1 \quad (9)$$

$$\propto = 0.9 + 2.6\psi\frac{\sigma_{ys}}{\sigma_u} \quad (10)$$

$$c = 0.13 - 0.21\psi\frac{\sigma_{ys}}{\sigma_u} \quad (11)$$

$$\overline{S}_T = 2 - 0.83\psi\frac{\sigma_{ys}}{\sigma_u} \quad (12)$$

The analysis of the dependencies proposed by different authors for the calculation of structures and elements under elastoplastic deformation conditions has shown that durability N_c and N_f and parameters of the generalized diagram of cyclic deformation A_1, A_2, $\propto$, c, $\overline{S}_T$ are often linked by the dependencies that are used for the calculation of mechanical characteristics.

However, the scientific literature reviewed did not investigate the level of correlation between these parameters. Therefore, the present study includes investigation of the correlation relations between N_c, N_f and σ_{ys}, σ_u, σ_f, ψ, ψ_u, $\psi\frac{\sigma_{ys}}{\sigma_u}$. The results were processed according to the known methods of mathematical statistics [35].

The coefficient of correlation between two correlating values was determined as follows:

$$r = \frac{m_{xy}}{\sigma_x\sigma_y} \quad (13)$$

where the second central mixed moment was determined according to the following dependency:

$$m_{xy} = \frac{1}{n}\left[\sum_{1}^{n} x_i y_i - \frac{\sum\limits_{1}^{n} x_i \sum\limits_{1}^{n} y_i}{n}\right] \quad (14)$$

while root mean square deviations of the investigated correlating quantities were determined according to the following dependencies:

$$\sigma_x = \sqrt{\frac{1}{n}\sum_{1}^{n}\left(x_i^2 - \overline{x}^2\right)},\ \sigma_y = \sqrt{\frac{1}{n}\sum_{1}^{n}\left(y_i^2 - \overline{y}^2\right)} \quad (15)$$

The mean arithmetic values of the investigated correlating quantities were determined according to the following dependencies:

$$\overline{x} = \frac{1}{n}\sum_{1}^{n} x_i,\ \overline{y} = \frac{1}{n}\sum_{1}^{n} y_i \quad (16)$$

The values of coefficients of regression b between the correlating quantities were determined on the basis of the following expressions:

$$b_{x/y} = r\frac{\sigma_x}{\sigma_y},\ b_{y/x} = r\frac{\sigma_y}{\sigma_x} \quad (17)$$

In the course of calculation of the value of the coefficient of correlation r, it can approximately be assumed that the estimate thereof has been distributed normally. Therefore, the confidence interval of the valid values r is:

$$r \pm t_\gamma \frac{1 - r^2}{\sqrt{n}} \quad (18)$$

while in case of $\gamma = 0.90$ and $t_\gamma = 1.65$ [36]:

$$r \pm 1.65\frac{1 - r^2}{\sqrt{n}} \quad (19)$$

The linear regression equation can be brought into the following form:

$$(x - \overline{x}) = b_{x/y}(y - \overline{y}) \quad (20)$$

The analysis of the results of calculation of correlation coefficients has suggested that, under strain-controlled loading, durability N_c and N_f correlates very well (almost linearly) with mechanical characteristics σ_{ys}, σ_u, σ_f, ψ, ψ_u, $\psi\frac{\sigma_{ys}}{\sigma_u}$ for all the investigated materials with contrasting cyclic properties. For steel 15Cr2MoVA, there is almost linear correlation with durability N_c and N_f yield strength σ_{ys} and true fracture strength σ_f (Figure 5).

According to Figure 5, there is a close correlation between the durability and mechanical characteristics of the 15Cr2MoVA steel. Similar results were obtained for the regression coefficients of the resistance characteristics to cyclic deformation in relation to the mechanical characteristics.

For steel C45, at the loading level $\overline{e}_0 = 2.5$ the correlation is better with durability N_c and N_f—reduction of area ψ, while at the level $\overline{e}_0 = 4.0$, the correlation is better with true fracture strength σ_f and yield strength σ_{ys}. For the aluminium alloy D16T1, yield strength σ_u correlates directly to durability. Interestingly, for steels 15Cr2MoVA and C45, multiplier $\psi\frac{\sigma_{ys}}{\sigma_u}$ insignificantly increases the coefficient of correlation between N_c,N_f and $\psi\frac{\sigma_{ys}}{\sigma_u}$. For aluminium alloy D16T1, introduction of multiplier $\psi\frac{\sigma_{ys}}{\sigma_u}$ leads to certain reduction of the coefficient of correlation for the levels of strain-controlled loading $\overline{e}_0 = 1.0; 1.5$, while for level $\bar{e}_0 = 2.0$, the multiplier increases the coefficient of correlation, same as in the case of steels 15Cr2MoVA and C45.

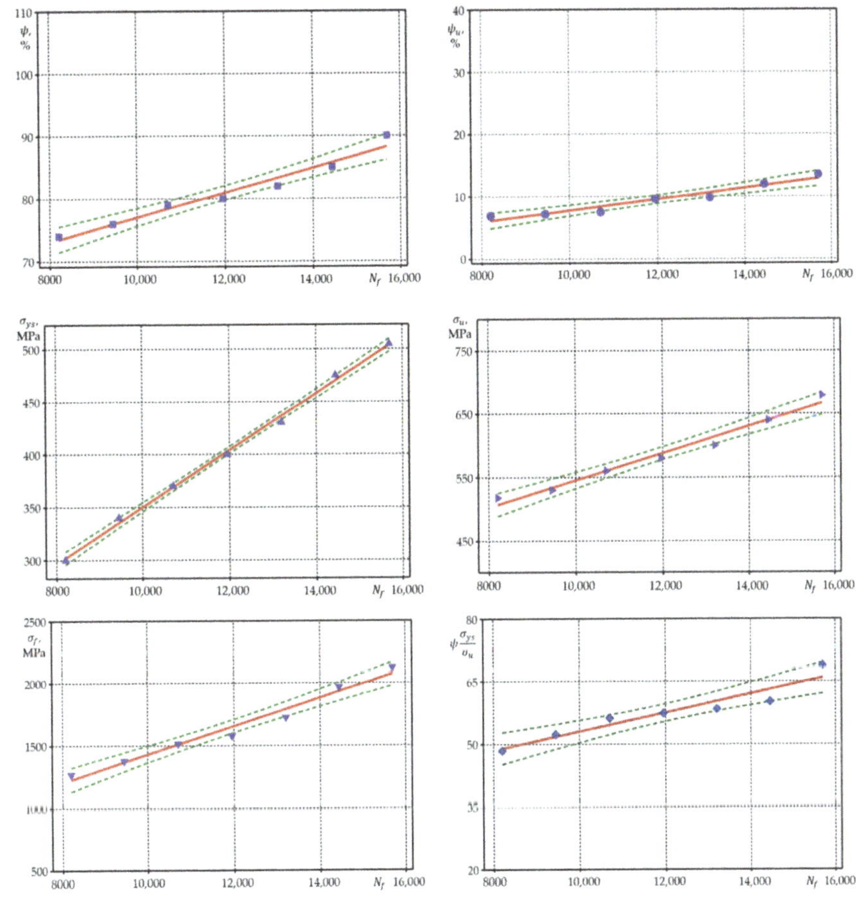

Figure 5. Regression lines under strain-controlled loading for steel 15Cr2MoVA ($\bar{e}_0 = 1.8$). Dots represent the experimental values, straight lines—the theoretical calculated dependencies, while dashed lines represent confidence intervals.

3.2. *Statistical Assessment of the Low-Cycle Fatigue Curves under Strain-Controlled Loading*

Until present, the low-cycle fatigue curves under strain-controlled loading have been built using Equations (1)–(6) or similar equations by applying the mechanical characteristics that correspond to a probability of 50%. Therefore, the curves calculated here correspond to the same failure probability. There are no systematic investigations into the building of calculated probabilistic low-cycle fatigue curves in the scientific literature.

In this study, probabilistic values of mechanical characteristics were determined. This enabled the authors to build the calculated low-cycle fatigue curves corresponding to different failure probabilities and to assess them from the probabilistic perspective.

Tables A1–A3 contain the values σ_{ys}, σ_u, ψ, e_{pr}, corresponding to 1%, 10%, 30%, 50%, 70%, 90%, 99% probabilities for all the materials investigated. The low-cycle fatigue curves under strain-controlled loading corresponding to 1%, 10%, 30%, 50%, 70%, 90%, 99% failure probabilities for the investigated steel 15Cr2MoVA were built using Equations (3), (5) and

(6) and the values of equal probability of mechanical characteristics (Figure 6a–c). The calculated curves were built using the absolute coordinates $\lg e_0 - \lg N_c$.

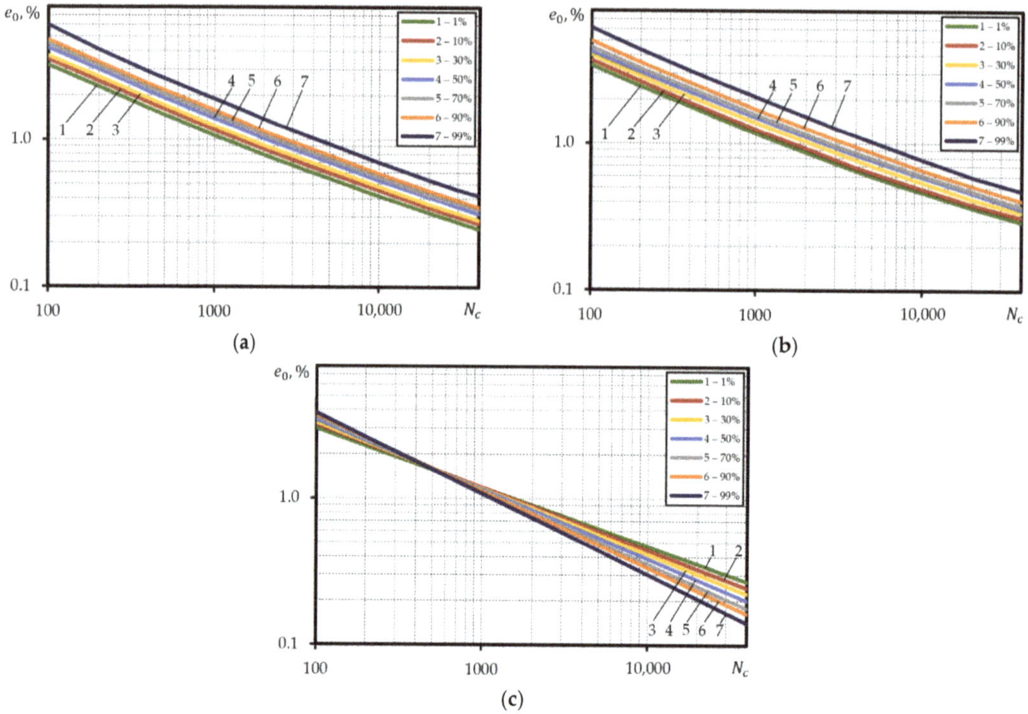

Figure 6. Calculated probabilistic low-cycle fatigue curves under strain-controlled loading for steel 15Cr2MoVA, built using the absolute coordinates, according to: (**a**)—Equation (3), (**b**)—Equation (5), (**c**)—Equation (6).

Based on Figure 6 the results of calculations using Equations (3) and (5), are little different from each other in terms of both the slope and occupied scatter band of the probabilistic low-cycle fatigue curves. The ratio between durability values for 99% and 1% curves is little dependent on the deformation level. The average ratio for the curves calculated using Equation (3) was about 3.2, for curves calculated using Equation (5)—about 3.3. Calculation of the probabilistic low-cycle fatigue curves using Equation (6) for steel 15Cr2MoVA generates a non-satisfactory result as the curves cross each other in case of $N_f = 200 - 400$ cycle durability or are positioned in reverse order in case of durabilities $N_f > 400$, i.e., the durability is the lowest in case of 99% failure probability and the highest in case of 1% failure probability. This is associated with dependency of α_{1p} and C_{1p} on the mechanical characteristics of the material. As follows from Equations (7) and (8), the probabilistic value of α_{1p} largely depends on ψ, as the probabilistic values of ratio σ_{ys}/σ_u differ very little (Table A1). Constant C_{1p} that depends on both α_{1p} and ψ changes within the range from 0.42 for probability curve 1% to 0.96 for probability curve 99%. Constant α_{1p} ranges from 0.41 (failure probability curve 1%) to 0.56 (failure probability curve 99%). The specified changes α_{1p} and C_{1p} lead to the positioning of the probabilistic curves as depicted in Figure 6c.

The dependencies built in the relative coordinates are used for calculation of the low-cycle fatigue of parts and structural elements. Hence, the probabilistic curves of low-cycle fatigue under strain-controlled loading for steel 15Cr2MoVA were also built in the relative coordinates using Equations (3), (5) and (6). The amplitude strains of curve 1%

were divided by probabilities e_{pr}—1%, strains of curve 10%—by probabilities e_{pr}—10%, etc. Hence, the designed calculated probabilistic curves of low-cycle fatigue in the coordinates $\lg\bar{e}_0 - \lg N_c$ are presented in Figure 7a–c.

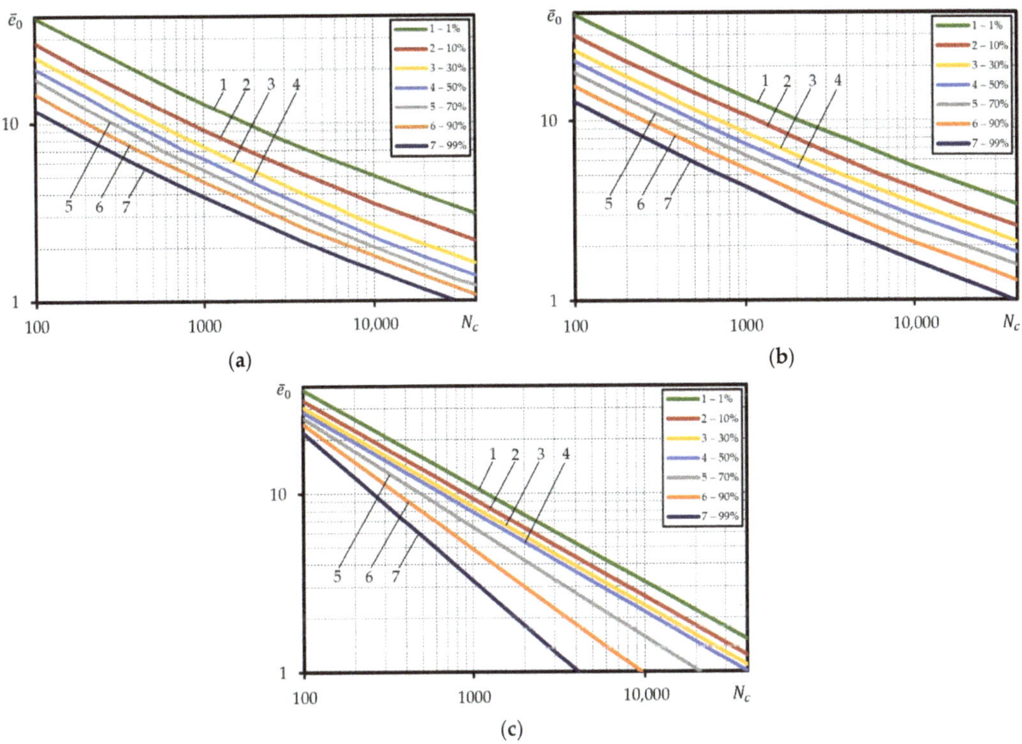

Figure 7. Calculated probabilistic low-cycle fatigue curves under strain-controlled loading for steel 15Cr2MoVA, built in the relative coordinates, according to: (**a**)—Equation (3), (**b**)—Equation (5), (**c**)—Equation (6).

Figure 7 suggests that the application of the relative coordinates for the design of probabilistic curves of low-cycle fatigue generates an implausible picture for steel 15Cr2MoVA. The implausibility of the mutual position of the calculated probabilistic curves lies in that curve 99% is characterized by the lowest durability, while curve 1%—the highest durability. The 'reverse' layout of the probabilistic curves of low-cycle durability is related to very vast scatter e_{pr} compared to the scatter of other mechanical characteristics, for example, ψ, that largely determine the durability. As suggested by Table A1, for steel 15Cr2MoVA, the ratio of values e_{pr} for probability 99% to 1% probability is 7.6, while the ratio of values ψ for probability 99% to 1% probability—1.4.

The sharp contrast in the values of the scatter between e_{pr} on one side and σ_{ys}, σ_u, ψ on the other side is likely to be due to higher sensitivity e_{pr} to thermal processing, hardening during mechanical processing, accuracy of the experiment, and other factors, in comparison to other mechanical characteristics. The conducted analysis of the calculated probabilistic curves for steel 15Cr2MoVA suggests that the probabilistic values of strain e_{pr}, cannot be used for the design of probabilistic curves of low-cycle fatigue in the relative coordinates as they distort the true layout of the curves.

To define more truthful layout of the calculated probabilistic curves of low-cycle fatigue under strain-controlled loading in the relative coordinates for steel 15Cr2MoVA, the percentage curve strain was divided by the mean arithmetic value of e_{pr} (Figure 8a–c).

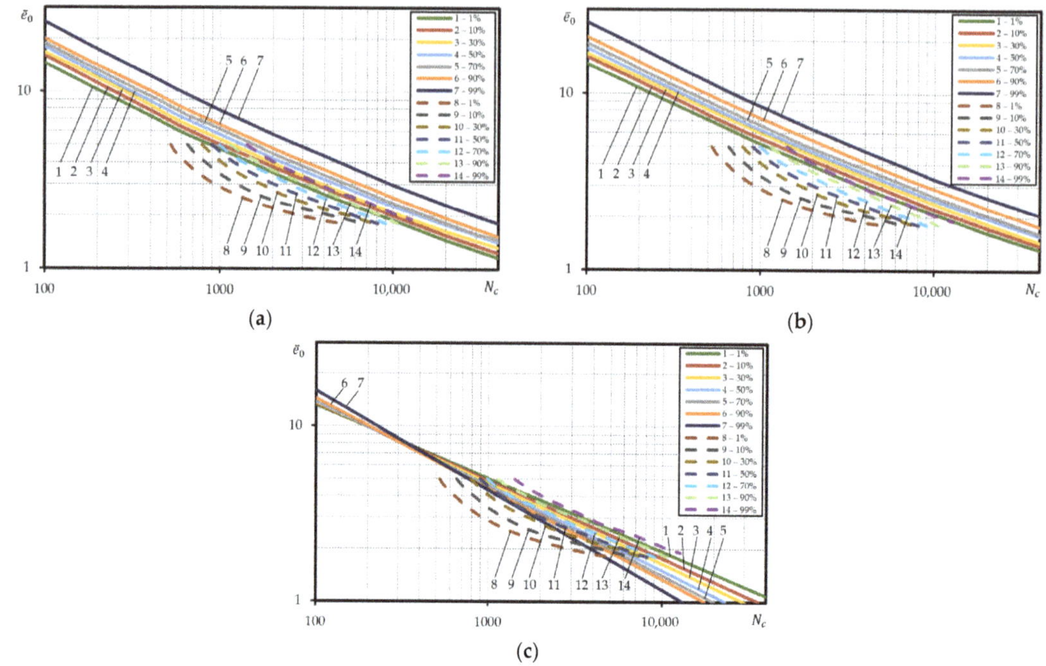

Figure 8. Comparison of the calculated probabilistic low-cycle fatigue curves under strain-controlled loading with the experimental ones for steel 115Cr2MoVA; straight lines—calculated, dashed lines—experiment; according to: (**a**)—Equation (3), (**b**)—Equation (5), (**c**)—Equation (6).

Here, as suggested by Figure 8a–c there is little difference in the layout of the probabilistic curves in case of the absolute coordinates (Figure 6). For the calculated curves built according to Equations (3) and (5), the ratio of durability for curve 99% and curve 1% was little dependent on the strain. The average ratio for the curves built according to Equation (3) was about 2.9, and for the curves built according to Equation (5)—3.2.

To validate the calculation for steel 15Cr2MoVA, the experimental curves of equal probability were compared to the calculated curves. Figure 8a,b suggests that the slope angle and the occupied scatter band of the experimental curves of equal probability is little different from those of the calculated curves designed according to Equations (3) and (5). Nonetheless, the experimental curves were located lower than the calculated curves. For example, at low durability, experimental curve 99% corresponded to calculated curve 30% (Equation (3)), while experimental curve 50% corresponded to calculated curve 1% (Figure 8a). Correspondence between the experimental curves and calculated curves according to Equation (5) was less accurate at high durability. In this case, experimental curve 99% was calculated above than calculated area 1% for $N_c > 3000$ cycle durability. In case of durabilities $N_c < 3000$, experimental curve 99% corresponded to calculated curve 1% (Figure 8b). Figure 8c suggests that the calculated probabilistic curves designed according to Equation (6) completely fall within the zone of the experimental curves. Nonetheless, the comparison renders the 'reverse' layout of the calculated probabilistic curves impossible in the area of $N_c > 200 - 400$ cycle durabilities. The reasons have already been covered above.

Similar analysis was conducted with the calculated and experimental probabilistic low-cycle fatigue curves under strain-controlled loading for steel C45. Equations (3), (5) and (6) were used to calculate the curves of equal probability by applying the probabilistic values of mechanical characteristics (Table A2). The obtained results were built using the absolute coordinates $\lg e_0 - \lg N_c$. For steel C45, same as for steel 15Cr2MoVA, the probabilistic calculated curves obtained according to Equations (3) and (5) were positioned in a similar way in terms of both the slope angle and the occupied scatter band. For probabilistic curves calculated according to Equation (6), increase accompanied by higher failure probability was observed, same as in the case of steel 15Cr2MoVA. For steel C45, however, the larger range of variation of the probabilistic value and smaller range of variation of probabilistic value C_{1p}, than for steel 15Cr2MoVA lead to regular layout of the probabilistic curves, i.e., curve 1% provides the lowest durability, while curve 99%—the highest durability.

The investigation of the durability scatter band for steel C45 has demonstrated that the ratio of durability for probabilistic curves 99% and 1% depends on the strain level. As suggested by the analysis performed, for curves designed according to Equations (3), (5) and (6), the ratio of durability of curves 99% and 1% at strain amplitude $e_0 = 0.9\%$ was 7.3; 8.2; 10.6 respectively, while at $e_0 = 0.4\%$ — 10; 18.7; 3.7. For steel C45, same as for steel 15Cr2MoVA, the use of probabilistic value e_{pr} for design of probabilistic calculated curves in the relative coordinates $\lg\bar{e}_0 - \lg N_c$ distorts their true layout in terms of all the dependencies applied Equations (3), (5) and (6), i.e., the curves are positioned in the reverse order. At strain amplitude $\bar{e}_0 = 4$, the ratio of durability of probabilistic curves 1% and 99% was respective 2.8; 3.5; 8.3, and at $\bar{e}_0 = 2$ — 4.0; 6.7; 20.7. To obtain a valid layout of the calculated probabilistic curves in the relative coordinates for steel 15Cr2MoVA and steel C45, strains e_0 were divided by mean arithmetic value e_{pr}. Hence, the obtained probabilistic curves are depicted in Figure 9a–c.

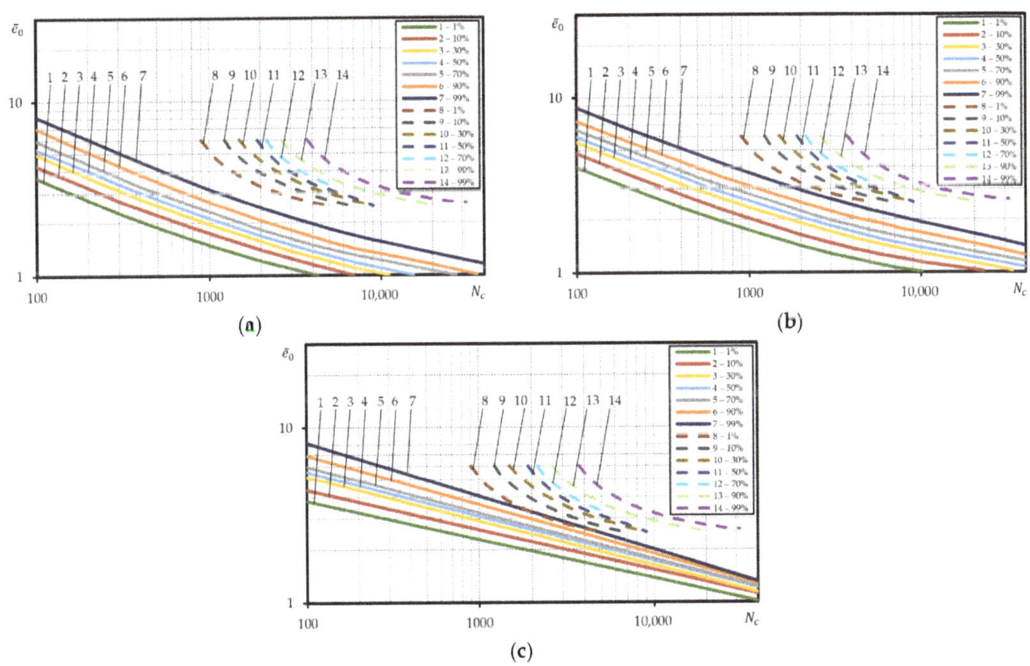

Figure 9. Comparison of the calculated probabilistic low-cycle fatigue curves under strain-controlled loading with the experimental ones for steel C45; straight lines—calculated, dashed lines—experiment; according to: (**a**)—Equation (3), (**b**)—Equation (5), (**c**)—Equation (6).

The calculated probabilistic curves are close to the curves in the absolute coordinates by the layout character and slope angle. For these curves, the durability ratio of the curves 99% and 1% depend on the strain level, the same as for the curves in the absolute coordinates, i.e., at strain amplitude $\bar{e}_0 = 4$, the ratio of durability according to Equations (3), (5) and (6) is 7.1; 7.6; 10.3, and at $\bar{e}_0 = 2$ — 8.4; 11.7; 5.3.

At the same time, the same figures also include the experimental low-cycle fatigue curves under strain-controlled loading for steel C45 (Figure 9a–c). Comparison of the experimental probabilistic curves with the calculated one has shown that the calculated probabilistic curves for steel C45 (Figure 9a–c) are located below the 1% experimental probabilistic curve. Of all the dependencies applied to the calculation of probabilistic low-cycle fatigue curves under strain-controlled loading, the calculated (Equation (6)) for steel C45 was the closest to reality as demonstrated by the investigation.

The calculations of the probabilistic low-cycle fatigue curves under strain-controlled loading for aluminum alloy D16T1 (Table A3) were performed according to Equation (6), as Equations (3) and (5) were designed for low-alloy steels used for energy purpose. The same methodology was used for the design of the calculated curves as for the steels 15Cr2MoVA and C45, that is, the curves were designed in the absolute coordinates and in the relative coordinates by using probabilistic and mean arithmetic values e_{pr}.

Investigating the durability scatter band for the D16T1 aluminium alloy has demonstrated that the ratio for probabilistic curves 99% and 1% depends on the strain level (Figure 10).

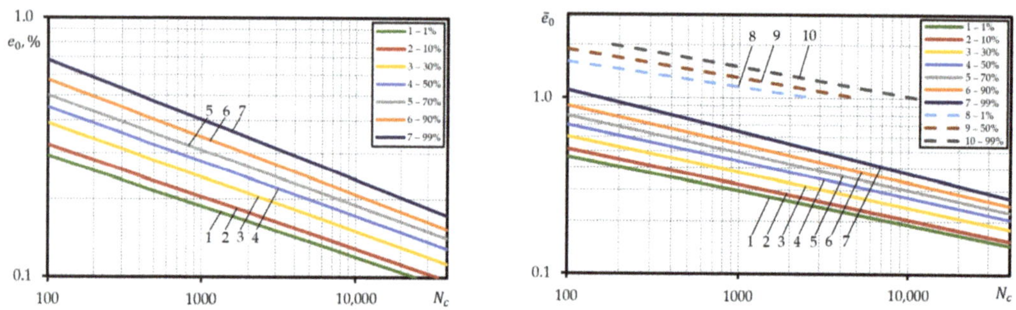

(a) (b)

Figure 10. Probabilistic low-cycle fatigue curves under strain-controlled loading for aluminium alloy D16T1 built using: the absolute coordinates (**a**); the relative coordinates (**b**); straight lines—calculated, dashed lines—experiment.

The conducted analysis has shown that the ratio of durability at strain amplitude $\bar{e}_0 = 0.3$ was 37, and at $\bar{e}_0 = 0.18\%$ — 24. At the same time, in the case of the calculated probabilistic curves, an increase in the slope angle with the increase in the failure probability has been observed. This is directly affected by the scatter of reduction of area ψ, same as for steels 15Cr2MoVA and C45. Application of relative strain $\bar{e}_0$, determined according to the probabilistic values of strain e_{pr} to the calculations leads to narrowing of the durability scatter band. In this case, the ratio of durability of the probabilistic curves 99% and 1% at strain amplitude $\bar{e}_0 = 4$ was 3.3, and at $\bar{e}_0 = 3 - 2.7$. The slope angles of the calculated probabilistic curves in the relative coordinates were reducing in comparison to the same curves in the absolute coordinates. However, the curve angle of the slope increased with increasing failure probability.

Same as for steels 15Cr2MoVA and C45, to calculate the strains $\bar{e}_0$, of aluminium alloy D16T1, mean arithmetic value e_{pr} was used. In this case, the ratio of durability of probabilistic curves 99% and 1% was close to the results of the ratio of durabilities of the probabilistic curves designed in the absolute coordinates, i.e., at $\bar{e}_0 = 4$, the ratio was 3.3, and at $\bar{e}_0 = 3 - 2.7$ (Figure 10b). Figure 10b also depicts the comparison of the calculated probabilistic curves with the experimental ones for aluminium alloy D16T1. As suggested by Figure 10b the correspondence of the experiment results with the calculated results is non-satisfactory, as the calculated curves are fully reflected in the elastic area.

Figures 11 and 12 for steels 15Cr2MoVA and C45 compare the experimental low-cycle fatigue curves under strain-controlled loading of failure probability 1%, 50%, 99% with the calculated curves designed according to Equations (3), (5) and (6) by using the normalized mechanical characteristics determined according to Equation (21) and mechanical characteristics taken from reference documents [37], and the low-cycle fatigue curves defined using the safety factor $n_N = 10$ for cycles and $n_e = 2$ for strain as used in the field of mechanical engineering.

$$K = \frac{x_{\max}}{x_{\min}} \tag{21}$$

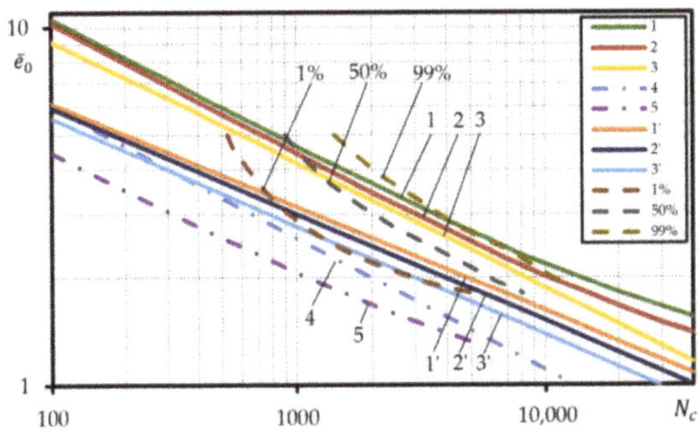

Figure 11. Comparison of the probabilistic low-cycle fatigue curves calculated under strain-controlled loading with the experimental ones for steel 15Cr2MoVA; straight lines—calculated, dashed lines–experiment; normalized characteristic according to: 1—Equation (3), 2—Equation (5), 3—Equation (6); reference characteristics, according to: 1′—Equation (3), 2′—Equation (5), 3′—Equation (6); calculated, according to: 4—Equation (3) $n_e = 2$, 5—Equation (3) $n_N = 10$.

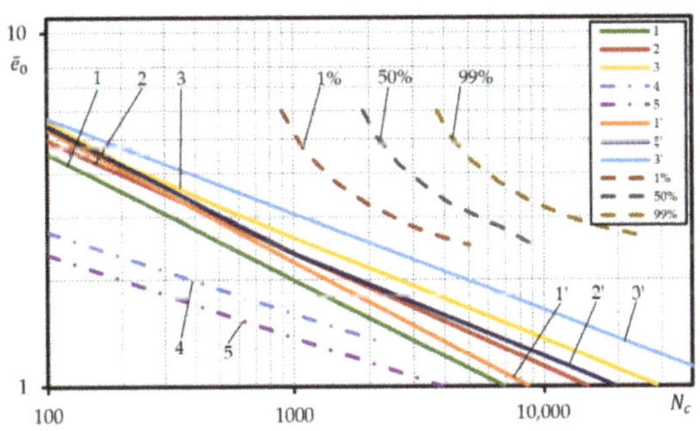

Figure 12. Comparison of the probabilistic low-cycle fatigue curves calculated under strain-controlled loading with the experimental ones for steel C45; straight lines—calculated, dashed lines– experiment; normalized characteristic according to: 1—Equation (3), 2—Equation (5), 3—Equation (6); reference characteristics, according to: 1′—Equation (3), 2′—Equation (5), 3′—Equation (6); calculated, according to: 4—Equation (3) if $n_e = 2$, 5—Equation (3) if $n_N = 10$.

Table A4 provides the values of ratios K of the highest and lowest mechanical characteristics $\sigma_{pr}, \sigma_{ys}, \sigma_u, \sigma_f, \psi, \psi_u$.

The design of the last curves employed the low-cycle fatigue curves designed according to the reference mechanical characteristics and the dependencies providing the most accurate description of experimental durability, i.e., Equation (3) for steel 15Cr2MoVA and Equation (6) for steel C45.

As suggested in Figure 11, for steel 15Cr2MoVA, the low-cycle fatigue curves determined according to Equations (3), (5) and (6) using normalized mechanical characteristics are above the experimental curve of probability of failure 99%. The same curves designed by using the reference mechanical characteristics are located in the durability band between curves 1% and 50%. This was predictable, as the normalized mechanical characteristics of the steel 15Cr2MoVA are close to the experimental mechanical characteristics with a probability of 12% to 25%, while the reference mechanical characteristics are located in the probability band of the experimental characteristics of 0.0003 to 74%. Due to the 'high' layout of the calculated low-cycle fatigue curves compared to the experimental ones, the curves designed using safety factors $n_e = 2$ and $n_N = 10$ are also positioned fairly high. The low-cycle fatigue curve designed using $n_e = 2$ virtually corresponds to the experimental curve of failure probability 1%, while the curve designed using $n_N = 10$ is located below.

Another situation is presented in Figure 12 that shows the listed low-cycle fatigue curves for steel C45. In this case, the calculated curves designed using both the normalized and the reference mechanical characteristics are located considerably lower than the experimental ones. However, this is the consequence of poor correspondence of the calculated curves designed according to Equation (6) with the experimental curves for the steel C45 (Figure 9c).

3.3. Case Study Objective

To Determine the Probabilistic Values of Cumulative Durability Damage (at the Crack Initiation Phase) for the Zone of Sleeve Connection to the Vessel Body under Hydraulic Forging at Temperature 20°C. The Nominal Strain Range for the Outer Surface of the Sleeve Connection under Hydraulic Forging: $\varepsilon_{1n} = 0$, $\varepsilon_{2n} = 0.54\%$, $\varepsilon_{3n} = -0.97\%$. Concentration Factor (Theoretical) of Elastic Stress $\alpha_\sigma = 3$, Strain-Controlled Loading Mode, Vessel Material—Grade 15Cr2MoVA Steel.

Mechanical and cyclic characteristics determined during the course of the study (Table A1 and Table A5) were used in the calculations.

All calculations were carried out for the failure probability of 1%, 10%, 30%, 50%, 70%, 90%, 99%. The case study presents the calculation for probability 1%. For other probabilities, the calculated values are presented in A6. The calculation was carried out as follows.

Hardening rate m_0 (Table A5) was used to design the tensile stress-strain diagrams $\overline{\sigma} - \overline{e}$. Their linear approximation resulted in following relative linear hardening modules:

$$G_T = \frac{\overline{\sigma} - 1}{\overline{e} - 1}, \tag{22}$$

$$G_T = \frac{2.06 - 1}{40 - 1} = 0.0272$$

Parameter χ_1 characterizing the sensitivity to cycle asymmetry was determined according to the following dependency:

$$\chi_1 = 2.7\psi\frac{\sigma_{ys}}{\sigma_u} - 1, \tag{23}$$

$$\chi_1 = 2.7 \cdot 0.74\frac{300}{500} - 1 = 0.1988$$

$$\begin{gathered} \chi_2 = 0.23(\tfrac{\sigma_u}{\sigma_{ys}} - 1) + \chi_1, \\ \chi_2 = 0.23(\tfrac{500}{300} - 1) + 0.1988 = 0.3521 \end{gathered} \tag{24}$$

The coefficient of the cycle asymmetry intensity of the nominal stresses was assumed to be $r_{\sigma_n} = -1.05$, and the range of intensity of nominal stresses for the first semi-cycle of loading was determined according to the following dependency:

$$\overline{S}_{inK} = \frac{2G_T\left[\overline{\varepsilon}_{inK}\overline{S}_T - B \cdot A_{1;2} \cdot F(k)\right]}{\overline{S}_T[2G_T + p_{1;2} \cdot A_{1;2} \cdot F(k)]}, \tag{25}$$

$$\overline{S}_{inK} = \frac{2 \cdot 0.0272[9.37 \cdot 1.16 - 37.19 \cdot 0.23 \cdot 1]}{1.15[2 \cdot 0.0272 + 1.0048 \cdot 0.23 \cdot 1]}$$

where

$$p_{1;2} = 1 + \chi_{1;2}\frac{1 + r_{\sigma_n}}{1 - r_{\sigma_n}}, \tag{26}$$

$$p_1 = 1 + 0.1988\frac{1 + 1.05}{1 - 1.05} = 1.0048,$$

$$B = 1 - \frac{1}{G_T} - \frac{\overline{S}_T}{2}, \tag{27}$$

$$B = 1 - \frac{1}{0.0272} - \frac{1.15}{2} = -37.19$$

Hence, to calculate the Poisson's ratio in normal section μ_n for the first semi-cycle of loading, it was necessary to use the value of the relative linear hardening module g_{n1}, and in the first approximation after replacement of the sign of main strains with the opposite sign, the following was assumed:

$$\overline{\varepsilon}_{in1} = \overline{\varepsilon}_{1n1},\ \overline{\varepsilon}_{1n1} = \frac{\varepsilon_{in1}}{e_{pr}\overline{S}_T}, \tag{28}$$

$$\overline{\varepsilon}_{in1} = \frac{0.97}{0.09 \cdot 1.15} = 9.37$$

Linear hardening module:

$$g_{n1} = \frac{\overline{S}_{in1} - 1}{\overline{\varepsilon}_{in1} - 1}, \tag{29}$$

$$g_{n1} = \frac{3.0231 - 1}{9.37 - 1} - 0.2417$$

and, accordingly, the Poisson's ratio:

$$\mu_{n1} = 0.5 - 0.2\frac{1 - g_{n1}(\overline{\varepsilon}_{in1} - 1)}{\overline{\varepsilon}_{in1}}, \tag{30}$$

$$\mu_{n1} = 0.5 - 0.2\frac{1 - 0.2417(9.37 - 1)}{9.37} = 0.4355$$

Intensity of nominal strains of the first approximation were determined according to dependency [34]:

$$\varepsilon_{in1} = \frac{\sqrt{2}}{2(1 + \mu_{n1})}\sqrt{(e_1 - e_2)^2 + (e_2 - e_3)^2 + (e_3 - e_1)^2}, \tag{31}$$

$$\varepsilon_{in1} = \frac{\sqrt{2}}{2(1 + 0.4355)}\sqrt{(0.97 - 0.54)^2 + (0.54)^2 + (-0.97)^2} = 0.5764$$

In relative units:

$$\varepsilon_{in1} = \frac{0.5864}{0.09 \cdot 1.15} = 5.6657$$

The obtained value and dependencies were used to repeat the calculation and determine final values $\overline{S}_{in1}$, g_{n1}, μ_{n1} and $\overline{\varepsilon}_{in1}$ (Table A5). If symmetrical cycle of strain intensity in the nominal section was accepted for the task considered, then:

$$\overline{e}_{in} = \frac{\overline{\varepsilon}_{in1}\overline{S}_T}{1 - r_{en}} \tag{32}$$

The coefficient of asymmetry of the intensity cycle of nominal strain was assumed as $r_{en} = -1$, then:

$$\bar{e}_{in} = \frac{4.8279 \cdot 1.15}{1+1} = 2.7761$$

The range of main stress in the first semi-cycle of loading in the nominal section was determined according to the following dependencies:

$$S_{1n1} = \frac{\varepsilon_{1n1} + \mu_{n1}\varepsilon_{2n1}}{1-\mu_{n1}^2} E'_1, S_{2n1} = \frac{\varepsilon_{2n1} + \mu_{n1}\varepsilon_{1n1}}{1-\mu_{n1}^2} E'_1 \quad (33)$$

$$E'_1 = \frac{S_{in1}}{\varepsilon_{in1}} = \frac{2.3744 \cdot 205}{4.8279 \cdot 0.0009} = 112,023\text{MPa}$$

$$S_{1n1} = \frac{0.0097 + 0.4151 \cdot 0.0054}{1 - 0.4151^2} \cdot 112,023 = 1616\text{MPa},$$

$$S_{2n1} = \frac{0.054 + 0.4151 \cdot 0.0097}{1 - 0.4151^2} \cdot 112,023 = 1276\text{MPa}$$

In accordance with the ranges of main stress available, the strain-controlled stiffness coefficients were determined according to the dependency:

$$D_e = \frac{\sqrt{(S_{1n1} - S_{2n1})^2 + (S_{2n1} - S_{3n1})^2 + (S_{3n1} - S_{1n1})^2}}{\sqrt{2}(\sigma_1 + \sigma_2 + \sigma_3)}, \quad (34)$$

$$D_e = \frac{\sqrt{(1616 - 1276)^2 + (1276)^2 + (-1616)^2}}{\sqrt{2}(1616 + 1276)} = 0.5103$$

By using $\bar{e}_{in}, \alpha_B, r_{en}, \chi_1, \chi_2, G_T$ (Table A5) and $A_1, A_2, \overline{S}_T, \beta$ as well as the Matlab programme of calculations designed according to the dependencies in papers [32], the data characterizing the stress-strain state in the zone of maximum concentration were determined (Table A6). In the durability calculation, the same as in relation to the durability calculation norms [33], the safety factor for maximum deformations $n_e = 2.$ was accepted. Then, the cumulative quasi-static damage [34]:

$$d_k = \frac{2(\bar{e}_i - \bar{\sigma}_i)}{\bar{e}_B D_e}, \quad (35)$$

Based on study [38], $\bar{e}_B = m_0$ was assumed:

$$\bar{e}_B = \frac{m_0}{e_{pr}} = \frac{0.15}{0.0009} = 166.6$$

then

$$d_k = \frac{2 \cdot (15.9 - 1.3405)}{166.6 \cdot 0.5103} = 0.3424, \text{ and } d_y = 1 - d_k = 0.6576$$

The number of semi-cycles k_0, within which fatigue damage $d_y = 0.6576$ would be cumulated was determined according to the following dependencies:

$$d_y = \frac{\sum_1^{k_0} \frac{\bar{\delta}_{ik}}{D_e} \left(\frac{\bar{\delta}_{ik}}{D_e} + \overline{S}_{ik}\overline{S}_T \right)^{\alpha_3}}{C_2 C_3^{\alpha_3}} \quad (36)$$

Durability k_0 calculations were conducted by using the following two techniques:

(a) in the calculations, probabilistic values of characteristics $A_1, A_2, \chi_1, \chi_2, \overline{S}_T, G_T, \beta, D_e, C_2, C_3$ and α_3 were used:

$$0.7259 = \frac{\sum_{1}^{k_0} \frac{2\overline{\delta}_{ik}}{0.5103}\left(\frac{2\overline{\delta}_{ik}}{0.5103} + \overline{S}_{ik} \cdot 1.28\right)^{0.98}}{500 \cdot 286^{0.98}}, \; k = 12.5 \text{ or } N = 6$$

(b) in the calculations, the values of probability 50% of characteristics $A_1, A_2, \chi_1, \chi_2, \overline{S}_T, G_T, \beta, D_e$ and probabilistic values of parameters C_2, C_3 and α_3 were used:

$$0.6576 = \frac{\sum_{1}^{k_0} \frac{\overline{\delta}_{ik}}{0.5103}\left(\frac{2\overline{\delta}_{ik}}{0.5103} + \overline{S}_{ik} \cdot 1.15\right)^{0.98}}{500 \cdot 286^{0.98}},$$

The results of probabilistic calculations are presented in Table 4.

Table 4. Results of probabilistic calculation of life in the concentration zone.

Parameter	Probability, %						
	1	10	30	50	70	90	99
k_0	12.5	21.5	52.0	86.0	201.0	287.0	389.0
k_0	59.5	66.5	78.5	86.0	98.5	123.0	171.0

The total number of start-stop operations of the system for the vessel considered expected during the lifetime is equal to 25 [33]. Hence, fatigue cracks may appear in the sleeve-to-body connection during the lifetime of the vessel considered at failure probability 30% according to calculation (a) and at failure probability lower than 1% according to calculation (b).

4. Conclusions

The methodology of conduction of an integrated experiment in a probabilistic setting to investigate the parameters of resistance to cyclic deformation and failure for the materials representing major types of cyclic properties (hardening, softening, stabilization).

The investigations presented in the paper point to the correlation relationship between mechanical characteristics and durability as well as the cyclic deformation that is close to the linear correlation. This supports the correctness and physical rationale of the mathematical dependencies proposed by different authors for describing the low-cycle deformation and failure process under strain-controlled loading. The regression coefficients allow for the calculation of preliminary cyclic characteristics and durability using the available mechanical characteristics.

The analysis of mutual layout of the calculated and experimental probabilistic low-cycle fatigue curves conducted in the study once again has demonstrated that the design of low-cycle fatigue curves according to the both analytical dependencies for specific materials may lead to considerable errors. Hence, in calculations of critical structures, it is necessary to have at least an experimental curve of failure probability 50% to conduct a reliable assessment of the strength and durability of the structure considered.

Investigations have pointed at the presence of a stable correlation relationship between mechanical characteristics and durability as well as the parameters of cyclic deformation parameters. This supports the physical rationale of the phenomenological dependencies proposed by different authors for the description of the low-cycle deformation and failure process under strain-controlled loading by using mechanical characteristics.

The above investigation has suggested that probabilistic curves should not be built in the relative coordinates using the probabilistic e_{pr}, if scatter e_{pr} is considerably larger than scatter σ_{ys}, σ_u and ψ. Equations (3), (5) and (6) are not universal, since, for steel

15Cr2MoVA, the best correspondence is provided by Equations (3) and (5), while for steel C45 Equation (6).

The statistical investigations conducted in the paper have shown that that phenomenological dependencies used at present for the description of low-cycle fatigue curves on the basis of mechanical characteristics are not universal for the materials with contrasting cyclic properties, and a reliable assessment of the durability of structure requires the probabilistic experimental curve 50%. For example, the calculated low-cycle fatigue curves under strain-controlled loading defined using the safety factor [31] ($n_e = 2,\ N_n = 10$) and the calculated curves that provide more accurate description of the experimental data provide the following strength safety margins in comparison to the experimental curves: $n_e = 0.45,\ N_n = 4.10$ (for steel 15Cr2MoVA) and $n_e = 4.85,\ N_n = 100$ (for steel C45).

Author Contributions: Conceptualization, methodology, validation, writing—original draft preparation, investigation, writing—review and editing Ž.B. and V.L.; software, visualization V.L.; formal analysis, data curation, resources project administration Ž.B. All authors have read and agreed to the published version of the manuscript.

Funding: This research received no external funding.

Acknowledgments: All co-authors have contributed equally. The authors declare that they have no known competing financial interests or personal relationships that could have appeared to influence the work reported in this paper.

Conflicts of Interest: The authors declare no conflict of interest.

Nomenclature

Latin symbols

$A_1,\ A_2$	constants of the low–cycle fatigue curve under strain-controlled loading describing the first and second semi-cycle form respectively;
B	parameter $(B = 1 - \frac{1}{G_T} - \frac{\bar{S}_T}{2})$;
$b, b_{\frac{x}{y}},\ b_{\frac{y}{x}}$	regression coefficients;
$C_2,\ C_3, \propto,\ c$	constants of the low–cycle fatigue curve under strain-controlled loading;
$C_{1p},\ C_\psi$	constants of the Coffin–Manson equation;
D_e	strain-controlled stiffness coefficient;
d_k	cumulated quasi-static damage;
d_y	cumulated fatigue damage;
E	modulus of elasticity (MPa);
E_1'	secant modulus for the first semi-cycle diagram (MPa);
$E_T\ , E_{Tk}$	linear hardening modules for single and cyclic (k-th semi-cycle) loading respectively (GPa);
e	strain intensity under linear strain-controlled state (%);
e_1, e_2, e_3	principal strain (%);
e_i	strain intensity (%);
e_{in}	nominal strain intensity (%);
$\bar{e},\ \bar{e}_i,\ \bar{e}_{in}$	respective deformations in relative units $(\bar{e}_i = e_i/e_{pr},\ \bar{e}_{in} = e_{in}/e_{pr})$, (%);
e_0	intensity of maximum strain at the initial loading under the linear strain-controlled state (%);
$\bar{e}_0$	intensity of maximum strain at the initial loading normalized to proportional limit strain;
e_a	total strain amplitude (%);
e_y	amplitude of elastic strain (%);
e_p	amplitude of plastic strain (%);
e_{pr}	proportional limit strain (%);
e_t	material plasticity indicator determined by assessment of the variation in the cross -section area of the standard cylindrical specimen subjected to tension;
$i = 1 \ldots n$	specimen ranks in the rank order;
$F(k)$	function of k semi-cycles [36];

G_T, g_{ni}	relative linear hardening modules for single and cyclic loading diagrams respectively ($G_T = E_T/E$, $g_k = E_{Tk}/E$), (GPa);
k	number of loading semi-cycle;
k_0	number of semi-cycles of the loading cycles to fatigue crack initiation;
K	coefficient K values of relative measures of the key mechanical properties;
m	constant of the Coffin–Manson equation;
m_0	hardening parameter for the tensile diagram under its stepped approximation;
m_{xy}	second central mixed moment;
N_0	number of loading cycles to fatigue crack initiation;
N_c	durability (number of load cycles) until crack initiation;
N_f	durability (number of load cycles) till the cracks propagated to complete rupture;
n	number of of measurements;
n_e	safety factor of strength by maximum deformations;
n_N	safety factor of strength by the number of cycles;
$p_{1;2}$	parameter ($p_{1;2} = 1 + \chi_{1;2}\frac{1+r_{en}}{1-r_{\sigma n}}$);
r	coefficient of correlation;
r_{en}	coefficient of asymmetry of the intensity cycle of nominal;
r_{σ_n}	coefficient of asymmetry of the intensity cycle of nominal stress;
S	cyclic stress intensity under linear strain-controlled state, calculated from the start of unloading (MPa);
S_i	cyclic stress intensity (MPa);
S_{in}	nominal cyclic stress intensity (MPa);
S_k, S_{ik}, S_{ink}	respective cyclic strain for the k-th semi-cycle of loading (MPa);
$\overline{S}_k, \overline{S}_{ik}, \overline{S}_{ink}$	respective cyclic strains in relative units ($\overline{S}_k = S_k/S_T$, $\overline{S}_{ik} = S_{ik}/S_T$, $\overline{S}_{ik} = S_{ink}/S_T$), (MPa);
S_T	cyclic limit of proportionality calculated from the start of unloading (MPa);
$\overline{S}_T$	cyclic stress normalized respectively to proportional limit stress ($\overline{S}_T = S_T/\sigma_{pr}$), (MPa);
t_γ	Student's t-distribution;
$\overline{x}, \overline{y}$	statistical mean;
x_i, y_i	independent variables;
x_{max}	maximum values of the material mechanical properties in the bins (statistical intervals);
x_{min}	minimum values of the material mechanical properties in the bins (statistical intervals);
Greek symbols	
α, β	constants of the low–cycle fatigue curve under strain-controlled loading characterizing materials hardening or softening;
α_{1p}	constant of the Coffin–Manson equation;
α_σ	theoretical coefficient of concentration of elastic stresses;
γ	reliability of normal distribution;
δ	intensity of cyclic plastic deformations under linear strain-controlled states or width of the elastoplastic hysteresis loop;
$\varepsilon_{1n}, \varepsilon_{2n}, \varepsilon_{3n}$	range of main nominal strains (%);
ε	cyclic elastoplastic strain intensity under linear strain-controlled state, calculated from the start of unloading (%);
ε_i	cyclic elastoplastic strain intensity (%);
ε_{in}	nominal cyclic elastoplastic strain intensity (%);
$\varepsilon_k, \varepsilon_{ik}, \varepsilon_{ink}$	respective cyclic strains for the k-th semi-cycle of loading (%);
$\overline{\varepsilon}_k, \overline{\varepsilon}_{ik}, \overline{\varepsilon}_{ink}$	respective cyclic strain in relative units ($\overline{\varepsilon}_k = \varepsilon_k/\varepsilon_T$, $\overline{\varepsilon}_{ik} = \varepsilon_{ik}/\varepsilon_T$, $\overline{\varepsilon}_{ink} = \varepsilon_{ink}/\varepsilon_T$);
ε_T	strain corresponding to S_T (%);
$\overline{\varepsilon}_T$	strain corresponding to $\overline{S}_T$ (%);
μ_{ni}	Poisson's ratio for elastoplastic deformation;
σ_x, σ_y	root mean square deviation;
σ_a^*	conditional elastic stresses ($\sigma_a^* = e_0 E e_t$), (MPa);
σ_{pr}	proportional limit stress (MPa);
σ_{ys}	elastic limit or yield strength, the stress at which 0.2% plastic strain occurs (MPa);
σ_u	ultimate tensile stress (MPa);
σ_f	fracture strength (MPa);

$\overline{\sigma}$	normalized to proportional limit stress (MPa), $\overline{\sigma} = \sigma/\sigma_{pr}$;
χ_1, χ_2	parameter characterizing the sensitivity to cycle asymmetry;
ψ	cross-sectional narrowing (%);
ψ_u	continuous cross-sectional narrowing (%).

Appendix A

Table A1. Probabilistic values of mechanical characteristics for steel 15Cr2MoVA.

Characteristic	Probability, %						
	1	**10**	**30**	**50**	**70**	**90**	**99**
σ_{ys}, MPa	300	340	370	400	430	475	535
σ_u, MPa	500	530	560	580	600	640	680
ψ, %	74	76	79	80	82	85	90
e_{pr}, %	0.090	0.130	0.170	0.200	0.245	0.320	0.475

Table A2. Probabilistic values of mechanical characteristics for steel C45.

Characteristic	Probability, %						
	1	**10**	**30**	**50**	**70**	**90**	**99**
σ_{ys}, MPa	220	265	300	340	360	420	500
σ_u, MPa	620	700	750	800	850	900	1020
ψ, %	28	32	37	39	42	47	54
e_{pr}, %	0.140	0.180	0.225	0.260	0.300	0.360	0.480

Table A3. Probabilistic values of mechanical characteristics for aluminium alloy D16T1.

Characteristic	Probability, %						
	1	**10**	**30**	**50**	**70**	**90**	**99**
σ_{ys}, MPa	260	300	320	350	370	405	460
σ_u, MPa	580	620	650	680	700	750	800
ψ, %	9.5	11.3	12.8	14.0	15.5	17.5	21.0
e_{pr}, %	0.46	0.52	0.56	0.60	0.64	0.70	0.78

Table A4. Coefficient K values of relative measures of the key mechanical properties.

Material	σ_{pr}	σ_{ys}	σ_u	σ_f	ψ	ψ_u
	MPA				%	
15Cr2MoVa	2.06	1.88	1.52	1.78	1.24	4.19
C45	2.41	2.41	1.57	1.62	1.71	3.09
D16T1	1.72	1.66	1.45	1.43	2.24	1.38

Table A5. Probabilistic values of parameters of cyclic deformation resistance for steel 15Cr2MoVA.

Parameter	Probability, %						
	1	**10**	**30**	**50**	**70**	**90**	**99**
A_1	0.23	0.87	1.32	1.60	1.94	2.40	3.04
A_1	0.30	0.94	1.34	1.64	2.00	2.43	3.06
$\overline{S}_T$	1.15	1.20	1.25	1.28	1.30	1.35	1.40
m_0	0.15	0.17	0.19	0.21	0.23	0.26	0.30

Table A6. Calculation data in the concentration zone for failure probabilities 1%, 10%, 30%,50%, 70%, 90%, 99%.

Parameter	Probability, %						
	1	**10**	**30**	**50**	**70**	**90**	**99**
G_T	0.0272	0.0328	0.0400	0.0451	0.0497	0.0605	0.0754
χ_1	0.1988	0.3164	0.4093	0.4896	0.5867	0.7033	0.9118
χ_2	0.3521	0.4449	0.5274	0.5931	0.6776	0.7832	0.9741
p_1	1.0048	1.0077	1.0099	1.0119	1.0143	1.0171	1.0222
B	−37.19	−30.08	−24.62	−21.81	−20.06	−16.20	−12.96
$\overline{\varepsilon}_{1n1}$	9.37	6.16	4.56	3.79	3.04	2.26	1.46
$\overline{S}_{1n1}$	3.0231	1.9699	1.7305	1.6435	1.5871	1.4779	1.3771
g_{n1}	0.2417	0.1879	0.2049	0.2300	2.2870	0.3786	0.8223
ε_{in1}	0.4355	0.4360	0.4242	0.4133	0.3958	0.3693	0.3111
$\overline{e}_{in}$	2.7761	0.23073	1.9606	1.5412	1.2836	1.0122	0.7181
β	0.00004	0.00016	0.00042	0.00085	0.0018	0.0048	0.0190
E'_1	112,023	92,705	87,001	90,992	94,974	98,169	105,772
S_{1n1}	1616	1315	1199	1226	1236	1220	1227
S_{2n1}	1276	1031	929	939	931	895	858
D_e	0.5103	0.5109	0.5119	0.5130	0.5146	0.5174	0.5230
α_σ	3	3	3	3	3	3	3
r_{en}	−1	−1	−1	−1	−1	−1	−1
d_k	0.3424	0.3443	0.2976	0.2741	0.2551	0.1232	0.2197
d_y	0.6576	0.6557	0.7024	0.7259	0.7449	0.7802	0.8768

References

1. Makhutov, N.A.; Gadenin, M.M. Integrated Assessment of the Durability, Resources, Survivability, and Safety of Machinery Loaded under Complex Conditions. *J. Mach. Manuf. Reliab.* **2020**, *49*, 292–300. [CrossRef]
2. Makhutov, N.A.; Panov, A.N.; Yudina, O.N. The development of models of risk assessment complex transport systems. In *IOP Conference Series: Materials Science and Engineering, Proceedings of the V International Scientific Conference, Survivability and Structural Material Science (SSMS 2020), Moscow, Russia, 27–29 October 2020*; Institute of Physics Publishing (IOP): Moscow, Russia, 2020; Volume 1023. Available online: https://scholar.google.lt/scholar?hl=lt&as_sdt=0%2C5&q=The+development+of+models+of+risk+assessment+complex+transport+systems&btnG= (accessed on 16 July 2022).
3. Sekhar, A.P.; Nandy, S.; Bakkar, M.A.; Ray, K.K.; Das, D. Low cycle fatigue response of differently aged AA6063 alloy: Statistical analysis and microstructural evolution. *Materialia* **2021**, *20*, 101219. [CrossRef]
4. Jiang, Z.; Han, Z.; Li, M. A probabilistic model for low-cycle fatigue crack initiation under variable load cycles. *Int. J. Fatigue* **2022**, *155*, 106528. [CrossRef]
5. Li, X.-K.; Chen, S.; Zhu, S.-P.; Ding Liao, D.; Gao, J.-W. Probabilistic fatigue life prediction of notched components using strain energy density approach. *Eng. Fail. Anal.* **2021**, *124*, 105375. [CrossRef]
6. Weibull, W. *Fatigue Testing and Analysis of Results*; Pergamon Press: New York, NY, USA, 1961. Available online: https://books.google.lt/books?hl=lt&lr=&id=YM4gBQAAQBAJ&oi=fnd&pg=PP1&dq=Weibull,+W.+Fatigue+Testing+and+Analysis+of+Results&ots=VIVGA6VzJY&sig=UsrDkGvsP8go6hPS6IuYK3q9-H0&redir_esc=y#v=onepage&q=Weibull%2C%20W.%20Fatigue%20Testing%20and%20Analysis%20of%20Results&f=false (accessed on 20 July 2022).

7. Weibull, W.; Rockey, K.C. Fatigue Testing and Analysis of Results. *J. Appl. Mech.* **1962**, *29*, 607. [CrossRef]
8. Coffin, L.F., Jr. A study of the effects of cyclic thermal stresses on a ductile metal. *Trans. Metall. Soc. ASME* **1954**, *76*, 931–950. [CrossRef]
9. Manson, S.S. Fatigue: A complex subject—Some simple approximations. *Exp. Mech.* **1965**, *7*, 193–225. [CrossRef]
10. Langer, B.F. Design of Pressure Vessels for Low-Cycle Fatigue. *ASME J. Basic Eng.* **1962**, *84*, 389–399. [CrossRef]
11. Iida, K.; Inoue, H. Evaluation of low cycle fatigue design curve based on life distribution shape. *J. Soc. Nav. Archit. Jpn.* **1973**, *133*, 235–247. [CrossRef]
12. Freudenthal, A.M.; Gumbel, E.J. On the statistical interpretation of fatigue tests. *Proc. R. Soc. Lond. Ser. A* **1953**, *216*, 309–332. [CrossRef]
13. Freudenthal, A.M.; Gumbel, E.J. Physical and Statistical Aspects of Fatigue. *Adv. Appl. Mech.* **1956**, *4*, 117–158. [CrossRef]
14. Serensen, S.V.; Shneiderovich, R.M. Deformations and rupture criteria under low-cycles fatigue. *Exp. Mech.* **1966**, *6*, 587–592. [CrossRef]
15. Serensen, S.V.; Kogayev, V.P.; Shneiderovich, R.M. *Load Carrying Ability and Strength Evaluation of Machine Components*, 3rd ed.; Mashinostroeniya: Moscow, Russia, 1975; pp. 255–311. (In Russian)
16. Stepnov, M.N. *Statistical Methods for Computation of the Results of Mechanical Experiments, Mechanical Engineering*; Mashinostroeniya: Moscow, Russia, 2005. (In Russian)
17. Makhutov, N.A.; Zatsarinny, V.V.; Reznikov, D.O. Fatigue prediction on the basis of analysis of probabilistic mechanical properties. *AIP Conf. Proc.* **2020**, *2315*, 040025. [CrossRef]
18. Daunys, M.; Šniuolis, R. Statistical evaluation of low cycle loading curves parameters for structural materials by mechanical characteristics. *Nucl. Eng. Des.* **2006**, *236*, 13. [CrossRef]
19. Daunys, M.; Bazaras, Z.; Timofeev, B.T. Low cycle fatigue of materials in nuclear industry. *Mechanics* **2008**, *73*, 12–17. Available online: https://scholar.google.lt/scholar?hl=lt&as_sdt=0%2C5&q=28.%09Daunys%2C+M.%3B+Bazaras%2C+Z.%3B+Timofeev%2C+B.+T.+Low+cycle+fatigue+of+materials+in+nuclear+industry&btnG= (accessed on 22 December 2021).
20. Timofeev, B.T.; Bazaras, Z.L. Cyclic strength of the equipment of nuclear power plants made of 22K steel. *Mater. Sci.* **2005**, *41*, 680–685. [CrossRef]
21. Raslavičius, L.; Bazaras, Ž.; Lukoševičius, V.; Vilkauskas, A.; Česnavičius, R. Statistical investigation of the weld joint efficiencies in the repaired WWER pressure vessel. *Int. J. Press. Vessel. Pip.* **2021**, *189*, 1–9. [CrossRef]
22. Guo, S.; Liu, R.; Jiang, X.; Zhang, H.; Zhang, D.; Wang, J.; Pan, F. Statistical Analysis on the Mechanical Properties of Magnesium Alloys. *Materials* **2017**, *10*, 1271. [CrossRef]
23. Zhu, S.-P.; Huang, H.-Z.; Smith, R.; Ontiveros, V.; He, L.-P.; Modarres, M. Bayesian framework for probabilistic low cycle fatigue life prediction and uncertainty modeling of aircraft turbine disk alloys. *Probabilistic Eng. Mech.* **2013**, *34*, 114–122. [CrossRef]
24. Strzelecki, P. Determination of fatigue life for low probability of failure for different stress levels using 3-parameter Weibull distribution. *Int. J. Fatigue* **2021**, *145*, 106080. [CrossRef]
25. Fekete, B. New energy-based low cycle fatigue model for reactor steels. *Mater. Des.* **2015**, *79*, 42–52. [CrossRef]
26. Bazaras, Ž.; Lukoševičius, V. Statistical Assessment of Low-Cycle Fatigue Durability. *Symmetry* **2022**, *14*, 1205. [CrossRef]
27. Chen, L.; Wang, D.S.; Shi, F.; Sun, Z.G. Low-Cycle Fatigue Properties of Austenitic Stainless Steel S30408 under Large Plastic Strain Amplitude. *Adv. Steel Constr.* **2022**, *18*, 517–527. [CrossRef]
28. Pelegatti, M.; Lanzutti, A.; Salvati, E.; Srnec Novak, J.; De Bona, F.; Benasciutti, D. Cyclic Plasticity and Low Cycle Fatigue of an AISI 316L Stainless Steel: Experimental Evaluation of Material Parameters for Durability Design. *Materials* **2021**, *14*, 3588. [CrossRef] [PubMed]
29. Lee, S.J.; Theerthagiri, J.; Nithyadharseni, P.; Arunachalam, P.; Balaji, D.; Kumar, A.M.; Madhavan, J.; Mittal, V.; Choi, M.Y. Heteroatom-doped graphene-based materials for sustainable energy applications: A review. *Renew. Sustain. Energy Rev.* **2021**, *143*, 110849. [CrossRef]
30. Lee, Y.; Yu, Y.; Das, H.T.; Theerthagiri, J.; Lee, S.J.; Min, A.; Kim, G.-A.; Choi, H.C.; Choi, M.Y. Pulsed laser-driven green synthesis of trimetallic AuPtCu nanoalloys for formic acid electro-oxidation in acidic environment. *Fuel* **2023**, *332 Pt 2*, 126164. [CrossRef]
31. *GOST 25502-79 Standard*; Strength Analysis and Testing in Machine Building. Methods of Metals Mechanical Testing. Methods of Fatigue Testing. Standardinform: Moscow, Russia, 1993.
32. *GOST 22015–76 Standard*; Quality of Product. Regulation and Statistical Quality Evaluation of Metal Materials and Products on Speed-torque Characteristics. Standardinform: Moscow, Russia, 2010.
33. Regularities and Norms in Nuclear Power Engineering (PNAE) No. G-7-002-89. In *Rules of Equipment and Pipelines Strength Calculation of Nuclear Power Plant*; Energoatomizdat: Moscow, Russia, 1989.
34. Daunys, M. *Cycle Strength and Durability of Structures*; Technologija: Kaunas, Lithuania, 2005. (In Lithuanian)
35. Montgomery, D.C.; Runger, G.C. *Applied Statistics and Probability for Engineers*, 7th ed.; Wiley: Hoboken, NJ, USA, 2018.
36. Čekanavičius, V.; Murauskas, G. *Statistics and Its Applications: I Part*; TEV: Vilnius, Lithuania, 2002; pp. 129–133. (In Lithuanian)
37. Pavaras, A.; Žvinys, J. *Steels*; Technologija: Kaunas, Lithuania, 1995. (In Lithuanian)
38. Makhutov, N.A. Strength analysis under long-term static and cyclic loading. In *Structural Durability, Resource and ASafety*; Nauka: Novosibirsk, Russia, 2005; Volume 1, pp. 223–256. (In Russian)

MDPI
St. Alban-Anlage 66
4052 Basel
Switzerland
Tel. +41 61 683 77 34
Fax +41 61 302 89 18
www.mdpi.com

Materials Editorial Office
E-mail: materials@mdpi.com
www.mdpi.com/journal/materials

www.ingramcontent.com/pod-product-compliance
Lightning Source LLC
Chambersburg PA
CBHW061558220926
51850CB00005B/369
9783036576008